Wireless Sensor Networks

The *Ian F. Akyildiz Series in Communications and Networking* offers a comprehensive range of graduate-level text books for use on the major graduate programmes in communications engineering and networking throughout Europe, the USA and Asia. The series provides technically detailed books covering cutting-edge research and new developments in wireless and mobile communications, and networking. Each book in the series contains supporting material for teaching/learning purposes (such as exercises, problems and solutions, objectives and summaries etc), and is accompanied by a website offering further information such as slides, teaching manuals and further reading.

Titles in the series:

Akyildiz and Wang: *Wireless Mesh Networks*, 978-0470-03256-5, January 2009
Akyildiz and Vuran: *Wireless Sensor Networks*, 978-0470-03601-3, August 2010
Akyildiz, Lee and Chowdhury: *Cognitive Radio Networks*, 978-0470-68852-6 (forthcoming, 2011)
Ekici: *Mobile Ad Hoc Networks*, 978-0470-68193-0 (forthcoming, 2011)

Wireless Sensor Networks

Ian F. Akyildiz

Georgia Institute of Technology, USA

Mehmet Can Vuran

University of Nebraska-Lincoln, USA

A John Wiley and Sons, Ltd, Publication

Registered office
John Wiley & Sons Ltd, The Atrium, Southern Gate, Chichester, West Sussex, PO19 8SQ,
United Kingdom

For details of our global editorial offices, for customer services and for information about how to apply
for permission to reuse the copyright material in this book please see our website at www.wiley.com.

Library of Congress Cataloging-in-Publication Data

Akyildiz, Ian Fuat.
   Wireless sensor networks / Ian F. Akyildiz, Mehmet Can Vuran.
     p. cm.
   Includes bibliographical references and index.
   ISBN 978-0-470-03601-3 (cloth)
1. Wireless sensor networks. I. Vuran, Mehmet Can. II. Title.
   TK7872.D48A38 2010   681'.2–dc22
                                              2010008113

A catalogue record for this book is available from the British Library.

ISBN 978-0-470-03601-3 (H/B)

Set in 9/11pt Times by Sunrise Setting Ltd, Torquay, UK.

*To my wife Maria and
children Celine, Rengin
and Corinne for their continous
love and support...*

IFA

*Hemşerim'e...
To the loving memory of my
Dad, Mehmet Vuran...*

MCV

Contents

About the Series Editor

 **Ian F. Akyildiz** is the Ken Byers Distinguished Chair Professor with the School of Electrical and Computer Engineering, Georgia Institute of Technology; Director of Broadband Wireless Networking Laboratory and Chair of the Telecommunications Group. Since June 2008 he has been an Honorary Professor with the School of Electrical Engineering at the Universitat Politècnica de Catalunya, Barcelona, Spain. He is the Editor-in-Chief of *Computer Networks Journal* (Elsevier), is the founding Editor-in-Chief of the *Ad Hoc Networks Journal* (Elsevier) in 2003 and is the founding Editor-in-Chief of the *Physical Communication (PHYCOM) Journal* (Elsevier) in 2008. He is a past editor for *IEEE/ACM Transactions on Networking* (1996–2001), the *Kluwer Journal of Cluster Computing* (1997–2001), the *ACM-Springer Journal for Multimedia Systems* (1995–2002), *IEEE Transactions on Computers* (1992–1996) as well as the *ACM-Springer Journal of Wireless Networks (ACM WINET)* (1995–2005).

Dr Akyildiz was the Technical Program Chair and General Chair of several IEEE and ACM conferences including ACM MobiCom'96, IEEE INFOCOM'98, IEEE ICC'03, ACM MobiCom'02 and ACM SenSys'03, and he serves on the advisory boards of several research centers, journals, conferences and publication companies. He is an IEEE Fellow (1996) and an ACM Fellow (1997), and served as a National Lecturer for ACM from 1989 until 1998. He received the ACM Outstanding Distinguished Lecturer Award for 1994, and has served as an IEEE Distinguished Lecturer for IEEE COMSOC since 2008. Dr Akyildiz has received numerous IEEE and ACM awards including the 1997 IEEE Leonard G. Abraham Prize award (IEEE Communications Society), the 2002 IEEE Harry M. Goode Memorial award (IEEE Computer Society), the 2003 Best Tutorial Paper Award (IEEE Communications Society), the 2003 ACM SIGMOBILE Outstanding Contribution Award, the 2004 Georgia Tech Faculty Research Author Award and the 2005 Distinguished Faculty Achievement Award.

Dr Akyildiz is the author of two advanced textbooks entitled *Wireless Mesh Networks* and *Wireless Sensor Networks*, published by John Wiley & Sons in 2010.

His current research interests are in Cognitive Radio Networks, Wireless Sensor Networks, Wireless Mesh Networks and Nano-networks.

Preface

Wireless sensor networks (WSNs) have attracted a wide range of disciplines where close interactions with the physical world are essential. The distributed sensing capabilities and the ease of deployment provided by a wireless communication paradigm make WSNs an important component of our daily lives. By providing distributed, real-time information from the physical world, WSNs extend the reach of current cyber infrastructures to the physical world.

WSNs consist of tiny sensor nodes, which act as both data generators and network relays. Each node consists of sensor(s), a microprocessor, and a transceiver. Through the wide range of sensors available for tight integration, capturing data from a physical phenomenon becomes standard. Through on-board microprocessors, sensor nodes can be programmed to accomplish complex tasks rather than transmit only what they observe. The transceiver provides wireless connectivity to communicate the observed phenomena of interest. Sensor nodes are generally stationary and are powered by limited capacity batteries. Therefore, although the locations of the nodes do not change, the network topology dynamically changes due to the power management activities of the sensor nodes. To save energy, nodes aggressively switch their transceivers off and essentially become disconnected from the network. In this dynamic environment, it is a major challenge to provide connectivity of the network while minimizing the energy consumption. The energy-efficient operation of WSNs, however, provides significantly long lifetimes that surpass any system that relies on batteries.

In March 2002, our survey paper "Wireless sensor networks: A survey" appeared in the Elsevier journal *Computer Networks*, with a much shorter and concise version appearing in *IEEE Communications Magazine* in August 2002. Over the years, both of these papers were among the top 10 downloaded papers from Elsevier and IEEE Communication Society (ComSoc) journals with over 8000 citations in total.[1] Since then, the research on the unique challenges of WSNs has accelerated significantly. In the last decade, promising results have been obtained through these research activities, which have enabled the development and manufacture of sophisticated products. This, as a result, eventually created a brand-new market powered by the WSN phenomenon. Throughout these years, the deployment of WSNs has become a reality. Consequently, the research community has gained significant experience through these deployments. Furthermore, many researchers are currently engaged in developing solutions that address the unique challenges of the present WSNs and envision new WSNs such as wireless underwater and underground sensor networks. We have contributed to this research over the years through numerous articles and four additional survey/roadmap papers on wireless sensor actor networks, underwater acoustic networks, wireless underground sensor networks, and wireless multimedia sensor networks which were published in different years within the last decade.

In summer 2003, we started to work on our second survey paper on WSNs to revisit the state-of-the-art solutions since the *dawn* of this phenomenon. The large volume of work and the interest in both academia and industry have motivated us to significantly enhance this survey to create this book, which is targeted at teaching graduate students, stimulating them for new research ideas, as well as providing academic and industry professionals with a thorough overview and in-depth understanding of the state-of-the-art in wireless sensor networking and how they can develop new ideas to advance this technology as well as support emerging applications and services. The book provides a comprehensive coverage of

---

[1] According to Google Scholar as of October 2009.

the present research on WSNs as well as their applications and their improvements in numerous fields. This book covers several major research results including the authors' own contributions as well as all standardization committee decisions in a cohesive and unified form. Due to the sheer amount of work that has been published over the last decade, obviously it is not possible to cover every single solution and any lack thereof is unintentional.

The contents of the book mainly follow the TCP/IP stack starting from the physical layer and covering each protocol layer in detail. Moreover, cross-layer solutions as well as services such as synchronization, localization, and topology control are discussed in detail. Special cases of WSNs are also introduced. Functionalities and existing protocols and algorithms are covered in depth. The aim is to teach the readers what already exists and how these networks can further be improved and advanced by pointing out *grand research challenges* in the final chapter of the book.

Chapter 1 is a comprehensive introduction to WSNs, including sensor platforms and network architectures. Chapter 2 summarizes the existing applications of WSNs ranging from military solutions to home applications. Chapter 3 provides a comprehensive coverage of the characteristics, critical design factors, and constraints of WSNs. Chapter 4 studies the physical layer of WSNs, including physical layer technologies, wireless communication characteristics, and existing standards at the WSN physical layer. Chapter 5 presents various medium access control (MAC) protocols for WSNs, with a special focus on the basic carrier sense multiple access with collision avoidance (CSMA/CA) techniques used extensively at this layer, as well as distinct solutions ranging from CSMA/CA variants, time division multiple access (TDMA)-based MAC, and their hybrid counterparts. Chapter 6 focuses on error control techniques in WSNs as well as their impact on energy-efficient communication. Along with Chapter 5, these two chapters provide a comprehensive evaluation of the link layer in WSNs. Chapter 7 is dedicated to routing protocols for WSNs. The extensive number of solutions at this layer are studied in four main classes: data-centric, hierarchical, geographical, and quality of service (QoS)-based routing protocols. Chapter 8 firstly introduces the challenges of transport layer solutions and then describes the protocols. Chapter 9 introduces the cross-layer interactions between each layer and their impacts on communication performance. Moreover, cross-layer communication approaches are explained in detail. Chapter 10 discusses time synchronization challenges and several approaches that have been designed to address these challenges. Chapter 11 presents the challenges for localization and studies them in three classes: ranging techniques, range-based localization protocols, and range-free localization protocols. Chapter 12 is organized to capture the topology management solutions in WSNs. More specifically, deployment, power control, activity, scheduling, and clustering solutions are explained. Chapter 13 introduces the concept of wireless sensor–actor networks (WSANs) and their characteristics. In particular, the coordination issues between sensors and actors as well as between different actors are highlighted along with suitable solutions. Moreover, the communication issues in WSANs are discussed. Chapter 14 presents wireless multimedia sensor networks (WMSNs) along with their challenges and various architectures. In addition, the existing multimedia sensor network platforms are introduced, and the protocols are described in the various layers following the general structure of the book. Chapter 15 is dedicated to underwater wireless sensor networks (UWSNs) with a major focus on the impacts of the underwater environment. The basics of underwater acoustic propagation are studied and the corresponding solutions at each layer of the protocol stack are summarized. Chapter 16 introduces wireless underground sensor networks (WUSNs) and various applications for these networks. In particular, WUSNs in soil and WUSNs in mines and tunnels are described. The channel properties in both these cases are studied. Furthermore, the existing challenges in the communication layers are described. Finally, Chapter 17 discusses the grand challenges that still exist for the proliferation of WSNs.

It is a major task and challenge to produce a textbook. Although usually the authors carry the major burden, there are several other key people involved in publishing the book. Our foremost thanks go to Birgit Gruber from John Wiley & Sons who initiated the entire idea of producing this book. Tiina Ruonamaa, Sarah Tilley, and Anna Smart at John Wiley & Sons have been incredibly helpful, persistent, and patient. Their assistance, ideas, dedication, and support for the creation of this book will always be greatly appreciated. We also thank several individuals who indirectly or directly contributed to our book. In particular, our sincere thanks go to Özgur B. Akan, Tommaso Melodia, Dario Pompili, Weilian Su, Eylem Ekici, Cagri Gungor, Kaushik R. Chowdhury, Xin Dong, and Agnelo R. Silva for their help.

I (MCV) would like to specifically thank the numerous professors who have inspired me throughout my education in both Bilkent University, Ankara, Turkey and Georgia Institute of Technology, Atlanta, GA. I would like to thank my colleagues and friends at the University of Nebraska–Lincoln and the Department of Computer Science and Engineering for the environment they created during the development of this book. I am especially thankful to my PhD advisor, Professor Ian F. Akyildiz, who introduced me to the challenges of WSNs. I wholeheartedly thank him for his strong guidance, friendship, and trust during my PhD as well as my career thereafter. I would also like to express my deep appreciation to my wife, Demet, for her love, exceptional support, constructive critiques, and her sacrifices that made the creation of this book possible. I am thankful to my mom, Ayla, for the love, support, and encouragement that only a mother can provide. Finally, this book is dedicated to the loving memory of my dad, Mehmet Vuran (or *Hemşerim* as we used to call each other). He was the greatest driving force for the realization of this book as well as many other accomplishments in my life.

I (IFA) would like to specifically thank my wife and children for their support throughout all these years. Without their continuous love, understanding, and tolerance, none of these could have been achieved. Also my past and present PhD students, who became part of my family over the last 25 years, deserve the highest and sincerest thanks for being in my life and letting me enjoy the research to the fullest with them. The feeling of seeing how they developed in their careers over the years is indescribable and one of the best satisfactions in my life. Their research results contributed a great deal to the contents of this book as well.

Ian F. Akyildiz and Mehmet Can Vuran

1

Introduction

With the recent advances in *micro electro-mechanical systems* (MEMS) technology, wireless communications, and digital electronics, the design and development of low-cost, low-power, multifunctional sensor nodes that are small in size and communicate untethered in short distances have become feasible. The ever-increasing capabilities of these tiny sensor nodes, which include sensing, data processing, and communicating, enable the realization of wireless sensor networks (WSNs) based on the collaborative effort of a large number of sensor nodes.

WSNs have a wide range of applications. In accordance with our vision [18], WSNs are slowly becoming an integral part of our lives. Recently, considerable amounts of research efforts have enabled the actual implementation and deployment of sensor networks tailored to the unique requirements of certain sensing and monitoring applications.

In order to realize the existing and potential applications for WSNs, sophisticated and extremely efficient communication protocols are required. WSNs are composed of a large number of sensor nodes, which are densely deployed either inside a physical phenomenon or very close to it. In order to enable reliable and efficient observation and to initiate the right actions, physical features of the phenomenon should be reliably detected/estimated from the collective information provided by the sensor nodes [18]. Moreover, instead of sending the raw data to the nodes responsible for the fusion, sensor nodes use their processing capabilities to locally carry out simple computations and transmit only the required and partially processed data. Hence, these properties of WSNs present unique challenges for the development of communication protocols.

The intrinsic properties of individual sensor nodes pose additional challenges to the communication protocols in terms of energy consumption. As will be explained in the later chapters, WSN applications and communication protocols are mainly tailored to provide high energy efficiency. Sensor nodes carry limited power sources. Therefore, while traditional networks are designed to improve performance metrics such as throughput and delay, WSN protocols focus primarily on power conservation. The deployment of WSNs is another factor that is considered in developing WSN protocols. The position of the sensor nodes need not be engineered or predetermined. This allows random deployment in inaccessible terrains or disaster relief operations. On the other hand, this random deployment requires the development of self-organizing protocols for the communication protocol stack. In addition to the placement of nodes, the density in the network is also exploited in WSN protocols. Due to the short transmission ranges, large numbers of sensor nodes are densely deployed and neighboring nodes may be very close to each other. Hence, multi-hop communication is exploited in communications between nodes since it leads to less power consumption than the traditional single hop communication. Furthermore, the dense deployment coupled with the physical properties of the sensed phenomenon introduce correlation in spatial and temporal domains. As a result, the spatio-temporal correlation-based protocols emerged for improved efficiency in networking wireless sensors.

Wireless Sensor Networks   Ian F. Akyildiz and Mehmet Can Vuran

In this book, we present a detailed explanation of existing products, developed protocols, and research on algorithms designed thus far for WSNs. Our aim is to provide a contemporary look at the current state of the art in WSNs and discuss the still-open research issues in this field.

1.1   Sensor Mote Platforms

WSNs are composed of individual embedded systems that are capable of (1) interacting with their environment through various sensors, (2) processing information locally, and (3) communicating this information wirelessly with their neighbors. A sensor node typically consists of three components and can be either an individual board or embedded into a single system:

- **Wireless modules** or **motes** are the key components of the sensor network as they possess the communication capabilities and the programmable memory where the application code resides. A mote usually consists of a microcontroller, transceiver, power source, memory unit, and may contain a few sensors. A wide variety of platforms have been developed in recent years including Mica2 [3], Cricket [2], MicaZ [3], Iris [3], Telos [3], SunSPOT [9], and Imote2 [3].

- **A sensor board** is mounted on the mote and is embedded with multiple types of sensors. The sensor board may also include a prototyping area, which is used to connect additional custom-made sensors. Available sensor boards include the MTS300/400 and MDA100/300 [3] that are used in the Mica family of motes. Alternatively, the sensors can be integrated into the wireless module such as in the Telos or the SunSPOT platform.

- **A programming board**, also known as the gateway board, provides multiple interfaces including Ethernet, WiFi, USB, or serial ports for connecting different motes to an enterprise or industrial network or locally to a PC/laptop. These boards are used either to program the motes or gather data from them. Some examples of programming boards include the MIB510, MIB520, and MIB600 [3]. Particular platforms need to be connected to a programming board to load the application into the programmable memory. They could also be programmed over the radio.

While the particular sensor types vary significantly depending on the application, a limited number of wireless modules have been developed to aid research in WSNs. Table 1.1 captures the major characteristics of popular platforms that were designed over the past few years in terms of their processor speed, programmable and storage memory size, operating frequency, and transmission rate. The timeline for these platforms is also shown in Figure 1.1. As can be observed, the capabilities of these platforms vary significantly. However, in general, the existing platforms can be classified into two based on both their capabilities and the usage. Next, we overview these existing platforms as *low-end* and *high-end* platforms. Moreover, several standardization efforts that have been undertaken for the proliferation of application development will be explained in Section 1.1.3. Finally, the software packages that have been used within these devices are described.

1.1.1   Low-End Platforms

The low-end platforms are characterized by their limited capabilities in terms of processing, memory, and communication. These platforms are usually envisioned to be deployed in large numbers in a WSN to accomplish sensing tasks as well as providing a connectivity infrastructure. The following platforms have been mostly used in developing communication protocols recently:

Mica family: The Mica family of nodes consist of Mica, Mica2, MicaZ, and IRIS nodes and are produced by Crossbow [3]. Each node is equipped with 8-bit Atmel AVR microcontrollers with a speed of 4–16 MHz and 128–256 kB of programmable flash. While the microcontrollers are similar, the Mica family of nodes have been equipped with a wide range of transceivers. The Mica node includes a 916 or 433 MHz transceiver at 40 kbps, while the Mica2 platform is equipped with a 433/868/916 MHz

Table 1.1   Mote hardware.

Mote type	CPU speed (MHz)	Prog. mem. (kB)	RAM (kB)	Radio freq. (MHz)	Tx. rate (kbps)
Berkeley [3]					
WeC	8	8	0.5	916	10
rene	8	8	0.5	916	10
rene2	8	16	1	916	10
dot	8	16	1	916	10
mica	6	128	4	868	10/40
mica2	16	128	4	433/868/916	38.4 kbaud
micaz	16	128	4	2.4 GHz	250
Cricket [3]	16	128	4	433	38.4 kbaud
EyesIFX [17]	8	60	2	868	115
TelosB/Tmote [3]	16	48	10	2.4 GHz	250
SHIMMER [16]	8	48	10	BT/2.4 GHz[a]	250
Sun SPOT [9]	16–60	2 MB	256	2.4 GHz	250
BTnode [1]	8	128	64	BT/433–915[a]	Varies
IRIS [3]	16	128	8	2.4 GHz	250
V-Link [15]	N/A	N/A	N/A	2.4 GHz	250
TEHU-1121 [7]	N/A	N/A	N/A	0.9/2.4 GHz	N/A
NI WSN-3202 [6]	N/A	N/A	N/A	2.4 GHz	250
Imote [3]	12	512	64	2.4 GHz (BT)	100
Imote2 [3]	13–416	32 MB	256	2.4 GHz	250
Stargate [3]	400	32 MB	64 MB SD	2.4 GHz	Varies[b]
Netbridge NB-100 [3]	266	8 MB	32 MB	Varies[b]	Varies[b]

[a] BTnode and SHIMMER motes are equipped with two transceivers: Bluetooth and a low-power radio.
[b] The transmission rate of the Stargate board and the Netbridge depends on the communication device connected to it (MicaZ node, WLAN card, etc.).

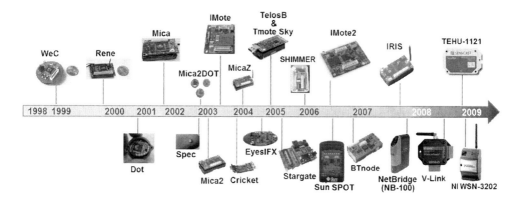

Figure 1.1   Timeline for the sensor mote platforms.

transceiver at 40 kbps. On the other hand, the MicaZ and IRIS nodes are equipped with IEEE 802.15.4 compliant transceivers, which operate at 2.4 GHz with 250 kbps data rate. Each platform has limited memory in terms of RAM (4–8 kB) and data memory (512 kB). Moreover, each version is equipped with a 51-pin connector that is used to connect additional sensor boards and programming boards to the mote.

Telos/Tmote: An architecture similar to the MicaZ platform has been adopted for the Telos motes from Crossbow and Tmote Sky motes from Sentilla (formerly Moteiv). While the transceiver is kept intact, Telos/Tmote motes have larger RAM since an 8 MHz TI MSP430 microcontroller with 10 kB RAM is used. Furthermore, Telos/Tmote platforms are integrated with several sensors including light, IR, humidity, and temperature as well as a USB connector, which eliminates the need for additional sensor or programming boards. Moreover, 6- and 10-pin connectors are included for additional sensors.

EYES: The EYES platform has been designed as a result of a 3-year European project and is similar to the Telos/Tmote architectures. A 16-bit microcontroller with 60 kB of program memory and 2 kB data memory is used in EYES [24]. Moreover, the following sensors are embedded with the mote: compass, accelerometer, and temperature, light, and pressure sensors. The EYES platform includes the TR1001 transceiver, which supports transmission rates up to 115.2 kbps with a power consumption of 14.4 mW, 16.0 mW, and 15.0 μW during receive, transmit, and sleep modes, respectively. The platform also includes an RS232 serial interface for programming.

In addition to these platforms, several low-end platforms have been developed with similar capabilities as listed in Table 1.1 and shown in Figure 1.1. An important trend to note is the appearance of proprietary platforms from the industry such as V-Link, TEHU, and the National Instruments motes in recent years (2008–2009).

The low-end platforms are used for sensing tasks in WSNs and they provide a connectivity infrastructure through multi-hop networking. These nodes are generally equipped with low-power microcontrollers and transceivers to decrease the cost and energy consumption. As a result, they are used in large numbers in the deployment of WSNs. It can be observed that wireless sensor platforms generally employ the *Industrial, Scientific, and Medical* (ISM) bands, which offer license-free communication in most countries. More specifically, most recent platforms include the CC2420 transceiver, which operates in the 2.4 GHz band and is compatible with the IEEE 802.15.4 standard. This standardization provides heterogeneous deployments of WSNs, where various platforms are used in a network. Most of the communication protocols discussed in this book are developed using these platforms.

1.1.2   High-End Platforms

In addition to sensing, local processing, and multi-hop communication, WSNs require additional functionalities that cannot be efficiently carried out by the low-end platforms. High-level tasks such as network management require higher processing power and memory compared to the capabilities of these platforms. Moreover, the integration of WSNs with existing networking infrastructure requires multiple communication techniques to be integrated through *gateway* modules. Furthermore, in networks where processing or storage hubs are integrated with sensor nodes, higher capacity nodes are required. To address these requirements, high-end platforms have been developed for WSNs.

Stargate: The Stargate board [8] is a high-performance processing platform designed for sensing, signal processing, control, and sensor network management. Stargate is based on Intel's PXA-255 Xscale 400 MHz RISC processor, which is the same processor found in many handheld computers including the Compaq IPAQ and the Dell Axim. Stargate has 32 MB of flash memory, 64 MB of SDRAM, and an on-board connector for Crossbow's Mica family motes as well as PCMCIA Bluetooth or IEEE 802.11 cards. Hence, it can work as a wireless gateway and computational hub for in-network processing algorithms.

When connected with a webcam or other capturing device, it can function as a medium-resolution multimedia sensor, although its energy consumption is still high [22].

Stargate NetBridge was developed as a successor to Stargate and is based on the Intel IXP420 XScale processor running at 266 MHz. It features one wired Ethernet and two USB 2.0 ports and is equipped with 8 MB of program flash, 32 MB of RAM, and a 2 GB USB 2.0 system disk, where the Linux operating system is run. Using the USB ports, a sensor node can be connected for gateway functionalities.

Imote and Imote2: Intel has developed two prototypal generations of wireless sensors, known as Imote and Imote2 for high-performance sensing and gateway applications [3]. Imote is built around an integrated wireless microcontroller consisting of an 8-bit 12 MHz ARM7 processor, a Bluetooth radio, 64 kB RAM, and 32 kB flash memory, as well as several I/O options. The software architecture is based on an ARM port of TinyOS.

The second generation of Intel motes, Imote2, is built around a new low-power 32-bit PXA271 XScale processor at 320/416/520 MHz, which enables DSP operations for storage or compression, and an IEEE 802.15.4 ChipCon CC2420 radio. It has large on-board RAM and flash memories (32 MB), additional support for alternate radios, and a variety of high-speed I/O to connect digital sensors or cameras. Its size is also very limited, 48 × 33 mm, and it can run the Linux operating system and Java applications.

1.1.3   Standardization Efforts

The heterogeneity in the available sensor platforms results in compatibility issues for the realization of envisioned applications. Hence, standardization of certain aspects of communication is necessary. To this end, the IEEE 802.15.4 [14] standards body was formed for the specification of low-data-rate wireless transceiver technology with long battery life and very low complexity. Three different bands were chosen for communication, i.e., 2.4 GHz (global), 915 MHz (the Americas), and 868 MHz (Europe). While the PHY layer uses binary phase shift keying (BPSK) in the 868/915 MHz bands and offset quadrature phase shift keying (O-QPSK) in the 2.4 GHz band, the MAC (Medium Access Control) layer provides communication for star, mesh, and cluster tree-based topologies with controllers. The transmission range of the nodes is assumed to be 10–100 m with data rates of 20 to 250 kbps [14]. Most of the recent platforms developed for WSN research comply with the IEEE 802.15.4 standard. Actually, the IEEE 802.15.4 standard, explained in Chapter 4, acquired a broad audience and became the *de facto* standard for PHY and MAC layers in low-power communication. This allows the integration of platforms with different capabilities into the same network.

On top of the IEEE 802.15.4 standard, several standard bodies have been formed to proliferate the development of low-power networks in various areas. It is widely recognized that standards such as Bluetooth and WLAN are not well suited for low-power sensor applications. On the other hand, standardization attempts such as ZigBee, WirelessHART, WINA, and SP100.11a, which specifically address the typical needs of wireless control and monitoring applications, are expected to enable rapid improvement of WSNs in the industry. In addition, standardization efforts such as 6LoWPAN are focused on providing compatibility between WSNs and existing networks such as the Internet.

Next, three major standardization efforts will be described in detail: namely, ZigBee [13], WirelessHART [12], and 6LoWPAN [4]. In addition, other standardization efforts will be summarized.

ZigBee

The ZigBee [13] standard has been developed by the ZigBee Alliance, which is an international, non-profit industrial consortium of leading semiconductor manufacturers and technology providers. The ZigBee standard was created to address the market need for cost-effective, standard-based wireless networking solutions that support low data rates, low power consumption, security, and reliability

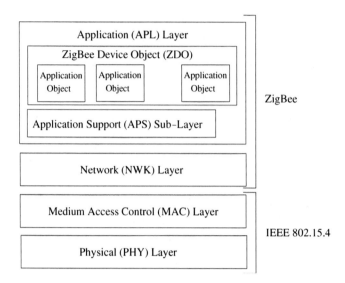

Figure 1.2   IEEE 802.15.4 and the ZigBee protocol stack [13].

through wireless personal area networks (WPANs). Five main application areas are targeted: home automation, smart energy, building automation, telecommunication services, and personal health care.

The ZigBee standard is defined specifically in conjunction with the IEEE 802.15.4 standard. Therefore, both are usually confused. However, as shown in Figure 1.2, each standard defines specific layers of the protocol stack. The PHY and MAC layers are defined by the IEEE 802.15.4 standard while the ZigBee standard defines the network layer (NWK) and the application framework. Application objects are defined by the user. To accommodate a large variety of applications, three types of traffic are defined, Firstly, *periodic data traffic* is required for monitoring applications, where sensors provide continuous information regarding a physical phenomenon The data exchange is controlled through the network controller or a router. Secondly, *Intermittent data traffic* applies to most event-based applications and is triggered through either the application or an external factor. This type of traffic is handled through each router node. To save energy, the devices may operate in disconnected mode, whereas they operate in sleep mode most of the time. Whenever information needs to be transmitted, the transceiver is turned on and the device associates itself with the network. Finally, *repetitive low-latency data traffic* is defined for certain communications such as a mouse click that needs to be completed within a certain time. This type of traffic is accommodated through the polling-based frame structure defined by the IEEE 802.15.4 standard.

The ZigBee network (NWK) layer provides management functionalities for the network operation. The procedures for establishing a new network and the devices to gain or relinquish membership of the network are defined. Furthermore, depending on the network operation, the communication stack of each device can be configured. Since ZigBee devices can be a part of different networks during their lifetime, the standard also defines a flexible addressing mechanism. Accordingly, the network coordinator assigns an address to the devices as they join the network. As a result, the unique ID of each device is not used for communication but a shorter address is assigned to improve the efficiency during communication. In a tree architecture, the address of a device also identifies its parent, which is used for routing purposes. The NWK layer also provides synchronization between devices and network controllers. Finally, multi-hop routes are generated by the NWK layer according to defined protocols.

As shown in Figure 1.2, the ZigBee standard also defines certain components in the application layer. This layer consists of the APS sub-layer, the ZigBee device object (ZDO), and the manufacturer-defined application objects [13]. The applications are implemented through these manufacturer-defined application objects and implementation is based on requirements defined by the standard. The ZDO defines functions provided by the device for network operation. More specifically, the role of devices such as a network coordinator or a router is defined through the ZDO. Moreover, whenever a device needs to be associated with the network, the binding requests are handled through the ZDO. Finally, the APS sub-layer provides discovery capability to devices so that the neighbors of a device and the functionalities provided by these neighbors can be stored. This information is also used to match the binding requests of the neighbors with specific functions.

WirelessHART

WirelessHART [12] has been developed as a wireless extension to the industry standard Highway Addressable Remote Transducer (HART) protocol. HART is the most used communication protocol in the automation and industrial applications that require real-time support with a device count around 20 million [12]. It is based on superimposing a digital FSK-modulated signal on top of the 4–20 mA analog current loop between different components. HART provides a master/slave communication scheme, where up to two masters are accommodated in the network. Accordingly, devices connected to the system can be controlled through a permanent system and handheld devices for monitoring and control purposes.

The WirelessHART standard has been released as a part of the HART 7 specification as the first open wireless communication standard specifically designed for process measurement and control applications [12]. WirelessHART relies on the IEEE 802.15.4 PHY layer standard for the 2.4 GHz band. Moreover, a TDMA-based MAC protocol is defined to provide several messaging modes: one-way publishing of process and control values, spontaneous notification by exception, ad hoc request and response, and auto-segmented block transfers of large data sets.

The network architecture of the WirelessHART standard is shown in Figure 1.3. Accordingly, five types of components are defined: *WirelessHART field devices* (WFDs) are the sensor and control elements that are connected to process or plant equipment. *Gateways* provide interfaces with wireless portions of the network and the wired infrastructure. As a result, host application and the controller can interact with the WFDs. The *network manager* maintains operation of the network by scheduling communication slots for devices, determining routing tables, and monitoring the health of the network. In addition to the three main components, the *WirelessHART adapters* provide backward compatibility by integrating existing HART field devices with the wireless network. Finally, *handhelds* are equipped with on-board transceivers to provide on-site access to the wireless network and interface with the WFDs.

Based on these components, a full protocol stack has been defined by the WirelessHART standard. As explained above, at the PHY layer, the IEEE 802.15.4 standard is employed and a TDMA-based MAC protocol is used at the data link layer. In addition, the network topology is designed as a mesh network and each device can act as a source or a router in the network. This network topology is very similar to what is generally accepted for WSNs.

At the network layer, table-based routing is used so that multiple redundant paths are established during network formation and these paths are continuously verified. Accordingly, even if a communication path between a WFD and a gateway is corrupted, alternate paths are used to provide network reliability greater than 3σ (99.7300204%). In addition to established paths, source routing techniques are used to establish ad hoc communication paths. Moreover, the network layer supports dynamic bandwidth management by assigning allocated bandwidth to certain devices. This is also supported by the underlying TDMA structure by assigning appropriate numbers of slots to these devices. The bandwidth is allocated on a demand basis and can be configured when a device joins the network.

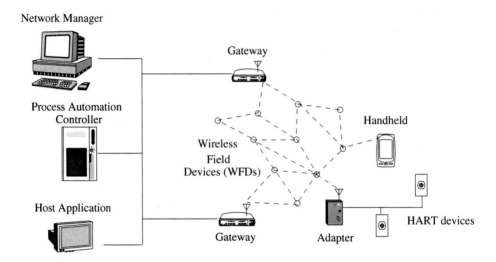

Figure 1.3   WirelessHART architecture and components [12].

The transport layer of the WirelessHART standard provides reliability on the end-to-end path and supports TCP-like reliable block transfers of large data sets. End-to-end monitoring and control of the network are also provided. Accordingly, WFDs continuously broadcast statistics related to their communication success and neighbors, which is monitored by the network manager to establish redundant routes and improve energy efficiency.

Finally, the application layer supports the standard HART application layer, where existing solutions can be implemented seamlessly.

6LoWPAN

The existing standards enable application-specific solutions to be developed for WSNs. Accordingly, stand-alone networks of sensors can be implemented for specific applications. However, these networks cannot be easily integrated with the Internet since the protocols based on IEEE 802.15.4 are not compliant with the IP. Therefore, sensors cannot easily communicate with web-based devices, servers, or browsers. Instead, gateways are required to collect the information from the WSN and communicate with the Internet. This creates single-point-of-failure problems at the gateways and stresses the neighbors of the gateway.

To integrate WSNs with the Internet, the Internet Engineering Task Force (IETF) is developing the IPv6 over Low-power Wireless Personal Area Network (6LoWPAN) standard [4]. This standard defines the implementation of the IPv6 stack on top of IEEE 802.15.4 to let any device be accessible from and with the Internet.

The basic challenge in integrating IPv6 and WSNs is the addressing structure of IPv6, which defines a header and address information field of 40 bytes. However, IEEE 802.15.4 allows up to 127 bytes for the *whole* packet including header and payload information. Accordingly, straightforward integration of both standards is not efficient. Instead, 6LoWPAN adds an adaptation layer that lets the radio stack and IPv6 communications operate together. A *stacked header* structure has been proposed for the 6LoWPAN standard [23], where, instead of a single monolithic header, four types of headers are utilized according to the type of packet being sent. In addition, stateless compression techniques are used to decrease the size of the header from 40 bytes to around 4 bytes, which is suitable for WSNs.

The four header types are as follows:

- **Dispatch header (1 byte):** This header type defines the type of header following it. The first 2 bits are set to 01 for the dispatch header and the remaining 6 bits define the type of header following it (uncompressed IPv6 header or a header compression header).
- **Mesh header (4 bytes):** This header is identified by 10 in the first 2 bits and is used in mesh topologies for routing purposes. The first 2 bits are followed by additional 2 bits that indicate whether the source and destination addresses are 16-bit short or 64-bit long addresses. A 4-bit hop left field is used to indicate the number of hops left. Originally, 15 hops are supported but an extra byte can be used to support 255 hops. Finally, the remaining fields indicate the source and the destination addresses of the packet. This information can be used by the routing protocols to find the next hop.
- **Fragmentation header (4–5 bytes):** IPv6 can support payloads up to 1280 bytes whereas this is 102 bytes for IEEE 802.15.4. This is solved by fragmenting larger payloads into several packets and the fragmentation header is used to fragment and reassemble these packets. The first fragment includes a header of 4 bytes, which is indicated by 11 in the first 2 bits and by 000 in the next 3 bits. This is followed by the datagram size and datagram tag fields. The following fragment header uses 11100 in the first 5 bits followed by the datagram size, tag, and the datagram offset.
- **Header compression header (1 byte):** Finally, the 40-byte IPv6 header is compressed into 2 bytes including the header compression header. This compression exploits the fact that IEEE 802.15.4 packet headers already include the MAC addresses of the source and destination pairs. These MAC addresses can be mapped to the lowest 64 bits of an IPv6 address. As a result, the source and destination addresses are completely eliminated from the IPv6 header. Similar techniques are used to eliminate the unnecessary fields for each communication and allow these fields to be inserted when the packet reaches a gateway to the Internet.

Header compression is not the only challenge for WSN–Internet integration. The ongoing efforts in the development of the 6LoWPAN standard aim to address some of these challenges including routing and transport control to provide seamless interoperation of WSNs and the Internet.

Other Standardization Efforts

In addition to the above efforts, several additional platforms have been engaged with defining standards for WSN applications. The ISA SP100.11a standard [5] also centers around the process and factory automation and is being developed by the Systems and Automation Society (ISA). Moreover, the Wireless Industrial Networking Alliance (WINA) [11] was formed in 2003 to stimulate the development and promote the adoption of wireless networking technologies and practices to help increase industrial efficiency. As a first step, this ad hoc group of suppliers and end-users is working to define end-user needs and priorities for industrial wireless systems. The standardization attempts such as ZigBee, WirelessHART, WINA, and SP100.11a, which specifically address the typical needs of wireless control and monitoring applications, are expected to enable rapid improvement of WSNs in the industry.

WSN applications have gained significant momentum during the past decade with the acceleration in research in this field. Although existing applications provide a wide variety of possibilities where the WSN phenomenon can be exploited, there exists many areas waiting for WSN empowerment. Moreover, further enhancements in WSN protocols will open up new areas of applications. Nevertheless, commercialization of these potential applications is still a major challenge.

1.1.4   Software

In addition to hardware platforms and standards, several software platforms have also been developed specifically for WSNs. Among these, the most accepted platform is the TinyOS [10], which is an open-source operating system designed for wireless embedded sensor networks. TinyOS incorporates a component-based architecture, which minimizes the code size and provides a flexible platform for implementing new communication protocols. Its component library includes network protocols, distributed services, sensor drivers, and data acquisition tools, which can be further modified or improved based on the specific application requirements. TinyOS is based on an event-driven execution model that enables fine-grained power management strategies.

Most of the existing software code for communication protocols today is written for the TinyOS platform. Coupled with TinyOS, a TinyOS mote simulator, TOSSIM, has been introduced to simplify the development of sensor network protocols and applications [21]. TOSSIM provides a scalable simulation environment and compiles directly from the TinyOS code. It simulates the TinyOS network stack at the bit level, allowing experimentation with low-level protocols in addition to top-level application systems. It also provides a graphical user interface tool, TinyViz, in order to visualize and interact with running simulations.

In addition to TinyOS, several software platforms and operating systems have been introduced recently. LiteOS [19] is a multi-threading operating system that provides Unix-like abstractions. Compared to TinyOS, LiteOS provides multi-threaded operation, dynamic memory management, and command-line shell support. The shell support, LiteShell, provides a command-line interface at the user side, i.e., the PC, to provide interaction with the sensor node to be programmed.

Contiki [20] is an open-source, multitasking operating system developed for use on a variety of platforms including microcontrollers such as the TI MSP430 and the Atmel AVR, which are used in the Telos, Tmote, and Mica families. Contiki has been built around an event-driven kernel but it is possible to employ preemptive multithreading for certain programs as well as dynamic loading and replacement of individual programs and services. As a result, compared to TinyOS, which is statically linked at compile-time, Contiki allows programs and drivers to be replaced during run-time and without relinking. Moreover, TCP/IP support is also provided through the μIP stack.

The recent SunSPOT platform [9] does not use an operating system but runs a Java virtual machine (VM), Squawk, on the bare metal, which is a fully capable Java ME implementation. The VM executes directly out of flash memory.

While several operating systems with additional capabilities have become available, TinyOS is still being widely used in WSN research. One of the main reasons for this popularity is the vast code space built throughout the development of WSN solutions. Clearly, it is hard to port existing applications and communication protocols to these new operating systems. This calls for platforms that support interoperability for existing code space so that additional flexibility and capabilities are provided to both the research community and industry.

1.2 WSN Architecture and Protocol Stack

The sensor nodes are usually scattered in a *sensor field* as shown in Figure 1.4. Each of these scattered sensor nodes has the capability to collect data and route data back to the *sink/gateway* and the end-users. Data are routed back to the end-user by a multi-hop infrastructureless architecture through the sink as shown in Figure 1.4. The sink may communicate with the *task manager/end-user* via the Internet or satellite or any type of wireless network (like WiFi, mesh networks, cellular systems, WiMAX, etc.), or without any of these networks where the sink can be directly connected to the end-users. Note that there may be multiple sinks/gateways and multiple end-users in the architecture shown in Figure 1.4.

In WSNs, the sensor nodes have the dual functionality of being both data originators and data routers. Hence, communication is performed for two reasons:

- **Source function:** Source nodes with event information perform communication functionalities in order to transmit their packets to the sink.
- **Router function:** Sensor nodes also participate in forwarding the packets received from other nodes to the next destination in the multi-hop path to the sink.

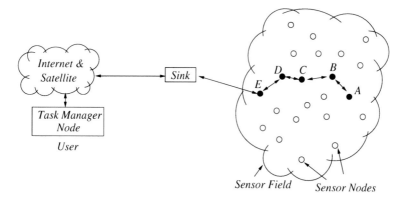

Figure 1.4   Sensor nodes scattered in a sensor field.

The protocol stack used by the sink and all sensor nodes is given in Figure 1.5. This protocol stack combines power and routing awareness, integrates data with networking protocols, communicates power efficiently through the wireless medium, and promotes cooperative efforts of sensor nodes. The protocol stack consists of the *physical layer, data link layer, network layer, transport layer, application layer*, as well as *synchronization plane, localization plane, topology management plane, power management plane, mobility management plane*, and *task management plane*. The physical layer addresses the needs of simple but robust modulation, transmission, and receiving techniques. Since the environment is noisy and sensor nodes can be mobile, the link layer is responsible for ensuring reliable communication through error control techniques and manage channel access through the MAC to minimize collision with neighbors' broadcasts. Depending on the sensing tasks, different types of application software can be built and used on the application layer. The network layer takes care of routing the data supplied by the transport layer. The transport layer helps to maintain the flow of data if the sensor network application requires it. In addition, the power, mobility, and task management planes monitor the power, movement, and task distribution among the sensor nodes. These planes help the sensor nodes coordinate the sensing task and lower the overall power consumption.

The power management plane manages how a sensor node uses its power. For example, the sensor node may turn off its receiver after receiving a message from one of its neighbors. This is to avoid getting duplicated messages. Also, when the power level of the sensor node is low, the sensor node broadcasts to its neighbors that it is low in power and cannot participate in routing messages. The remaining power is reserved for sensing. The mobility management plane detects and registers the movement of sensor nodes, so a route back to the user is always maintained, and the sensor nodes can keep track of their neighbors. By knowing these neighbor sensor nodes, the sensor nodes can balance their power and task usage. The task management plane balances and schedules the sensing tasks given to a specific region. Not all sensor nodes in that region are required to perform the sensing task at the same time. As a result, some sensor nodes perform the task more than others, depending on their power level. These management planes are needed so that sensor nodes can work together in a power-efficient way, route data in a mobile sensor network, and share resources between sensor nodes. Without them, each sensor node will just work individually. From the standpoint of the whole sensor network, it is more efficient if sensor nodes can collaborate with each other, so the lifetime of the sensor networks can be prolonged.

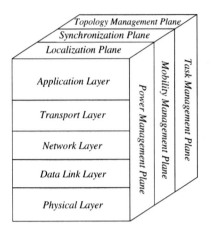

Figure 1.5   The sensor network protocol stack.

1.2.1   Physical Layer

The physical layer is responsible for frequency selection, carrier frequency generation, signal detection, modulation, and data encryption. Frequency generation and signal detection have more to do with the underlying hardware and transceiver design and hence are beyond the scope of our book. More specifically, we focus on signal propagation effects, power efficiency, and modulation schemes for sensor networks.

1.2.2   Data Link Layer

The data link layer is responsible for the multiplexing of data streams, data frame detection, and medium access and error control. It ensures reliable point-to-point and point-to-multipoint connections in a communication network. More specifically, we discuss the medium access and error control strategies for sensor networks.

MAC

The MAC protocol in a wireless multi-hop self-organizing sensor network must achieve two goals. The first goal is creation of the network infrastructure. Since thousands of sensor nodes can be densely scattered in a sensor field, the MAC scheme must establish communication links for data transfer. This forms the basic infrastructure needed for hop-by-hop wireless communication and provides the self-organizing capability. The second objective is to fairly and efficiently share communication resources between sensor nodes. These resources include time, energy, and frequency. Several MAC protocols have been developed for WSNs to address these requirements over the last decade.

Regardless of the medium access scheme, energy efficiency is of utmost importance. A MAC protocol must certainly support the operation of power saving modes for the sensor node. The most obvious means of power conservation is to turn the transceiver off when it is not required. Though this power saving method seemingly provides significant energy gains, it may hamper the connectivity of the network. Once a transceiver is turned off, the sensor node cannot receive any packets from its neighbors, essentially becoming disconnected from the network. Moreover, turning a radio on and off has an overhead in terms of energy consumption due to the startup and shutdown procedures required for both hardware and software. In fact, if the radio is blindly turned off during each idling slot, over a period of time the sensor may end up expending more energy than if the radio had been left on. As a result, operation in a power saving mode is energy efficient only if the time spent in that mode is greater than a certain threshold. There can be a number of such useful modes of operation for the wireless sensor node, depending on the number of states of the microprocessor, memory, A/D converter, and the transceiver. Each of these modes can be characterized by its power consumption and the latency overhead, which is the transition power to and from that mode.

Error Control

Another important function of the data link layer is the error control of transmission data. Two important modes of error control in communication networks are forward error correction (FEC) and automatic repeat request (ARQ), and hybrid ARQ. The usefulness of ARQ in sensor network applications is limited by the additional retransmission cost and overhead. On the other hand, decoding complexity is greater in FEC, as error correction capabilities need to be built in. Consequently, simple error control codes with low-complexity encoding and decoding might present the best solutions for sensor networks. In the design of such a scheme, it is important to have a good knowledge of the channel characteristics and implementation techniques.

1.2.3   Network Layer

Sensor nodes are scattered densely in a field either close to or inside the phenomenon as shown in Figure 1.4. The information collected relating to the phenomenon should be transmitted to the sink, which may be located far from the sensor field. However, the limited communication range of the sensor nodes prevents direct communication between each sensor node and the sink node. This requires efficient multi-hop wireless routing protocols between the sensor nodes and the sink node using intermediate sensor nodes as relays. The existing routing techniques, which have been developed for wireless ad hoc networks, do not usually fit the requirements of the sensor networks. The networking layer of sensor networks is usually designed according to the following principles:

- Power efficiency is always an important consideration.
- Sensor networks are mostly data-centric.
- In addition to routing, relay nodes can aggregate the data from multiple neighbors through local processing.
- Due to the large number of nodes in a WSN, unique IDs for each node may not be provided and the nodes may need to be addressed based on their data or location.

An important issue for routing in WSNs is that routing may be based on data-centric queries. Based on the information requested by the user, the routing protocol should address different nodes that would provide the requested information. More specifically, the users are more interested in querying an attribute of the phenomenon rather than querying an individual node. For instance, "the areas where the temperature is over 70 °F (21 °C)" is a more common query than "the temperature read by node #47."

One other important function of the network layer is to provide internetworking with external networks such as other sensor networks, command and control systems, and the Internet. In one scenario, the sink nodes can be used as a gateway to other networks, while another scenario is to create a backbone by connecting sink nodes together and making this backbone access other networks via a gateway.

1.2.4   Transport Layer

The transport layer is especially needed when the network is planned to be accessed through the Internet or other external networks. TCP, with its current transmission window mechanisms, does not address the unique challenges posed by the WSN environment. Unlike protocols such as TCP, the end-to-end communication schemes in sensor networks are not based on global addressing. These schemes must consider that addressing based on data or location is used to indicate the destinations of the data packets. Factors such as power consumption and scalability, and characteristics like data-centric routing, mean sensor networks need different handling in the transport layer. Thus, these requirements stress the need for new types of transport layer protocols.

The development of transport layer protocols is a challenging task because the sensor nodes are influenced by hardware constraints such as limited power and memory. As a result, each sensor node cannot store large amounts of data like a server in the Internet, and acknowledgments are too costly for sensor networks. Therefore, new schemes that split the end-to-end communication probably at the sinks may be needed where UDP-type protocols are used in the sensor network.

For communication inside a WSN, transport layer protocols are required for two main functionalities: reliability and congestion control. Limited resources and high energy costs prevent end-to-end reliability mechanisms from being employed in WSNs. Instead, localized reliability mechanisms are necessary. Moreover, congestion that may occur because of the high traffic during events should be mitigated by the transport layer protocols. Since sensor nodes are limited in terms of processing, storage, and energy consumption, transport layer protocols aim to exploit the collaborative capabilities of these sensor nodes and shift the intelligence to the sink rather than the sensor nodes.

1.2.5 Application Layer

The application layer includes the main application as well as several management functionalities. In addition to the application code that is specific for each application, query processing and network management functionalities also reside at this layer.

The layered architecture stack has been initially adopted in the development of WSNs due to its success with the Internet. However, the large-scale implementations of WSN applications reveal that the wireless channel has significant impact on the higher layer protocols. Moreover, resource constraints and the application-specific nature of the WSN paradigm leads to *cross-layer* solutions that tightly integrate the layered protocol stack. By removing the boundaries between layers as well as the associated interfaces, increased efficiency in code space and operating overhead can be achieved.

In addition to the communication functionalities in the layered stack, WSNs have also been equipped with several functionalities that aid the operation of the proposed solutions. In a WSN, each sensor device is equipped with its own local clock for internal operations. Each event that is related to operation of the sensor device including sensing, processing, and communication is associated with timing information controlled through the local clock. Since users are interested in the collaborative information from multiple sensors, timing information associated with the data at each sensor device needs to be consistent. Moreover, the WSN should be able to correctly order the events sensed by distributed sensors to accurately model the physical environment. These timing requirements have led to the development of *time synchronization* protocols in WSNs.

The close interaction with physical phenomena requires location information to be associated in addition to time. WSNs are closely associated with physical phenomena in their surroundings. The gathered information needs to be associated with the location of the sensor nodes to provide an accurate view of the observed sensor field. Moreover, WSNs may be used for tracking certain objects for monitoring applications, which also requires location information to be incorporated into the tracking algorithms. Further, location-based services and communication protocols require position information. Hence, *localization* protocols have been incorporated into the communication stack.

Finally, several *topology management* solutions are required to maintain the connectivity and coverage of the WSN. The topology management algorithms provide efficient methods for network deployment that result in longer lifetime and efficient information coverage. Moreover, topology control protocols help determine the transmit power levels as well as the activity durations of sensor nodes to minimize energy consumption while still ensuring network connectivity. Finally, clustering protocols are used to organize the network into clusters to improve scalability and improve network lifetime.

The integration of each of the components for efficient operation depends on the applications running on the WSN. This application-dependent nature of the WSNs defines several unique properties compared to traditional networking solutions. Although the initial research and deployment of WSNs have focused on data transfer in wireless settings, several novel application areas of WSNs have also emerged. These include *wireless sensor and actor networks*, which consist of actuators in addition to sensors that convert sensed information into actions to act on the environment, and *wireless multimedia sensor networks*, which support multimedia traffic in terms of visual and audio information in addition to scalar data. Furthermore, recently the WSN phenomenon has been adopted in constrained environments such as underwater and underground settings to create *wireless underwater sensor networks* and *wireless underground sensor networks*. These new fields of study pose additional challenges that have not been considered by the vast number of solutions developed for *traditional* WSNs.

The flexibility, fault tolerance, high sensing fidelity, low cost, and rapid deployment characteristics of sensor networks create many new and exciting application areas for remote sensing. In the future, this wide range of application areas will make sensor networks an integral part of our lives. However, realization of sensor networks needs to satisfy the constraints introduced by factors such as fault tolerance, scalability, cost, hardware, topology change, environment, and power consumption. Since these constraints are highly stringent and specific for sensor networks, new wireless ad hoc networking techniques are required. Many researchers are currently engaged in developing the technologies needed for different layers of the sensor network protocol stack. Commercial viability of WSNs has also been shown in several fields. Along with the current developments, we encourage more insight into the problems and more development of solutions to the open research issues as described in this book.

References

[1] BTnode platform. http://www.btnode.ethz.ch/.

[2] Cricket indoor location system. http://nms.lcs.mit.edu/projects/cricket.

[3] Crossbow technology. http://www.xbow.com.

[4] IPv6 over low power WPAN working group. http://tools.ietf.org/wg/6lowpan/.

[5] ISA SP100.11a working group. http://www.isa.org/ISA100.

[6] National Instruments ni wsn-3202 wireless sensor. http://sine.ni.com/nips/cds/view/p/lang/en/nid/206921.

[7] Sensicast TEHU-1121 wireless sensor. http://www.sensicast.com/wireless_sensors.php.

[8] The Stargate platform. http://www.xbow.com/Products/Product_pdf_files/Wireless_pdf/Stargate_Datasheet.pdf.

[9] SunSPOT mote specifications. http://www.sunspotworld.com.

[10] TinyOS. Available at http://www.tinyos.net/.

[11] Wireless Industrial Networking Alliance (WINA). http://www.wina.org.

[12] WirelessHART communications protocol. http://www.hartcomm2.org/index.html.

[13] ZigBee Alliance. http://www.zigbee.org/.

[14] IEEE standard for information technology – telecommunications and information exchange between systems – local and metropolitan area networks – specific requirement part 15.4: wireless medium access control (MAC) and physical layer (PHY) specifications for low-rate wireless personal area networks (WPANS). *IEEE Std. 802.15.4a-2007 (Amendment to IEEE Std. 802.15.4-2006)*, pp. 1–203, 2007.

[15] Microstrain V-Link wireless sensor. http://www.microstrain.com/v-link.aspx, 2009.

[16] Shimmer project. http://www.eecs.harvard.edu/ konrad/projects/shimmer/, 2009.

[17] EYES project. http://www.eyes.eu.org/, March 2002–February 2005.

[18] I. F. Akyildiz, W. Su, Y. Sankarasubramaniam, and E. Cayirci. Wireless sensor networks: a survey. *Computer Networks*, 38(4):393–422, March 2002.

[19] Q. Cao, T. Abdelzaher, J. Stankovic, and T. He. The LiteOS operating system: towards Unix-like abstractions for wireless sensor networks. In *Proceedings of IPSN'08*, pp. 233–244, St. Louis, Missouri, USA, April 2008.

[20] A. Dunkels, B. Grönvall, and T. Voigt. Contiki – a lightweight and flexible operating system for tiny networked sensors. In *Proceedings of IEEE EmNets'04*, Tampa, Florida, USA, November 2004.

[21] P. Levis, N. Lee, M. Welsh, and D. Culler. Tossim: accurate and scalable simulation of entire TinyOS applications. In *Proceedings of ACM SenSys'03*, pp. 126–137, Los Angeles, CA, USA, November 2003.

[22] C. B. Margi, V. Petkov, K. Obraczka, and R. Manduchi. Characterizing energy consumption in a visual sensor network testbed. In *Proceedings of the IEEE/Create-Net International Conference on Testbeds and Research Infrastructures for the Development of Networks and Communities (TridentCom'06)*, Barcelona, Spain, March 2006.

[23] G. Mulligan and the 6LoWPAN Working Group. The 6LoWPAN architecture. In *Proceedings of EmNets'07*, pp. 78–82, Cork, Ireland, June 2007.

[24] L. F. W. van Hoesel, S. O. Dulman, P. J. M. Havinga, and H. J. Kip. Design of a low-power testbed for wireless sensor networks and verification. Technical report, University of Twente, 2003.

2

WSN Applications

The emergence of the WSN paradigm has triggered extensive research on many aspects of it. However, the applicability of sensor networks has long been discussed with emphasis on potential applications that can be realized using WSNs. In this chapter, we present an overview of the existing commercial and academic applications developed for WSNs.

WSNs may consist of many different types of sensors including seismic, magnetic, thermal, visual, infrared, acoustic, and radar, which are able to monitor a wide variety of ambient conditions that include the following [15, 26]: temperature, humidity, pressure, speed, direction, movement, light, soil makeup, noise levels, the presence or absence of certain kinds of objects, and mechanical stress levels on attached objects. As a result, a wide range of applications are possible. This spectrum of applications includes homeland security, monitoring of space assets for potential and human-made threats in space, ground-based monitoring of both land and water, intelligence gathering for defense, environmental monitoring, urban warfare, weather and climate analysis and prediction, battlefield monitoring and surveillance, exploration of the Solar System and beyond, monitoring of seismic acceleration, strain, temperature, wind speed and GPS data.

These ever-increasing applications of WSNs can be mainly categorized into five categories. In this chapter, several examples are described for military (Section 2.1), environmental (Section 2.2), health (Section 2.3), home (Section 2.4), and industry (Section 2.5) applications as illustrated in Figure 2.1.

2.1 Military Applications

WSNs can be an integral part of military *command, control, communications, computing, intelligence, surveillance, reconnaissance, and targeting* (C4ISRT) systems. The rapid deployment, self-organization, and fault tolerance characteristics of sensor networks make them a very promising sensing technique for military C4ISRT. Since sensor networks are based on the dense deployment of disposable and low-cost sensor nodes, destruction of some nodes by hostile action does not affect a military operation as much as the destruction of a traditional sensor, which makes the sensor network concept a better approach for battlefields. Some of the military applications of sensor networks are monitoring friendly forces, equipment, and ammunition; battlefield surveillance; reconnaissance of opposing forces and terrain; targeting; battle damage assessment; and nuclear, biological, and chemical (NBC) attack detection and reconnaissance [5, 9, 15, 29, 37]. Three of the representative military applications are discussed next.

2.1.1 Smart Dust

Smart Dust was one of the earliest applications of the WSN phenomenon. The main goal of this DARPA-funded project is to provide technologies for sensor networks that will be used for military operations in hostile environments. The use of WSNs extend the reach by deploying these networks in areas where

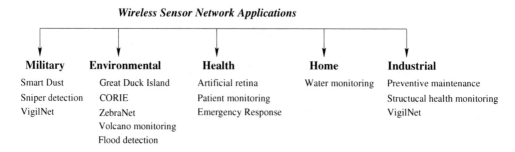

Figure 2.1   Main categories of WSN applications and the specific examples discussed in this chapter.

it is too dangerous for humans to operate continuously. WSNs obtain information needed to assess critical situations by dropping a robust, self-configuring, self-organizing WSN onto the battlefield. The military applications include collecting information from enemy movements, hazardous chemicals, and infrastructure stability. The main goal of the Smart Dust project has been developing sensor platforms in cubic millimeter packages. Instead of focusing on a particular sensor, the transceiver and microcontroller unit (MCU) design are performed.

The resulting sensor node is a 100 cubic millimeter prototype with two chips: a MEMS corner cube optical transmitter array and a CMOS ASIC with an optical receiver, charge pump, and simple digital controller. The Smart Dust project has resulted in the creation of Dust Networks Inc. [9], to commercialize this technology. Accordingly, the Smart Dust platform has been utilized in military applications such as battlefield surveillance, treaty monitoring, transportation monitoring, and scud hunting. In addition, other commercial applications are also viable with this technology. As an example, a virtual keyboard is developed, where each dust mote is attached to each fingernail and the acceleration information is used to detect movements and convert them into key strokes. Similarly, the Smart Dust node is used as a part of inventory control, where each item is equipped with a unique wireless sensor node that can communicate with the palette, truck, warehouse, and the Internet. Finally, Smart Dust nodes are used in product quality monitoring, where the temperature and humidity of meat, produce, and dairy products are monitored.

2.1.2   Sniper Detection System

The Boomerang sniper detection system [5] has been developed for accurate sniper location detection by pinpointing small-arms fire from the shooter. It has been used by the military, law-enforcement agencies, and municipalities. This application has been realized through two distinct architectures.

The countersniper system shown in Figure 2.2 consists of an array of microphones and can be mounted on a vehicle or worn by a soldier [5]. The system uses passive acoustic sensors to detect incoming fire. The detected audio from the microphones is processed to estimate the relative position of the shooter. The Boomerang system is helpful in urban settings, where a soldier or a vehicle can be subjected to fire from any location. The vehicle-mounted sensors can detect the shooter even while moving.

This centralized concept has been enhanced through a distributed network of acoustic sensors for a countersniper system as shown in Figure 2.3 to combat the effects of multi-path in sound detection [37]. The network consists of Mica2 nodes equipped with a custom-made sensor board. The sensor board is embedded with a high-power DSP to provide real-time detection, classification, and correlation of acoustic events. Once a shot has been detected by several sensors, the time and location information of these sensors is used to determine the trajectory of the bullet and estimate the location of the shooter. As a result, the distributed system improves the accuracy of the centralized system. This system is suitable for law-enforcement agencies and municipalities to provide protection during events such as speeches.

Figure 2.2   Boomerang sniper detection system [5].

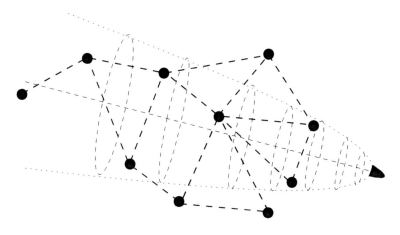

Figure 2.3   Distributed sniper detection system [37].

2.1.3   VigilNet

VigilNet [29] is a large-scale surveillance network designed for energy-efficient and stealthy target tracking in harsh environments. The design and implementation of the network were based on 70 Mica2 nodes equipped with magnetic sensors that detect the magnetic field generated by the movement of vehicles and magnetic objects. The main goal of this application is to provide energy-efficient surveillance support through distributed sensor nodes.

The lifetime of VigilNet is improved through a hierarchical structure of nodes in the network. Certain nodes denoted as *sentries* are responsible for maintaining the multi-hop routes and managing event-based operation of the *non-sentry* nodes, which remain in a low-power state until an event occurs. In case of an event, e.g., a vehicle passing a group of nodes, the network is reorganized into clusters including

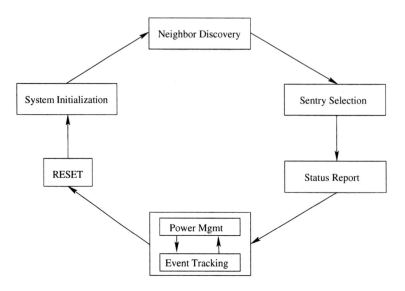

Figure 2.4   Operation phases of VigilNet [29].

the associated group of nodes with the help of sentries. This scheme provides event-based collaborative tracking with a long network lifetime.

The network operation of VigilNet is organized into several phases that are executed in a cycle as shown in Figure 2.4. The five phases are as follows:

1. *System initialization:* This phase includes synchronization, multi-hop backbone setup, and reconfiguration. Network-wide synchronization is essential for the accuracy of the tracking application. This is accomplished by broadcast messages from the sink node sent during the initialization phase. During the propagation of the broadcast messages, a backbone of sentries is also formed throughout the network. This backbone provides a stable path between the nodes and the sink and allows the nodes that are not part of the backbone to sleep. To avoid the diverse effects of asymmetric links between any pair of nodes, each node performs *link symmetry detection* (LSD) to filter out its neighbors with highly symmetric links. The backbone is then built using these links. The time synchronization packets are also used to reconfigure the network on demand. Accordingly, the operating parameters such as cycle length, duty cycle, sampling rate, and aggregation parameters are piggybacked on the synchronization packets. Finally, localization is performed manually during this phase. During deployment, an operator equipped with a GPS unit and a transceiver programs each node with its location based on the GPS reading on site. Apart from localization, each initialization service is performed at the beginning of each cycle.

2. *Neighbor discovery:* The initialization phase is followed by local broadcast of HELLO messages by each node. This message includes information about the node including its ID, whether it is a sentry or not, the number of sentries it is connected to, and its location. This information is used for local message exchanges and sentry selection.

3. *Sentry selection:* The sentries remain active during the network operation and are selected at the beginning of each cycle. Each node determines its duty as a sentry if it is part of the backbone or none of its neighbors is a sentry. This duty is communicated to other nodes using a broadcast message after a certain broadcast duration. Consequently, each node is guaranteed to be covered by at least one sentry.

4. *Status report:* This phase is used by each node to communicate its status to the base station using the backbone. The status report phase can be omitted, but is used for visualization and debugging purposes.

5. *Power management and event tracking:* The final phase of the operating cycle constitutes the majority of the lifetime of the network. Throughout the hierarchical structure, which consists of sentry and non-sentry nodes, a significant amount of energy is saved. The sleep and wakeup cycles of the non-sentry nodes are controlled by the sentry nodes through proactive or reactive control schemes. For proactive control, each sentry node broadcasts sleep beacons that let each non-sentry node switch to a sleep state. In case of an event, the sleep beacons are not sent and the non-sentry nodes participate in delivering information. On the other hand, for reactive control, the sentry nodes send beacons only when an event occurs to wake up the non-sentry nodes, which, otherwise, follow a sleep/wakeup schedule. When an event occurs, the nodes that sense the event form *event-based clusters* and a leader is selected. The sensed information is locally collected by the leader, which forwards this information to the sink only after a certain confidence level has been satisfied. The collected information consists of the location of the node and the timestamp related to the sensed event. Using this information the location and speed of the moving object are estimated.

Each cycle is periodically initiated by a reset phase followed by the system initialization phase. The system spends a very short time, e.g., 2 minutes, during the first four phases (system initialization, neighbor discovery, sentry selection, and status report), which update the system-wide parameters to be adopted in the dynamics of the network. The fifth phase (power management and event tracking) constitutes actual application operation, where the event-based tracking takes place. This phase may last up to a day until the next cycle. This sequential operation also helps isolate one system management task from another as well as the actual tracking. This helps conserve the limited resources and the wireless bandwidth.

VigilNet illustrates how several functionalities of a WSN are integrated into a coherent framework in a practical setting. The experimental evaluations reveal that the local aggregation and the hierarchical sleep/wakeup schemes significantly contribute to the energy efficiency of a tracking application in military settings.

In addition to these specific projects, WSNs have been widely used in battlefield surveillance applications, where exposure of military personnel to hazardous materials and enemy attack are minimized. Furthermore, WSNs are used for target tracking, where the location of a known or a mobile object is detected by a group of sensors and reported to the command control stations.

2.2 Environmental Applications

The autonomous coordination capabilities of WSNs are utilized in the realization of a wide variety of environmental applications. Some environmental applications of WSNs include tracking the movements of birds, small animals, and insects; monitoring environmental conditions that affect crops and livestock; irrigation; macroinstruments for large-scale Earth monitoring and planetary exploration; chemical/biological detection; precision agriculture; biological, Earth, and environmental monitoring in marine, soil, and atmospheric contexts; forest fire detection; meteorological or geophysical research; flood detection; biocomplexity mapping of the environment; and pollution studies [6, 10, 14, 17, 18, 19, 21, 30, 31, 32, 33, 34, 38, 49, 50, 51, 52, 54].

2.2.1 *Great Duck Island*

The Great Duck Island project [38, 50] has been developed as a collaboration between College of the Atlantic and the Intel Research Laboratory at Berkeley to study the distribution and abundance

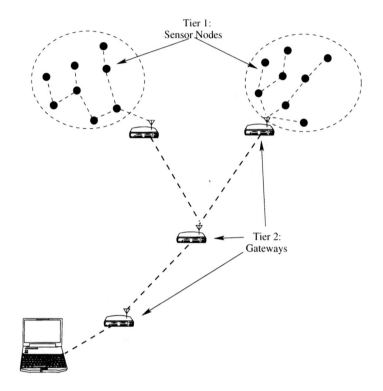

Figure 2.5   Two-tier architecture of Great Duck Island habitat monitoring network [38].

of seabirds on Great Duck Island (GDI), Maine. A network of Mica sensors is used to measure the occupancy of nesting burrows and the role of microclimatic factors in the habitat selection of seabirds, i.e., petrels. More specifically, the occupancy of the burrows during 1–3 days is monitored as well as the environmental changes during the breeding season and the corresponding changes in bird behavior.

The GDI habitat monitoring network employs a two-tier hierarchical architecture as shown in Figure 2.5. In the first tier, groups of sensors are used to gather information. For this tier, Mica motes equipped with Mica Weather Boards are used, which include temperature, photoresistor, barometric pressure, humidity, and thermopile sensors. More specifically, these sensor nodes are used for two different purposes: *burrow motes* are deployed in the burrows to detect occupancy using non-contact infrared thermopiles and temperature/humidity sensors, while *weather motes* are used to monitor the surface microclimate. Each group of sensors is connected to a gateway that collects these data and sends them to the second tier. The second tier consists of gateways that provide connectivity between the sensor groups in the field and the remote base station on the island through long-haul point-to-point communication. Two different platforms are used for this purpose: an embedded Linux system equipped with an IEEE 802.11 card and a 12 dBi omnidirectional antenna with a range of 1000 feet (30.5 m). Moreover, the Mica nodes are equipped with a 916 MHz directional antenna for a range of 1200 feet (365.75 m). The base station is also connected to the Internet by satellite, which enables remote access to the WSN.

The GDI project was conducted in 2002 and provided essential information on the characteristics of seabirds. In addition, the GDI network stands as one of the first proof-of-concept deployments of the WSN phenomenon for environmental applications.

2.2.2   CORIE

The Columbia River Ecosystem (CORIE) [6] is an example of an environmental observation and forecasting system (EOFS) built by the Center for Coastal and Land Margin Research at the Oregon Graduate Institute. It consists of sensor stations in the Columbia River estuary that carry various environmental sensors. In all, 24 stations in or around the Columbia River and other Oregon estuaries are equipped with various environmental sensors to measure water velocity, water temperature, salinity, and depth. Moreover, several meteorological stations are deployed to measure wind and air properties, and determine solar radiation at different wavelengths. The information gathered from these sensors is used for online control of vessels, marine research and rescue, and ecosystem research and management. The stations communicate via a Freewave DGR-115 spread-spectrum wireless network. In addition, the transmission of signals is accomplished by an ORBCOMM LEO satellite in case of disruptions in line of sight. Accordingly, the observations are displayed via the World Wide Web, in near real time.

The Columbia River estuary system includes the lower river, the estuary, and the near ocean. The goal of the CORIE system is to combine the observed information from the sensors with numerical data models to characterize the physical state of this system. More specifically, the goal is to characterize the complex circulation and mixing processes and predict any changes [24]. Since a large area of the river is covered by a limited number of sensors, the CORIE network represents significantly different characteristics than traditional sensor networks. Compared to low-cost sensor motes, the sensors used in CORIE are large complex systems that are expensive to deploy and maintain. Therefore, highly accurate data gathering solutions with a minimum number of stations are required. This necessitates estimation methods tailored to the data models used by the domain scientists.

2.2.3   ZebraNet

ZebraNet [33, 54] is an animal tracking system developed to investigate the long-term movement patterns of zebras, their interactions within and between species, as well as the impacts of human development. The system was deployed in Kenya to track two species of zebras.

ZebraNet consists of collars attached to the zebras to collect location information. Each collar includes a GPS unit, associated microcontroller, two types of radios (short range and long range), a lithium-ion polymer high-energy-density battery, and a solar array for recharging the battery as shown in Figure 2.6. The attached collar logs location information through the GPS unit every 3 minutes, which is shown through biological studies to result in statistically independent location samples that can be used to develop mobility models. This information is collected through a base station, which is kept by the researchers and used intermittently during their trips into the field. Therefore, ZebraNet can be characterized as a highly mobile sensor network without a static sink. Accordingly, the location information collected at each node is transmitted in a *delay-tolerant* manner, i.e., when the opportunity to communicate with the base station arises.

Since a particular node may not be within communication range of the mobile sink for a very long time, a data sharing policy is adopted in ZebraNet so that each sensor node shares the collected information with its neighbors. As a result, when one of the nodes establishes a connection with the sink, the location information of this node as well as its previous neighbors is downloaded. For data sharing, two different delay-tolerant routing protocols are used. Firstly, a simple flooding protocol is used, where each node sends its data to its neighbor whenever it detects a new neighbor. This improves the chance to deliver the data to the sink but increases the data overload in the network and limits the storage space at each node. The second protocol is based on the successful sink communication history of the nodes and assigns each node a hierarchy level [33]. If a node has recently communicated with a node it is assigned a higher level. Each node shares its location information with a node at a higher hierarchy level with higher probability.

Figure 2.6   The ZebraNet sensor collar attached to a zebra [54].

The location information of the zebras collected by a group of sensors provides the domain scientists with valuable information to model the movement of these animals. In addition to its contribution to the domain sciences, this movement information is also used to fine-tune the developed protocols according to the mobility patterns of the zebras. The network deployment has provided an important insight into the operation of these networks and resulted in the development of communication protocols such as Z-MAC [47] that was devised to address the unique challenges of this application.

2.2.4   Volcano Monitoring

WSNs have also been used in extreme environments, where continuous human access is impossible. Volcano monitoring is an example of these extreme applications, where a network of sensors can be easily deployed near active volcanoes to continuously monitor their activities and provide data at a scale and resolution not previously possible with existing tools.

Two case studies were conducted on two volcanoes in Ecuador during 2004–2005 as a proof of concept of WSN applications in volcano monitoring [52]. In 2004, a small network of three sensor nodes, equipped with microphones, was used to monitor an erupting volcano, i.e., Volcán Tangurahua in central Ecuador. In 2005, 16 TMote Sky nodes equipped with seismic and acoustic sensors were used for a duration of 19 days to monitor the activities of an active volcano, Volcán Reventador, in northern Ecuador. The sensor nodes were equipped with higher gain external antennas to improve the communication range and three long-haul communication nodes were used to transmit the data to a central controller covering a 3 km array. A laptop equipped with a directional antenna was used as a sink to collect information and manage the network remotely.

The main goal of the application was to collect seismic information based on earthquakes that occur near the volcanoes. Since these earthquakes usually last less than 60 seconds, a high sampling rate is employed for the seismic sensors, i.e., 100 Hz, which limits the amount of locally stored information to 20 minutes of volcanic activity. Each sensor node uses the local storage as a cyclic buffer and filters the collected information through a short-term average/long-term average threshold detector to determine the events related to volcanic activities. When an event is generated, it is reported to the sink. If high confidence of an event is established through multiple event reports, the sink sends back a data collection command to the network. As a response to this command, each node transmits the information stored in its memory. Consequently, a large volume of data is transmitted to the sink only if an event of interest is detected by a smaller number of nodes. This reduces the bandwidth wastage and energy consumption at individual nodes.

The sensor nodes are synchronized using the Flooding Time Synchronization protocol (FTSP) [40] and a GPS unit equipped with a MicaZ node to provide location information. The results of this application provide important information related to the physical processes at work within a volcano's interior.

2.2.5 Early Flood Detection

WSNs are also used for early warning flood detection in developing countries [17]. The main challenge with existing systems is the involvement of personnel for continuous monitoring of river beds. Instead, model-based prediction systems can be used by exploiting the statistical properties of data collected by a network of sensor nodes. Such a system has been developed at MIT and tested in Honduras, where frequent floods significantly affect urban life.

Flood monitoring requires a large area to be covered with sensors. Considering the limited communication range of individual sensors, a two-tier network topology is used as shown in Figure 2.7. Three different sensors are used in the lower tier of the network for measuring rainfall, air temperature, and water flow data. These types of data are required for the prediction model. Each closely located sensor forms a group and is connected to the second-tier computation nodes. Data collection and information processing are performed at the computation nodes, which inform the third tier, i.e., control centers, in case of a potential flood. Overall, four different types of nodes are used in the system. Each node has the same base system that includes a microcontroller and a transceiver but is enhanced with sister-boards according to its function:

- **Sensing node:** This node is equipped with one of the sensors monitoring rainfall, air temperature, and water flow and collects information over a period on the order of minutes [17]. Each node is equipped with a 900 MHz transceiver and reports its data to the computation node in its cluster. While the rainfall and temperature sensors are located on the ground, the water flow data requires underwater measurements. The water level can be easily converted into water flow. To measure the water level, a water pressure sensor is interfaced with the base board and deployed under water.

- **Computation node:** Each group of closely located sensing nodes is maintained by a computation node as shown in Figure 2.7. The computation node has two main duties in the network. Firstly, the data set collected from the group of sensors is fed into the prediction model, which estimates the river flow. In case the prediction confidence is not high, the computation node requests additional information from the sensing nodes. Secondly, the computation node provides long-haul communication between the group and the control and monitoring nodes through a 144 MHz transceiver. The information collected by different groups of sensing nodes is collected at the interface nodes. Through long-haul communication techniques, the computation nodes can directly reach the interface nodes. This two-tier approach improves the scalability of the system.

- **Government office interface node:** This node provides a user interface to the network by visualizing data and enabling network maintenance tasks. Larger scale predictions can also be performed using the data received from multiple sensors located at different locations.

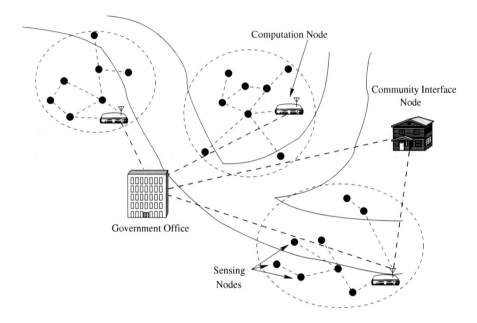

Figure 2.7   Sensor network architecture for flood detection [17].

- **Community interface node:** Finally, the flood prediction system will be connected to community interface nodes that inform the interested community about the results of the prediction. This allows the necessary action to be taken timely in the case of severe floods.

The prediction model has been validated using 7-year data collected from the Blue River, Oklahoma, through the Distributed Model Intercomparison project run by the National Oceanic and Atmospheric Administration. Moreover, the lower tier of the system has been tested in Charles River, Massachusetts, during October–November 2007 using a group of three sensors. Finally, the flood prediction system was deployed in Honduras over a long period of time between January 2004 and March 2007.

2.3   Health Applications

The developments in implanted biomedical devices and smart integrated sensors make the usage of sensor networks for biomedical applications possible. Some of the health applications for sensor networks are the provision of interfaces for the disabled; integrated patient monitoring; diagnostics; drug administration in hospitals; monitoring the movements and internal processes of insects or other small animals; telemonitoring of human physiological data; and tracking and monitoring doctors and patients inside a hospital [4, 28, 34, 39, 41, 46, 51, 53].

2.3.1   Artificial Retina

The Artificial Retina (AR) project supported by the US Department of Energy aims to build a chronically implanted artificial retina for visually impaired people [4]. More specifically, the project focuses on addressing two retinal diseases, namely age-related macular degeneration (AMD) and retinitis pigmentosa (RP). AMD is an age-related disease and results in severe vision loss at the center of the retina caused by fluid leakage or bleeding in people aged 60 and above. RP, on the other hand, affects

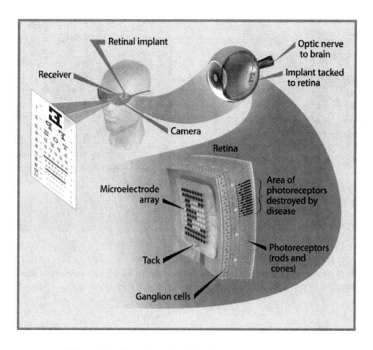

Figure 2.8 Operating principle of the artificial retina [4].

the photoreceptor or rod cells, which results in a loss of peripheral vision [4]. A healthy photoreceptor stimulates the brain through electric impulses when light is illuminated from the external world. When damaged, vision is blocked at the locations of the photoreceptors. The AR project aims to replace these damaged photoreceptors with an array of microsensors.

The receptors along the retina have different functionalities. The periphery is specialized on temporal events, whereas the center of the macula is specialized on spatially oriented information. In order to mimic the functionality of the retina, the distribution and the transmission principles of smart sensors on each segment of the retina must be tailored.

The operating principle of the AR is shown in Figure 2.8. A tiny camera placed in eyeglasses is used to capture visual information from the external world. This information is wirelessly transmitted to a microprocessor, which converts it into electrical signals. These signals are then sent to an array of microelectrodes, which are implanted along the retina at the locations of the photoreceptors. The microelectrodes then send electric impulses to the brain via the optic nerve.

So far, three models have been developed as part of the AR project. The first model, Argus I, has been completely tested and implanted into six patients between 2002 and 2004. This model consists of a 16-electrode array and helps the patients to detect whether lights are on or off, describe the motion of an object, count individual items, and locate objects. The second model, Argus II, consists of a 60-electrode array and is undergoing clinical trials. Finally, the third model, which is under development, will include more than 200 electrodes. The ultimate goal for the prosthetic device is to create a lasting device that will enable facial recognition and the ability to read large print [4].

(a)   (b)

Figure 2.9 Medical sensor nodes used in CodeBlue: (a) Telos mote with EMG sensor and (b) Mica2 mote with pulse oximeter sensor [7].

2.3.2 Patient Monitoring

The CodeBlue project at Harvard University focuses on wearable sensors that monitor vital signs of patients throughout their daily lives [39]. To this end, sensor boards with pulse oximeter, electrocardiograph (EKG), and electromyograph (EMG) circuitry have been designed for MicaZ and Telos motes as shown in Figure 2.9. Accordingly, pulse rate, blood oxygen saturation, electrical activities of the heart, patient movements, and muscular activity can be monitored continuously. The CodeBlue software platform enables these nodes to be operated in a networked setting, where medical personnel can monitor patients through a PDA.

The architecture of the CodeBlue network is shown in Figure 2.10 in a hospital setting. Accordingly, sensor motes with various sensors are attached to patients for monitoring. Additional motes are deployed inside the hospital as an infrastructure to provide message delivery to/from the motes. The medical personnel can access the network through either PDAs or computers. The network is based on a *publish/subscribe* mechanism, where the sensor motes publish the available data and the medical personnel subscribe to this data based on certain conditions. The publish/subscribe mechanism is based on the Adaptive Demand-Drive Multicast Routing (ADMR) protocol, which maintains routing information based on the published information. Whenever a mote publishes data, each node in the network is informed about the best path to this node. Route discovery between nodes is performed periodically through limited flooding to adapt to the changes in the wireless channel.

When a sensor is connected to the network, its capabilities are communicated to the network through a discovery protocol. Subscribing to the published data is performed through the CodeBlue query (CBQ) layer, which specifies a query through a tuple $\langle S, \tau, chan, \rho, C, p \rangle$, where S is the set of subscribed node IDs, τ is the subscribed sensor type, *chan* is the channel type used for routing, ρ is the sampling rate, and C is the optional maximum number of samples to be reported. To model the *interest*s of the users, a filter predicate, p, is also used in the query, according to which the sensor information to be reported is filtered. Accordingly, a doctor may specify a sensor connected to a particular patient and request to receive heartbeat information only if it exceeds normal limits.

In addition to the publish/subscribe mechanism, CodeBlue also provides a localization service through a set of nodes that periodically broadcast beacon messages. A mobile node can then estimate its location based on the received RF signal signature from these beacon nodes. This information is essential in

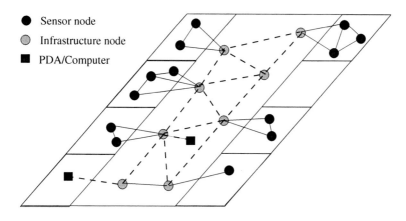

Figure 2.10 CodeBlue architecture in a hospital setting.

hospitals to locate doctors, nurses, patients, and important equipment in a timely fashion. Finally, a graphical user interface (GUI) is provided to users so that medical personnel can easily utilize the system.

ALARM-NET provides similar monitoring capabilities for elderly people in assisted-living and residential monitoring applications [53]. A hierarchical network of sensor nodes monitors environmental and physiological data. At the lowest layer, a body sensor network (BSN) is worn by the resident(s), which collects vital signs. Moreover, *emplaced* sensors are deployed in the living space to sense information related to temperature, dust, motion, and light. Finally, the heterogeneous set of sensor nodes reports to *AlarmGates*, which are connected through an IP network. Accordingly, the daily activities of residents can be monitored and reported to the medical personnel.

2.3.3   Emergency Response

WSNs are also used in emergency response situations to automate monitoring and assessment of the vital signs of casualties [28]. The miTag platform has been developed to include several medical sensors including GPS, pulse oximetry, blood pressure, temperature, and ECG. Using these sets of sensors, the emergency responders can monitor the vital signs of casualties in an accident. The miTag architecture consists of two tiers of networks. The first tier is a body sensor network including the various sensors and the second tier is a relay network consisting of nodes dispensed by responders on the scene. Accordingly, an ad hoc network can be formed rapidly to collect information from multiple casualties in a large disaster or an accident.

2.4   Home Applications

As technology advances, smart sensor nodes and actuators can be buried in appliances such as vacuum cleaners, microwave ovens, refrigerators, and DVD players [43] as well as water monitoring systems [35]. These sensor nodes inside domestic devices can interact with each other and with the external network via the Internet or satellite. They allow end-users to more easily manage home devices both locally and remotely. Accordingly, WSNs enable the interconnection of various devices at residential places with convenient control of various applications at home.

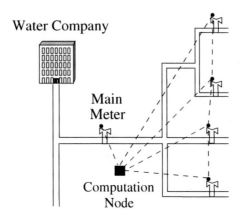

Water Company

Main
Meter

Computation
Node

Figure 2.11   Non-intrusive Autonomous Water Monitoring System (NAWMS) architecture [35].

2.4.1   Water Monitoring

The Nonintrusive Autonomous Water Monitoring System (NAWMS) for homes has recently been developed using WSNs [35]. The main goal of NAWMS is to localize the wastage in water usage and inform tenants about more efficient usage. Since the water utility companies only provide total water usage in a house, it is not easy to determine the individual sources that contribute to that total. Using a distributed WSN, the water usage in each pipe of the house's plumbing system can be monitored at a low cost.

The operating principle of NAWMS is based on the fact that the water flow in a particular pipe can be estimated by measuring the vibrations of that pipe because of the proportional relationship between the two. Accordingly, wireless sensor nodes are attached to the water pipes to measure the vibrations through accelerometers. Because of the nonlinear relationship between vibration and water flow, however, each sensor node needs to be calibrated to determine the optimal set of parameters that relate acceleration information to water flow. Instead of manual calibration that can be performed by individually installing water flow sensors at each pipe, in NAWMS this calibration is performed automatically with the help of the main water meter.

The architecture of NAWMS is illustrated in Figure 2.11, which consists of three types of components. A wireless sensor node is attached to the main water meter to collect actual water flow information and communicate it to the rest of the network. Since the water meter sensor is already calibrated by the utility company, this node serves as a *ground truth* for the rest of the system. The second type of component is the vibration sensors, which are installed in each pipe to monitor vibrations. The distributed vibration information is then sent to the central computation node, which automatically calibrates the sensors and determines the water usage associated with each pipe. Calibration is performed through an online optimization algorithm, which is based on the fact that the sum of the water flows at each non-calibrated sensor should be equal to that of the water meter, which is known to be accurate. Accordingly, the vibrations are used in a microscopic flow model to estimate the flow rate at spatially distributed locations. The system provides real-time water usage information at different locations of the water pipe system, which can be utilized to improve the efficiency of homes in the future.

2.5   Industrial Applications

Networks of wired sensors have long been used in industrial fields such as industrial sensing and control applications, building automation, and access control. However, the cost associated with the deployment of wired sensors limits the applicability of these systems. Moreover, even if a sensor system were deployed in an industrial plant, upgrading this system would cost almost as much as a new system. In addition to sensor-based monitoring systems, manual monitoring has also been used in industrial applications for preventive maintenance [36]. Manual monitoring is generally performed by experienced personnel using handheld analyzers that are collected from a central location for analysis. While sensor-based systems incur high deployment costs, manual systems have limited accuracy and require personnel. Instead, WSNs are a promising alternative solution for these systems due to their ease of deployment, high granularity, and high accuracy provided through battery-powered wireless communication units.

Some of the commercial applications are monitoring material fatigue; building virtual keyboards; managing inventory; monitoring product quality; constructing smart office spaces; environmental control of office buildings; robot control and guidance in automatic manufacturing environments; interactive toys; interactive museums; factory process control and automation; monitoring disaster areas; smart structures with embedded sensor nodes; machine diagnosis; transportation; factory instrumentation; local control of actuators; detecting and monitoring car theft; vehicle tracking and detection; instrumentation of semiconductor processing chambers, rotating machinery, wind tunnels, and anechoic chambers; and distributed spectrum sensing to help realize cognitive radio networks [1, 2, 11, 12, 13, 14, 16, 20, 22, 23, 25, 26, 36, 27, 34, 42, 44, 45, 46, 48, 51].

2.5.1   Preventive Maintenance

Preventive maintenance has been utilized in many large industrial plants since it provides cost-effective solutions to long-term operation of expensive equipment. However, existing systems cannot gain wide acceptance since the cost associated with the deployment of a preventive maintenance system outweighs the gains. WSNs have been shown to provide cost-effective yet accurate maintenance capabilities in two case studies: a semiconductor fabrication plant and an oil tanker [36].

The "health" of equipment can be monitored through *vibration analysis* techniques that require accelerometer sensors attached to the equipment. In *Intel's semiconductor fabs*, thousands of sensors track the vibrations of various pieces of equipment. Based on the established science that maps a particular signature to a well-functioning device, the machines are monitored continuously. However, the data from the sensors are collected manually by employees. In an effort to automate the data collection, a network of Mica2 and Intel Mote sensors along with Stargate gateways, namely *FabApp*, has been deployed.

The FabApp architecture is shown in Figure 2.12, where the hierarchical three-tier network is illustrated. The lowest tier consists of either Mica2 or Intel motes connected to multiple accelerometers to collect vibration information. This tier is organized into clusters and each cluster is controlled by a Stargate gateway node to improve the scalability of the network. The data collection is controlled by each gateway node through a duty cycle scheme. The gateways form the second tier, i.e., an IEEE 802.11 mesh network of high-end nodes that is overlaid on the first-tier sensor network. The data collected by each gateway are transmitted to a *root node* over the higher data-rate 802.11 connections. The root node is also connected to the enterprise network through a cable, which forms the third tier. Using this architecture, the data collected by the entire network are sent to the enterprise server for analysis.

FabApp illustrates a practical implementation of the cluster-based protocols for industrial applications. The sensor node activities are controlled by the gateway to decrease the processing required at each node. Accordingly, the gateway notifies the report and sleep durations to the sensor nodes and each sensor node wakes up at the same time to report its vibration information and switch to sleep state for the rest of the

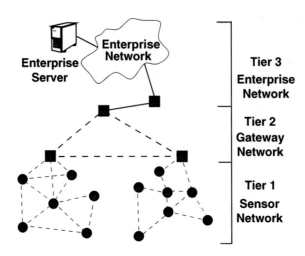

Figure 2.12   FabApp architecture [36].

cycle. In addition, both sensor nodes and gateway are programmed such that the software is reset at the end of each reset cycle. This removes any state dependency for each cycle and ensures long-term unattended operation in the network. Furthermore, several watchdog timers are implemented to ensure the embedded devices operate properly.

The same tiered structure is also deployed in an oil tanker to monitor equipment through a WSN. While the application goals are similar to FabApp, the environmental differences significantly affect the design choices. Since the ship consists of steel floors and bulkheads and the compartments are divided by hatchways, signal propagation and wireless communication are major challenges [36]. This requires pre-deployment analysis to determine suitable locations for the gateways. Moreover, a more robust data delivery protocol is required so that data are not lost whenever hatchways are closed and the network is partitioned. Both deployments have been operated for several days with valuable information collected by distributed sensors and provide a proof-of-concept for industrial applications of WSNs.

2.5.2   Structural Health Monitoring

WSNs have also been used in structural health monitoring (SHM) applications, where distributed sensors track the spatio-temporal patterns of vibrations induced throughout the structures [23]. Accordingly, potential damage can be localized and its extent can be estimated in almost real time. Wisden [42] has been developed to address the limitations of existing SHM techniques which rely on either periodic visual inspections or expensive wired data acquisition systems. To this end, actuators, which apply forced excitation to the structure, are networked with sensors, which detect the effects of these excitations. Accordingly, the challenges of intolerance to data loss, strict synchronization requirements, and large volumes of raw data are addressed through a distributed WSN. The system has been deployed in a SHM testbed for development as well as in an abandoned building for evaluation. This system has also been extended to the netSHM [22], which provides a two-level architecture such that the details of the underlying WSN operation are hidden from the application. This provides the flexibility of using higher level languages such as MATLAB or C, with which domain experts are much more comfortable.

2.5.3   Other Commercial Applications

Wireless automatic meter reading (AMR) is one of the fastest growing markets for short-range radio devices. Wireless collection of utility meter data (electricity, water, gas) is a very cost-efficient way of gathering consumption data for the billing system [3, 8]. *Chipcon* produces low-cost, low-power radio chips and transceivers for wireless AMR applications [1].

Heating, ventilating, and air-conditioning (HVAC) applications are another field where WSNs have an important impact. In commercial buildings, it is common to control multiple spaces or rooms by a single HVAC unit. Hence, systems configured this way are most commonly controlled by a single sensor in one of the rooms. However, low-cost wireless sensor technology offers the opportunity to replace this single sensor with a network of sensors where there is at least one sensor per room. *ZenSys* produces wireless RF-based communications technology designed for residential and light commercial control and status reading applications such as meter reading, lighting, and appliance control [13].

Sensicast is producing the *H900 Sensor Network Platform*, a wireless mesh networking system, which can be used in many industrial applications [11]. Moreover, *SYS Technologies* (formerly Xsiology) produces systems for real-time monitoring of a wide variety of remote industrial applications including wastewater, oil and gas, utilities, and railroads [12].

Soflinx provides a perimeter security system for the real-time detection of hazardous explosive, nuclear, biological, and chemical warfare agents. *Soflinx* uses its *Datalinx* technology which pushes the intelligence of the network closer to the source of the data by using gateways in the network edges and enabling the individual components to automate responses to these data. Moreover, a transportable security system for the detection of the above hazardous agents has also been produced [2].

References

[1] Available at http://www.chipcon.com.

[2] Available at http://www.soflinx.com.

[3] Altera Corp. http://www.altera.com/.

[4] Artificial Retina project. Available at http://artificialretina.energy.gov.

[5] Boomerang shooter detection system. http://bbn.com/boomerang.

[6] Center for Coastal Margin Observation and Prediction. http://www.ccalmr.ogi.edu/corie.

[7] CodeBlue project. Available at http://fiji.eecs.harvard.edu/CodeBlue.

[8] DataRemote Inc. http://www.dataremote.com/.

[9] Dust Networks Inc. http://www.dust-inc.com.

[10] ENSCO Inc. http://www.ensco.com.

[11] Sensicast systems. Available at http://www.sensicast.com.

[12] SYS Technologies (formerly Xsiology). http://www.systechnologies.com.

[13] Zensys, Inc. http://www.zen-sys.com.

[14] J. Agre and L. Clare. An integrated architecture for cooperative sensing networks. *IEEE Computer Magazine*, 33(5):106–108, May 2000.

[15] I. F. Akyildiz, W. Su, Y. Sankarasubramaniam, and E. Cayirci. Wireless sensor networks: a survey. *Computer Networks*, 38(4):393–422, March 2002.

[16] I.F. Akyildiz, W.-Y. Lee, M. C. Vuran, and S. Mohanty. Next generation/dynamic spectrum access/cognitive radio wireless networks: a survey. *Computer Networks*, 50(13):2127–2159, September 2006.

[17] E. A. Basha, S. Ravela, and D. Rus. Model-based monitoring for early warning flood detection. In *Proceedings of ACM SenSys'08*, pp. 295–308, Raleigh, NC, USA, November 2008.

[18] M. Bhardwaj, T. Garnett, and A. P. Chandrakasan. Upper bounds on the lifetime of sensor networks. In *Proceedings of IEEE ICC'01*, volume 3, pp. 785–790, Helsinki, Finland, June 2001.

[19] P. Bonnet, J. Gehrke, and P. Seshadri. Querying the physical world. *IEEE Personal Communications*, 7(5):10–15, October 2000.

[20] N. Bulusu, D. Estrin, L. Girod, and J. Heidemann. Scalable coordination for wireless sensor networks: self-configuring localization systems. In *Proceedings of the International Symposium on Communication Theory and Applications (ISCTA'01)*, Ambleside, UK, July 2001.

[21] A. Cerpa, J. Elson, M. Hamilton, and J. Zhao. Habitat monitoring: application driver for wireless communications technology. In *Proceedings of ACM SIGCOMM'01*, pp. 20–41, Costa Rica, April 2001.

[22] K. Chintalapudi, J. Paek, O. Gnawali, T. Fu, K. Dantu, J. Caffrey, R. Govindan, and E. Johnson. Structural damage detection and localization using NetSHM. In *Proceedings of IPSN/SPOTS'06*, pp. 475–482, Nashville, TN, USA, April 2006.

[23] K. Chintalapudi, J. Paek, N. Kothari, S. Rangwala, J. Caffrey, R. Govindan, E. Johnson, and S. Masri. Monitoring civil structures with a wireless sensor network. *IEEE Internet Computing*, 10(2):26–34, March/April 2006.

[24] T. Dang, S. Frolov, N. Bulusu, W. Feng, and A. Baptista. Near optimal sensor selection in the COlumbia RIvEr (CORIE) observation network for data assimilation using genetic algorithms. In *Proceedings of DCOSS'07*, pp. 253–266, Santa Fe, NM, USA, June 2007.

[25] D. Estrin, R. Govindan, and J. Heidemann. Embedding the internet. *Communications of the ACM*, 43(5):38–41, May 2000.

[26] D. Estrin, R. Govindan, J. Heidemann, and S. Kumar. Next century challenges: scalable coordination in sensor networks. In *Proceedings of ACM/IEEE MobiCom'99*, pp. 263–270, Seattle, WA, USA, August 1999.

[27] V. Fodor, I. Glaropoulos, and L. Pescosolido. Detecting low-power primary signals via distributed sensing to support opportunistic spectrum access. In *Proceedings of IEEE ICC'09*, Dresden, Germany, June 2009.

[28] T. Gao, C. Pesto, L. Selavo, Y. Chen, J. G. Ko, J.H. Lim, A. Terzis, A. Watt, J. Jeng, B. r. Chen, K. Lorincz, and M. Welsh. Wireless medical sensor networks in emergency response: implementation and pilot results. In *Proceedings of IEEE International Conference on Technologies for Homeland Security*, pp. 187–192, Waltham, MA, USA, May 2008.

[29] T. He, S. Krishnamurthy, J. A. Stankovic, T. Abdelzaher, L. Luo, R. Stoleru, T. Yan, L. Gu, G. Zhou, J. Hui, and B. Krogh. VigilNet: an integrated sensor network system for energy-efficient surveillance. *ACM Transactions on Sensor Networks*, 2(1):1–38, February 2006.

[30] W. R. Heinzelman, J. Kulik, and H. Balakrishnan. Adaptive protocols for information dissemination in wireless sensor networks. In *Proceedings of MobiCom'99*, pp. 174–185, Seattle, WA, USA, August 1999.

[31] C. Intanagonwiwat, R. Govindan, and D. Estrin. Directed diffusion: a scalable and robust communication paradigm for sensor networks. In *Proceedings of MobiCom'00*, pp. 56–67, Boston, MA, USA, August 2000.

[32] C. Jaikaeo, C. Srisathapornphat, and C. Shen. Diagnosis of sensor networks. In *Proceedings of IEEE ICC'01*, volume 5, pp. 1627–1632, Helsinki, Finland, June 2001.

[33] P. Juang, H. Oki, Y. Wang, M. Martonosi, L. S. Peh, and D. Rubenstein. Energy-efficient computing for wildlife tracking: design tradeoffs and early experiences with ZebraNet. *ACM SIGOPS Operating Systems Review*, 36(5):96–107, 2002.

[34] J. M. Kahn, R. H. Katz, and K. S. J. Pister. Next century challenges: mobile networking for "smart dust". In *Proceedings of ACM MobiCom'99*, pp. 271–278, Seattle, WA, USA, August 1999.

[35] Y. Kim, T. Schmid, Z. M. Charbiwala, J. Friedman, and M. B. Srivastava. NAWMS: Nonintrusive Autonomous Water Monitoring System. In *Proceedings of ACM SenSys'08*, pp. 309–322, Raleigh, NC, USA, November 2008.

[36] L. Krishnamurthy, R. Adler, P. Buonadonna, J. Chhabra, M. Flanigan, N. Kushalnagar, L. Nachman, and M. Yarvis. Design and deployment of industrial sensor networks: experiences from a semiconductor plant and the North Sea. In *Proceedings of SenSys'05*, pp. 64–75, San Diego, CA, USA, 2005.

[37] Á. Lédeczi, A. Nádas, P. Völgyesi, G. Balogh, B. Kusy, J. Sallai, G. Pap, S. Dóra, K. Molnár, M. Maróti, and G Simon. Countersniper system for urban warfare. *ACM Transactions on Sensor Networks*, 1(2):153–177, November 2005.

[38] A. Mainwaring, J. Polastre, R. Szewczyk, D. Culler, and J. Anderson. Wireless sensor networks for habitat monitoring. In *Proceedings of ACM WSNA'02*, Atlanta, GA, USA, September 2002.

[39] D. Malan, T. Fulford-Jones, M. Welsh, and S. Moulton. CodeBlue: an Ad Hoc sensor network infrastructure for emergency medical care. In *Proceedings of Workshop on Applications of Mobile Embedded Systems (WAMES 2004)*, Boston, MA, USA, June 2004.

[40] M. Maroti, B. Kusy, G. Simon, and A. Ledeczi. The flooding time synchronization protocol. In *Proceedings of ACM SenSys'04*, Baltimore, MD, USA, November 2004.

[41] N. Noury, T. Herve, V. Rialle, G. Virone, E. Mercier, G. Morey, A. Moro, and T. Porcheron. Monitoring behavior in home using a smart fall sensor. In *IEEE–EMBS Special Topic Conference on Microtechnologies in Medicine and Biology*, pp. 607–610, Lyon, France, October 2000.

[42] J. Paek, K. Chintalapudi, J. Cafferey, R. Govindan, and S. Masri. A wireless sensor network for structural health monitoring: performance and experience. In *Proceedings of the 2nd IEEE Workshop on Embedded Networked Sensors (EmNetS-II)*, Sydney, Australia, May 2005.

[43] E. M. Petriu, N. D. Georganas, D. C. Petriu, D. Makrakis, and V. Z. Groza. Sensor-based information appliances. *Instrumentation & Measurement Magazine*, 3(4):31–35, December 2000.

[44] G. J. Pottie and W. J. Kaiser. Wireless integrated network sensors. *Communications of the ACM*, 43:51–58, May 2000.

[45] N. Priyantha, A. Chakraborty, and H. Balakrishnan. The cricket location-support system. In *Proceedings of ACM MobiCom'00*, Boston, MA, USA, pp. 32–43, August 2000.

[46] J. M. Rabaey, M. J. Ammer, J. L. da Silva Jr., D. Patel, and S. Roundy. Pico radio supports ad hoc ultra-low power wireless networking. *IEEE Computer Magazine*, 33(7):42–48, 2000.

[47] I. Rhee, A. Warrier, M. Aia, and J. Min. Z-mac: a hybrid mac for wireless sensor networks. In *Proceedings of ACM SenSys'05*, pp. 90–101, San Diego, CA, USA, 2005.

[48] E. Shih, S. H. Cho, N. Ickes, R. Min, A. Sinha, A. Wang, and A. Chandrakasan. Physical layer driven protocol and algorithm design for energy-efficient wireless sensor networks. In *Proceedings of ACM MobiCom'01*, pp. 272–287, Rome, Italy, 2001.

[49] S. Slijepcevic and M. Potkonjak. Power efficient organization of wireless sensor networks. In *Proceedings of IEEE ICC'01*, Helsinki, Finland, June 2001.

[50] R. Szewczyk, E. Osterweil, J. Polastre, M. Hamilton, A. Mainwaring, and D. Estrin. Application driven systems research: habitat monitoring with sensor networks. *Communications of the ACM Special Issue on Sensor Networks*, 47(6):34–40, June 2004.

[51] B. Warneke, B. Liebowitz, and K. S. J. Pister. Smart Dust: communicating with a cubic-millimeter computer. *IEEE Computer*, 34(1):2–9, January 2001.

[52] G. Werner-Allen, K. Lorincz, M. Ruiz, O. Marcillo, J. Johnson, J. Lees, and M. Welsh. Deploying a wireless sensor network on an active volcano. *IEEE Internet Computing*, 10(2):18–25, March/April 2006.

[53] A. Wood, G. Virone, T. Doan, Q. Cao, L. Selavo, Y. Wu, L. Fang, Z. He, S. Lin, and J. Stankovic. ALARM-NET: wireless sensor networks for assisted-living and residential monitoring. Technical report CS-2006-11, Department of Computer Science, University of Virginia, 2006.

[54] P. Zhang, C. Sadler, S. Lyon, and M. Martonosi. Hardware design experiences in ZebraNet. In *Proceedings of ACM SenSys'04*, Baltimore, MD, USA, November 2004.

3

Factors Influencing WSN Design

The design of WSNs requires ample knowledge of a wide variety of research fields including wireless communication, networking, embedded systems, digital signal processing, and software engineering. This is motivated by the close coupling between several hardware and software entities of wireless sensor devices as well as the distributed operation of a network of these devices. Consequently, several factors exist that significantly influence the design of WSNs. In this chapter, the major factors are described including the *hardware constraints*, *fault tolerance*, *scalability*, *production costs*, *sensor network topology*, *transmission media*; and *power consumption*. These factors have been addressed by many researchers in a wide range of areas concerning the design and deployment of WSNs. Moreover, the integration of the solutions for these factors is still a major challenge because of the interdisciplinary nature of this research area.

3.1 Hardware Constraints

The general architecture and the major components of a wireless sensor device (node) are illustrated in Figure 3.1. A wireless sensor device is generally composed of four basic components: a *sensing unit*, a *processing unit*, a *transceiver unit* and a *power unit*. Moreover, additional components can also be integrated into the sensor node depending on the application. These components as shown by the dashed boxes in Figure 3.1 include: a *location finding system*, a *power generator*, and a *mobilizer*. Next, each component is described in detail:

- **Sensing unit:** The sensing unit is the main component of a wireless sensor node that distinguishes it from any other embedded system with communication capabilities. The sensing unit may generally include several *sensor units*, which provide information gathering capabilities from the physical world. Each sensor unit is responsible for gathering information of a certain type, such as temperature, humidity, or light, and is usually composed of two subunits: a sensor and an analog-to-digital converter (ADC). The analog signals produced by the sensor based on the observed phenomenon are converted to digital signals by the ADC, and then fed into the processing unit.

- **Processing unit:** The processing unit is the main controller of the wireless sensor node, through which every other component is managed. The processing unit may consist of an on-board memory or may be associated with a small storage unit integrated into the embedded board. The processing unit manages the procedures that enable the sensor node to perform sensing operations, run associated algorithms, and collaborate with the other nodes through wireless communication.

- **Transceiver unit:** Communication between any two wireless sensor nodes is performed by the transceiver units. A transceiver unit implements the necessary procedures to convert bits to be transmitted into radio frequency (RF) waves and recover them at the other end. Essentially, the WSN is connected to the network through this unit.

Wireless Sensor Networks   Ian F. Akyildiz and Mehmet Can Vuran
© 2010 John Wiley & Sons, Ltd

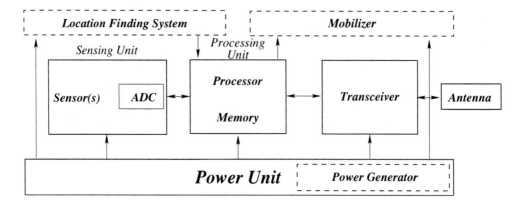

Figure 3.1   General hardware architecture of a sensor node.

- **Power unit:** One of the most important components of a wireless sensor node is the power unit. Usually, battery power is used, but other energy sources are also possible. Each component in the wireless sensor node is powered through the power unit and the limited capacity of this unit requires energy-efficient operation for the tasks performed by each component.

- **Location finding system:** Most of the sensor network applications, sensing tasks, and routing techniques need knowledge of the physical location of a node. Thus, it is common for a sensor node to be equipped with a location finding system. This system may consist of a GPS module for a high-end sensor node or may be a software module that implements the localization algorithms that provide location information through distributed calculations.

- **Mobilizer:** A mobilizer may sometimes be needed to move sensor nodes when it is necessary to carry out the assigned tasks. Mobility support requires extensive energy resources and should be provided efficiently. The mobilizer can also operate in close interaction with the sensing unit and the processor to control the movements of the sensor node.

- **Power generator:** While battery power is mostly used in sensor nodes, an additional power generator can be used for applications where longer network lifetime is essential. For outdoor applications, solar cells are used to generate power. Similarly, energy scavenging techniques for thermal, kinetic, and vibration energy can also be used [22].

These components should also fit into a matchbox-sized embedded system [11]. For certain applications, the actual size may be smaller than even a cubic centimeter [19] and weigh light enough to remain suspended in the air. These sophisticated features and the size requirements place additional constraints on the design of wireless sensor nodes. In addition to size, there are also some other stringent constraints for sensor nodes. These nodes must [12]: consume *extremely low power*; operate in *high density*; have *low production cost* and be dispensable; be *autonomous* and operate unattended; and be *adaptive* to the environment.

The main concern for the operation of WSNs is the energy consumption. For most applications, the WSN is inaccessible or it is not feasible to replace the batteries of the sensor nodes. With limited battery power, the lifetime of the network, which is the maximum time for which network is operational, is also limited. Because of size and cost limitations of sensor nodes, the power is a scarce resource in WSNs. As an example, the capacity of the *Smart Dust mote* is 33 mA h [19] as powered by a single size-5 hearing aid battery. For MicaZ and Mica2 nodes [2], which use two AA batteries, the node capacity is limited to 1400–3400 mA h. Similarly, the recent SunSPOT platform uses a 750 mA h lithium-ion battery [5]. It is possible to extend the lifetime of the WSNs through energy scavenging [21, 22], which is extracting energy from the environment. Solar cells are a practical and low-cost example of energy scavenging. However, the amount of energy that can be extracted is still limited. Hence, WSNs are severely constrained in terms of available energy. Moreover, a WSN is required to operate for a long time over months or even years. Therefore, energy-efficient operation is the most important factor for the design of WSNs.

Among the components described above, the transceiver unit is the most important part of a sensor node because it consumes the most energy and provides connectivity to the rest of the network. The transceiver may be a passive or active optical device as in *Smart Dust motes* [19] or an RF device as found in the Mica family of nodes [2], TelosB [3], or SunSPOT [5]. RF communications require modulation, bandpass filtering, demodulation, and multiplexing circuitry, which makes them more complex and expensive as well as energy consuming. Moreover, RF waves experience high path loss, which is proportional to the fourth-order exponent of the distance between two nodes [19, 29]. Consequently, the communication range of a low-power RF transceiver is limited to tens to hundreds of meters. Nevertheless, RF communication is preferred as the *de facto* standard for most sensor node prototypes. This is based on the fact that small packets can be efficiently sent through RF signals with low data rates and the limited transmission range can be exploited for frequency reuse. Furthermore, the low startup and power-off delays of current sensor transceivers provide extensive capabilities to communication protocols, where low-duty-cycle operations can be implemented. In other words, a sensor transceiver can be turned off for most of the time and turned on whenever there is a packet to be sent or received. One of the major factors for transceivers is the design of *low-cost*, energy-efficient, and low-duty-cycle radio circuits, which is still technically challenging.

In addition to the transceiver, the sensor node is also constrained in terms of processing and memory. Although higher computational power is made available with smaller processors at low cost, the processing power of the current sensor nodes is significantly lower than many other embedded systems because of the cost and size constraints. As an example, earlier platforms such as the Smart Dust mote prototype had a 4 MHz Atmel AVR 8535 microcontroller with 8 kB instruction flash memory, 512 bytes of RAM, and 512 bytes of EEPROM [17]. These capabilities have been increased with the recent SunSPOT and Imote2 platforms. SunSPOT is equipped with a 180 MHz 32-bit ARM920T core processor with 512 kB RAM and 4 MB flash [5], while the higher end Imote2 platform has a 416 MHz Marvell PXA271 XScale microcontroller with 256 kB SRAM, 32 MB flash, and 32 MB SDRAM [1]. Although the capabilities of sensor node hardware are increasing, these values are still greatly below the current capabilities of embedded devices such as PDAs or cell phones. Consequently, the software designed for WSNs should be lightweight and the computational requirements of the algorithms should be low for efficient operation in WSNs.

WSNs closely interact with their environment to gather data about various physical phenomena. Most sensing tasks require knowledge of position. Since sensor nodes are generally deployed randomly and run unattended, a location finding system is required. Location finding systems are also required by many of the routing protocols developed for WSNs as explained in Chapter 7. Location information can easily be provided by GPS, which provides an accuracy up to 10 m through the recent GPS units developed for WSNs [4]. However, the cost of these units is significantly higher than a single sensor node and, hence, equipping all sensor nodes with GPS is not viable. Instead, a limited number of nodes, which use GPS or other means to identify their location, are used to help the other nodes determine their locations. The localization solutions are discussed in Chapter 12.

3.2 Fault Tolerance

The hardware constraints lead sensor nodes to frequently fail or be blocked for a certain amount of time. These faults may occur because of a lack of power, physical damage, environmental interference, or software problems. The failure of a node results in disconnection from the network. Since the WSN is interested in information regarding the physical phenomenon instead of information from a single sensor, the failure of a single node should not affect the overall operation of the network. The level of failures that is allowed by the network to adequately continue its functions defines its *fault tolerance*.

More specifically, fault tolerance is the ability to sustain sensor network functionalities without any interruption due to sensor node failures [10, 15, 23]. Firstly, the hardware and software components of a node affect the failure rate. Since the sensor nodes are embedded with low-cost devices, the majority of the failures are caused by hardware problems. Moreover, because of the limited memory space and processing capabilities, the software may result in a node to halt. In addition to the internal problems, the deployment environment may also affect how a sensor works. Indoor applications result in less interference with the sensor nodes and may not increase the failure rate. On the other hand, in applications where sensor nodes are deployed outdoors, sensor node failures also occur because of environmental interference.

The protocols and algorithms designed for WSNs aim to address the frequent failures of sensor nodes through redundancy. The fault tolerance of a network can be improved by relying on more than a single node in the broadcast range of a node. As a result, even if a sensor node fails, other nodes in the broadcast range can be utilized for connectivity to the network.

The fault tolerance of a network also depends on the application it is built for. If the environment, where the sensor nodes are deployed, has little interference, then, the protocols can be more relaxed. For example, if sensor nodes are being deployed in a house to keep track of humidity and temperature levels, the fault tolerance requirement may be low since this kind of sensor network is not easily damaged or affected by environmental noise. On the other hand, if sensor nodes are being deployed on a battlefield for surveillance and detection, then the fault tolerance has to be high because the sensed data are critical and sensor nodes can be destroyed by hostile action. As a result, the fault tolerance level depends on the application of the sensor networks, and the protocols or algorithms should be developed accordingly.

3.3   Scalability

While high-density deployment of sensor nodes in a WSN provides redundancy and improves the fault tolerance of the network, this also creates scalability challenges. The number of sensor nodes deployed for sensing a physical phenomenon may be on the order of hundreds or thousands. Therefore, the networking protocols developed for these networks should be able to handle these large numbers of nodes efficiently. The density can range from a few to hundreds of sensor nodes in a region, which can be less than 10 m in diameter [8]. The node density depends on the application for which the sensor nodes are deployed.

3.4   Production Costs

Since the sensor networks consist of a large number of sensor nodes, the cost of a single node is very important to justify the overall cost of the networks. If the cost of the network is more expensive than deploying traditional single sensor devices, then the sensor network will not be cost justified. As a result, the cost of each sensor node has to be kept low. The cost for Bluetooth is usually less than $10 [21]. The cost of a sensor node should be less than $1 in order for sensor networks to be practically feasible [20]. Current prices for sensor devices are much more higher than even for Bluetooth. Furthermore, a sensor node may also have some additional units, e.g., for sensing and processing as described in Section 3.1. Also it may be equipped with a location finding system, mobilizer, or power generator depending on the applications of the sensor networks. These units all add to the cost of the sensor devices. As a result, the cost of a sensor node is a very challenging issue given the number of functionalities.

3.5   WSN Topology

The large number of inaccessible and unattended sensor nodes which are prone to frequent failures make topology maintenance a challenging task. The major challenge is the deployment of these sensor nodes in the field so that the phenomenon of interest can be monitored efficiently. This constitutes the pre-deployment and deployment phase. Topology maintenance is also important after the initial deployment, i.e., the post-deployment phase, where the protocol parameters and operations can be adopted according to the network topology. Finally, the re-deployment phase may be necessary if several nodes fail or deplete their energy to prolong the network lifetime. Overall, densely deploying a high number of nodes requires careful handling of topology maintenance. We describe next issues related to topology maintenance and change in these three phases.

3.5.1   Pre-deployment and Deployment Phase

Sensor nodes can be either thrown *en masse* or placed one by one in the sensor field. They can be deployed by dropping from an aircraft; delivering by an artillery shell, rocket, or missile; placing in a factory; and placing one by one either by a human or robot. Although the sheer number of sensors and their unattended deployment usually preclude their placing according to a carefully engineered plan, the schemes for initial deployment must (1) *reduce* the installation *cost*; (2) *eliminate* the need for any pre-organization and *preplanning*; (3) *increase* the *flexibility* of arrangement; and (4) *promote* self-organization and *fault tolerance*.

3.5.2   Post-deployment Phase

After the deployment phase, the topology may vary due to changes in sensor conditions [11, 13]. In mobile WSNs, the movements of sensors affect the topology of the network. Accordingly, significant changes may occur in the topology for long times. Moreover, the connectivity of the nodes can change because of jamming, interference, noise, or moving obstacles. These factors also affect the topology of the network for a short time. Another cause of topology changes after deployment is due to node failures, which result in permanent changes. Finally, the topology of the network may change periodically according to the sensing tasks and the application, when certain nodes need to be turned off for a specific amount of time.

   These changes result in a different operation than the initial deployment of the network. Consequently, the networking protocols should be able to adapt to these short-term, periodic, and long-term changes in the topology.

3.5.3   Re-deployment Phase of Additional Nodes

The post-deployment phase changes may require additional nodes to be deployed if the connectivity and fault tolerance of the network are severely affected by the changes in the topology. Accordingly, additional sensor nodes can be re-deployed at any time to replace the malfunctioning nodes or due to changes in task dynamics. The addition of new nodes poses a need to reorganize the network. Coping with frequent topology changes in an ad hoc network that has myriads of nodes and very stringent power consumption constraints requires special routing protocols. This issue is examined in detail in Chapter 7.

3.6   Transmission Media

The successful operation of a WSN relies on reliable communication between the nodes in the network. In a multi-hop sensor network, nodes can communicate through a wireless medium creating *links* between each other. These links can be formed by radio, infrared, optical, acoustic or magneto-inductive links. To enable interoperability and global operation of these networks, the chosen transmission medium must be available worldwide.

Table 3.1 Frequency bands available for ISM applications.

Frequency band	Center frequency
6765–6795 kHz	6780 kHz
13 553–13 567 kHz	13 560 kHz
26 957–27 283 kHz	27 120 kHz
40.66–40.70 MHz	40.68 MHz
433.05–434.79 MHz	433.92 MHz
902–928 MHz	915 MHz
2400–2500 MHz	2450 MHz
5725–5875 MHz	5800 MHz
24–24.25 GHz	24.125 GHz
61–61.5 GHz	61.25 GHz
122–123 GHz	122.5 GHz
244–246 GHz	245 GHz

A popular option for radio links is the use of ISM bands, which offer license-free communication in most countries. The *International Table of Frequency Allocations*, contained in Article S5 of the Radio Regulations (Volume 1), specifies some frequency bands that may be made available for ISM applications; they are listed in Table 3.1.

Some of these frequency bands are already being used for communication in cordless phone systems and *wireless local area networks* (WLANs). For WSNs, a small-sized, low-cost, ultra low power transceiver is required. Hardware constraints and the tradeoff between antenna efficiency and power consumption limit the choice of a carrier frequency for such transceivers to the *ultra high frequency* (UHF) range [18]. Consequently, earlier sensor nodes supported the 433 MHz ISM band in Europe and the 915 MHz ISM band in North America. More recent sensor nodes use the 2.4 GHz band, which is also supported by the new IEEE 802.15.4 standard [6]. The IEEE 802.15.4 standard will be explained in detail in Chapter 5.

The main advantages of using the ISM bands are the free radio, huge spectrum allocation, and global availability. Communication in the ISM band is not bound to a particular standard, thereby giving more freedom for the implementation of energy-efficient networking protocols for WSNs. On the other hand, there are various rules and constraints, such as power limitations and harmful interference from existing applications. Furthermore, since the ISM band is not regulated or assigned to a particular type of user, it can be used by any wireless network. This increases the chance of interference to WSNs, which typically use low-power communication techniques in this spectrum band.

Much of the current hardware for sensor nodes is based upon RF circuit design. The earlier μAMPS wireless sensor node uses a Bluetooth-compatible 2.4 GHz transceiver with an integrated frequency synthesizer [24]. The MicaZ, TelosB, and SunSPOT nodes all use the same transceiver chip (CC2420), which operates in the 2.4 GHz band and supports direct sequence spread-spectrum techniques as defined by the IEEE 802.15.4 standard. It can be observed that with the wide acceptance of the IEEE 802.15.4 standard, the transmission media used by WSNs are also being standardized. This provides extensive capabilities in terms of interoperability and ease of development for sensor node design.

Another possible mode of internode communication in sensor networks is infrared. Infrared communication is license free and robust to interference from electrical devices. Infrared-based transceivers are cheaper and easier to build. Many of today's laptops, PDAs, and cell phones offer an *Infrared Data Association* (IrDA) interface. The main drawback, on the other hand, is the requirement of a line of sight between the sender and the receiver. This makes infrared a reluctant choice for the transmission medium in the sensor network scenario. Nevertheless, infrared can be used in harsh environments, where RF signals suffer from high attenuation, such as underwater links, as will be described in Chapter 16.

The unusual application requirements of sensor networks make the choice of transmission media more challenging. For instance, marine applications may require the use of an aqueous transmission medium. Here, one would like to use long-wavelength radiation that can penetrate the water surface. Inhospitable terrain or battlefield applications might encounter error-prone channels and greater interference. Moreover, a sensor antenna might not have the height and radiation power of those in other wireless devices. Hence, the choice of transmission medium must be supported by robust coding and modulation schemes that efficiently model these vastly different channel characteristics. Considering these facts, acoustic communication techniques have recently been adopted for underwater sensor network applications instead of RF waves, which experience high attenuation in this environment.

3.7   Power Consumption

A wireless sensor node can only be equipped with a limited power source (<0.5 A h, 1.2 V) due to the several hardware constraints described in Section 3.1. Moreover, for most applications, replenishment of power resources is impossible. WSN lifetime, therefore, shows a strong dependence on battery lifetime. Thus the sources that consume energy during the operation of each node should be analyzed and maintained efficiently.

In a multi-hop ad hoc sensor network, each node plays two separate and complementary roles:

- **Data originator:** Each sensor node's primary role is to gather data from the environment through the various sensors. The data generated from *sensing* the environment need to be processed and transmitted to nearby sensor nodes for multi-hop delivery to the sink.

- **Data router:** In addition to originating data, each sensor node is responsible for relaying the information transmitted by its neighbors. The low-power communication techniques in WSNs limit the communication range of a node. In a large network, multi-hop communication is required so that nodes relay the information sent by their neighbors to the data collector, i.e., the sink. Accordingly, the sensor node is responsible for receiving the data sent by its neighbors and forwarding these data to one of its neighbors according to the routing decisions.

The operations related to each role affect how the energy is consumed in a sensor node. Moreover, as explained in Section 3.5, the failure of a few nodes can cause significant topological changes and might require rerouting of packets and reorganization of the network. Hence, power conservation and power management are integral parts of any communication protocol in WSNs. Consequently, the design of power-aware protocols and algorithms for WSNs is of paramount importance.

The main task of a sensor node in a sensor field is to detect events, perform local data processing, and then transmit the data. Power consumption can hence be divided into three domains, *sensing*, *communication*, and *data processing*, which are performed by the sensors, the CPU, and the radio, respectively. A breakdown of the power consumption of a MicaZ sensor node is shown in Figure 3.2 [2]. It can be seen that, among these three, a sensor node expends maximum energy for *data communication*. The three sources of energy consumption are discussed next.

3.7.1   Sensing

The sensing unit and its components were introduced in Section 3.1. Sensing power varies with the nature of applications and the specific sensors used. Sporadic sensing might consume less power than constant event monitoring. The complexity of event detection also plays a crucial role in determining energy expenditure. Higher ambient noise levels might cause significant corruption and increase detection complexity.

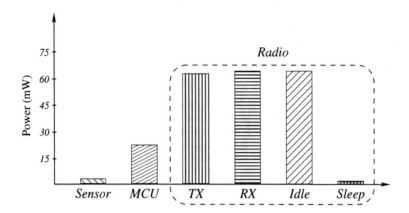

Figure 3.2   The breakdown of power consumption of a MicaZ node.

While the energy consumption for sensing varies significantly with the type of sensor used, the sensor system is generally associated with an ADC subsystem. This subsystem usually consists of the sensor, a low-noise preamplifier, an anti-aliasing filter, an ADC, and a DSP [7]. The energy consumption of an ADC depends on the performance of each of these components; however, in general the power dissipated by an ADC depends on two major factors [26]:

$$P \propto F_S \cdot 2^{ENOB} \tag{3.1}$$

where F_S is the sampling rate and *ENOB* is the effective number of bits, i.e., the resolution of the sensor. As a result, based on the application requirements, the accuracy of the sensed data can be controlled by adjusting the sampling rate or the resolution. Increased sampling rate provides better temporal resolution of the sensed data. However, the physical properties of the sensed phenomenon may not require high sampling rates. As an example, for a temperature sensing application, a sensing rate of 1 ms will not be required since the temperature usually changes over tens of minutes or even hours. As a result, the energy consumption of sensing can be minimized by adjusting the sampling rate according to the requirements of the application and the sensed phenomenon. The resolution, on the other hand, defines the level of information gathered through a single sensing. Increasing the resolution from an 8-bit to 10-bit ADC provides more accurate results. However, this increase is possible through more sophisticated ADC circuitry and an increased number of operations for a single sensing. Accordingly, the energy consumption can be increased by approximately a factor of 4 [7]. In addition to adjusting sampling frequency and resolution, the energy management of sensors should also include sleep states. Whenever a sensor is not required for a certain amount of time, it should be switched to a sleep state, which consumes power only in proportion to the leakage current. However, the overhead in turning on the sensor and the energy consumption associated with it should also be considered to prevent frequent switches.

3.7.2   Data Processing

The energy expenditure for data processing is similar to that for sensing. However, computation requires much less energy compared to data communication as shown in Figure 3.2. As an example, assuming Rayleigh fading and fourth-power distance loss, the energy cost of transmitting a 1 kb packet over a distance of 100 m is approximately equal to executing 3 million instructions by a typical microprocessor [19]. This drastic difference between communication and computation signifies the importance of local data processing in minimizing power consumption in a multi-hop sensor network.

A sensor node has built-in computational abilities and is capable of interacting with its surroundings through the transceiver. The computational abilities are provided through a microprocessor, which is based on *complementary metal oxide semiconductor* (CMOS) technology because of cost and size limitations. While CMOS technology enables low-cost processors to be embedded into sensor platforms, this technology has inherent limitations in terms of energy efficiency. A CMOS transistor pair draws power every time it is switched. This switching power is proportional to the clock frequency, device capacitance (which further depends on the area), and square of the voltage swing. Moreover, the processor dissipates energy due to leakage currents between the power and ground. As a result, the energy consumption for data processing (E_p) for a single task can be represented as a sum of two components as follows [27]:

$$E_p = N \cdot C \cdot V_{dd}^2 + V_{dd}(I_0 e^{V_{dd}/n \cdot V_T})(N/f) \tag{3.2}$$

where the first term is energy dissipation as a result of transistor switches and N is the number of clock cycles per task, C is the total switching capacitance, and V_{dd} is the supply voltage. The second term in (3.2) is the power loss due to leakage currents, where I_0 is the leakage current, n is a constant related to the processor hardware, V_T is the threshold voltage, and f is the clock frequency.

The energy consumption for data processing, as shown in (3.2), depends on the supply voltage, V_{dd}, and the clock frequency, f, which can be controlled, in addition to other parameters that depend on the microprocessor architecture. More specifically, the energy consumption decreases quadratically as the supply voltage is decreased. On the other hand, the gate delay is also dependent on the supply voltage as follows [25, 27]:

$$T_g = \frac{V_{dd}}{K(V_{dd} - V_{th})^a} \tag{3.3}$$

where K and a are microprocessor-dependent variables with $a \sim 2$. Therefore, decreasing the supply voltage increases the gate delay which can be controlled to reduce the idle time of the microprocessor. If the microprocessor is operating at a clock frequency, f, this corresponds to a gate switch for each $T_0 = 1/f$ seconds during which the processor has a single task to process. If the gate delay, T_g, is much less than the gate switch duration T_0, then the processor is kept idle from when the task is finished to when the next task is assigned. Therefore, the gate delay can be increased by decreasing the supply voltage according to (3.3) such that $T_g \le T_0$ or [27]

$$f \le \frac{K(V_{dd} - V_{th})^a}{V_{dd}}. \tag{3.4}$$

As shown in (3.4), for each clock frequency value, there exists a minimum supply voltage level. Reducing the supply voltage to this level is hence an effective means of lowering power consumption without hampering the performance. This is referred to as *dynamic voltage scaling* (DVS), [14, 16], where the processor power supply, V_{dd}, and the clock frequency, f, are adapted to match workloads.

When a microprocessor handles time-varying computational load, simply reducing the clock frequency during periods of reduced activity only affects the power loss due to leakage currents as shown in (3.2). In addition, the supply voltage can also be reduced in conjunction with the clock frequency. This results in almost quadratic savings in energy consumption and reduces the leakage current as well. While DVS provides computation energy savings, decreasing the supply voltage compromises the peak performance of the processor. Significant energy gains can be obtained by recognizing that the peak performance is not always desired and, therefore, the processor's operating voltage and frequency can be dynamically adapted to instantaneous processing requirements.

As shown in Figure 3.2, the power consumption for data processing is significantly smaller than that for communication. This motivates the importance of energy-efficient communication protocols. Nevertheless, each communication task is associated with a computational component. Hence, further energy reductions are possible by combining energy-efficient communication techniques with adaptive computation such as DVS.

3.7.3   Communication

Of the three domains, a sensor node expends the maximum energy in data communication as shown in Figure 3.2. Communication is performed by the transceiver circuitry during both receiving and transmitting data. The breakdown of the transceiver power consumption in Figure 3.2 shows that a transceiver expends a similar amount of energy for transmitting and receiving, as well as when it is idle. Moreover, a significant amount of energy can be saved by turning off the transceiver to a sleep state whenever the sensor node does not need to transmit or receive any data. This provides energy savings of up to 99.99% (from 59.1 mW to 3 μW).

A transceiver circuit consists of a mixer, frequency synthesizer, voltage-controlled oscillator (VCO), phase-locked loop (PLL), demodulator, and power amplifiers, all of which consume valuable power. For a transmitter–receiver pair, the power consumption for data communication, P_c, is a combination of power consumed by these components and can simply be modeled as

$$P_c = P_o + P_{tx} + P_{rx} \tag{3.5}$$

where P_o is the output transmit power and P_{tx} and P_{rx} are the power consumed in the transmitter and receiver electronics, respectively. In other words, the first two terms in (3.5) model the power consumption of the transmitting party, while the last term is for the receiving party.

The major difference between transmitting and receiving is the usage of power amplifiers for transmitting data, i.e., P_o. However, for short-range communication with low radiation power ($\sim$0 dBm), transmission and reception energy costs are nearly the same. Furthermore, with the development of more sophisticated modulation schemes, decoding energy dominates encoding energy and the receive power is higher than the transmit power for MicaZ nodes. The transmission power consumption may change according to the transmit power level set by the sensor node software, which can range from 0 to -25 dBm. Accordingly, if the transmit power level is set to very low values, then the receiving power dominates data communication in WSNs.

While the power consumption model in (3.5) includes the basic factors for energy consumption, more sophisticated models are required to accurately analyze communication energy consumption. In addition to the transmitting and receiving modes, a transceiver can be switched to sleep mode to save energy during inactive periods. However, the transition between different modes of the transceiver is not instantaneous and consumes additional energy. The energy consumption due to the transition between the sleep mode and active (transmit or receive) modes is called the *startup energy consumption*, which is attributed to the startup delays of the frequency synthesizer and the VCO as well as the lock time of the PLL in the transceiver. The startup time, which is on the order of hundreds of microseconds, makes the startup energy consumption non-negligible. As the packet size is reduced, the startup power consumption starts to dominate the active power consumption. As a result, it is inefficient to turn the transceiver on and off frequently, due to the large amount of power spent in turning the transceiver back on each time. The startup energy consumption, E_{st}, can be modeled as follows [28]:

$$E_{st} = P_{LO} \cdot t_{st} \tag{3.6}$$

where P_{LO} is the power consumption of the circuitry including the synthesizer and the VCO; t_{st} is the time required to start up all the transceiver components.

In addition to the energy consumed for transitioning between the sleep and active modes, energy is also consumed when the transceiver switches from the transmit mode to receive mode. This energy consumption, E_{sw}, is given as

$$E_{sw} = P_{LO} \cdot t_{sw} \tag{3.7}$$

where t_{sw} is the switching time. Finally, the energy consumption for receive and transmit modes is represented as follows. In the receive mode, the transceiver uses the synthesizer, VCO, low-noise amplifier, mixer, intermediate-frequency amplifier, and demodulator components. Since the power consumption of the synthesizer and the VCO are denoted as P_{LO}, the energy consumption while receiving is given as [28]

$$E_{rx} = (P_{LO} + P_{RX})t_{rx} \tag{3.8}$$

where P_{RX} is the power consumption of the remaining active components, which is constant for the data-rate values used for WSNs, and t_{rx} is the time it takes to receive a packet.

When the transceiver is switched to transmit mode, in addition to the frequency synthesizer and VCO, it uses the modulator and the power amplifier. Since the modulator's energy consumption is negligible, the energy consumption for transmitting is given as

$$E_{tx} = (P_{LO} + P_{PA})t_{tx} \tag{3.9}$$

where P_{PA} is the power amplifier's power consumption. Contrary to the receive mode energy consumption, which is constant, transmit energy consumption varies with the transmit power, i.e., RF output power. As the desired RF output power level is increased by the system software, the power amplifier consumes more energy, which increases the energy consumption. Consequently,

$$P_{PA} = \frac{1}{\eta} P_{out} \tag{3.10}$$

where η is the power efficiency of the power amplifier and P_{out} is the desired RF output power level. As an example, for the MicaZ node [2], the transceiver consumes 52.2 mW and 33 mW for a RF output power of 0 dBm and -10 dBm, respectively. The RF output power level is usually controlled by higher layer protocols such as the MAC or routing protocols to provide an acceptable communication success at a given distance d. Accordingly, the power amplifier's power consumption can also be written in terms of this distance d as follows [28]:

$$P_{PA} = \frac{1}{\eta} \cdot \gamma_{PA} \cdot r \cdot d^n \tag{3.11}$$

where r is the data rate, n is the path loss exponent of the channel, and γ_{PA} is a factor which depends on the antenna gains, wavelength, thermal noise power spectral density, as well as the desired signal-to-noise ratio (SNR) at the distance d. Since it is a function of distance d, the power amplifier's power consumption, P_{PA} is usually referred to as the *distance-dependent component* for energy consumption, while the other components for transmit, receive, startup, and switch energy consumption are referred to as the *distance-independent components*.

According to the above definitions, a detailed model for the energy consumed in communication can be derived. A communication cycle, where a node transmits a packet to a neighbor node and receives a response back, includes the startup of the transceiver, packet transmission, switching from transmit mode to receive mode, and packet reception. As a result, the overall energy consumption can be given as [28]

$$E_c = E_{st} + E_{rx} + E_{sw} + E_{tx} \tag{3.12}$$

$$= P_{LO} \cdot t_{st} + (P_{LO} + P_{RX})t_{rx} + P_{LO} \cdot t_{sw} + (P_{LO} + P_{PA})t_{tx}. \tag{3.13}$$

Assuming the transmit and receive durations can be represented as $t_{rx} = t_{tx} = l_{PKT}/r$, where l_{PKT} is the packet length, and using (3.11), the overall energy consumption is

$$E_c = P_{LO}(t_{st} + t_{sw}) + (2P_{LO} + P_{RX})\frac{l_{PKT}}{r} + \frac{1}{\eta} \cdot \gamma_{PA} \cdot d^n t_{tx} \cdot l_{PKT} \tag{3.14}$$

It can be seen from (3.14) that the energy consumption for communication has three main components. The first component is constant and depends on the transceiver circuitry. The second component is independent of the communication distance, d, but can be controlled through the packet size or the transmission rate. The first two terms in (3.14) constitute the distance-independent components of energy consumption. The final component depends on the communication distance as well as the packet length and can be controlled through higher layer protocols such as the MAC and routing protocols.

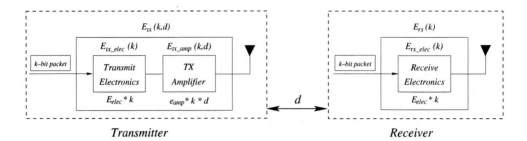

Figure 3.3 The simplified energy model.

Simplified Energy Model

The energy consumption model can be simplified for a transmitter–receiver pair a distance d apart as follows:

$$E_c = E_{tx}(k, d) + E_{rx}(k) \tag{3.15}$$

where $E_{tx}(k, d)$ and $E_{rx}(k)$ are the energy consumption of the transmitter and the receiver, respectively. This simplified model is illustrated in Figure 3.3 for the transmitter–receiver pair. Accordingly, the energy consumption at the transmitter is divided into the transmit electronics and transmitter amplifier while the receiver energy consumption depends on only the receiver electronics. Then, the transmitter and receiver energy consumptions are

$$E_{tx}(k, d) = E_{tx\text{-}elec} \cdot k + e_{amp} \cdot k \cdot d^n \tag{3.16}$$

$$E_{rx}(k) = E_{rx\text{-}elec} \cdot k \tag{3.17}$$

respectively, where $E_{tx\text{-}elec}$ and $E_{rx\text{-}elec}$ are the energy consumption per bit for the transmitter and receiver circuitry, respectively, and e_{amp} is the energy consumption per bit per distance for the power amplifier.

Detailed Energy Model

Energy consumption for communication depends on various factors such as the hardware profile, packet size, transmit power level, and distance. Since energy efficiency is the most important parameter in WSNs, these various factors should be carefully chosen. It can be seen from the overall energy consumption in (3.14) that the energy consumption directly depends on the duration over which the transceiver is active. Consequently, one way to increase the energy efficiency of communication is to reduce the transmission time of the radio. This can be achieved by sending multiple bits per symbol, i.e., by using M-ary modulation [9]. Consequently, the effective packet size, l_{PKT}, is decreased, which reduces the transmit and receive times.

   More specifically, to send a packet of l_{PKT} bits, a binary modulation scheme uses l_{PKT} symbols. Given the symbol duration as $T_s = 1/r$, the transmit time of a packet is $L \cdot T_s$ for binary modulation. This duration can be reduced through an M-ary modulation scheme, which uses b bits per symbol, where $b = \log_2 M$. Therefore, the effective packet length is $L_s = l_{PKT}/b$ and the transmit time can be decreased by a factor of b. While M-ary modulation decreases the transmit duration, the circuit complexity and power consumption of the radio are increased. Moreover, for M-ary modulation, the efficiency of the power amplifier is also reduced. This implies that more power is needed to obtain reasonable levels of transmit output power for a single symbol. Therefore, using M-ary modulation, the energy consumption of the power amplifier can be decreased at the cost of increasing the energy consumption of the transceiver circuitry. As a result, an optimal modulation level exists for a certain WSN, where the activity time is decreased without compromising the circuit energy.

Based on the above discussion, and assuming a multilevel quadrature amplitude modulation (MQAM) scheme is used, the overall energy consumption of a pairwise communication can be represented as [9]

$$E_c = (1 + \alpha)\frac{4}{3}N_f\sigma^2\frac{2^b - 1}{b}\ln\frac{4(1 - 2^{-b/2})}{bP_b}G_d + \frac{P_cT_{on} + 2P_{syn}T_{tr}}{l_{PKT}} \tag{3.18}$$

where

$$\alpha = \frac{\xi}{\eta} - 1, \quad \xi = 3\frac{\sqrt{M} - 1}{\sqrt{M} + 1}, \quad M = 2^{L/(B \cdot T_{on})} \tag{3.19}$$

and L is the packet length, B is the channel bandwidth, N_f is the receiver noise figure, σ^2 is the power spectrum energy, P_b is the probability of bit error, G_d is the power gain factor, P_c is the circuit power consumption, P_{syn} is the frequency synthesizer power consumption, T_{tr} is the frequency synthesizer settling time, T_{on} is the transceiver on time, and M is the modulation parameter.

The detailed energy model in (3.18) captures several factors that affect the energy consumption for communication. Compared to the simplified energy model in (3.16), this model provides more detail and flexibility for analyzing the energy consumption. However, for certain applications, this level of detail is not necessary. Therefore, the selection of the most appropriate energy model depends on the focus of the analysis.

Overall Energy Consumption

In addition to the energy consumed in a single cycle of communication as given in (3.14), its simplified version in (3.16), and the detailed version in (3.18), the communication energy also depends on the *rate* at which the transceiver is used. This rate depends on many factors including the application type, transport protocol, routing protocol, as well as the MAC layer. In a monitoring application, where frequent information retrieval is necessary, the transceiver is more frequently used compared to an event-based application, where data are generated only if certain conditions are met and the network is idle otherwise. Similarly, the transport protocol throttles the rate at which the data should be injected into the network, which determines how frequently the transceiver should be turned on. The routing protocols, on the other hand, may increase the communication rate by sensing control messages among neighbors to gather neighborhood information. The next hop selected by the routing protocol also determines the communication rate since multiple retransmissions may be required if a node with poor channel conditions is chosen. Finally, the MAC layer controls the channel access rate through which the node gains access to the broadcast channel. Depending on the MAC type, this rate may change drastically. Overall, the communication energy consumption can be modeled as follows [24]:

$$E_c = N_T[P_{LO}(t_{tx} + t_{st}) + P_{PA}(t_{tx})] + N_R[P_R(t_{rx} + t_{st})] \tag{3.20}$$

where P_{LO} is the power consumed by the synchronizer and the VCO, P_{PA} is the output power of the transmitter, $P_R = (P_{LO} + P_{RX})$ is the power consumed by the receiver, t_{tx} and t_{rx} are the transmitter and receiver on time, and t_{st} is the transmitter or receiver startup time. Finally, N_T and N_R are the number of times the transmitter and receiver are switched on per unit time, respectively, which depends on the application, transport, network, and MAC layers.

References

[1] Crossbow Imote2 mote specifications. http://www.xbow.com.

[2] Crossbow MicaZ mote specifications. http://www.xbow.com.

[3] Crossbow TelosB mote specifications. http://www.xbow.com.

[4] Mts420 environment sensor board. Available at http://www.xbow.com.

[5] SunSPOT mote specifications. http://www.sunspotworld.com.

[6] Wireless medium access control (MAC) and physical layer (PHY) specifications for low-rate wireless personal area networks (LR-WPANs). ANSI/IEEE Standard 802.15.4, October 2003.

[7] B. H. Calhoun, D. C. Daly, N. Verma, D. F. Finchelstein, D. D. Wentzloff, A. Wang, S.-H. Cho, and A. P. Chandrakasan. Design considerations for ultra-low energy wireless microsensor nodes. *IEEE Transactions on Computers*, 54(6):727–740, June 2005.

[8] S. Cho and A. P. Chandrakasan. Energy efficient protocols for low duty cycle wireless microsensor networks. In *Proceedings of IEEE International Conference on Acoustics, Speech, and Signal Processing (ICASSP'01)*, volume 4, pp. 2041–2044, Salt Lake City, UT, USA, January 2001.

[9] S. Cui, A. J. Goldsmith, and A. Bahai. Energy-constrained modulation optimization. *IEEE Transactions on Wireless Communications*, 4(5):2349–2360, September 2005.

[10] G. Hoblos, M. Staroswiecki, and A. Aitouche. Optimal design of fault tolerant sensor networks. In *Proceedings of IEEE International Conference on Control Applications*, pp. 467–472, Anchorage, AK, USA, September 2000.

[11] C. Intanagonwiwat, R. Govindan, and D. Estrin. Directed diffusion: a scalable and robust communication paradigm for sensor networks. In *Proceedings of MobiCom'00*, pp. 56–67, Boston, MA, USA, August 2000.

[12] J. M. Kahn, R. H. Katz, and K. S. J. Pister. Next century challenges: mobile networking for "smart dust." In *Proceedings of ACM MobiCom'99*, pp. 271–278, Seattle, WA, USA, August 1999.

[13] S. Meguerdichian, F. Koushanfar, M. Potkonjak, and M. B. Srivastava. Coverage problems in wireless ad-hoc sensor networks. In *Proceedings of IEEE INFOCOM 2001*, volume 3, pp. 1380–1387, Anchorage, AK, USA, April 2001.

[14] R. Min, T. Furrer, and A. Chandrakasan. Dynamic voltage scaling techniques for distributed microsensor networks. In *Proceedings of ACM MobiCom'95*, Berkeley, CA, USA, August 1995.

[15] D. Nadig and S. S. Iyengar. A new architecture for distributed sensor integration. In *Proceedings of IEEE Southeastcon'93*, Charlotte, NC, USA, April 1993.

[16] T. Pering, T. Burd, and R. Brodersen. The simulation and evaluation of dynamic voltage scaling algorithms. In *Proceedings of International Symposium on Low Power Electronics and Design (ISLPED'98)*, pp. 76–81, Monterey, CA, USA, August 1998.

[17] A. Perrig, R. Szewczyk, V. Wen, D. Culler, and J. D. Tygar. Spins: security protocols for sensor networks. In *Proceedings of ACM MobiCom'01*, pp. 189–199, Rome, Italy, 2001.

[18] A. Porret, T. Melly, C. C. Enz, and E. A. Vittoz. A low-power low-voltage transceiver architecture suitable for wireless distributed sensors network. In *Proceedings of IEEE International Symposium on Circuits and Systems'00*, volume 1, pp. 56–59, Geneva, Switzerland, 2000.

[19] G. J. Pottie and W. J. Kaiser. Wireless integrated network sensors. *Communications of the ACM*, 43:51–58, May 2000.

[20] J. Rabaey, J. Ammer, J. L. da Silva Jr., and D. Patel. Picoradio: ad-hoc wireless networking of ubiquitous low-energy sensor/monitor nodes. In *Proceedings of the IEEE Computer Society Annual Workshop on VLSI (WVLSI'00)*, pp. 9–12, Orlando, FL, USA, April 2000.

[21] J. M. Rabaey, M. J. Ammer, J. L. da Silva Jr., D. Patel, and S. Roundy. Pico radio supports ad hoc ultra-low power wireless networking. *IEEE Computer Magazine*, 33(7):42–48, 2000.

[22] S. Roundy, P. K. Wright, and J. M. Rabaey. *Energy Scavenging for Wireless Sensor Networks: with Special Focus on Vibrations*. Springer, 2003.

[23] C. Shen, C. Srisathapornphat, and C. Jaikaeo. Sensor information networking architecture and applications. *IEEE Personal Communications*, 8(4):52–59, August 2001.

[24] E. Shih, S. H. Cho, N. Ickes, R. Min, A. Sinha, A. Wang, and A. Chandrakasan. Physical layer driven protocol and algorithm design for energy-efficient wireless sensor networks. In *Proceedings of ACM MobiCom'01*, pp. 272–287, Rome, Italy, 2001.

[25] A. Sinha and A. Chandrakasan. Dynamic power management in wireless sensor networks. *IEEE Design and Test of Computers*, 18(2):62–74, 2001.

[26] R. H. Walden. Analog-to-digital converter survey and analysis. *IEEE Journal of Selected Areas in Communication*, 17(4):539–550, April 1999.

[27] A. Wang and A. Chandrakasan. Energy-efficient DSPs for wireless sensor networks. *IEEE Signal Processing Magazine*, 43(5):68–78, July 2002.

[28] A. Y. Wang and C. G. Sodini. A simple energy model for wireless microsensor transceivers. In *Proceedings of IEEE Globecom'04*, pp. 3205–3209, Dallas, TX, USA, November 2004.

[29] M. Zuniga and B. Krishnamachari. Analyzing the transitional region in low power wireless links. In *Proceedings of IEEE SECON'04*, pp. 517–526, Santa Clara, CA, USA, October 2004.

4

Physical Layer

The physical (PHY) layer is responsible for the conversion of bit streams into signals that are best suited for communication across the wireless channel. More specifically, the physical layer is responsible for frequency selection, carrier frequency generation, signal detection, modulation, and data encryption. The reliability of the communication depends also on the hardware properties of the nodes, such as antenna sensitivity and transceiver circuitry.

Most of the unique advantages of WSNs are provided through wireless communication. Ease of deployment, infrastructure-free networking, and broadcast communication are some of these advantages. However, wireless communication also brings several challenges in terms of limited communication range, frequent errors, and interference.

It is well known that long-distance wireless communication can be expensive, in terms of both energy consumption and implementation complexity. While designing the physical layer for sensor networks, energy minimization assumes significant importance, over and above the decay, scattering, shadowing, reflection, diffraction, multi-path, and fading effects. In general, the minimum output power required to transmit a signal over a distance d is proportional to d^n, where $2 < n < 4$. The exponent n is closer to 4 for low-lying antennas and near-ground channels [25, 27], as is typical in sensor network communication. Multi-hop communication in a sensor network can effectively overcome shadowing and path loss effects, if the node density is high enough. Similarly, while propagation losses and channel capacity limit data reliability, this very fact can be used for spatial frequency reuse.

In this chapter, physical layer issues in WSNs are discussed. In Section 4.1, major PHY layer technologies used for wireless communication are described. An overview of RF communication, which is most frequently used in WSNs, is provided in Section 4.2. Based on this overview, channel coding (Section 4.3) and modulation (Section 4.4) techniques are explained. Moreover, the effects of wireless communication and wireless channel models for WSNs are discussed in Section 4.5. Finally, the IEEE 802.15.4 PHY layer standard and the platforms used in WSNs are described in Section 4.6.

4.1 Physical Layer Technologies

The wireless medium used in WSNs is one of most important factors, since the unique properties of different media place several constraints on the capabilities of the physical layer. In general, wireless links can be formed by RF, optical, acoustic, or magnetic induction techniques. While RF communication is generally adopted, the other communication techniques also have particular application scenarios. These physical layer techniques are described next.

4.1.1 RF

Most of the current hardware for sensor nodes is based on RF circuit design. RF communication takes place through electromagnetic (EM) waves that are transmitted on the RF bands, which span the 3 Hz to 300 GHz spectrum. As explained in Chapter 3, for RF links, one option is to use the ISM bands, which offer license-free communication in most countries. Other frequency bands have also been used for this type of communication.

The main types of technologies used for RF communication in WSNs can be classified into three as *narrow-band*, *spread-spectrum*, and *ultra-wide-band* (UWB) techniques. Narrow-band technologies aim to optimize bandwidth efficiency by using M-ary modulation schemes in a narrow band. Spread spectrum and UWB, on the other hand, use a much higher bandwidth and spread the information onto this higher bandwidth. Spread-spectrum techniques use chip codes of higher rate for spreading the spectrum. On the other hand, UWB accomplishes communication by relative positioning of UWB pulses with respect to a reference time [33]. These three main RF communication techniques are described next.

Narrow-Band Communication

The earlier platforms developed for WSNs employ narrow-band communication techniques. As an example, the Mica2 platform uses the CC1000 transceiver, which operates in 433, 868, and 915 MHz bands with a bandwidth of up to 175 kHz and data rate of 76 kbps. In addition, the low-power sensor device [32] uses a single channel RF transceiver operating at 916 MHz with a data rate of 10 kbps. The *wireless integrated network sensors* architecture [23] also uses narrow-band RF techniques for communication.

Spread Spectrum

Spread-spectrum techniques have recently been used for RF communication to improve the data rate and resistance to interference. A narrow-band signal is transmitted using a spectrum that is much larger than the frequency content of the signal. In other words, a signal of limited bandwidth is *spread* over a much larger band using spread-spectrum techniques. Accordingly, the signal at a particular band is observed as noise, which improves the resilience to interference from other signals. Two types of spread-spectrum techniques exist: frequency hopping spread spectrum (FHSS) and direct sequence spread spectrum (DSSS).

FHSS relies on a frequency hopping scheme, where the wide-band spectrum is divided into frequency channels. Accordingly, the transmitter and receiver pair hops through these channels based on a predefined hopping scheme. FHSS has been mainly used in the Bluetooth standard [3]. The μAMPS wireless sensor node [26] uses a Bluetooth-compatible 2.4 GHz transceiver with an integrated frequency synthesizer.

On the other hand, DSSS is based on pseudo-noise (PN) codes that are called *chips*. A stream of chips is used to modulate the information bits to be sent. The chips have a much smaller duration compared to a bit and, hence, each bit is modulated with a number of chips. Since the chip rate is much higher than the bit rate, the narrow-band information is again spread over a much larger bandwidth. By communicating the sequence of PN codes to the receiver, a DSSS signal can be decoded. The DSSS technique has become the *de facto* standard for WSNs with the establishment of the IEEE 802.15.4 standard [4]. Accordingly, many recent sensor platforms including MicaZ, TelosB, and Imote2 employ the DSSS technique. The CC2420 transceiver chip, which is used in these platforms, operates at the 2.4 GHz band with a chip rate of 2 Mchips/s and a bit rate of 250 kbps.

Ultra Wide Band

The UWB or impulse radio has also been used as communication technology in WSN applications, especially in indoor wireless networks [22]. UWB employs baseband transmission and, thus, requires no intermediate or radio carrier frequencies. Generally, pulse position modulation (PPM) is used. The main advantage of UWB is its resilience to multi-path fading [10, 19]. Hence, increased reliability is possible by exploiting the UWB techniques in sensor networks along with low transmission power and simple transceiver circuitry. Since UWB uses bandwidth modulation, implementation costs are significantly lower than DSSS systems. UWB will be described in detail in Chapter 15 as a potential physical layer technique for wireless multimedia sensor networks (WMSNs).

A comparison of the three RF communication techniques, i.e., narrow band, spread spectrum, and UWB, shows that narrow-band communication performs poorly in WSNs [31]. The main reason is that these techniques trade bandwidth efficiency for energy efficiency. More importantly, as the modulation level is increased, energy efficiency decreases further although bandwidth efficiency is increased. On the other hand, spread spectrum and UWB enable low-power communication with robustness against multi-path effects. In particular, spread-spectrum techniques provide robustness to other narrow-band signals since the communication takes place over a larger spectrum using very low power. Moreover, for a given communication success requirement, spread-spectrum techniques can afford lower transmit power levels compared to narrow-band techniques, which is important for WSNs.

UWB, on the other hand, improves the advantages of spread-spectrum techniques since much larger spectrum bands are used for transmission. Furthermore, UWB can be transmitted at the baseband without requiring intermediate-frequency (IF) transmission, which significantly decreases the complexity of the transceiver circuitry. However, UWB communication is limited to small communication ranges (< 10 m), which limits its applicability for WSN deployment. Furthermore, since a carrier does not exist, well-established carrier sensing (CS) techniques cannot be employed for MAC.

Compared to UWB, DSSS can provide a better packet error rate if equal bandwidth occupancies are considered for binary modulation [33]. UWB performance is comparable to DSSS only for higher modulation schemes, which, on the other hand, increases the complexity required for a UWB transceiver and its cost. However, when multi-path effects are considered, UWB provides higher resilience when compared to DSSS. It can be observed that spread-spectrum techniques provide a fine balance between system complexity and interference mitigation compared to other RF communication techniques.

4.1.2   Other Techniques

RF communication techniques have been mostly used in WSNs. However, these techniques suffer from limited bandwidth and susceptibility to interference from other wireless sources. In addition to RF communication techniques, several other solutions have also been used for certain scenarios in WSNs. These techniques are optical (infrared), acoustic, and magnetic induction communication.

Optical Communication

Optical (or infrared) communication takes place in the band that is at the lower end of the visible spectrum with a wavelength of 750 to 1000 nm. Infrared communication has been used mainly for short-range communication in portable devices such as cell phones, laptops, and PDAs. A transmitter consisting of a light-emitting diode (LED) or a laser diode (LD) is used to emit infrared light, which can be captured by a photodiode [9, 17]. The transmitter modulates the intensity of the infrared according to the information content. At the receiver end, the current at the photodiode is used to decode the transmitted information, resulting in an intensity modulation with direct detection (IM/DD) system [9].

Infrared communication can be performed in two ways as shown in Figure 4.1. The most commonly used communication type is *point-to-point* communication, where the transmitter and receiver are

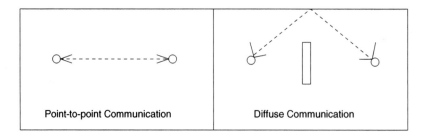

Figure 4.1   Infrared communication techniques.

directed to each other. Point-to-point communication requires a line of sight (LoS) between the transmitter and the receiver. By using narrow beams, communication distances longer than 10 m can be achieved provided that the LoS path exists. The second type of infrared communication is *diffuse* communication, which does not require LoS. Instead, the communication is conducted based on reflections of the light from surrounding objects. To increase success in diffuse communication, generally wide beams at the transmitter and a wide field of view at the receiver are used. Hence, the communication range of diffuse systems is limited.

In addition to the infrared band, optical communication has also been performed over a larger spectrum band for communication among sensor nodes. An example is the Smart Dust mote [16], which is an autonomous sensing, computing, and communication system that uses the optical medium for transmission. The main goal of the Smart Dust project is the development of a cubic-millimeter sensor node [29]. Hence, size is an extremely important design parameter for this platform. RF transceivers, which provide sufficient data rates with a form factor on the order of millimeters, are not feasible. Instead, optical transmitters and receivers are used. Accordingly, Smart Dust supports two types of optical communication techniques: passive reflective and active-steered laser systems.

The passive reflective system is based on a corner-cube retroreflector (CCR), which consists of three micro-scale mirrors that are placed perpendicular to each other as the corner of a cube. One of the mirrors is controlled by the node to modulate an incoming light beam to transmit information to a base station. Using this technique a data rate of 4 kbps at a range of 180 m is possible [30]. The Smart Dust node not only includes a passive CCR system, but also integrates an active transmitter. The active transmitter includes a light source and a steered micro-mirror controlled by a MEMS controller. The light illuminated at the mirror is reflected in the direction selected by the controller to perform communication between different motes. The system is shown to operate at a distance of 21.4 km using 3.5 mW transmission power.

Compared to RF communication, infrared communication technologies do not suffer from electro-magnetic interference. Hence, interoperability with existing wireless systems is not an issue. Moreover, LoS communication enables simultaneous communication between multiple closely located transmitter–receiver pairs, which decreases the complexity required by MAC protocols. Although not affected by existing electromagnetic waves, external light sources create interference to optical communication. For outdoor deployment, sunlight is an important source of interference that requires noise reduction techniques at the receiver. On the other hand, indoor deployment is generally influenced by the periodic noise generated by fluorescent lights, which should be combated through appropriate filters at the receivers [9]. While the optical medium can enable ultra-low-power communication with the help of passive devices and mirrors on the sensor nodes, LoS requirements and robustness problems with node position changes constitute difficulties for the deployment of WSNs. This severely limits the applicability of optical communication technology as a transmission medium for WSNs. On the other hand, infrared transmitter and receivers can be used as a sensor type.

Acoustic Communication

The particular characteristics of the deployment environment also dictate the wireless communication technique to be deployed for the WSN. As an example, for underwater wireless sensor networks (UWSNs), the transmission medium is the water. Since the communication ranges of RF and optical communication techniques are severely limited in water, an alternative communication technique is required. Instead, acoustic waves that can propagate in water over very long distances are suitable for the transmission of digital signals.

Underwater acoustic communication requires efficient microphones and receivers to establish wireless links in water. Communication success is influenced by path loss, noise, multi-path, Doppler spread, and propagation delay, all of which depend on the properties of the water, such as temperature, depth, and composition. Consequently, the available bandwidth and the propagation delay significantly depend on time and frequency. While the data rate provided by acoustic communication varies according to the target communication range, most communication is limited to the order of tens of kilobits per second for existing devices. Furthermore, because of the three-dimensional nature of the underwater topology, acoustic links are also classified as *vertical* and *horizontal*, according to the direction of the sound ray with respect to the ocean bottom. Details of acoustic communication are deferred to Chapter 16 in the explanation of UWSNs.

Magnetic Induction Communication

WSNs have also been recently deployed in underground settings, where communication is severely affected by underground obstacles such as soil, rocks, and tunnels. This led to the development of wireless underground sensor networks (WUSNs). Traditional signal propagation techniques using EM waves encounter three major problems in the soil medium [7]. Firstly, EM waves experience high levels of attenuation due to absorption by soil, rock, and water underground. Secondly, the path loss is highly dependent on numerous soil properties such as water content, soil makeup (sand, silt, or clay), and density, and can change dramatically with time and space. Thirdly, operating frequencies in the megahertz or lower ranges are necessary to achieve a practical transmission range [6]. To efficiently transmit and receive signals at that frequency, the antenna size would be too large to be deployed in the soil.

Magnetic induction (MI) is an alternative signal propagation technique for underground wireless communication, which addresses the challenges of dynamic channel condition and large antenna size in the EM wave techniques. In particular, dense media such as soil and water cause little variation in the attenuation rate of magnetic fields from that of air, since the magnetic permeabilities of each of these materials are similar. Details of communication through magnetic induction techniques are deferred to Chapter 17 in the explanation of WUSNs.

4.2   Overview of RF Wireless Communication

Despite the various short-range, low-power wireless communication techniques explained in Section 4.1, RF communication techniques are generally used in WSNs. In the rest of the chapter, we will focus on the details of RF communication techniques.

An illustration of RF wireless communication and the general components is given in Figure 4.2. Accordingly, the following are performed to transmit information between a transmitter and a receiver:

- **Source coding (data compression):** At the transmitter end, the information source is first encoded with a source encoder, which exploits the information statistics to represent the source with a fewer number of bits, i.e., source codeword. Source coding is also referred to as *data compression*. Source (en/de)coding is performed at the application layer as shown in Figure 4.2.

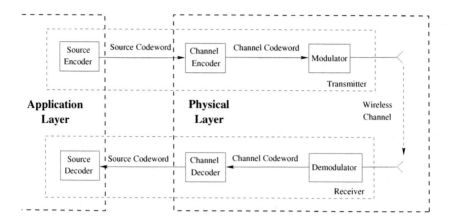

Figure 4.2   Overview of RF communication blocks.

- **Channel coding (error control coding):** The source codeword is then encoded by the channel encoder to address the wireless channel errors affecting the transmitted information. Channel coding is also referred to as *error control coding*.
- **Interleaving and modulation:** The encoded channel symbols are then interleaved to combat the bursty errors that can affect a large number of consecutive bits. The channel coding and the interleaving mechanism help the receiver either to (1) identify bit errors to initiate retransmission or (2) correct a limited number of bits in case of errors. Then, an analog signal (or a set thereof) is modulated by the digital information to create the waveform that will be sent over the channel. Finally, the waveforms are transmitted through the antenna to the receiver.
- **Wireless channel propagation:** The transmitted waveform, which is essentially an EM wave, travels through the channel. Meanwhile, the waveform is attenuated and distorted by several wireless channel effects.

At the receiver end, symbol detection is performed first to lock into the sent waveform. The receiver signal is demodulated to extract the channel symbols, which are then de-interleaved and decoded by the channel and the source decoders to determine the information sent by the transmitter.

The success of RF wireless communication depends on the techniques used for each block shown in Figure 4.2 as well as the wireless channel effects and operating parameters such as frequency, antenna properties, and ambient noise. In addition to these fundamental parameters, wireless communication for WSNs is also influenced by energy efficiency. The limited energy resources and the low cost requirements of the WSN paradigm result in low-power communication techniques being employed. Furthermore, high-complexity channel codes, modulation schemes, or antenna technologies cannot be employed. These limitations exacerbate the effects of the wireless channel. Consequently, providing energy-efficient wireless communication is a major challenge for the PHY layer of WSNs.

The first step in information delivery in wireless systems is *source coding (or data compression)*. These solutions rely on the statistical characteristics of information for more efficient representation. Accordingly, redundant information is compressed to decrease the data volume while preserving the information content. Source coding is generally performed at the application layer and will be explained in Chapter 9.

In the remainder of this chapter, the concepts of channel coding and modulation are discussed. Furthermore, the *log-normal channel model*, which adequately models the effects of the wireless channel for low-power communication, is described. Finally, the PHY layer standards and the transceiver types used in WSN applications are introduced.

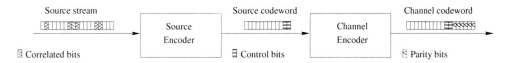

Figure 4.3 Source and channel coding.

4.3 Channel Coding (Error Control Coding)

Source coding techniques aim to decrease the amount of data to be transmitted by exploiting the statistical properties of the information, as will be explained in Chapter 9. Once encoded, this information needs to be transmitted reliably over the wireless channel. However, the transmitted information can be easily corrupted due to the adverse effects of the wireless channel. To combat these effects, channel coding schemes have long been investigated in the context of wireless communication theory. The main goal of these channel coding approaches is to exploit the statistical properties of the channel to inject redundancy into the information to be sent. Consequently, the received information can be decoded successfully even if certain portions of it are distorted. Channel coding is also referred to as error control coding (ECC) or forward error correction (FEC).

The notions of source and channel coding can be more clearly seen in Figure 4.3. When a certain number of information bits are provided to the source encoder, an output that consists of a smaller number of bits is created. Although compressed, if this source codeword is successfully received at the source decoder, the original set of information bits can be decoded. In order to successfully transmit the source codeword, the channel encoder creates an output that consists of a larger number of bits compared to the source codeword. The redundant bits (or parity bits) are added to the source codeword to create the channel codeword, which helps combat the wireless channel errors. Note that the resulting channel word may be longer or shorter than the initial information since source and channel coding are performed independently. As a result, the reduction due to source coding depends on the properties of information content whereas the increase due to channel coding depends on the channel properties.

There exist several powerful channel codes that have been developed for communication systems. These codes constitute the basis of error control techniques such as automatic repeat request (ARQ), forward error correction (FEC), and hybrid ARQ schemes as we will describe in Chapter 6. Next, we present some of the common channel codes that are being used in WSNs.

4.3.1 Block Codes

Block codes are generally preferred in WSNs due to their relatively simpler implementation and smaller memory requirements. A block code transforms an input message $\mathbf{u}$ of k bits into an output message $\mathbf{v}$ of n bits, $n > k$ [20, 21]. The output message $\mathbf{v}$ is denoted as the *channel codeword*. Depending on the complexity of the code, each code is capable of correcting up to t bits in error. Accordingly, a block code is identified by a tuple (n, k, t).

The error detection and error correction capabilities of block codes are determined by the *minimum distance* of the code. The Hamming distance between two codewords is defined as the number of places they differ [20]. Accordingly, the minimum Hamming distance, $d_{\min}$, is the minimum distance between any two words in a code. A code with minimum distance $d_{\min}$ can detect up to $d_{\min} - 1$ errors and correct up to t errors such that $2t + 1 \leq d_{\min} \leq 2t + 2$.

Three main types of block codes are used in WSNs in general.

BCH Codes

A popular example of block codes is the Bose, Chaudhuri, and Hocquenghem (BCH) code. BCH codes have been used in many different applications and provide both error detection and correction capabilities. A BCH code is identified by a tuple (n, k, t), where n is the block length, k is the information length, and t is the maximum number of errors that can be corrected. The following hold true for any BCH code:

$$n = 2^m - 1$$
$$n - k \leq mt \tag{4.1}$$
$$d_{min} \geq 2t + 1$$

The encoding and decoding operations of BCH codes are performed in a finite field $GF(2^m)$, called a Galois field, which has 2^m elements [20].

RS Codes

RS (Reed–Solomon) codes are a family of BCH codes that are non-binary, i.e., the operations are performed in $GF(q)$, where q is a prime number. While they retain the properties of binary BCH codes, the following hold true for all RS codes:

$$n = q - 1$$
$$n - k = 2t \tag{4.2}$$
$$d_{min} = 2t + 1$$

CRC Codes

CRC (Cyclic Redundancy Check) codes are a special family of BCH codes that are used to detect errors in a packet. Irrespective of whether an error control code is used within communication, CRC codes are used in almost any communication system. The automatic repeat request (ARQ) scheme (see Chapter 6), which relies on retransmissions for reliability, is based on CRC codes. In particular, CRC codes are BCH codes with $d_{min} = 4$. Upon decoding, CRC codes *detect* whether a packet is received in error or not. However, they cannot *correct* these errors.

Block codes are easy to implement because of the relatively simpler encoder and decoder structures. Consequently, the encoding complexity of block codes is negligible [20]. Hence, only the decoding complexity is considered. Accordingly, assuming software implementation, block codes can be decoded with $(2nt + 2t^2)$ additions and $(2nt + 2t^2)$ multiplications. A more detailed description of the benefits of block codes in communication in WSNs is provided in Chapter 6.

4.3.2   Joint Source–Channel Coding

Traditional information theory separates source and channel coding as shown in Figure 4.3. This is based on the fact that, generally, information and channel exhibit independent characteristics. Accordingly, information content is first compressed using source coding techniques that exploit the statistical redundancy within it. Secondly, redundant parity bits are added by channel coding techniques to the compressed bits according to the channel characteristics. In a typical communication system, the information generated by the users is directly related to the applications on their devices. Hence, separation of source and channel coding is optimum for these systems.

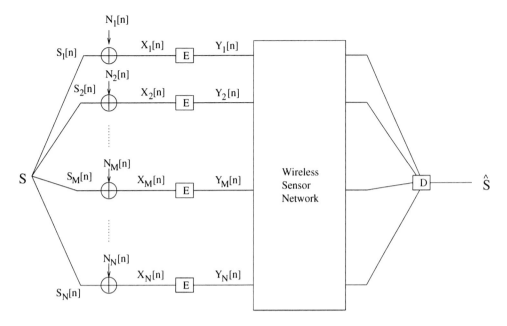

Figure 4.4 Correlation model and architecture.

In WSNs, however, the applications generally observe physical phenomena. Since the information gathered by sensor nodes follows the physical properties of the sensed phenomenon, the characteristics of the source can be closely matched with the channel characteristics. Accordingly, the efficiency of joint source–channel coding has recently been investigated in the context of WSNs [12].

In a sensor field, each sensor observes the noisy version of a physical phenomenon. The sink is interested in observing the physical phenomenon using the observations from sensor nodes with the highest accuracy [28]. The physical phenomenon of interest can be modeled as a spatio-temporal process $s(t, x, y)$ as a function of time t and location (x, y).

The model for the information gathered by N sensors in an event area is illustrated in Figure 4.4. The sink is interested in estimating the event source, S, according to the observations of the sensor nodes, n_i, in the event area. Each sensor node n_i observes $X_i[n]$, the noisy version of the event information, $S_i[n]$. The event information $S_i[n]$ is also correlated with the event source, S. In order to communicate this observation to the sink through the WSN, each node has to encode its observation. The encoded information, $Y_i[n]$, is then sent to the sink through the WSN. The sink, at the other end, decodes this information to get the estimate, $\hat{S}$, of the event source S. The encoders and the decoders are labeled as E and D in Figure 4.4, respectively.

Each observed sample, $X_i[n]$, of sensor i at time n is represented as

$$X_i[n] = S_i[n] + N_i[n] \tag{4.3}$$

where the subscript i denotes the spatial location of node n_i, i.e. (x_i, y_i), $S_i[n]$ is the realization of the space–time process $s(t, x, y)$ at time $t = t_n$ and $(x, y) = (x_i, y_i)$, and $N_i[n]$ is the observation noise. $\{N_i[n]\}_n$ is a sequence of i.i.d. Gaussian random variables of zero mean and variance σ_N^2. We further assume that the noise each sensor node encounters is independent of another, i.e., $N_i[n]$ and $N_j[n]$ are independent for $i \neq j$ and $\forall n$.

As shown in Figure 4.4, each observation $X_i[n]$ is then encoded into $Y_i[n]$ by the source coding at the sensor node as

$$Y_i[n] = f_i(X_i[n]) \tag{4.4}$$

where $f_i(\cdot)$ is the encoding function. The $Y_i[n]$ are then sent through the network to the sink. The sink decodes the received data to reconstruct an estimate $\hat{S}$ of the source S

$$\hat{S} = g(Y_1[n_1], \ldots, Y_1[n_\tau]; \ldots; Y_N[n_1], \ldots, Y_N[n_\tau]) \tag{4.5}$$

based on the data received from N nodes in the event area over a time period $\tau = t_{n_\tau} - t_{n_1}$. The sink is interested in reconstructing the source S according to a distortion constraint

$$D = E[d(S, \hat{S})]. \tag{4.6}$$

In distributed networks, where the information about an event is more important than the individual readings of each sensor, joint source–channel coding results in optimal results [13]. More specifically, in the case when source–channel coding is performed separately, the distortion in (4.6) can only be decreased by $1/\log N$, where N is the number of sensors. However, using joint source–channel coding, the observation $X_i[n]$ can be encoded using a scalar power constraint as follows:

$$Y_i[n] = \sqrt{\frac{P_E}{\sigma_S^2 + \sigma_N^2}} X_i[n] \tag{4.7}$$

where P_E is the power constraint. Accordingly, using the MMSE estimator at the sink, each sample is estimated as

$$Z[n] = \frac{E[S[n]|Y[n]]}{E[Y^2[n]]} Y[n]. \tag{4.8}$$

For (4.7) and (4.8), the distortion in (4.6) decreases with $1/N$ as the number of sensors, N, that observe the phenomenon increases. For the case of Gaussian sensor networks, where the source and noise statistics follow a Gaussian distribution, in addition to outperforming separation techniques, joint source–channel coding also leads to the *optimum* communication solution [14].

Joint source–channel coding provides a better performance compared to traditional separation techniques, in theory. However, existing communication platforms still follow the source–channel separation principle since they are generally implemented in the digital domain. On the other hand, joint source–channel coding techniques require analog communication techniques, which are generally nonlinear and harder to implement [15]. Therefore, practical implementations of joint source–channel coding techniques have not yet been realized. The existing digital source and channel coding techniques perform well for point-to-point communication. However, the optimality of these techniques is not well known for networked settings and especially for sensor networks, where the information is also related to the physical phenomenon. Developing practical joint source–channel communication solutions for WSNs is an important challenge in this field.

4.4 Modulation

Source and channel coding schemes are generally operated in the digital domain, where the information bits that will be communicated over the wireless channel are determined. Wireless communication, however, takes place in the analog domain, where waveforms generated by the transmitter antenna are radiated. The conversion from the bit streams to the waveforms is accomplished by modulation techniques.

In a typical modulation scheme, a waveform is a sinusoid of the form

$$s(t) = r(t) \cos[2\pi f_c t + \psi(t)] \tag{4.9}$$

which has three main components:

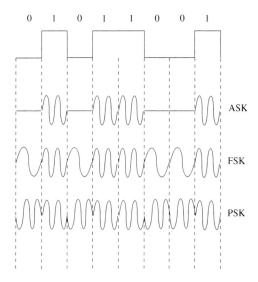

Figure 4.5   The three basic modulation schemes: amplitude shift keying (ASK), frequency shift keying (FSK), and phase shift keying (PSK).

- **Amplitude $r(t)$:** The amplitude or envelope of the waveform.
- **Frequency f_c:** The center frequency of the waveform.
- **Phase $\psi(t)$:** The phase of the waveform.

Digital information is transmitted by modifying one of these components according to the bits to be transmitted. This leads to the three main digital modulation schemes that constitute the bases of several other modulation schemes used for wireless communication. The representation of these schemes is shown in Figure 4.5 and described as follows:

- **Amplitude shift keying (ASK):** As shown in Figure 4.5, the ASK modulation scheme is based on modulating the amplitude of the waveform according to the bit to be sent. The simplest form of ASK is ON–OFF keying (OOK), which transmits a waveform for a digital 1 and keeps silent for the duration of a digital 0. A simple signal amplitude comparison is performed at the receiver to detect the transmitted bits.

- **Frequency shift keying (FSK):** FSK is one of the most frequently used modulation schemes for wireless communication. As shown in Figure 4.5, the frequency, f_c, of the waveform is varied based on the information bit to be transmitted. By choosing distant frequency values, different signals and hence bits can be detected at the receiver.

- **Phase shift keying (PSK):** PSK is based on modifying the phase of the waveform, $\psi(t)$, according to the transmitted bits as shown in Figure 4.5. Accordingly, the changes in the phase of the received signal can be mapped to the transmitted bits at the receiver.

The modulation schemes used for wireless communication are generally derived from the three main schemes shown in Figure 4.5. Next, we describe two main modulation schemes used for WSN platforms: frequency shift keying (FSK) and offset quadrature phase shift keying (OQPSK). The energy efficiency of modulation schemes is also discussed.

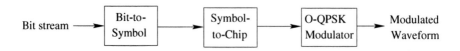

Figure 4.6 The modulation structure in IEEE 802.15.4 [4].

4.4.1 FSK

Binary FSK has been used by early WSN platforms such as Mica2 [1]. As shown in Figure 4.5, the modulation is based on using two distinct frequency values to represent digital 1 and 0. At the receiver side, the center frequency of the received waveform is estimated to decide whether a 1 and/or 0 is received.

At the transmitter side, the appropriate waveform is selected to transmit the bits sequentially. The transmitter ensures that the phase is preserved between different bit transmissions. At the receiver side, the waveform is passed through two matched filters that operate at frequencies f_1 and f_2. The output of each filter represent the signal content at the particular center frequency, which is input to an envelope detector. The results of the envelope detector are compared for a decision.

The bit error rate of the binary FSK modulation scheme is given by

$$p_b^{FSK} = \frac{1}{2} \exp\left(-\frac{E_b}{2N_0}\right) \tag{4.10}$$

where E_b/N_0 is the ratio of energy per bit to noise spectral density, which represents the quality of the received signal at the receiver.

4.4.2 QPSK

The QPSK modulation scheme has been adopted by the IEEE 802.15.4 [4] standard to modulate the chips sent for each bit as a part of the DSSS scheme as explained in Section 4.1.1. More specifically, modulation of bits in an IEEE 802.15.4 transceiver uses offset QPSK (O-QPSK) with DSSS. The modulation structure is shown in Figure 4.6, which consists of the conversion of bits to symbols, conversion of symbols to chips, and O-QPSK modulation for the chips.

In the IEEE 802.15.4 standard, a byte is represented by two symbols with 4 bits per symbol, which results in a 16-ary modulation. Each 16 symbol is represented by a combination of 32 chips. Then, each chip is transmitted using O-QPSK modulation.

The modulation scheme used in MicaZ nodes is O-QPSK with DSSS. The bit error rate of this scheme is given by [18]

$$p_b^{OQPSK} = Q(\sqrt{(E_b/N - 0)_{DS}}) \tag{4.11}$$

where

$$(E_b/N_0)_{DS} = \frac{2N \times E_b/N_0}{N + 4E_b/N_0(K - 1)/3} \tag{4.12}$$

where N is the number of chips per bit, and K is the number of simultaneously transmitting users.

4.4.3 Binary vs. M-ary Modulation

The modulation schemes shown in Figure 4.5 are denoted as *binary* modulation schemes, since two types of waveforms are used to represent digital 1 and 0. In addition to binary schemes, a higher number of waveforms can be used to improve the efficiency of the modulation scheme. These schemes are referred to as *M-ary modulation* schemes, where $M = 2^n$ is the number of waveforms used by the modulation scheme and n is the number of bits represented by each waveform. Using M-ary modulation, multiple bits can be sent through a single symbol, i.e., waveform. This is accomplished by parallelizing the input data and using these parallel data as inputs to a digital-to-analog converter (DAC). As a result, the parallel input levels provide the in-phase and quadrature components of the modulated signal.

Binary and M-ary modulation schemes have different advantages for WSNs. An M-ary scheme can send multiple bits per symbol, which reduces the time it takes to transmit a given set of bits. Accordingly, the transmitter can be kept on for a shorter time, which can reduce the *energy consumption*. On the other hand, compared to binary modulation, M-ary modulation schemes require higher complexity for transceiver circuitry to process the transmitted and received waveforms. Moreover, compared to the binary scheme, where two different waveforms are used to represent 1 and 0, M-ary modulation schemes use a higher number of waveforms. As the number of waveforms increases, it becomes harder to distinguish any received distorted waveform from other waveforms, which increases the decoding errors. To combat the increased decoding errors and to achieve the same symbol error rate, M-ary modulation schemes require a higher transmit power compared to binary schemes [26]. Furthermore, with M-ary modulation, the efficiency of the power amplifier at the transmitter decreases as M increases. Consequently, to achieve the same output power level, higher power consumption is required at the amplifier. As a result, M-ary schemes decrease communication *duration* while increasing *power*, which may or may not decrease the overall energy consumption.

The overall energy consumptions of the binary and M-ary modulation schemes are compared next, considering the tradeoffs in terms of circuit power and transmit duration. According to the energy consumption factors discussed in Chapter 3, for a binary modulation scheme, the energy consumption is modeled as follows [26]:

$$E_2 = (P_{mod\text{-}B} + P_{FS\text{-}B})T_{on} + P_{FS\text{-}B}T_{st} + P_{out\text{-}B}T_{on} \qquad (4.13)$$

where $P_{mod\text{-}B}$ is the power consumption for binary modulation circuitry, $P_{FS\text{-}B}$ is the power consumed at the frequency synthesizer, and $P_{out\text{-}B}$ is the output transmit power consumed at the amplifier. The durations T_{on} and T_{st} are the transmit on time and the transceiver circuitry startup latency. While the transmit on time, T_{on}, depends on the number of bits to be sent, T_{st} is fixed and platform specific.

For an M-ary modulation scheme, the transmit on time is shorter than that for binary modulation and is proportional to the number of bits per symbol, $n = \log_2 M$. On the other hand, the transmit power levels are higher. Accordingly, the energy consumption for M-ary modulation schemes is [26]

$$E_M = (P_{mod\text{-}M} + P_{FS\text{-}M})\frac{T_{on}}{n} + P_{FS\text{-}M}T_{st} + P_{out\text{-}M}\frac{T_{on}}{n} \qquad (4.14)$$

where $P_{mod\text{-}M}$, $P_{FS\text{-}M}$, and $P_{out\text{-}M}$ are the corresponding power consumption at the modulation circuitry, frequency synthesizer, and amplifier, respectively. While the transmit on time is scaled by n, such a clear relationship between the power consumption levels cannot be found easily. Instead, the ratios of power consumption at the modulation circuitry and the frequency synthesizer are represented as follows:

$$\alpha = \frac{P_{mod\text{-}M}}{P_{mod\text{-}B}} \qquad (4.15)$$

$$\beta = \frac{P_{FS\text{-}M}}{P_{FS\text{-}B}}. \qquad (4.16)$$

Then, M-ary modulation is more energy efficient, i.e., $E_M < E_2$ if the following inequality holds:

$$\alpha < n\left[1 + \frac{P_{FS\text{-}B}[(1 - \beta/n)T_{on} + (1 - \beta)T_{st}]}{P_{mod\text{-}B}T_{on}}\right] \qquad (4.17)$$

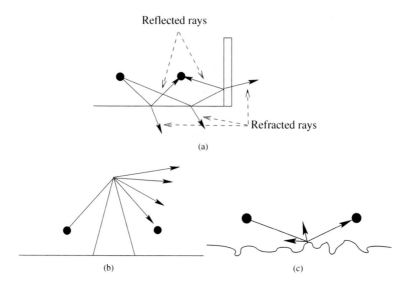

Figure 4.7 Sources of distortion in wireless communication: (a) reflection and refraction, (b) diffraction, and (c) scattering.

considering the power consumption of the amplifier is small compared to that of the frequency synthesizer. According to (4.17), M-ary modulation is more energy efficient when the overhead of the modulation scheme, i.e., α and the startup power, T_{st}, are small. However, under startup power dominant conditions, i.e., $T_{st} \gg T_{on}$, the effect of decreased on time is not pronounced. Instead, the overall energy consumption increases because of the difference in the power consumed at the frequency synthesizer for M-ary modulation, i.e., β. In this case, the binary modulation scheme is more energy efficient irrespective of the values of α. Consequently, the efficiency of M-ary modulation schemes partly depends on the circuit parameters.

4.5   Wireless Channel Effects

The modulation schemes discussed in Section 4.4 convert a stream of bits into waveforms. These waveforms are transmitted from the transceiver via EM signals through the air and received at the receiver antenna. During this transmission, the EM waves are distorted due to several external factors and hence may not be correctly decoded at the receiver. This results in *wireless channel errors*. Understanding the circumstances under which wireless channel errors occur requires accurate wireless channel models. Since the adverse effects of these errors influence each protocol in WSNs, capturing the effects of the wireless channel is essential in WSNs. In the following, the wireless channel effects and the techniques for modeling these effects for WSNs will be described.

The wireless channel *distorts* signals transmitted from a transceiver. The sources of this distortion can be classified into four main phenomena:

- **Attenuation:** As the signal wave propagates through air, the signal strength is attenuated. The attenuation is proportional to the distance traveled over the air. This results in path loss for radio waves in air.
- **Reflection and refraction:** When a signal wave is incident at a boundary between two different types of material, a certain fraction of the wave bounces off the surface, which is called *reflection*. Depending on the properties of the two materials, a certain fraction of the wave may also

propagate through the boundary, which is called *refraction*. Reflection and refraction are usually observed on the ground or the walls of a building as shown in Figure 4.7(a). As a result the received signal may *fade* based on the constructive or destructive effects of multiple waves that are received at the receiver.

- **Diffraction:** A signal wave may also propagate through sharp edges such as the tip of a mountain or a building. As shown in Figure 4.7(b), this causes the sharp edge to act as a *source*, where new waves are generated. In effect, the signal strength is distributed to the new generated waves.

- **Scattering:** Signal waves do not generally encounter perfect boundaries. Instead, when a signal wave is incident at a rough surface, it scatters in different directions as shown in Figure 4.7(c).

Next, we explain in detail the effects of each phenomenon on low-power wireless communication for WSNs.

4.5.1 Attenuation

Attenuation in air results in *path loss* for the EM waves. More specifically, the signal strength decreases as a function of distance. This decrease also defines the *transmission range* of a node. If the signal strength decreases below a certain threshold, the receiver cannot correctly detect the EM wave and decode the signal. This threshold is referred to as the *receiver sensitivity*. The distance at which the signal strength drops below the receiver sensitivity can be regarded as the maximum communication distance or the transmission range of a node.

Attenuation is generally modeled based on empirical evaluations, which show that signal strength decreases as a function of the distance, d, between the transmitter and the receiver. Consequently, the attenuation of the wireless channel can be generalized by representing the path loss $PL(d)$ as the ratio of the transmitted power, P_t, and the received power, P_r:

$$\frac{P_t}{P_r(d)} = PL(d) = PL(d_0)\left(\frac{d_0}{d}\right)^{-\eta} \tag{4.18}$$

where $PL(d)$ is the *path loss* at a distance d, $PL(d_0)$ is the received signal strength at a reference distance d_0, and η is the *attenuation constant* or *path-loss exponent*. Path loss is generally represented in logarithmic form (in dB):

$$PL(d)[\text{dB}] = PL(d_0)[\text{dB}] + 10\eta \log_{10}\left(\frac{d}{d_0}\right) \tag{4.19}$$

where $PL(d_0)[\text{dB}]$ is the path loss at the reference distance, d_0, in dB and η is the path-loss exponent. Accordingly, the relation between the transmit power, P_t, and the received signal power, $P_r(d)$, at a distance d is given below:

$$P_r(d) = P_t - PL(d_0) - 10\eta \log_{10}\left(\frac{d}{d_0}\right) \tag{4.20}$$

From this equation, it can be seen that as the communication range increases, the required transmit power level should be increased for successful communication. Therefore, long-distance wireless communication is expensive, in terms of both energy and cost. However, WSNs are generally deployed over large areas, where information from a single sensor node needs to be transmitted at the sink, which may be a long distance from the node. Since the signal strength decreases with d^η, where $\eta \geq 2$, it is more energy efficient to use multi-hop communication with smaller node distances. Multi-hop communication not only limits the maximum transmit power that needs to be used at a sensor node, but also decreases the overall energy consumption compared to long-haul communication.

4.5.2   Multi-path Effects

The path-loss model in (4.19) captures only the attenuation of the signal waves with distance. However, as explained in Section 4.5, additional factors such as reflection and scattering also affect communication in the wireless channel. Due to the relatively short distances used for communication, diffraction is not a major factor in low-power, short-range communication. However, the effects of reflection and scattering should be considered.

Reflection and scattering from a surface generally result in multiple copies of the signal being received at the receiver. This is mainly related to the omnidirectional propagation of signal waves from the transmitter as shown in Figure 4.7(a). Accordingly, one can visualize multiple rays propagating away from the transmitter in all directions. Because of the reflection and scattering, multiple paths exist through which these rays can reach the receiver. The direct path between the transmitter and receiver is called the *line-of-sight* (LoS) path. In addition to the LoS path between a transmitter and the receiver, a group of rays may also reflect or scatter from surrounding surfaces and reach the receiver. These paths are referred to as the *non-line-of-sight* (NLoS) paths. Those NLoS rays that encounter attenuation comparable to the LoS path also contribute to the success of the communication. These effects are called *multi-path* effects because of the multiple rays that are received at the receiver from different paths.

The effects of NLOS paths on communication depend on both the surrounding and the operating frequency. Moreover, these effects vary with time since the surroundings may vary with time. Therefore, the effects cannot be accurately modeled by the attenuation formula in (4.20). Instead, experimental evaluations show that the multi-path effects are random and can be represented by several probability distribution functions depending on the communication scale, frequency, and location. For short-range communication such as in WSNs, the multi-path effects are modeled by a *log-normal random variable* [34]. As a result, the path loss formula can be rewritten as

$$PL(d) = PL(d_0)[\text{dB}] + 10\eta \log_{10}\left(\frac{d}{d_0}\right) + X_\sigma \qquad (4.21)$$

where X_σ is a normal random variable, $\mathcal{N}(0, \sigma)$, with zero mean and variance σ^2.

Multi-path effects effectively introduce randomness into the wireless communication. Consequently, the success of the communication no longer depends on the distance between the transmitter and receiver. Depending on the path-loss exponent, η, and the variance due to multi-path effects, σ^2, this may lead to a highly non-deterministic wireless communication success, which should be considered in the design of higher layer protocols.

4.5.3   Channel Error Rate

The channel model developed so far captures the effects of signal propagation from a single transmitter. Using (4.18) and (4.21), the received signal strength at a receiver can be found. However, the success of wireless communication depends not only on the received signal strength from a single transmitter but also on the effects of transceiver noise and interference from other users. To be able to capture the channel error rate in terms of bits that are transmitted (bit error rate (BER)) and packets that are transmitted (packet error rate (PER)), the multi-user wireless environment should be considered. Accordingly, the following two additional sources of distortion should also be considered:

- **Noise:** The receiver electronics as well as external sources result in non-negligible noise that affects the accuracy of signal reception. This is generally referred to as the *receiver sensitivity*, which is the minimum signal strength required for the receiver to decode the waveform. Noise generally depends on the particular receiver architecture, the environment, as well as the temperature.

- **Interference:** The wireless channel is a multi-user environment due to its broadcast nature. As a result, more than one device may transmit at the same time, which creates *co-channel* *interference.* Furthermore, other devices, which emit energy at a band close to the sensor node operating frequency, may cause interference with the ongoing communication and create *adjacent-channel interference.* While adjacent-channel interference can be minimized through appropriate filter design, it is still an important source of distortion for WSNs. Common sources of adjacent-channel interference for WSNs are WLANs and microwave ovens.

The success of wireless communication depends on decoding the received signal waveform according to the modulation scheme as explained in Section 4.4. In addition to the particular modulation scheme, this also depends on the received signal strength compared to the distortion caused by the noise and interference. Accordingly, the *signal-to-noise and interference ratio (SNIR)* in dB is defined as follows:

$$SNIR\,[\text{dB}] = 10\log_{10}\left(\frac{P_r}{N_0 + \sum_{i=1}^{k} I_i}\right) \tag{4.22}$$

where P_r is the received signal strength, N_0 is the noise power, I_i is the interference from node i, and k is the number of neighbors that contribute to the interference. The higher the SNIR, the better the wireless channel quality.

In WSNs, the communication attempts of multiple nodes are controlled through the MAC layer solutions as we will see in Chapter 5. Generally, these MAC protocols limit the simultaneous communication effects so that the interference from different nodes is minimized. Hence, co-channel interference can be neglected for WSNs. Similarly, adjacent-channel interference can be regarded as random and, hence, modeled as additional noise. Accordingly, a simplified representation is possible through the definition of signal-to-noise ratio (SNR) as follows:

$$SNR(d) = \psi(d) = P_t - PL(d) - P_n \tag{4.23}$$

where P_n is the noise power (or noise floor) in dB. The noise floor depends on the environment (indoor, outdoor, factory, etc.) as well as time.

As shown in Section 4.4, the BER of a modulation scheme depends on the ratio of energy per bit to noise spectral density, E_b/N_0, which is generally expressed in terms of SNR as follows:

$$E_b/N_0 = \psi\frac{B_N}{R} \tag{4.24}$$

where ψ is the received SNR, B_N is the noise bandwidth, and R is the data rate. Accordingly, the BER for FSK and O-QPSK modulation schemes used for Mica2 and MicaZ can be expressed as in (4.10) and (4.11), respectively.

Based on the BER p_b, the PER can be calculated for different channel coding schemes as explained in Section 4.3. For ARQ, the CRC-16 error detection mechanism is deployed. Assuming all possible errors in a packet can be detected, the PER of a single transmission of a packet with payload of l bits is given by

$$PER^{CRC}(l) = 1 - (1 - p_b)^l. \tag{4.25}$$

For BCH codes, assuming perfect interleaving at the transceiver, the block error rate (BLER) is given by

$$BLER(n, k, t) = \sum_{i=t+1}^{n} \binom{n}{i} p_b^i (1 - p_b)^{n-i}. \tag{4.26}$$

Since a packet can be larger than the block length n, especially where small block lengths are used, the PER for FEC is given by

$$PER^{FEC}(l, n, k, t) = 1 - (1 - BLER(n, k, t))^{\lceil l/k \rceil} \tag{4.27}$$

where $\lceil l/k \rceil$ is the number of blocks required to send l bits and $\lceil \cdot \rceil$ is the ceiling function.

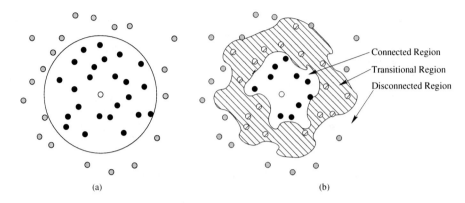

Figure 4.8   Channel models for WSNs: (a) unit disc graph model and (b) statistical channel model.

4.5.4   Unit Disc Graph vs. Statistical Channel Models

The PHY layer specializes in providing reliable wireless links for multi-hop networking in WSNs. The wireless channel properties affect almost all operations at the higher layers. Therefore, an accurate model of the wireless channel is essential. So far, wireless channel models in WSNs have been considered in two main contexts: unit disc graph models and statistical channel models.

Unit Disc Graph Model

Founded on graph theory principles, the unit disc graph (UDG) model represents the range of a node as a disc centered around the node with a radius of 1 as shown in Figure 4.8(a). Accordingly, the node can communicate with its neighbors inside the unit disc perfectly and it is disconnected from the nodes outside the circle. This model leads to the definition of *communication range* of a node as a fixed distance, r_{comm}. Accordingly, the BER for the unit disc graph model, p_b^{udg}, is represented as

$$p_b^{udg} = \begin{cases} 0 & \text{if } d \leq r_{comm} \\ 1 & \text{if } d > r_{comm} \end{cases} \tag{4.28}$$

This model is extremely useful for simplifying the analysis of large graphs, which represent the connections between sensor nodes. Accordingly, several protocols have been created and analyzed based on the UDG model.

Statistical Channel Models

As shown in Section 4.5.2, wireless communication exhibits non-deterministic characteristics due to the random multi-path effects. This randomness lends itself to statistical channel models as explained in Section 4.5. Compared to the UDG model, statistical channel models capture the randomness in the wireless channel more accurately.

Using the log-normal channel model, the SNR is given by [34]

$$\psi(d) = P_t - PL(d_0)[\text{dB}] - 10\eta \log_{10}\left(\frac{d}{d_0}\right) + X_\sigma - P_n \tag{4.29}$$

which can be represented as a Gaussian random variable as follows:

$$\Psi(d) = \mathcal{N}(\beta(d, \eta), \sigma) \tag{4.30}$$

where

$$\beta(d, \eta) = P_t - PL(d_0)|\text{dB}| - 10\eta \log_{10}\left(\frac{d}{d_0}\right) - P_n \qquad (4.31)$$

is the mean and σ^2 is the variance of the Gaussian random variable ψ.

Through (4.10), (4.11), (4.24), and (4.25)–(4.27), we know that the PER is a function of the SNR:

$$\pi(d) = f(\psi(d)), \qquad (4.32)$$

where $f(\cdot)$ depends on the path loss, receiver electronics, modulation, and error control coding schemes. Due to the randomness in SNR, the PER is also a random variable at a given distance, d. Accordingly, the statistical channel model can be represented as in Figure 4.8(b) through three main regions [34]:

- **Connected region:** This is the region in which the nodes have a high probability p_c of having a high PER ($> \pi_c$).
- **Disconnected region:** In this region the nodes have a high probability, p_d, of having a low PER ($< \pi_d$).
- **Transitional region:** This region resides between the connected and disconnected regions, where the variance of the PER is high. Consequently, nodes residing in this region may observe high-quality links for a fraction of the time.

Compared to the UDG model, where only connected and disconnected regions exist, the statistical channel model introduces the concept of a *transitional region*, where the success of the communication has a high variance. According to the definition of PER in (4.32), the lower (d_l) and higher (d_h) bounds for the transitional region can be defined.

The connected region is defined according to the following condition:

$$Pr(\pi > \pi_c) = Pr(\psi > \psi_c) = p_c \qquad (4.33)$$

$$Q\left(\frac{\psi_c - \beta(d_c^{\max})}{\sigma}\right) = p_c \qquad (4.34)$$

where $\pi_c = f(\psi_c)$, $Q(x) = 1/\sqrt{2\pi}(\int_x^\infty e^{-(t^2/2)})\, dt$, and $d_c^{\max}$ is the maximum distance that defines the boundary between the connected and transitional regions as shown in Figure 4.8(b). Similarly, the disconnected region is defined as follows:

$$Pr(\pi < \pi_d) = Pr(\psi < \psi_d) = p_d \qquad (4.35)$$

$$Q\left(\frac{\psi_d - \beta(d_d^{\min})}{\sigma}\right) = (1 - p_d) \qquad (4.36)$$

where $d_d^{\min}$ is the minimum distance that defines the boundary between the transitional and disconnected regions as shown in Figure 4.8(b). Accordingly, the critical distances $d_c^{\max}$ and $d_d^{\min}$ are given as follows [34]:

$$d_c^{\max} = 10^{e_c}, \quad e_c = \frac{\psi_c - \sigma Q^{-1}(p_c) - P_t + P_n + PL(d_0)}{-10\eta} \qquad (4.37)$$

$$d_d^{\min} = 10^{e_d}, \quad e_d = \frac{\psi_d - \sigma Q^{-1}(1 - p_d) - P_t + P_n + PL(d_0)}{-10\eta}. \qquad (4.38)$$

Since communication is still possible in the transitional region, its relation to the connected region is of interest for communication and routing in WSNs. As can be seen from Figure 4.8(b), the connected region has a limited range for multi-hop routing in WSNs. Exploiting the transitional region for communication is important since this leads to longer communication distances. In a multi-hop scenario, a fewer number of hops can be utilized to transfer information from sensors to the sink. Accordingly, the *transitional region coefficient*, Γ, is defined as follows:

$$\Gamma = \frac{d_d^{\min} - d_c^{\min}}{d_c^{\min}} \tag{4.39}$$

According to (4.37)–(4.38), the transitional region coefficient decreases with the attenuation constant η and increases with the variance, σ^2, due to multi-path effects. In other words, in an environment where the attenuation constant is high, the signal strength falls rapidly so that the effect of the transitional region is small. However, as the randomness in the channel increases due to multi-path effects, the transitional region constitutes most of the communication range. Depending on η and σ^2, the transitional region coefficient, Γ, can be as high as 10, i.e., the transitional region length is 10 times the connected region. This fact highlights the importance of exploiting the transitional region in communication for WSNs.

It can be observed that the UDG model is a special case of the statistical channel model, where the length of the transitional region is zero, i.e., $\Gamma = 0$. In cases where $\Gamma \gg 1$, the UDG model leads to highly pessimistic results since the communication range of a sensor node is confined to only the connected region, which is only a fraction of the potential communication range. However, because of the dependency on the random SNR values, exploiting the transitional region requires channel-aware routing protocols. Accordingly, links can be selected according to the instantaneous channel quality, i.e., SNR, and much more efficient multi-hops can be created by using longer links. This requires cross-layer solutions as we will discuss in Chapter 10.

For the design of the physical layer for WSNs, energy minimization assumes significant importance, over and above the attenuation, reflection, refraction, diffraction, and scattering effects. For instance, multi-hop communication in a sensor network can effectively overcome shadowing and path-loss effects, if the node density is high enough. Similarly, while propagation losses and channel capacity limit data reliability, this very fact can be used for spatial frequency reuse. Moreover, network layer protocols are usually developed to provide the shortest hop routes to the packet. Although these routes may be optimal for the network layer, the high error rates due to increased transmission range result in a penalty [11]. However, with the aid of physical layer information, fewer error-prone links can be chosen providing a cross-layer energy optimization in WSNs. Energy-efficient and reliable physical layer solutions are currently being pursued by researchers. Although some of these topics have been addressed in the literature, the area still remains a vastly unexplored domain of WSNs, especially in terms of power-efficient transceiver design, new modulation techniques, and the efficient implementation of low-power communication solutions for sensor network applications.

4.6 PHY Layer Standards

Standardization in WSNs has been a major challenge. Over the years, various types of platforms have been developed, mainly for research purposes. However, due to the independent development of these platforms, interdependency has been a problem among them. Recently, the IEEE has undertaken a worldwide effort to develop a standard for low-power wireless communication. As a result, the IEEE 802.15.4 [4] standard was developed. Since its establishment, this standard has been adopted by the majority of the platforms and industrial applications as the *de facto* standard for WSNs. In this section, we describe the IEEE 802.15.4 standard and the existing platforms used for WSNs.

4.6.1 IEEE 802.15.4

The IEEE 802.15.4 [4] standards body was formed for the specification of low-data-rate wireless transceiver technology with long battery life and very low complexity. The IEEE 802.15.4 standard can be viewed as the low-power counterpart of the IEEE 802.11 standard developed for WLANs [2].

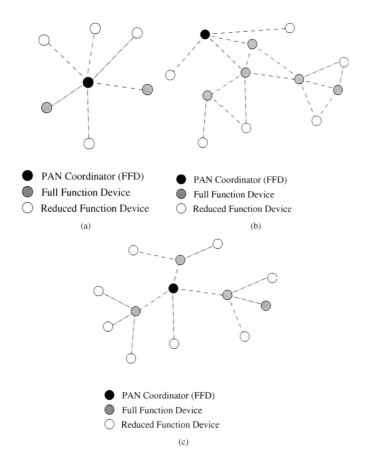

Figure 4.9 Topology structures for IEEE 802.15.4 networks: (a) star, (b) mesh, and (c) cluster tree.

The IEEE 802.15.4 standard defines the PHY and MAC layers and provides flexibility for higher layer solutions. More specifically, the wireless spectrum to be used, wireless communication techniques, and MAC algorithms are defined. This allows for compliant transceivers to communicate with each other even though they may have been produced by different vendors.

At the PHY layer, three different bands are chosen for communication, i.e., 2.4 GHz (global), 915 MHz (the Americas), and 868 MHz (Europe), in addition to larger bands reserved for UWB communication; 47 channels are distributed among these bands. One channel is associated with the 868 MHz range with a data rate of 20–250 kbps; 30 channels are defined for the 915 MHz range and 16 are used in the 2.4 GHz range. The transmission range of the nodes is assumed to be 10–100 m with data rates from 20 to 250 kbps [4]. Based on this channel allocation, six different PHY layers are defined. Among these, three PHY layer specifications are based on DSSS. In the 868/915 MHz bands, BPSK and O-QPSK are used. O-QPSK is also used in the 2.4 GHz band. In addition to these three specifications, the parallel sequence spread spectrum (PSSS) is used to define an additional PHY layer in the 868/915 MHz bands. With the recent amendments, the chirp spread spectrum (CSS) is used as an additional PHY layer definition for highly mobile communication devices in the 2.4 GHz band. Finally, direct sequence UWB is defined to operate in the 1, 3–5, and 6–10 GHz bands as a part of the IEEE 802.15.4a standard [4].

Table 4.1   Existing transceivers used in WSNs.

	Radio				
	RFM TR1000	**Infineon TDA5250**	**TI CC1000**	**TI CC2420**	**Zeevo ZV4002**
Platforms	WeC, Rene Dot, Mica	eyesIFX	Mica2Dot, Mica2 BTnode	MicaZ, TelosB SunSPOT, Imote2	Imote BTnode
Standard	N/A	N/A	N/A	IEEE 802.15.4	Bluetooth
Data rate (kbps)	2.4–115.2	19.2	38.5	250	723.2
Modulation	OOK/ASK	ASK/FSK	FSK	O-QPSK	FHSS–GFSK
Radio frequency (MHz)	916	868	315/433/868/915	2.4 GHz	2.4 GHz
Supply voltage (V)	2.7–3.5	2.1–5.5	2.1–3.6	2.1–3.6	0.85–3.3
TX max (mA/dBm)	12/−1	11.9/9	26.7/10	17.4/0	32/4
TX min (mA/dBm)	N/A	4.9/−22	5.3/−20	8.5/−25	N/A
RX (mA)	1.8–4.5	8.6–9.5	7.4–9.6	18.8	32
Sleep (μA)	5	9	0.2–1	0.02	3.3 mA
Startup time (ms)	12	0.77–1.43	1.5–5	0.3–0.6	N/A

The IEEE 802.15.4 standard also defines a MAC layer, which is based on a superframe structure and relies on carrier sense multiple access with collision avoidance (CSMA/CA) techniques as explained in Chapter 5. The MAC layer provides communication for star, mesh, and cluster tree-based topologies with controllers. Each of these topologies is shown in Figure 4.9. As part of these topologies, two types of devices are defined as a part of the standard. *Full function devices* (FFDs) are implemented with all the functionalities defined in the standard. As shown in Figure 4.9, they can function in any topology and can be used as a network coordinator or a router. An FFD can communicate with any other device in the network. On the other hand, *reduced function devices* (RFDs) are defined for very simple implementation in the network. RFDs are used only as part of the star topology as shown in Figure 4.9(a). Because of the simple implementation, they cannot perform some of the functionalities defined in the standard, such as routing or network coordinator operations. An RFD can only communicate with a network coordinator.

Applications for the IEEE 802.15.4 standard include sensor networks, industrial sensing and control devices, building and home automation products, and even networked toys. Most of the recent platforms developed for WSN research comply with the IEEE 802.15.4 standard.

4.6.2   Existing Transceivers

The major transceiver architectures used for WSNs and their properties are listed in Table 4.1. It can be observed that a wide variety of data rates, operating frequencies, and energy consumption levels exist for the transceivers. Moreover, the first four columns of the table illustrate the timeline of low-power transceivers. A steady increase in data rate from 10 up to 250 kbps can be observed. This is, however, achieved at the cost of increased power consumption because of higher complexity modulation schemes as well as transceiver circuitry used in the recent architectures. Nevertheless, since the transmit durations are decreased through higher data rates, recent transceiver architectures are much more energy efficient compared to earlier ones. An important trend is the change in the tradeoff between transmit and receive power. In earlier transceivers such as the RFM TR1000, transmit power dominated energy consumption by 4 to 1, which is mainly because of the relatively simpler receiver electronics. In this case, amplifier power consumption is higher compared to the energy consumption of the electronics. However, in more recent platforms such as the CC1000, which is used in Mica2 motes, the transmit–receive energy is almost equal. This tradeoff has recently tilted toward receive energy consumption in the CC2420, which has been used in many platforms and complies with the IEEE 802.15.4 standard. Due to the increased complexity with spread-spectrum techniques, receiver electronics dominate amplifier energy consumption, making receive energy consumption higher than that for transmit. This fundamental shift in energy consumption also influences the way MAC protocols have evolved as we will discuss in Chapter 5. Another important trend in transceiver design is the decrease in sleep power as can be observed from the first four columns in Table 4.1. The significant difference in energy consumption between sleep and idle modes ($\sim$1:1 000 000 for the CC2420) motivates more energy-efficient communication protocols that leave the node in sleep mode for most of the time. Furthermore, the startup time for the transceiver has considerably reduced during the last decade, which reduces the overhead in switching to sleep state too often. These changes in transceiver capabilities also drive the changes in communication approaches for WSNs [5, 8, 24].

References

[1] Crossbow Mica2 mote specifications. http://www.xbow.com.

[2] IEEE Std. 802.11b-1999/cor 1-2001. 2001.

[3] Specification of the Bluetooth system – version 1.1b, specification volume 1 & 2. Bluetooth SIG, February 2001.

[4] IEEE standard for information technology – telecommunications and information exchange between systems – local and metropolitan area networks – specific requirement part 15.4: wireless medium access control (MAC) and physical layer (PHY) specifications for low-rate wireless personal area networks (WPANS). *IEEE Std. 802.15.4a-2007 (Amendment to IEEE Std. 802.15.4-2006)*, pp. 1–203, 2007.

[5] A. A. Abidi, G. J. Pottie, and W. J. Kaiser. Power-conscious design of wireless circuits and systems. *Proceedings of the IEEE*, 88(10):1528–1545, October 2000.

[6] I. F. Akyildiz and E. P. Stuntebeck. Wireless underground sensor networks: research challenges. *Ad Hoc Networks Journal*, 4:669–686, July 2006.

[7] I. F. Akyildiz, M. C. Vuran, and Z. Sun. Signal propagation techniques for wireless underground communication networks. *Physical Communication Journal*, 2(3):167–183, September 2009.

[8] B. H. Calhoun, D. C. Daly, N. Verma, D. F. Finchelstein, D. D. Wentzloff, A. Wang, S.-H. Cho, and A. P. Chandrakasan. Design considerations for ultra-low energy wireless microsensor nodes. *IEEE Transactions on Computers*, 54(6):727–740, June 2005.

[9] J. B. Carruthers. Wireless infrared communications. Chapter in *Encyclopedia of Telecommunications*. John Wiley & Sons, Inc., 2002.

[10] R. J.-M. Cramer, M. Z. Win, and R. A. Scholtz. Impulse radio multipath characteristics and diversity reception. In *Proceedings of IEEE ICC'98*, volume 3, pp. 1650–1654, Atlanta, GA, USA, 1998.

[11] D. S. J. De Couto, D. Aguayo, J. Bicket, and R. Morris. A high-throughput path metric for multi-hop wireless routing. *Wireless Networks*, 11(4):419–434, 2005.

[12] M. Gastpar. Distributed source–channel coding for wireless sensor networks. In *Proceedings of IEEE ICASSP'04*, volume 3, pp. 829–832, Montreal, Quebec, Canada, May 2004.

[13] M. Gastpar and B. Rimoldi. Source–channel communication with feedback. In *Proceedings of IEEE Information Theory Workshop'03*, pp. 279–282, Paris, France, March 2003.

[14] M. Gastpar and M. Vetterli. Source–channel communication in sensor networks. In *Proceedings of 2nd International Workshop on Information Processing in Sensor Networks (IPSN'03)*, pp. 162–177, Palo Alto, CA, USA, April 2003.

[15] M. Gastpar, M. Vetterli, and P. L. Dragotti. Sensing reality and communicating bits: a dangerous liaison. *IEEE Signal Processing Magazine*, 23(4):70–83, July 2006.

[16] J. M. Kahn, R. H. Katz, and K. S. J. Pister. Next century challenges: mobile networking for "smart dust". In *Proceedings of ACM MobiCom'99*, pp. 271–278, Seattle, WA, USA, August 1999.

[17] J. M. Kahn and J. R. Barry. Wireless infrared communications. *Proceedings of the IEEE*, 85(2):265–298, February 1997.

[18] M. A. Landolsi and W. E. Stark. On the accuracy of Gaussian approximations in the error analysis of DS-CDMA with OQPSK modulation. *IEEE Transactions on Communications*, 50(12):2064– 2071, December 2002.

[19] H. Lee, B. Han, Y. Shin, and S. Im. Multipath characteristics of impulse radio channels. In *Proceedings of IEEE VTC 2000*, volume 3, pp. 2487–2491, Tokyo, Japan, May 2000.

[20] S. Lin and D. J. Costello Jr. *Error Control Coding: Fundamentals and Applications*. Prentice Hall, 1983.

[21] A. M. Michelson. *Error-Control Techniques for Digital Communication*. John Wiley & Sons, Inc., 1984.

[22] F. R. Mireles and R. A. Scholtz. Performance of equicorrelated ultra-wideband pulse-position-modulated signals in the indoor wireless impulse radio channel. *Proceedings of IEEE Communications, Computers and Signal Processing*, 2:640–644, August 1997.

[23] G. J. Pottie and W. J. Kaiser. Wireless integrated network sensors. *Communications of the ACM*, 43:51–58, May 2000.

[24] J. Rabaey, J. Ammer, B. Otis, F. Burghardt, Y. H. Chee, N. Pletcher, M. Sheets, and H. Qin. Ultra-low-power design. *IEEE Circuits & Devices Magazine*, 22(4):23–29, July/August 2006.

[25] T. Rappaport. *Wireless Communications: Principles and Practice*. Prentice Hall, 1996.

[26] E. Shih, S. H. Cho, N. Ickes, R. Min, A. Sinha, A. Wang, and A. Chandrakasan. Physical layer driven protocol and algorithm design for energy-efficient wireless sensor networks. In *Proceedings of ACM MobiCom'01*, pp. 272–287, Rome, Italy, 2001.

[27] K. Sohrabi, B. Manriquez, and G. Pottie. Near-ground wideband channel measurements. In *Proceedings of the IEEE Vehicular Technology Conference (VTC)*, volume 1, pp. 571–574, New York, USA, 1999.

[28] M. C. Vuran, O. B. Akan, and I. F. Akyildiz. Spatio-temporal correlation: theory and applications for wireless sensor networks. *Computer Networks*, 45(3):245–261, June 2004.

[29] B. Warneke, B. Liebowitz, and K. S. J. Pister. Smart Dust: communicating with a cubic-millimeter computer. *IEEE Computer*, 34(1):2–9, January 2001.

[30] B. A. Warneke, M. D. Scott, B. S. Leibowitz, L. Zhou, C. L. Bellew, J. A. Chediak, J. M. Kahn, B. Boser, and K. Pister. An autonomous $16\,mm^3$ solar-powered node for distributed wireless sensor networks. In *Proceedings of the International Conference on Sensors*, volume 2, pp. 1510–1515, Piscataway, NJ, USA, 2002.

[31] K. D. Wong. Physical layer considerations for wireless sensor networks. In *Proceedings of the IEEE International Conference on Networking, Sensing and Control*, volume 2, pp. 1201–1206, Taipei, Taiwan, March 2004.

[32] A. Woo and D. Culler. A transmission control scheme for media access in sensor networks. In *Proceedings of ACM MobiCom'01*, Rome, Italy, July 2001.

[33] R. Ziemer, M. Wickert, and T. Williams. A comparison between UWB and DSSS for use in a multiple access secure wireless sensor network. In *Proceedings of the IEEE Conference on Ultra Wideband Systems and Technologies, 2003*, pp. 428–432, Reston, VA, USA, November 2003.

[34] M. Zuniga and B. Krishnamachari. An analysis of unreliability and asymmetry in low-power wireless links. *ACM Transactions on Sensor Networks*, 3(2):1–34, June 2007.

5

Medium Access Control

The wireless channel exhibits a broadcast nature, where the transmission of a sensor node can be received by multiple sensor nodes surrounding it. This nature results in each sensor node sharing the wireless channel with the nodes in its transmission range. Since the communication has to be performed through this broadcast wireless channel, the design of *Medium Access Control* (MAC) protocols is of crucial importance in WSNs. The MAC protocols ensure communication in the wireless medium such that the communication links between nodes are established and connectivity is provided throughout the network. Moreover, the access to the channel should be coordinated such that collisions, which occur when two closely located nodes transmit at the same time, are minimized or eliminated.

In addition to the traditional requirements of MAC in wireless networks, additional challenges are posed because of the limited capabilities of each sensor node, the distributed nature of the WSNs, and the traffic properties of sensor applications. We begin this chapter by discussing these challenges and potential solutions in detail. Accordingly, the factors that affect the design of a MAC protocol for WSNs are explained.

In order to address the unique challenges of MAC for WSNs, a large number of MAC protocols have been developed in recent years [5, 12]. The approaches can be classified into three main classes: *contention-based medium access*, *reservation-based medium access*, and *hybrid solutions* that merge these two schemes. The three main classes and the respective protocols developed for WSNs are shown in Figure 5.1. In the following sections, the MAC protocols shown in bold in Figure 5.1 are discussed in detail.

In this section, we discuss only three types of MAC techniques: contention-based, reservation-based, and hybrid solutions. These solutions depend on two fundamental multiple access schemes: carrier sense multiple access (CSMA) and time division multiple access (TDMA). However, typical medium access schemes also include frequency division multiple access (FDMA) [26], code division multiple access (CDMA) [20, 22, 29], and orthogonal frequency division multiple access (OFDMA) [10]. These techniques are generally not employed in WSNs for various reasons.

FDMA techniques are not preferred due to the narrow-band nature of communication employed for WSNs. Furthermore, since the channel bandwidth is limited, FDMA schemes do not provide high efficiencies in WSNs. On the other hand, CDMA and OFDMA techniques that are successfully used in cellular networks are not generally preferred due to cost constraints. The complexity and energy consumption required for these schemes are high compared to the traditional requirements of WSNs. However, as the technology progresses and more cost- and energy-efficient solutions are developed, these solutions will also find their applications in WSNs.

5.1 Challenges for MAC

Several MAC protocols have already been developed and are being used in the general area of wireless networks. These protocols focus mainly on two important performance metrics: throughput

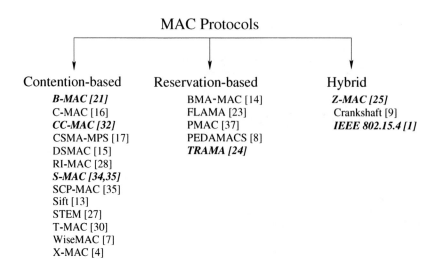

Figure 5.1 Overview of MAC protocols.

and latency. However, these metrics are of secondary importance for WSNs. As explained in Chapter 3, the stringent energy reserves of the sensor nodes make the energy consumption of primary importance. Hence, the MAC protocols designed for WSNs need to be developed according to this unique challenge. In addition to energy consumption, more challenges exist, which are pointed out next.

5.1.1   Energy Consumption

The low cost requirements and the distributed nature of the sensor nodes constrain the energy consumption of all the layers [2]. Hence, the energy efficiency is of primary importance for the MAC layer protocol design. The MAC layer protocol should ensure that nodes transmit their information with minimum energy consumption.

The sources of energy consumption in WSNs are the three main functionalities of sensing, processing, and communication. The sensor and the processing circuitry consume negligible amounts of power when compared to the radio. Hence, communication attempts need to be coordinated carefully to provide energy-efficient operation in WSNs. MAC layer design, in this respect, constitutes an important challenge since the MAC constitutes the core of the communication coordination. Accordingly, the major sources for energy consumption during a communication attempt can be classified as follows:

- **Idle listening:** This refers to the cases where the radio is operated and no useful data are retrieved from the channel. One of the major sources of idle listening is leaving the radio on during long idle times when no sensing event happens. This may result in overhearing. When a node receives a packet that is not destined for itself, it wastes energy while receiving this packet. All the sources of idle listening in WSNs should be minimized in the MAC layer design.

- **Collisions:** These occur when two or more closely located sensor nodes transmit packets to the same receiver at overlapping times. The overlapped information causes the receiver not to receive either of the packets, leading to packet *collision*. Collisions constitute a major source of energy consumption in WSNs. Since the half-duplex nature of the wireless channel prevents collision detection, collision avoidance techniques are usually exploited by MAC protocols.

- **Protocol overhead:** Another major source of energy consumption is the control overhead of the communication protocols. In order to coordinate communication in the wireless channel, MAC protocols require control packets to be transmitted. Although these control packets provide robust operation of MAC protocols, they need to be minimized to improve the energy efficiency.

- **Transmit vs. receive power:** Transmitting and receiving a packet constitute the major sources of energy consumption in WSNs. However, depending on the hardware architecture, transmitting or receiving may dominate the energy consumption. While in earlier hardware architectures, such as Mica2, transmission energy is higher than receiving energy, the opposite is true for MicaZ nodes. As a result, the MAC protocol should be tailored to these relations between transmitting and receiving.

An important way to save energy in WSNs is by turning off the transceiver circuitry. The radio of a sensor node can be operated in sleep mode, where all the circuitry related to the radio is switched off. This is especially desirable when a node has no packets to transmit or receive during a specific period. Because of this sleep mode, the node is said to be in "sleep" when the radio circuitry is switched off and "awake" when it is switched on. Ideally, a sensor node should be awake only during transmitting and receiving data packets. However, this ideal behavior is hard to achieve in a distributed fashion. Since a centralized controller that can manage the operations of all the nodes in the network is not practical, distributed MAC protocols are required. Therefore, these protocols introduce a control overhead in coordinating the communication between sensor nodes. The major goal of a MAC protocol is to minimize this overhead while ensuring efficient, collision-free, and reliable communication in the broadcast wireless channel.

5.1.2   Architecture

The topological awareness of the network is a major property that should be incorporated into MAC protocols. In WSNs, a large number of sensor nodes can be deployed [2]. The increasing density increases the number of nodes within reach of a sensor node, which can be viewed as both a disadvantage and an advantage. Higher network density results in an increase in the number of nodes contending for the wireless channel, which results in higher collision probability. On the other hand, the connectivity of the network can be improved without compromising the increased transmission power due to the large number of neighbor nodes. Moreover, the multi-hop nature of the network needs to be exploited at the MAC layer for improved delay and energy consumption performance.

5.1.3   Event-Based Networking

The application-oriented nature of the WSNs should be exploited in order to increase the performance of the MAC protocol. In traditional networks, per-node fairness is an important aspect of the MAC layer protocol due to the competitive nature of the nodes. In WSNs, however, the system is interested in the collective information provided by the sensors instead of the information sent by each node. Hence, MAC layer protocols should take a collaborative approach so that the application-specific information is exploited to enhance the performance. As an example, for monitoring applications where the traffic follows a periodic pattern, a reservation-based approach can be used to exploit the periodicity in the traffic. On the other hand, in event-based applications, where bursty traffic is generated only during events, an access mechanism that is adaptive to the generated traffic is necessary.

5.1.4   Correlation

Due the high density of the sensor nodes, the information gathered by each node is highly correlated [31, 32]. Intuitively, data from spatially separated sensors are more useful to the sink than highly correlated data from closely located sensors. Hence, it may not be necessary for every sensor node to transmit its data; instead, a smaller number of sensor measurements might be adequate to communicate the event features to the sink. Similarly, the nature of the sensed physical phenomenon results in sensed

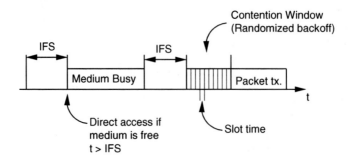

Figure 5.2 Basic CSMA protocol.

information being temporally correlated. Consequently, there is a fundamental limit in the rate at which sensor samples should be taken. Exploiting the correlation between sensor nodes at the MAC layer protocol can be a promising approach to further improve overall network performance.

5.2   CSMA Mechanism

Most of the MAC protocols proposed for WSNs rely on a conventional medium access scheme that has been introduced for WLANs. This scheme is the carrier sense multiple access (CSMA) mechanism. CSMA is used extensively in many types of MAC protocols in WSNs. In contention-based protocols, CSMA is used for basic data communication. Similarly, in reservation-based protocols, slot requests are generally performed through CSMA. Hence, a general knowledge of how CSMA works is necessary before explaining the specifics of each MAC protocols for WSNs. In this section, we overview the CSMA technique along with some modifications that improve the performance of basic CSMA.

CSMA, as its name implies, relies on carrier sense. Carrier sense refers to nodes listening to the channel for a specific amount of time to assess the activity on the wireless channel. In other words, CSMA is a *listen-before-transmit* method. The functionalities of the basic CSMA are shown in Figure 5.2. A node, first, listens to the channel for a specific time, which is generally referred to as the interframe space (IFS). Then, the node acts based on two conditions:

- *If the channel is idle* for the duration of the IFS, the node may transmit immediately.
- *If the channel becomes busy* during the IFS, the node defers transmission and continues to monitor the channel until the transmission is over.

■ **EXAMPLE 5.1**

The IFS duration ensures that nodes transmit only if the channel is idle and helps to prevent collisions. If the channel is sensed as busy during the IFS, it is determined that another node is transmitting a packet. Figure 5.3 illustrates this situation. When node A transmits a packet to node B, the nodes C, D, and E that can hear A's transmission will defer their transmissions if they have a packet to send. As an example, let us assume that all three nodes C, D, and E have a packet to transmit. Since they will sense the channel as busy during carrier sensing, they will defer their transmission until the end of node A's transmission. However, at the end of node A's transmission, all the three nodes will sense the channel as idle and will try to send their packets simultaneously. This, in turn, causes a collision at any node that can receive packets from any two of the nodes (node F). In order to prevent this collision, in CSMA nodes defer for a random amount of time before they transmit a packet. This is referred to as *backoff*.

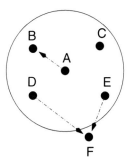

Figure 5.3   Collision in CSMA.

The backoff mechanism of CSMA works as follows. Once the current transmission is over, the station delays another IFS. If the medium remains idle for this period, the node picks up a random number of slots within a range of values to wait before transmitting its packet. This range of values is referred to as the *contention window*. As an example, in WLANs, an initial contention window of 32 is used, which means that a node can select a random number between 0 and 31 for its backoff. The backoff is performed through a timer, which reduces the backoff value for each specific duration referred to as a *slot*. After the nodes enter the backoff period, the first node with its clock expiring starts transmission. Other terminals sense the new transmission and freeze their backoff clocks, to be restarted after completion of the current transmission in the next contention period.

The random backoff mechanism of CSMA aims to prevent nodes from self-synchronizing at the end of a transmission and colliding with each other. However, it is clear that in the case of dense networks, there will be many nodes that will enter the backoff mechanism. As a result, with some probability, some nodes can select the same backoff period and collide with each other. In this case, the nodes that collide double their contention window, i.e., from 32 to 64, and select a new backoff period. This mechanism is referred to as the *binary exponential backoff* scheme. In the case of a successful communication, the contention window is set to its initial value, e.g., 32.

So far, only the basic CSMA mechanism has been described. However, in basic CSMA, the transmitter node has no way of knowing that a packet has been successfully transmitted. It is possible that a packet may be corrupted because of wireless channel errors or collide with another packet. In order for a node to be informed about its transmission, an acknowledgment mechanism is incorporated into CSMA. When a node receives a packet from the transmitter node, it waits for a smaller amount of time than the IFS, i.e., SIFS < IFS, and transmits an acknowledgment (ACK) packet back to the transmitter. On reception of the packet, the transmitter is informed that the packet has been received correctly. The lack of an ACK packet indicates an error in transmission.

One of the major shortcomings of the CSMA mechanism is its susceptibility to hidden terminal collisions. The *hidden terminal problem* is explained through the following example.

■ **EXAMPLE 5.2**

The hidden terminal problem is depicted in Figure 5.4. When node A is transmitting a packet to node B, nodes C, D, and E can hear this transmission and defer their transmission attempts. However, nodes G, H, and I, which can hear B but not A, are not aware of this transmission since they cannot sense the transmission of node A. Hence, if one of these nodes starts transmitting a packet, this packet may collide with the packet that is being sent by node A, at node B. This phenomenon is known as the *hidden terminal problem*.

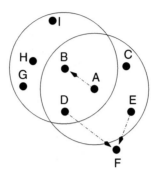

Figure 5.4   Hidden terminal problem.

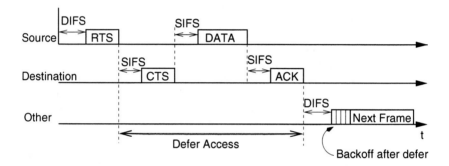

Figure 5.5   CSMA/CA mechanism.

If the data packet being sent is long, then the probability of hidden terminal collision is high. In order to solve this problem, the CSMA/CA scheme has been introduced in wireless networks, where CA stands for collision avoidance. In this scheme, nodes send small packets to reserve the wireless channel prior to data transmission. The operation of CSMA/CA is illustrated in Figure 5.5. This operation of exchanging four messages in CSMA/CA is generally referred to as *four-way handshaking*.

The difference between CSMA and CSMA/CA is the transmission of two reservation packets in the latter mechanism. The channel reservation in CSMA/CA is performed through two small packets, i.e., request to send (RTS) and clear to send (CTS). These packets are generally smaller than the actual DATA packets. The CSMA/CA mechanism is explained in the following:

■ **EXAMPLE 5.3**

As shown in Figure 5.5, the channel sensing and, if necessary, backoff procedure are performed prior to sending the RTS packet by the transmitter node. When a receiver node receives the RTS packet, it replies with a CTS packet, granting the right to send the packet. As shown in Figure 5.5, each node waits for a duration of SIFS before sending its packet in the four-way handshaking. Since SIFS < IFS, the four-way handshaking is given precedence over other transmission attempts. Other nodes have to wait for a duration of IFS for the channel to be idle. Hence, they cannot interfere with the exchange of control and data packets in CSMA/CA.

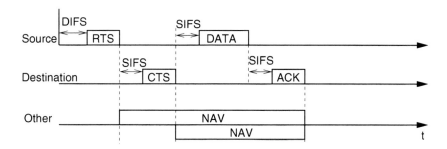

Figure 5.6   NAV mechanism in IEEE 802.11.

Now, let us refer to Example 5.2 again. It can be seen in Figure 5.3 that after the CTS packet is transmitted, the neighbors of node B are informed about the ongoing transmission. Hence, the hidden node problem is partly suppressed through the four-way handshaking. However, RTS collisions are still possible in CSMA/CA. This probability is reduced through binary exponential backoff.

The CSMA/CA mechanism is used in everyday life quite often. The reason for this wide use is because this mechanism is a part of the IEEE 802.11 MAC protocol, which is the MAC protocol of almost every WLAN card. In addition to physical channel sensing, in IEEE 802.11, virtual channel sensing is employed as well. Virtual sensing enables nodes to store a channel occupancy schedule in their local cache. Virtual sensing is accomplished through a local table called the *network allocation vector* (NAV), as explained next.

■ **EXAMPLE 5.4**

The NAV operation is illustrated in Figure 5.6 according to the setup in Example 5.3. When a node sends an RTS packet, the duration of the DATA packet that will be sent is also indicated in the RTS packet. Moreover, the receiver node replies to this RTS with a CTS packet, copying the duration of the DATA packet. As a result, nodes that hear either the RTS or CTS packet can determine the duration of the four-way handshaking. As a result, these nodes can refrain from continuously sensing the channel during the transmission. NAV consists of the duration of the transmission that is being performed and is reduced in every slot just as the backoff timer. Physical channel sensing is only performed when NAV expires.

One of the major problems of the CSMA/CA scheme is the requirement for continuous channel sensing. NAV helps, to some extent, to reduce the energy consumed for channel sensing by enabling the nodes to sleep during an ongoing transmission until NAV expires. The NAV concept is one of the most exploited ideas in MAC protocols for WSNs. In order to further reduce energy, many solutions aim to construct logical distributed schedules at each node that helps them to sleep more and be active only when required. CSMA/CA is the major building block of WSN MAC protocols. We will explore how it is exploited in the following sections.

5.3   Contention-Based Medium Access

One of the fundamental classes of MAC protocols is the contention-based protocols, which rely on controlled contention between nodes to establish communication links. Contention-based protocols provide flexibility since each node can independently perform contention decisions without the need for

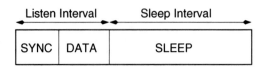

Figure 5.7   Listen and sleep intervals of S-MAC.

message exchanges. As a result, contention-based protocols generally do not require any infrastructure, which is important for many applications of WSNs. Instead, each node tries to access the channel based on the carrier sense mechanism. Contention-based protocols provide robustness and scalability to the network. On the other hand, the collision probability increases with increasing node density.

The IEEE 802.11 MAC protocol, which is based on the CSMA/CA technique explained in Section 5.2, constitutes the basic building block for many of the developed protocols for WSNs. However, CSMA/CA performs poorly in terms of energy efficiency since nodes have to listen to the channel for contention and before transmission. Nodes also consume energy during the idle listening period [34]. In addition, as the density of the network increases, the collision avoidance mechanism becomes ineffective due to the increased number of hidden nodes [33]. Hence, appropriate enhancements are required in the WSN scenario.

In this section, we describe the major MAC protocols developed for WSNs that address these enhancements to the CSMA/CA scheme to improve energy efficiency. Three main MAC protocols, i.e. Sleep MAC (S-MAC) [34, 35], Berkeley MAC (B-MAC) [21], and the Correlation-based Collaborative MAC (CC-MAC) [32], are explained in detail. We also briefly describe other examples of contention-based protocols including Dynamic Sensor MAC (DSMAC) [15], T-MAC [30], STEM [27], WiseMAC [7], CSMA-MPS [17], and Sift [13].

5.3.1   S-MAC

The CSMA/CA technique has the disadvantage of requiring nodes to continuously sense the channel for inactivity. This requirement results in significant energy consumption, especially when nodes do not have any packets to transfer. In order to address this challenge, *duty cycle operation* has been introduced through the S-MAC protocol [34, 35]. Using this operation, the activity of a node is scheduled according to a specific amount of time, called the *frame*. During this frame, which is shown in Figure 5.7, a node sleeps for a specific amount of time and listens to the wireless channel for the rest of the frame. The ratio of the listen interval and the total duration of the frame is denoted as the *duty cycle*. During the sleep interval, the radio of the node is switched off to save energy. In the meantime, the particular node is also detached from the network.

Periodic Listen and Sleep

S-MAC is a distributed contention-based MAC protocol which relies on coordinated sleep schedules to decrease the energy consumption while trading off throughput and latency. While the protocol is based on the CSMA/CA scheme, periodic sleep and listen cycles are introduced to reduce idle listening. The operation of each node is maintained during frames. Each frame consists of two intervals, *listen* and *sleep*, as shown in Figure 5.7. The listen interval is further divided into two intervals called SYNC and DATA. The main idea behind S-MAC is to construct virtual clusters of nodes that sleep and wake up at the same time. This goal is performed through periodic synchronization messages, abbreviated as SYNC messages. The SYNC portion of the listen interval is reserved for the exchange of these messages. Then, nodes try to find their intended receivers during the DATA interval.

Sender Node ID	Next Sleep Time

Figure 5.8   The structure of the S-MAC SYNC packets.

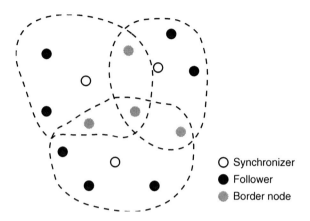

○ Synchronizer
● Follower
◉ Border node

Figure 5.9   Virtual clusters of the S-MAC protocol.

Now, let us illustrate how the virtual schedules are established using S-MAC. S-MAC ensures that nodes, which are in the transmission range of each other, synchronize according to a single sleep schedule. This is performed by exchanging periodic SYNC messages. The structure of a SYNC message is shown in Figure 5.8 and consists of the ID of the sender node and the remaining time until the sender switches to sleep mode. During protocol initialization, a node listens to the channel for a specific amount of time, long enough to receive any SYNC packets sent by its neighbors. If no SYNC packet is received during this interval, the node determines its own sleep schedule and broadcasts a SYNC packet. This particular node is referred to as the *synchronizer*. The nodes which receive this packet follow the sleep schedule of the synchronizer. The sleep schedule consists of the remaining time until the node will *go to sleep*. Since all the nodes are informed about the duty cycle and the frame size, this information is sufficient for other nodes to synchronize. However, inaccuracies in the clocks of a sensor node prevent actual time values being broadcast. Instead, a node broadcasts the remaining time to switch to sleep.

■ **EXAMPLE 5.5**

An example of the virtual cluster formation of S-MAC is shown in Figure 5.9. If a node receives a schedule from a neighbor before choosing its own schedule, it follows this neighbor's schedule, i.e. becomes a *follower*. Furthermore, the follower waits for a random delay and broadcasts this schedule. S-MAC does not aim to globally synchronize the network such that a single schedule is followed throughout the network. Instead, the nodes in close proximity are synchronized. As a result, it might happen that a node receives a neighbor's schedule after it has selected its own schedule. In these cases, this node is referred to as the *border node* as shown in Figure 5.9. Border nodes adapt to both schedules and wake up at the listen intervals of these two schedules. However, it is expected that a node adopts multiple schedules very rarely since every node tries to follow existing schedules before choosing an independent one.

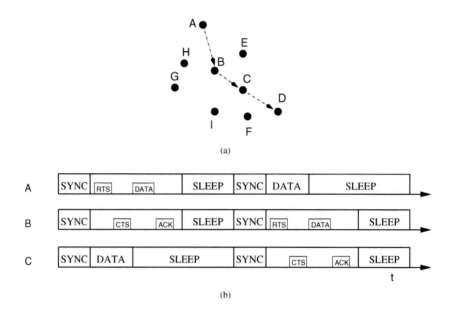

(a)

A	SYNC	RTS	DATA		SLEEP	SYNC	DATA		SLEEP	
B	SYNC		CTS	ACK	SLEEP	SYNC	RTS	DATA	SLEEP	
C	SYNC	DATA		SLEEP		SYNC		CTS	ACK	SLEEP

t

(b)

Figure 5.10   Multi-hop awareness problem: (a) topology and (b) MAC scheme.

Overhearing Avoidance

Once a schedule is established, data packet transmissions are performed during the DATA period of the listen interval. For this purpose, the CSMA/CA mechanism is used. Nodes that have a packet to send contend for the wireless medium during the DATA interval shown in Figure 5.7 through RTS–CTS exchanges. When a node transmits its RTS packet, the intended receiver sends back a CTS packet during the "for CTS" interval. After the RTS–CTS exchange, the transmitter node starts to transmit its DATA packet. Other nodes in the virtual cluster switch to sleep state until the end of the frame. This avoids wasting energy during idle listening and is referred to as *overhearing avoidance* [35]. The actual DATA packet transmission can continue during the sleep interval and only the two nodes that exchange DATA and ACK messages stay awake.

Multi-hop Awareness

A major disadvantage of the S-MAC protocol described so far is that it only controls the local interactions of the nodes in the network. More specifically, only a single hop operation is maintained. However, the multi-hop nature of WSNs requires various enhancements to the basic CSMA/CA mechanism, which was originally developed for single hop WLANs. While multi-hop communication is mainly handled at the network layer in the conventional layered structure, considering this fact for MAC design brings various advantages. This is commonly referred to as *multi-hop awareness*, which is not crucial for proper operation of the medium access scheme alone, but it is important for energy conservation and latency reduction due to cross-layer effects of the routing on the MAC.

■ **EXAMPLE 5.6**

The multi-hop awareness problem of the basic S-MAC scheme is illustrated in Figure 5.10, where node A tries to send a packet to node D, through nodes B and C.

When node A attempts to send a packet to node B it first performs carrier sensing. If node A successfully sends a packet to node B, the neighbors of nodes A and B, i.e., nodes C, H, and E, which overhear this transmission, switch to sleep state. However, in S-MAC, the RTS–CTS exchange is performed when all the nodes are awake. If a node does not receive any packets destined for itself during this interval, it switches to sleep state. As a result, the communication between nodes A and B continues as the nodes C, H, and E switch to sleep state. When node B successfully receives the packet from node A, it will try to find node C, which is the next hop to the destination node D. However, since node C is in sleep mode, node B has to wait for the subsequent listen interval for node C to wake up and receive an RTS packet. As a result, during a single frame, a packet can only travel a single hop. This results in an average delay proportional to the length of the path, which significantly affects the packet delivery delay of multi-hop networks.

The lack of multi-hop awareness results in medium access being performed for every hop independent of the multi-hop route of the packet. An ideal way would be for each next hop to wake up when the transmission of the previous hop is finished. This would be possible in Example 5.6 if node C wakes up at the time when the transmission from node A to node B is finished. As a result, node B can send its packet immediately to node C if there are no other nodes performing a transmission. Similarly, if node D wakes up at the end of the transmission from node B to node C, virtually no delay would be incurred due to the duty cycle operation of the MAC protocol.

The ideal solution for multi-hop awareness is hard to implement in a distributed network such as WSNs. Such a solution requires, firstly, a network-wide synchronization such that each node will wake up at the exact time and, secondly, knowledge of the route a packet will take. These requirements may not be feasible in a distributed WSN. Moreover, there will be other packets being transmitted in the network, which will contend with this packet. Instead, a localized solution, which exploits limited information, has been developed in S-MAC. This solution is referred to as *adaptive listening*, as explained next.

Adaptive Listening

The S-MAC protocol has been enhanced with the adaptive listening mechanism to provide multi-hop awareness [35]. Adaptive listening does not assume knowledge of the route, nor does it try to schedule all the nodes on the route of a packet that is sent. Instead, a best effort solution is provided. Adaptive listening allows nodes that overhear a packet transfer to wake up at the end of this transfer in case they become the next hop. An example of the operation of the adaptive listening mechanism is given next.

■ **EXAMPLE 5.7**

The adaptive listening mechanism of S-MAC is illustrated in Figure 5.11 based on the topology in Example 5.6 as shown in Figure 5.10(a). Node A transmits an RTS packet to node B. This transmission is also overheard by node C, which switches to sleep state to save energy during the communication. Node C is also informed about the duration of the transmission through the *duration* field in the RTS and CTS packets. Accordingly, node C sets up a timer such that it wakes up for a short period of time at the end of the transmission between nodes A and B. This allows node B to find its next hop by transmitting an RTS packet immediately. Since node C is awake, it can respond with a CTS packet and the packet can be transmitted one additional hop during a single frame.

Figure 5.11 shows the timeline for the adaptive listening operation. Note that node C wakes up when the communication between nodes A and B is finished. When node B transmits an RTS packet to node C, it can immediately receive the packet. This adaptive listening mechanism reduces the latency of the basic S-MAC protocol by half. However, note that the packet cannot progress any further when the transmission between nodes B and C finishes. As a result, node C has to wait for the next listen period in order to find the next hop (node D). Thus, adaptive listening

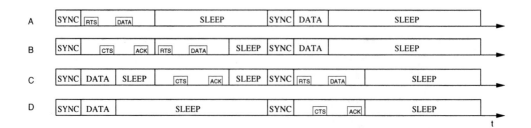

Figure 5.11   Adaptive listening mechanism of S-MAC.

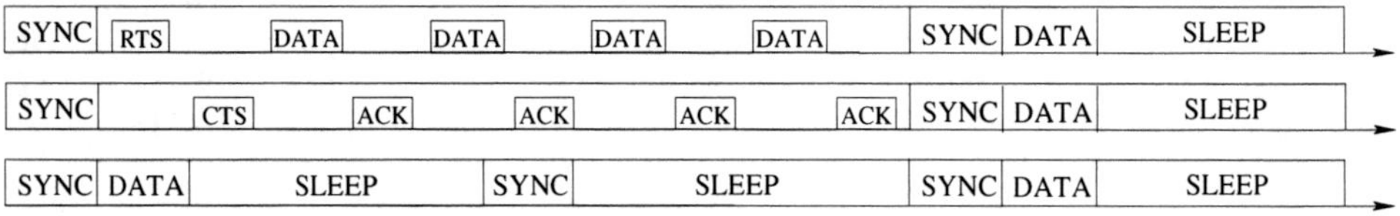

Figure 5.12   Message passing mechanism of S-MAC.

provides a best effort service to minimize latency in duty cycle-based MAC protocols. However, this scheme may not always decrease the latency. Instead, energy consumption may be increased because of the adaptive listening of all the neighbor nodes that overhear a transmission.

Message Passing

In certain applications, a sensor node may need to send a *burst* of packets to transmit a large amount of information that it has generated. If the default operation of S-MAC is used for these cases, a large overhead is incurred. The main reason for this overhead is the transmission of RTS–CTS packets before each DATA packet. This overhead is minimized in S-MAC through the *message passing* procedure. In this case, when a node has a burst of packets to send, it uses the RTS–CTS exchange only for the first packet. Each packet is followed by an acknowledgment from the receiver. Moreover, the remaining duration of the burst transfer is included in each packet sent by the sender and the receiver nodes. In this way, other nodes are prevented from accessing the channel. The main notion of the acknowledgment messages from the receiver is to prevent hidden terminal problems. The message passing procedure is explained in the following example.

■ EXAMPLE 5.8

A message passing example, which includes a sender node A and a receiver node B, is illustrated in Figure 5.12. Moreover, node C is a neighbor of node B and cannot hear the transmission of node A. Node A starts the transmission by broadcasting an RTS packet and node B responds with a CTS packet. If node C wakes up in the middle of the transmission, it receives the ACK packet from node B and is informed about the duration of the transmission. In this case, it sleeps until the end of the transmission. This procedure prevents collisions with long packet transmissions.

Qualitative Evaluation

S-MAC consumes much less energy compared to the CSMA/CA protocol as a result of energy savings through the duty cycle operation. Especially for lightly loaded networks, periodic listening plays a key role. At heavy load, on the other hand, idle listening rarely happens. As a result, energy savings from sleeping are very limited. S-MAC achieves energy savings by avoiding overhearing and efficiently transmitting long messages. Moreover, the synchronization mechanism of S-MAC results in the formation of virtual clusters in the network. Consequently, several cluster-based protocols developed for routing can be easily incorporated in S-MAC without explicit clustering mechanisms.

For the development of the S-MAC protocol, a network with periodic traffic is assumed. However, the fixed duty cycle operation of S-MAC cannot provide flexibility to bursty traffic since the sleep schedules are of fixed length. When the network traffic is low, each node has to wake up at the beginning of each frame to receive the SYNC packet sent by the synchronizer without sending any packets. As a result, S-MAC consumes a constant amount of energy due to the duty cycle structure. Moreover, if the network traffic increases because of an event, the listen interval may not be long enough to accommodate the traffic. This results in an increase in communication delay since a node will have to wait several frames to transmit its packets.

S-MAC trades off latency for reduced energy consumption. The virtual clustering mechanism results in all the nodes in a cluster sleeping at the same time and transmitting during the listen period. Consequently, the idle listening duration is reduced to save energy. However, this increases the end-to-end latency that a packet experiences, which makes S-MAC unsuitable for delay-sensitive data. Furthermore, in high-density networks or networks with high loads, S-MAC introduces an increased collision rate since all the nodes in the cluster are constrained to contend for the medium during the listen interval. Consequently, these types of networks may not be supported by S-MAC.

5.3.2   B-MAC

Energy consumption is the main performance metric that drives novel medium access protocols being developed for WSNs. As we have seen in Section 5.3.1, duty cycle operation is devised through the S-MAC protocol to decrease energy consumption by making nodes sleep during idle periods. In this section, we will describe the *preamble sampling* mechanism and the B-MAC (Berkeley MAC) protocol [21], which has been devised to overcome two main drawbacks of this mechanism.

Duty cycle operation requires sleep–wakeup schedules to be formed such that nodes in close proximity are active at the same time. This operation has two drawbacks in terms of energy inefficiency. Firstly, nodes need to send periodic messages, such as the SYNC packets used by S-MAC in each frame. Secondly, all the nodes need to be active during the listen period to wait for a possible incoming packet. As a result, even when there is no traffic, nodes consume energy at a rate at least equal to the duty cycle.

B-MAC has been designed to provide a core MAC protocol that is simple and reconfigurable by higher level protocols. To this end, B-MAC provides basic CSMA mechanism to the higher layers. Furthermore, an optional link level ACK mechanism is provided without any RTS–CTS message exchanges. The CSMA mechanism can be configured by higher layers by changing the backoff durations. Accordingly, a low-power operation and an effective collision avoidance mechanism is implemented. Furthermore, the simple implementation results in a small code space.

B-MAC is based on two mechanisms: sleep–wake scheduling using low-power listening (LPL) and carrier sensing using clear channel assessment (CCA). Both of these mechanisms improve the energy efficiency and channel utilization. Furthermore, B-MAC has been implemented in TinyOS and provides simple interfaces for the higher layer services to easily configure the underlying MAC operation. This enables rapid development of cross-layer solutions, which require the basic functionalities of a MAC protocol. Through the provided interfaces, the higher layer protocols can toggle the CCA and ACK

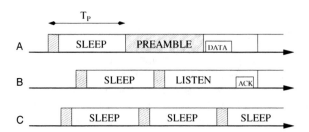

Figure 5.13   Preamble sampling.

services, set backoff parameters on a per-packet basis, and change the LPL mode for transmit and receive efficiently. Next, we explain the LPL and CCA functionalities of B-MAC.

LPL

The drawbacks of the fixed duty cycle operation of S-MAC and its derivatives can be addressed by removing the requirement of *schedule* for each node. Accordingly, all the nodes in the network do not need to wake up and sleep at the same time. Instead, each node can determine its sleep and wakeup schedule without any synchronization with the other nodes. This, however, requires transmitters to synchronize with their intended receivers when they have a packet to send. In other words, a transmitter needs either to *wake up* its intended receiver or wait for the intended receiver to wake up and send its data. This necessity is solved through preamble sampling [6], which is also referred to as LPL [21]. The main idea behind LPL is to send a preamble before each packet to *wake up* the intended receiver with the goal of minimizing the "listen cost" associate with the fixed duty cycle protocols. Accordingly, each node periodically wakes up, turns on its radio, and checks for activity on the channel. During this small period, if activity is detected on the channel, the node stays in the receive mode. However, if no activity is detected, then the node switches back to sleep state.

■ **EXAMPLE 5.9**

The operation of LPL is shown in Figure 5.13 for three nodes: transmitter node A, receiver node B, and neighbor node C. In this scheme, each node in the network determines its sleep schedule. The sleep schedule frame length is determined as T_P. At each T_P second, the node wakes up for a short amount of time to detect the activity of the channel. As shown in Figure 5.13, the wakeup time of each node is not synchronized with other nodes. When node A has a packet to send to node B, it first sends a preamble of length T_P to wake up the node. Note that a length of T_P is required for the preamble to wake up any node. When node B wakes up, it listens until the end of the preamble, determines that the packet is destined for itself, and does not switch to sleep state to wait for the subsequent packet. Node A then sends the DATA packet and node B replies with an ACK packet if the transmission is successful. Note that node C also wakes up during the preamble transmission. However, since the packet is not destined for itself, it switches back to sleep state without consuming more energy.

The LPL mechanism removes the need to send periodic SYNC messages to form virtual clusters. However, this is achieved at the cost of transmitting a long preamble before each DATA packet. This approach can be more energy consuming per packet transfer when compared to the sleep–wakeup schedule-based protocols. However, it is certainly more energy efficient for the durations with no traffic. For low traffic load, preamble sampling is, hence, more energy efficient. The channel listen interval,

T_P, is an important variable to determine for efficient operation. The traffic pattern defines the optimal listen interval. If the listen interval is too small, then the nodes waste energy by frequently waking up and listening to the channel. On the other hand, if the listen interval is too large, energy is wasted on transmissions since long preambles need to be transmitted before each communication attempt. However, since multiple nodes consume energy during listening, it is better to use a longer preamble than check for activity more often. The optimal listen interval value can be determined according to the traffic load.

CCA

The success of the LPL technique, in general, relies on the accuracy in sensing the activity on the channel. Each node samples the network for a short amount of time and goes back to sleep if it does not detect any activity such as the transmission of preamble packet(s). If a node assesses that there is activity on the channel and wakes up when there is no activity (false positive), precious energy is wasted. On the other hand, if a node cannot detect a preamble destined for itself (detection failure), the transmitter would waste energy because of the transmitted preamble and would have to wait for another sampling period to meet the receiver, which increases end-to-end latency. Furthermore, if a transmitter inaccurately assesses that the channel is idle when it is occupied, the transmission of a preamble would result in collisions, which reduces the capacity of the wireless channel and increases energy consumption. These problems are addressed by the B-MAC protocol in the context of CCA [21].

The main motivation behind the design of the CCA mechanism of B-MAC is as follows. In a wireless channel, the ambient noise may significantly change depending on the environment. This leads to frequent fluctuations in the received signal strength of a receiver even during a lack of packet transmissions. On the other hand, packet reception exhibits a fairly constant channel energy. This leads to the conclusion that, to accurately detect activity in the channel, several channel measurements should be taken over time. The main goal of the CCA mechanism is to differentiate between noise and a signal to accurately assess the channel activity. This is addressed through a software approach to estimate the noise floor before any decisions have been made.

The CCA mechanism of B-MAC consists of two phases: noise floor estimation and signal detection. Since the channel noise varies depending on the environment, B-MAC does not assume a fixed noise floor for CCA. Instead, signal strength samples are taken during times when the channel is assumed to be idle, which is usually right after a packet transmission. The noise floor estimation is performed using a first-in, first-out (FIFO) queue. Each sample is stored in the FIFO queue and the median, S_t, of the queue is added to an exponentially weighted moving average with decay. Accordingly, the noise floor estimate is calculated as follows:

$$A_t = aS_t + (1 - a)S_{t-1} \tag{5.1}$$

where a is a parameter which controls the effect of the decay. The moving average of the noise floor is then used for signal detection.

The traditional approach for signal detection in CSMA-based protocols is to take a single signal strength sample and compare this value to the noise floor. This is referred to as a *threshold* approach. However, the threshold approach leads to many false positives. Whenever the channel is idle, the fluctuations in the noise level may result in the receiver determining that the channel is occupied. As a result, the channel utilization is decreased. To address this issue, B-MAC follows an *outlier detection* approach, where the node searches for outliers in the received signal through multiple signal strength measurements. Since the packet transmission results in a constant signal strength, if a node finds outliers among the samples it determines that the channel is clear. On the other hand, if no outliers exist, then the channel is determined to be busy.

As a result, the operation of the B-MAC CCA mechanism works as follows. Following a transmission, a node first takes a sample of the channel to update its noise floor estimate. Then, before transmitting any packets, it takes several samples of the channel to determine outliers. If an outlier is found, the channel is assumed to be clear and the packet is sent. If no outlier is found, the channel is determined to be busy and the backoff mechanism is used.

B-MAC provides interfaces in TinyOS implementation so that several features can be controlled by higher layers. As a result, the CCA mechanism can be turned on or off by higher layer protocols. This allows TDMA-based MAC protocols to be implemented in conjunction with the B-MAC protocol, which provides basic functionalities. Similarly, the backoff mechanism of B-MAC can be controlled by higher layers through the provided interface. Whenever a channel is sensed to be busy, B-MAC sends a signal to the higher layers through this interface and the required backoff mechanism can be specified.

Qualitative Evaluation

The B-MAC protocol provides an efficient carrier sensing mechanism so that most of the false positives are eliminated. As a result, the channel efficiency can be improved. Furthermore, the noise floor estimation mechanism enables the MAC protocol to adapt to its surroundings. Moreover, B-MAC is a simple and lightweight protocol which does not require large memory space. As a result, most of the MAC functionalities can be provided without consuming scarce memory resources. Accordingly, the interfaces provided by the B-MAC protocol provide flexibility and interoperability for the development of higher layer protocols. Consequently, B-MAC can be used as a "core" MAC protocol, on which other schemes can be built.

B-MAC implements a simple CSMA-based MAC protocol. As a result, the hidden terminal problem is not addressed. Accordingly, the B-MAC protocol may lead to underutilization of the wireless channel in high-density and high-traffic networks. The LPL mechanism requires nodes to transmit long preambles, which creates significant overhead when the traffic load is high. Furthermore, this overhead increases if the duty cycle of the network is low and the nodes sleep for a longer time to save energy.

While trying to improve the accuracy of the carrier sense mechanism, B-MAC introduces additional complexity through the CCA mechanism. Each node needs to keep track of several channel measurements, which may increase memory usage. Furthermore, the CCA mechanism increases the channel access delay since multiple measurements need to be taken before the channel is determined to be clear.

5.3.3   CC-MAC

Typical WSN applications require spatially dense sensor deployment in order to achieve satisfactory coverage [18]. As a result, several sensor nodes record information about a single event in a sensor field. Due to the high density in the network topology, the sensor records may be spatially correlated subject to an event. Exploiting spatial correlation in the context of the collaborative nature of the WSNs can lead to a significant performance improvement in communication protocols. Spatial correlation increases when the distance between nodes decreases, making this phenomenon of local interest around a particular node. Since the local interactions among nodes are resolved at the MAC layer, it is natural to exploit this local redundancy at the MAC layer. The CC-MAC (Correlation-based Collaborative MAC) protocol aims to address this issue [32].

The operation of the CC-MAC protocol is based on a spatial correlation model, which will be explained next. Then, the details of CC-MAC operation are described.

Spatial Correlation Model

In a sensor field, each sensor observes the noisy version of a physical phenomenon. The sink is interested in observing this phenomenon using the observations from multiple sensor nodes with the highest accuracy. The physical phenomenon of interest can be modeled as a spatio-temporal process $s(t, x, y)$ as a function of time t and spatial coordinates (x, y).

The model for information delivery from N sensors in an event area is discussed in Chapter 4. The sink is interested in estimating the event source, S, according to the observations of a subset of M sensor nodes in the event area ($M \leq N$). Each sensor node n_i observes $X_i[n]$, the noisy version of the event information, $S_i[n]$, which is spatially correlated to the event source, S. This observation is then encoded in the form of $Y_i[n]$ and then sent to the sink through the WSN. The sink, at the other end, decodes this information to get the estimate, $\hat{S}$, of the event source S.

As explained above, the observations of each sensor node are correlated according to the properties of the physical phenomenon, which can be modeled as joint Gaussian random variables (JGRVs) at each observation point[1] as

$$E\{S_i\} = 0, \quad i = 1, \ldots, N$$

$$\text{var}\{S_i\} = \sigma_S^2, \quad i = 1, \ldots, N$$

$$\text{cov}\{S_i, S_j\} = \sigma_S^2 \, \text{corr}\{S_i, S_j\}$$

$$\text{corr}\{S_i, S_j\} = \rho_{i,j} = K_\vartheta(d_{i,j}) = \frac{E[S_i S_j]}{\sigma_S^2}$$

where $d_{i,j} = \|\mathbf{s}_i - \mathbf{s}_j\|$ is the distance between nodes n_i and n_j located at coordinates $\mathbf{s}_i$ and $\mathbf{s}_j$, respectively, and $K_\vartheta(\cdot)$ is the correlation model. The covariance function is assumed to be non-negative and to decrease monotonically with the distance $d = \|\mathbf{s}_i - \mathbf{s}_j\|$, with limiting values of 1 at $d = 0$ and of 0 at $d = \infty$.

According to this model, each observed sample, X_i, of sensor n_i is represented as

$$X_i = S_i + N_i, \quad i = 1, \ldots, N \tag{5.2}$$

where the subscript i denotes the location of node n_i, i.e. (x_i, y_i), S_i is the realization of the space–time process $s(x, y)$ at $(x, y) = (x_i, y_i)$, and N_i is the observation noise. $\{N_i\}_i$ is a sequence of i.i.d. Gaussian random variables of zero mean and variance σ_N^2.

Each observation $X_i[n]$ is then encoded into $Y_i[n]$ as

$$Y_i = \sqrt{\frac{P_E}{\sigma_S^2 + \sigma_N^2}} X_i, \quad i = 1, \ldots, N \tag{5.3}$$

where σ_S^2 and σ_N^2 are the variances of the event information S_i and the observation noise N_i, respectively. Y_i is then sent to the sink through the multi-hop network. Once the sink receives this information, it is decoded as Z_i as follows:

$$Z_i = \frac{E[S_i Y_i]}{E[Y_i^2]} Y_i. \tag{5.4}$$

The main goal of the sensing phenomenon is to accurately reconstruct the features of the physical phenomenon using information from a subset of M out of N sensors, where N is the total number of sensor nodes in the event area. Accordingly, $\hat{S}$, the estimate of S, is given as

$$\hat{S}(M) = \frac{1}{M} \sum_{i=1}^{M} Z_i. \tag{5.5}$$

Moreover, the level error in the reconstructions, i.e., the distortion, achieved by using M sensor nodes to estimate the event S is given as

$$D(M) = E[(S - \hat{S}(M))^2] \tag{5.6}$$

---

[1] Since the spatial correlation between nodes is considered, samples are assumed to be temporally independent.

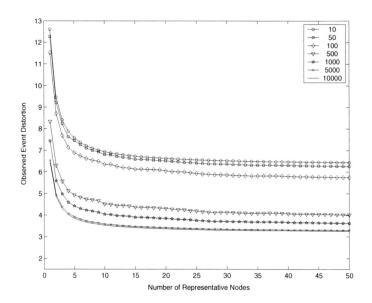

Figure 5.14   Average distortion for different θ_1 values according to changing representative node number.

where the mean-squared error is used as the distortion metric. Accordingly, the distortion function $D(M)$ is found to be [32]:

$$D(M) = \sigma_S^2 - \frac{\sigma_S^4}{M(\sigma_S^2 + \sigma_N^2)}\left(2\sum_{i=1}^{M}\rho_{(s,i)} - 1\right) + \frac{\sigma_S^6}{M^2(\sigma_S^2 + \sigma_N^2)^2}\sum_{i=1}^{M}\sum_{j \neq i}^{M}\rho_{(i,j)}. \qquad (5.7)$$

■ **EXAMPLE 5.10**

$D(M)$ shows the distortion achieved at the sink as a function of the number of nodes M that send information to the sink and correlation coefficients $\rho_{(i,j)}$ and $\rho_{(s,i)}$ between nodes n_i and n_j, and the event source S and node n_i, respectively. The change in distortion as a function of the number of nodes M that send information is shown in Figure 5.14 for a network of $N = 50$ nodes.

It can be observed that the minimum distortion is achieved when all the sensor nodes in the event area send information to the sink, i.e., $M = N = 50$. However, the achieved distortion at the sink can be preserved even though the number of representative nodes significantly decreases from 50 to 15. Furthermore, there are two factors affecting the distortion other than the number of representative nodes. Firstly, the distortion increases as the distance between the event source S and a node n_i increases. More specifically, if a representative sensor node is chosen distant from the source, it observes inaccurate data resulting in higher distortion at the sink. Secondly, as the distance between sensor nodes increases, the distortion decreases. Since sensor nodes that are further apart observe less correlated data, the distortion is decreased if these nodes are chosen as the representative nodes. Accordingly, the minimum distortion can be achieved by choosing these nodes such that (1) they are located as close to the event source S as possible, and (2) they are located as far apart from each other as possible.

As shown in Figure 5.15, the redundancy in the network allows a sensor node to act as a *representative node* for several other sensor nodes in its vicinity. These nodes are referred to as *correlation neighbors*

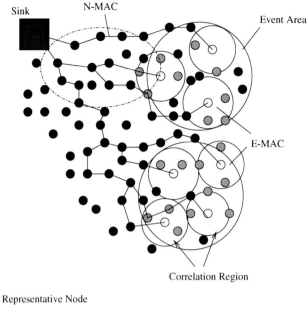

Figure 5.15   Two components of the CC-MAC protocol: E-MAC and N-MAC [32].

since the information they observe is highly correlated with that of the representative node. As a result, from the standpoint of the sink, it would be enough for only one node (representative node) to send its information. The area where the representative node and all its correlation neighbors reside is referred to as the *correlation region*. CC-MAC operation is based on the operation of two entities: the iterative node selection (INS) algorithm that runs at the sink and the distributed CC-MAC protocol that runs at each sensor node on the field.

Iterative Node Selection (at the sink):

The INS algorithm resides at the sink to determine the operating parameters of the distributed CC-MAC protocol. To this end, the statistical properties of the WSN topology are used as the input. These include the network density and the variance of the physical phenomenon samples and the noise. As the output, the INS algorithm provides the average correlation radius, r_{corr}, between the nodes to be used in the distributed protocol. This is the maximum distance, r_{corr}, at which two nodes are highly correlated, given a distortion requirement. More specifically, the goal is to find the minimum number of representative nodes that achieve the distortion constraint given by the sensor application:

$$M^* = \operatorname*{argmin}_{M}\{D_E(M) < D_{\max}\} \tag{5.8}$$

where $D_{\max}$ is the maximum distortion allowed by the sensor application. Given the density of the network, the distance, r_{corr}, is determined. The calculation of the correlation radius consists of two steps: (1) INS performs vector quantization (VQ) techniques to form Voronoi regions. (2) The average distance between the centers of these Voronoi regions are used as the correlation radius. This value is then communicated to the sensor nodes for distributed CC-MAC operation.

Given a vector source with known statistical properties and given a distortion constraint and the number of code vectors, VQ algorithms try to find a codebook and a partition which result in the smallest average distortion. In other words, VQ algorithms represent all possible codewords in a code space by a subset of codewords. In the case of the CC-MAC protocol, the goal is to represent all the sensor nodes in an event area with a smaller number of representative nodes. If we choose two-dimensional code vectors, the code space in the VQ approach can be mapped to the network topology with the node locations as the codeword spaces. Once the codebook is determined, VQ algorithms use Voronoi regions to determine the partition of a code such that any information in this partition is represented by the code vector. Voronoi regions determine the areas closest to the points representing the area.

■ **EXAMPLE 5.11**

The notions of codebook and partitions are illustrated in Figure 5.16 for two correlation radii, i.e., $r_{corr} = 90$ m in Figure 5.16(a) and $r_{corr} = 45$ m in Figure 5.16(b). Accordingly, the codebook represents the locations of the representative nodes (nodes filled with dots) and the partitions represent the Voronoi regions, i.e., areas for which the representative nodes are responsible.

VQ algorithms require only the statistical properties of the code space, i.e., only the statistical properties of the topology are required at the sink. Accordingly, the INS algorithm does not need the exact locations of the nodes but only the statistical properties, which are evident from initial network deployment.

In addition, the statistical properties of the sensed phenomenon should be estimated to determine the distortion in (5.7). This is performed during initialization of the protocol, when sensors send their observations for one cycle to the sink. Accordingly, the INS algorithm estimates the variance σ_S^2 from the collected data by variance estimation techniques. The correlation parameters are estimated similarly. More specifically, the empirical correlation between two nodes is estimated according to the samples $S_1, S_2, \ldots, S_N$, from each of the N nodes, as follows:

$$\rho(i,j) = \frac{S_i S_j}{\sigma_S} = e^{-d_{(i,j)}/\theta_1} \tag{5.9}$$

and the correlation coefficient, θ_1, is estimated accordingly. This value is used in (5.7) to calculate the distortion.

The optimum value of M^* in (5.8) is found first by setting $M = N$, the total number of nodes in the event area or the whole network. Then, the distortion, $D(M)$, given in (5.7) is calculated. If $D(M) < D_{max}$, then M is decreased as $M = M - k$. Using the updated M, the VQ algorithm is executed to generate multiple topologies with M nodes and find the minimum distortion, $D(M)$. This iteration is repeated as long as $D(M) < D_{max}$. The algorithm terminates when M^* is found and the resulting Voronoi regions are found as shown in Figure 5.16. Accordingly, the average distance, d_{avg}, between the representative nodes is found, where $d_{avg} = 2r_{corr}$.

CC-MAC Protocol (at the nodes)

The CC-MAC operation is performed through two components: Event MAC (E-MAC) and Network MAC (N-MAC). The correlation in the region where an event occurs is handled by the E-MAC component. E-MAC is initiated by a *first contention phase* (FCP), where nodes in close proximity to each other contend to become a representative node using the RTS/CTS/DATA/ACK mechanism. Each of these nodes sets the first-hop (FH) field of the RTS packet and tries to capture the medium for transmission.

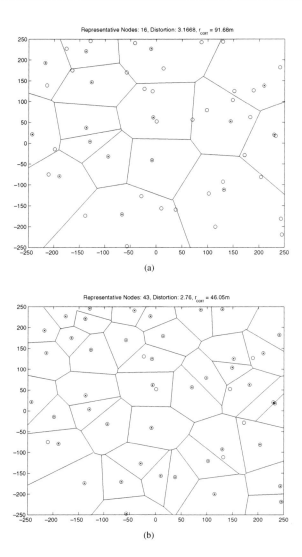

Figure 5.16 Representative nodes and Voronoi regions created by the INS algorithm of CC-MAC for (a) $r_{corr} = 90$ m and (b) $r_{corr} = 45$ m [32].

When a node n_i captures the channel after the first contention phase, it becomes the representative node of the area determined by the correlation radius r_{corr}. The node, n_i, continues to send information to the sink as the sole representative of its correlation region. This phase is similar to the clustering mechanisms explained in Chapter 13. Once a node becomes a representative node, it sets a *first-hop bit* in the transmitted packets to indicate to its correlated neighbors to drop their packets. Every other neighbor node, n_j, that listens to the RTS packet of the node n_i checks the FH field and determines that the transmission is related to a source functionality. In addition, each node n_j determines $d_{(i,j)}$, its distance to node n_i. If $d_{(i,j)}$ is found to be less than the correlation radius, r_{corr}, then the node, n_j, determines that it is a correlation neighbor of the node n_i and stops its transmission. If the node is outside the correlation region of node n_i, then it contends for the medium if it has a packet to send.

This operation constructs correlation clusters as shown in Figure 5.15. As a result, a lower number of communication attempts are performed, which leads to lower contention, energy consumption, and latency while achieving acceptable distortion for reconstructing event information at the sink.

Since the correlation is filtered in the event area, the transmitted packets will have a higher priority than newly generated packets. These packets are handled by the Network MAC (N-MAC) component of CC-MAC. The relay nodes clear the FH bit on receiving a packet and relay it to indicate that the transmitted packet is a "route-thru" packet. Consequently, a higher priority is given to the route-thru packets. A node in a correlation region with a route-thru packet listens to the channel for a time which is smaller than the *DCF interframe space* (DIFS) used by the nodes performing E-MAC. As a result, it can attempt to capture the channel earlier. Moreover, the router node uses a smaller backoff window size than the value used by the representative nodes. Such a principle increases the probability that the router node captures the channel since the router node begins backoff before the representative node of the correlation region. As a result, the route-thru packet is given precedence.

Exploiting spatial correlation for MAC operation results in high performance in terms of energy efficiency, packet drop rate, and latency compared to MAC layer protocols designed for WSNs. Consequently, exploiting the spatial correlation results in significant energy savings in WSNs.

Qualitative Evaluation

CC-MAC exploits the spatial correlation inherent in the sensor readings to significantly improve the energy efficiency. By restricting the communication attempts to only representative nodes, the efficiency of the MAC protocol is improved since local contentions are minimized. This operation does not affect the accuracy of event estimation due to the spatial correlation between closely located nodes. The E-MAC and N-MAC components of CC-MAC provide priority for the filtered traffic so that information from representative nodes is not affected by local contentions. This improves the reliability of data delivery. Furthermore, CC-MAC moves most of the complexity to the sink, where the computation-hungry INS algorithm is run. At the sensor nodes, a distributed protocol is implemented, which accepts only the correlation radius, r_{corr}, as a parameter from the sink.

CC-MAC is applicable to networks where spatial correlation is dominant. More specifically, in high-density networks, where the sensors observe correlated data, the energy savings are significant. However, CC-MAC may not provide high savings if nodes observe independent data. Similarly, the operation of CC-MAC requires locational knowledge of the neighbors of a node to determine correlation. This requires either on-board GPS devices or localization techniques as discussed in Chapter 12.

5.3.4   Other Contention-Based MAC Protocols

In addition to the S-MAC, B-MAC, and CC-MAC protocols, we briefly explain next other contention-based MAC protocols developed for WSNs. More specifically, the Dynamic Sensor MAC (DSMAC) [15], T-MAC [30], STEM [27], WiseMAC [7], CSMA-MPS [17], and Sift [13] protocols are described.

DSMAC

One of the basic enhancements for fixed duty cycle operation is provided through the DSMAC (Dynamic Sensor MAC) protocol [15]. The main motivation behind this protocol is to minimize the medium access delay that may occur due to high traffic rate. The S-MAC duty cycle allows a fixed number of packet transmissions during the listen period. If a node generates (or receives) more packets than it can immediately transmit, the delay that will be experienced by the packet will increase. The delay that is incurred by the MAC protocol in transmitting a packet is generally referred to as *medium access delay*. Static duty cycles may result in intolerable medium access delay, which builds up the packet queue. As will be explained in Chapter 8, increased queue length may cause congestion in the network, leading to

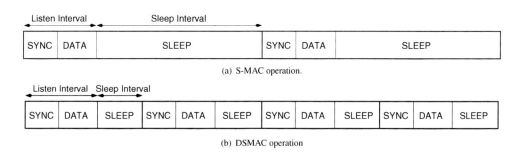

(a) S-MAC operation.

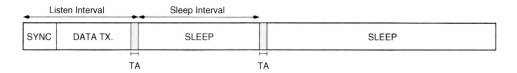

(b) DSMAC operation

Figure 5.17   Listen and sleep intervals of DSMAC.

Figure 5.18   Listen and sleep intervals of T-MAC.

some packets being lost. DSMAC aims to solve this problem. The solution is to double the duty cycle in case the medium access delay of a packet exceeds a pre-specified value. In order to conserve energy, a node also checks if its energy consumption is lower than a threshold value. The sleep schedule of the DSMAC protocol, when doubled due to high traffic load, is shown in Figure 5.17(b) along with the S-MAC sleep schedule in Figure 5.17(a). Doubling the duty cycle enables a node to receive or send more packets than nodes that perform the original schedule. Moreover, since every other listen interval coincides with the original schedule, connectivity is still established. When a node decides to double its duty cycle, it broadcasts this value inside the SYNC packet that is sent at the beginning of each original frame. The node also includes its intended receiver in the SYNC message. Accordingly, the receiver node, after receiving the SYNC packet, adjusts it duty cycle and wakes up at the specified time. Other nodes in the virtual cluster are not affected by this operation. As a result of doubling the duty cycle, the medium access delay can be reduced and the buffer length can be decreased.

T-MAC

Another drawback of static schedules is the waste of energy when the traffic load is low. T-MAC aims to solve this problem by introducing an adaptive duty cycle operation [30]. The nodes start to listen to the channel in each listen interval but time out when there is no traffic for a specific amount of time. In this way, the duration of the listen interval is reduced in cases where there is no node in the virtual cluster to transmit a packet. The sleep–wakeup schedule of the T-MAC protocol is shown in Figure 5.18. When there is no traffic, the listen time of the T-MAC protocol is lower than S-MAC's. The timeout value, denoted by TA, is an important parameter for the T-MAC protocol. This value is determined such that $TA > C + R + T$, where C is the length of the contention interval, R is the time for RTS packet transfer, and T is the turnaround time between the reception of the RTS packet and the transmission of the CTS packet. This ensures that potential hidden neighbors of the receiver nodes are aware of the transmission before switching to sleep. Although this awareness is not crucial for the communication taking place, it has implications for multi-hop awareness of the MAC protocol.

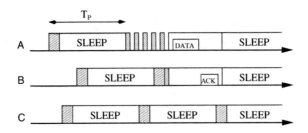

Figure 5.19   Wakeup scheme of STEM.

STEM

A problem with basic preamble sampling is that a node has to send the whole preamble even if the receiver node wakes up at the very beginning and is ready to receive the packet. An example of this problem is explained next.

■ EXAMPLE 5.12

As shown in Figure 5.13, after node A transmits the preamble, node B wakes up in the middle of the transmission. However, node B still has to listen to the remaining preamble before the DATA transmission can start. As a result, the long preamble wastes bandwidth as well as energy.

An alternative way and a solution to this problem is used in the STEM (Sparse Topology and Energy Management) protocol [27]. Although STEM is not a pure MAC protocol, here we will describe the wakeup scheme used in STEM. This wakeup scheme relies on successive small packets being transmitted instead of a single long preamble. Moreover, a separate signaling channel is reserved for wakeup packets. As a result, STEM requires two radios (main and wakeup radio) to be implemented on the sensor nodes. However, this requirement can also be relaxed by using a single radio.

After transmitting each wakeup packet, the transmitter node listens to the channel for a reply from the intended receiver. On the other side, each node in the network periodically listens to the channel as in the basic preamble sampling scheme. When a node hears a wakeup packet destined for itself, it replies with a small packet. After packet exchange, the transmitter starts to send the DATA packet as shown in Figure 5.19. In this way, no energy is wasted once the intended receiver wakes up. Let us denote the preamble length of the basic preamble sampling mechanism as T_P s. Using the basic preamble sampling mechanism, a transmitter node has to spend a duration of T_P s before each DATA packet. Accordingly, the wakeup scheme in STEM reduces this time to $T_P/2$ on average.

WiseMAC

Another source of energy wastage of the basic preamble sampling scheme is when a transmitter immediately starts to send the preamble once it has a DATA packet to send and the medium is idle. However, it still has to wait for a node to wake up before transmitting the DATA packet while sending the preamble. Since the transceiver consumes energy during this time, both bandwidth and energy are wasted. Recall that each receiver has a fixed wakeup schedule. This schedule can be exploited to schedule the start of the preamble packet and save energy. Hence, if a node is aware of the wakeup schedule of its intended receiver, it can wait until the intended receiver wakes up and then send the preamble packet. In this way, it will not waste energy by transmitting the preamble before its intended receiver can hear.

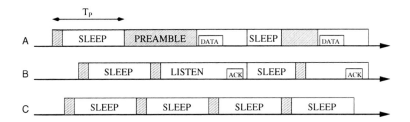

Figure 5.20 Wakeup scheme of WiseMAC.

WiseMAC [7] enhances the wakeup scheme of basic preamble sampling by exploiting the wakeup schedules of a node's neighbors, referred to as *local self-synchronization*. WiseMAC minimizes the length of the wakeup preamble by exploiting knowledge of the sampling schedule of direct neighbors, reducing transmit, receive, and overhearing overheads.

■ **EXAMPLE 5.13**

The operation of WiseMAC is shown in Figure 5.20 for nodes A (transmitter), B (intended receiver), and C (neighbor) similar to Figure 5.13. When node A has no knowledge of the wakeup schedule of node B, WiseMAC operates similar to basic preamble sampling. When node A sends a DATA packet to node B, node B replies with an ACK packet. In the ACK packet, the wakeup schedule information of node B is also indicated. This information consists of the next time node B will wake up and the value of the wakeup period. Accordingly, node A will know when node B will wake up in case it has more packets to send to it.

Sensor nodes use cheap clocks, which have a fairly significant drift. Hence, even if a node knows the wakeup schedule of its neighbor, it still may not reach the neighbor at the previously calculated wakeup time. As a result, the preamble has to be transmitted before the calculated wakeup time and last longer. More specifically, let us assume that node A sends a packet to node B and receives the ACK packet at time $t = 0$. At time $t = 0$, node A will know about node B's wakeup schedule. Let us further assume that node B's next wakeup is scheduled at $t = L$. If node A has more packets to send to B, it will wait L seconds and start to send the preamble. However, considering a clock drift of at most $\pm\delta$ per second, the actual wakeup time of a node may drift by $\pm 2\delta L$. As a result, WiseMAC adjusts the length of the preamble such that $T_P = \min(4\delta L, T_W)$, where T_W is the wakeup schedule. Consequently, as the time between subsequent transmission to a neighbor increases, a longer preamble is used to overcome the clock drift.

WiseMAC also addresses two of the drawbacks of the basic preamble sampling scheme. One is the possibility of a collision between two preambles. If two nodes start to send the preamble at the same time, the preambles will collide at the receivers. Hence, WiseMAC inserts a medium reservation preamble (MRP) in front of the preambles to resolve collisions just like the RTS packets in CSMA/CA. Another drawback of the basic preamble sampling scheme is the case when a node has more than one packet to send to the same receiver. In this case, the node has to wait until the next sampling time of the receiver for each subsequent packet. This adds latency to the end-to-end delivery performance of the protocol. However, in WiseMAC, nodes indicate subsequent packets by a "more" bit in the data header. As a result, at the end of each ACK packet, the receiver waits for the next DATA packet. This allows the transmission of bursty traffic with low delays. Moreover, the overhead due to transmitting a preamble is shared among multiple DATA packets.

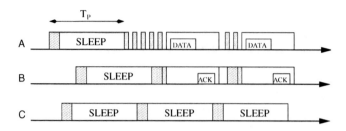

Figure 5.21   Wakeup scheme of CSMA-MPS.

CSMA-MPS

As we have seen above, STEM and WiseMAC reduce energy wastage at the two ends of the preamble. While STEM prevents preamble transmission time being longer than required, WiseMAC schedules the start of the preamble such that longer preambles are not sent. These two improvements are combined in CSMA-MPS [17]. The medium access mechanism is shown in Figure 5.21. CSMA-MPS uses multiple small wakeup packets as in STEM and starts the transmission of these packets according to the schedule information of the neighbors as in WiseMAC. Moreover, at the beginning of the preamble transmission, nodes introduce an additional random time T_{random} to avoid collisions with other nodes that are similarly synchronized to the receiving node and want to send at the same time.

Sift

The majority of MAC protocols proposed for WSNs rely on the CSMA/CA protocol as a building block. As a result, the backoff mechanism is not altered in many proposals and the binary exponential backoff mechanism, which is explained in Section 5.2, is deployed. One of the drawbacks of the binary exponential backoff occurs when the contention rate is increased, i.e., the number of nodes contending for the wireless channel is high. In this case, the backoff window of each node is increased due to collisions. Consequently, the medium access latency increases, leading to high end-to-end latency in the network. The Sift protocol has been designed to minimize the time taken to send packets in the network [13]. Sift uses a fixed contention window of length CW as opposed to the variable contention window used by CSMA/CA. The goal of the protocol is to minimize the time taken for $R \leq N$ nodes to send their packets without collisions, where N is the number of nodes contending for the channel at the same time. Instead of varying the contention window size, Sift uses a non-uniform probability distribution for picking a transmission slot in the contention window. The probability distribution for picking the slot $r \in [1, CW]$ is

$$p_r = \frac{(1-\alpha)\alpha^{CW}}{1-\alpha^{CW}}\alpha^{-r}, \quad \text{for } r = 1, \ldots, CW, \qquad (5.10)$$

where $0 < \alpha < 1$. This probability distribution increases exponentially with r, so the later slots have higher probability. The reasoning behind the probability distribution can be described as follows. When a transmission ends, each node that has a packet to send contends for the medium using the backoff strategy. Assuming there are multiple nodes that try to access the channel, nodes pick the earlier slots of the contention window with low probability. If there are a high enough number of nodes in the vicinity, with some probability, a node picks the earlier slot and wins the contention. However, when no slots are being chosen, the probability to pick later slots increases according to the distribution function (5.10). In other words, as the slots are not chosen, each node assumes that there are a fewer nodes contending for the medium and tries to access the channel more aggressively. As a result, access to the channel is

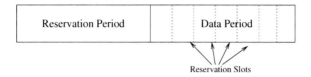

Figure 5.22   General frame structure for TDMA-based MAC protocols.

upper bounded by the fixed contention window length because, with a high probability, a node wins the contention at the end of CW.

5.3.5   *Summary*

Generally, contention-based protocols provide scalability and lower delay, when compared to the reservation-based protocols that will be explained next. On the other hand, the energy consumption is significantly higher than the reservation-based (or TDMA-based) approaches as will be explained in Section 5.4 due to collisions and collision avoidance schemes. Moreover, contention-based protocols are more adaptive to changes in the traffic volume and hence applicable to applications with bursty traffic such as event-based applications. Furthermore, the synchronization and clustering requirements of reservation-based protocols make contention-based ones more favorable in scenarios where such requirements cannot be fulfilled.

5.4   **Reservation-Based Medium Access**

Reservation-based protocols have the advantage of collision-free communication since each node transmits data during its reserved slot. Hence, the duty cycle of the nodes is decreased resulting in further energy efficiency. Recently, *time division multiple access* (TDMA)-based protocols have been proposed in the literature. Generally, these protocols follow common principles, where each node communicates according to a specific *superframe* structure. This superframe structure, which generally consists of two main parts, is illustrated in Figure 5.22. The *reservation period* is used by the nodes to reserve their time slots for communication through a central agent, i.e., *cluster head* or with other nodes. The *data period* consists of multiple time slots that are used by each sensor for transmitting information. Among the proposed TDMA schemes, the contention schemes for reservation protocols, the slot allocation principles, the frame size, and clustering approaches differ in each protocol. We explain each protocol along with these common features in the following.

5.4.1   *TRAMA*

The energy-efficient collision-free MAC protocol TRAMA is based on a time-slotted structure and uses a distributed election scheme based on the traffic requirements of each node [24]. As a result, the time slot that a node should use is determined such that any collisions with other nodes are prevented. TRAMA is a schedule-based MAC protocol, where no central entity is required for reservations. Rather, pairwise communications between neighbors are performed to schedule transmission slots for the communicating parties. Consequently, each node schedules the slots during which it will transmit or receive packets. Accordingly, the node can coordinate when it has to sleep or remain active in the network.

TRAMA consists of four main phases:

- **Neighborhood discovery:** In this phase, nodes need to be informed about their neighbors such that the potential receivers and transmitters can be determined.

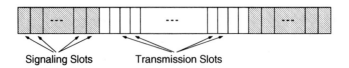

Figure 5.23   Frame structure of TRAMA.

- **Traffic information exchange:** During this phase, nodes inform their intended receivers about their traffic information. More specifically, if a node intends to send a packet to a node, it informs this particular node during this phase. Consequently, by collecting traffic information from other nodes, a node can form its schedule.

- **Schedule establishment:** Based on the traffic information from its neighbors, a node determines the slots for transmitting and receiving packets in a frame. These schedules are then exchanged between nodes.

- **Data transmission:** Based on the established schedule information, the nodes can switch to active mode and start communication in the specified slot.

The superframe structure of TRAMA is shown in Figure 5.23. The frame consists of *signaling slots* in the reservation period and *transmission slots* in the data period. TRAMA operation consists of three mechanisms: the neighbor protocol (NP), schedule exchange protocol (SEP), and adaptive election algorithm (AEA). In summary: (1) Each node gets information about its every two-hop neighbor using NP. (2) The traffic information of each node is gathered by SEP using signaling slots, i.e., the reservation period. Based on this information, (3) each node calculates its priority and decides using AEA which time slot to use. Nodes sleep during their allocated slots if they do not have any packets to send or receive. We explain these three mechanisms in detail next.

NP

NP of TRAMA propagates one-hop information among neighbors. This is achieved by each node broadcasting its neighborhood information using signaling packets during the signaling slots. The signaling packets indicate the list of one-hop neighbors that a node has and carry incremental neighbor updates to keep the size of the signaling packet small. Each node sends incremental updates about its one-hop neighborhood as a set of added and deleted neighbors. If there are no updates, signaling packets are still sent as "keep alive" beacons. As a result, signaling packets help maintain connectivity between neighbors. The signaling packets inform each node about its two-hop neighborhood. If a node does not hear from a node for a certain amount of time, the neighborhood information about that node is erased.

SEP

Using the neighborhood information gathered through NP, a node determines its intended transmission schedule based on the number of packets it has. This schedule is then communicated to the neighbors through SEP. Each node computes a *SCHEDULE_INTERVAL* (SCHED) based on the rate at which packets are produced. SCHED represents the number of slots for which the node can announce the schedule to its neighbors according to its current state. The node pre-computes the number of slots in the interval $[t, t+\text{SCHED}]$ for which it has the highest priority among its two-hop neighbors. The last winning slot is used for broadcasting the node's schedule for the next interval. The node announces the intended receivers for these slots. The intended receivers are indicated through a bitmap in the schedule packets. The format of the schedule packet is shown in Figure 5.24. If a node does not have data, it gives

Figure 5.24   TRAMA schedule packet format.

up the vacant slots by indicating this through the GIVEUP field of the schedule packet. The schedule information is also piggybacked to the DATA packets to maintain synchronization of the network.

AEA

The selection of the slots to transmit or receive a packet is determined through the distributed AEA of TRAMA. According to the schedule information obtained by SEP from neighbor nodes, AEA decides each neighbor node state as *transmit* (TX), *receive* (RX), or *sleep* (SL). Nodes without any data to send are removed from the election process, thereby improving channel utilization. After NP and SEP, each node knows its two-hop neighborhood and the current schedules of its one-hop neighbors. Using this information, the priority of each node in the schedule interval can be calculated.

This is performed as follows. A node uses identifier u and a globally known hash function h and computes its own priority

$$p(u, t) = h(u \oplus t) \tag{5.11}$$

for a time slot t. This priority is computed for the next k time slots for the node itself and its two-hop neighbors. This is also performed by the other nodes. Note that schedule information is not required to compute the priority of a node. Instead, the node identifier, u, determines the priority of nodes in different time instances, t, according to the hash function $h(\cdot)$. Then, a node uses the time slots for which it has the highest priority. For a given time slot t, the node is decided to be in TX state if it has the highest priority and it has a packet to send. Accordingly, each node determines the slot it should transmit and informs its intended receivers. Each intended receiver, when informed, is assigned to RX state for the selected slot. The other slots, when the node is not assigned to TX or RX state, are labeled as SL, during which the node can sleep.

TRAMA does not require complete information regarding the schedules of nodes in the network for slot assignment. Instead, local information exchanges are exploited for this purpose. As a result, collisions are possible since each node selects its slots according to limited neighborhood information, i.e., two-hop neighborhood. The cases for possible collisions are resolved by calculating the relative priorities of each neighbor among their respective neighbors. In a two-hop neighborhood of a node i, if a node j has the highest priority in a given slot, it is determined as the *absolute winner*. An absolute winner has the right to transmit at a slot. We explain next a case where the absolute winner may not always be selected.

■ EXAMPLE 5.14

In the network topology shown in Figure 5.25, according to node B, node D is the absolute winner. However, node A may not be aware of node D's priority. As a result, according to node A, it is the absolute winner in its two-hop neighborhood. If node D does not have a packet to send to node B, then node B may sleep during that particular slot. However, in this slot, node A is assigned as the absolute winner by itself and can send its packet to node B. For node B to be available at that slot, TRAMA introduces the concepts of *alternate winner* and *possible transmitter set*. If a node may be an absolute winner in its two-hop neighborhood it is labeled as an alternate winner by its neighbors. As a result, node B labels node A as an alternate winner and

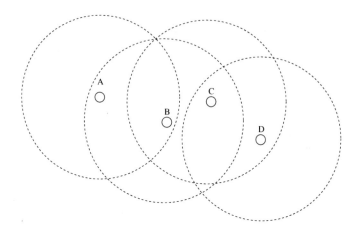

Figure 5.25   Absolute and alternate winners in TRAMA [24].

includes it in its possible transmitter set if it has a packet to send. When node *A* tries to send a packet to node *B*, then it will be available.

Qualitative Evaluation

TRAMA increases the energy efficiency compared to contention-based protocols by increasing the percentage of the time that nodes spend in sleep mode. While contention-based protocols include inefficiency because of random scheduling, the time-slotted communication structure of TRAMA decreases the collision rate. In addition to energy efficiency, the delivery rate of TRAMA is also high since collisions are minimized and time slots are assigned according to the traffic requirements of the nodes. The time slots are determined according to a distributed selection mechanism so that each node determines its own schedule. This eliminates the need for clustering algorithms and any central coordinators for slot allocation.

The time-slotted structure of TRAMA introduces significant delay for end-to-end communication in WSNs. Compared to contention-based protocols, this delay may increase by 3-4 orders of magnitude. Since the end-to-end delay is proportional to the frame length, determining the optimal frame length is crucial to minimizing this additional delay. Moreover, TRAMA requires frequent message exchanges with neighbor nodes for SEP. In high-density networks, this leads to increased overhead, since the traffic information should frequently be exchanged.

5.4.2   *Other Reservation-Based MAC Protocols*

In addition to TRAMA, there are other reservation-based MAC protocols developed for WSNs, which we explain next. More specifically, Pattern MAC (PMAC) [37], the energy-aware TDMA-based MAC protocol [3], bit-map-assisted (BMA) MAC [14], and the adaptive low-power reservation-based MAC protocol [19] are described.

Figure 5.26 Frame structure of PMAC.

PMAC

PMAC is an adaptive MAC protocol that relies on the exchange of activity patterns and schedule determination [37]. The time frame structure of PMAC is shown in Figure 5.26. Each frame consists of two periods, i.e., pattern repeat time frame (PRTF) and pattern exchange time frame (PETF), similar to data period and reservation period, respectively. Similar to TRAMA, in PMAC a node broadcasts a tentative schedule for active slots during the PETF. The tentative schedule is referred to as *patterns* in PMAC. The patterns consist of ones and zeros for each slot, where a 1 indicates that the node *intends* to be active during the slot, whereas a 0 indicates that it intends to switch to sleep state during the slot. For example, a pattern of 001 indicates that a node will be in sleep state for two consecutive slots and will be active during the third slot.

The *pattern* only indicates the intention of the node according to its traffic load. The actual slots where the node will be active or sleep are determined by its *schedule* which is determined after the pattern exchange. The PMAC pattern is a sequence of binary numbers with a maximum length of N, which is the period length. Moreover, the pattern is restricted to be in the form of 0^m1, where $m = 0, 1, \ldots, N-1$. Note that the length of a pattern can be less then N, in which case the pattern repeats itself. For example, for $N = 6$, a pattern of 001 corresponds to the tentative plan 001001.

A node's pattern starts with a 1 and zeros are added in front of it, if there is no traffic. This increment is performed in an exponential fashion. When a node does not have a packet to transmit, the pattern evolves as follows: $1, 01, 0^21, 0^41, \ldots, 0^\delta1, 0^\delta01, \ldots$. This approach of exponentially increasing the sleep time during light traffic allows the nodes to save a considerable amount of energy. Note that the exponential increase continues until the number of zeros reaches a predetermined value (δ). Then, zeros are added linearly to prevent oversleeping.

After each node exchanges its pattern during the PETF, the actual schedule determination is performed. This is done according to the patterns of the intended receiver and the number of packets that a node has to transmit. More specifically, considering the pattern value for node j at slot k as P^i_j, and the corresponding value for its intended receiver (in case a data packet is to be transmitted) as P^k_i, the schedule is determined as follows:

- $P^k_j = 1$, $P^k_i = 1$. If node j has a packet to send, it marks its schedule as 1 and transmits its packet at time slot k. If node j does not have a packet to send, it marks its schedule as $1-$ indicating that it will wake up at slot k, wait for a specific time for any incoming packet, and switch to sleep state if there is no activity.
- $P^k_j = 0$, $P^k_i = 1$. If node j has a packet to send, it marks its schedule as 1. If it does not have any packet to send, the schedule is marked as 0 and the node j sleeps during this slot.
- $P^k_j = 0$, $P^k_i = 0$. In this case, node j marks its schedule as 0 and waits for the next frame if it has a packet to send to node i.
- $P^k_j = 1$, $P^k_i = 0$. In this case, node j marks its schedule as $1-$ and wakes up at slot k to check for an incoming transmission. If it has a packet to send to node i, it waits for the next frame.

According to these rules the communication between each node is performed using CSMA/CA in each slot.

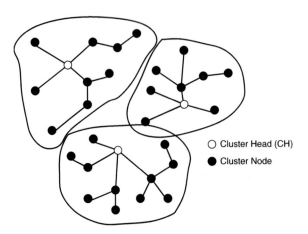

Figure 5.27   An example of the cluster structure in WSNs.

Energy-Aware TDMA-Based MAC Protocol

The reservation-based protocols described so far enable each node to exchange its schedule information with its neighbors to reserve slots for transmission. As a result, a distributed reservation approach is developed. This reservation approach, however, may lead to energy wastage due to collisions in message exchanges. Moreover, since a general view of even the close vicinity of nodes cannot be generated, suboptimal schedules can be generated distributively. In order to address these inefficiencies, cluster-based protocols have also been developed for WSNs. These protocols rely on a clustered hierarchy such that a fraction of the nodes in the network are part of a cluster. Moreover, a cluster head (CH) organizes the reservation of slots for efficient communication. An example of the cluster structure is also shown in Figure 5.27. In each cluster, the communication is organized by a CH.

The energy-aware TDMA-based MAC protocol [3] focuses on WSNs that are composed of clusters and gateways. Each gateway acts as a cluster-based centralized network manager and assigns slots in a TDMA frame based on the traffic requirements of the nodes in the cluster. Since, in a cluster, nodes may need multiple hops to reach the gateway, the traffic for routing messages is also incorporated into the TDMA frame structure. This protocol serves as another example of multi-hop awareness in WSNs. Similar to all cluster-based MAC protocols, the CH manages the slot assignments for other nodes in the cluster. This assignment informs the nodes about the slots during which they should either receive or transmit a packet. Since multi-hop routes can also exist in a cluster as shown in Figure 5.27, the nodes can also serve as routers in addition to data generators.

The structure of the TDMA frame is shown in Figure 5.28. Both the TDMA frame and the protocol have four phases for medium access. These four phases are *refresh*, *event-triggered rerouting*, *refresh-based rerouting*, and *data transfer* phases. The first three phases can be considered as the reservation period and the last phase is the data period in the context of the general structure of reservation-based protocols. The duration and period of each phase are fixed in the protocol. The main goal is to schedule the active times of slots based on routing duties and the generated packets of the nodes in a cluster. Accordingly, during the refresh phase, network-related information such as residual energy, location, and state of each node is collected. The refresh slot shown in Figure 5.28 is further slotted such that each node in the cluster has an assigned slot to transmit its information to the CH. During the event-triggered and refresh-based rerouting phases, the multi-hop routes from the nodes to the CH are updated and the nodes, which are on these routes, are assigned slots. Accordingly, the CH manages the times of communication at each hop as well as calculating paths.

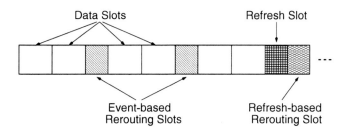

Figure 5.28   The frame structure of the energy-aware TDMA-based MAC protocol [3].

For slot assignment, two techniques, i.e., *breadth* and *depth*, are used. The breadth technique follows a breadth-first search technique to assign the slots to nodes. Accordingly, first nodes A and B send their packets to node C. Node C then sends these packets to node E. As a result, a node has to wait for all packets from its children before relaying packets to its next hop. On the other hand, the depth technique follows a depth-first search technique such that the packets from a node that is farther away from the sink are transmitted first. The CH informs each node of the slots it is going to receive or transmit information using one of these techniques.

Finally, the data phase is composed of data slots allocated to nodes with packets to send. The slot assignment performed in the rerouting phases is followed by the nodes in a cluster.

BMA-MAC

The BMA-MAC protocol is an intracluster communication protocol with an energy-efficient TDMA (E-TDMA) scheme [14]. BMA-MAC consists of a *cluster setup phase* and a *steady state phase*. In the cluster setup phase, the CH is selected based on the available energy of each node. If a node's remaining energy is above a certain threshold, then it can be elected as the CH. The elected nodes contend for becoming the CH by broadcasting an advertisement message using a simple CSMA scheme. The nodes that hear the advertisement message join the cluster of the CH. Once the clusters are built, the protocol enters the steady state phase.

In the steady state phase, an E-TDMA MAC scheme is used in each of the clusters formed by the CHs. The superframe structure of BMA-MAC is shown in Figure 5.29. In each superframe, the reservation period is slotted for contention. The data period is further divided into two periods, i.e., the data transmission period and idle period. The duration of the data period is fixed and the data transmission period is changed based on the traffic demands of the nodes. The slot assignment is performed by the CH according to the demands of the non-CHs during the reservation period. The reservation period is slotted and each node has a preassigned slot for its request. If a node has a packet to send, it sends a 1-bit control message or remains silent if it does not have any packets. According to the requests, the CH assigns slots for each node with a packet to send in the next data period. Depending on the number of nodes that request a slot, the duration of the data transmission period and the idle period changes. Each node transmits its packets to the CH during its assigned slot and sleeps during the other slots. During the idle slots, all the nodes in the cluster sleep. The BMA-MAC protocol performs the cluster setup phase after a specified number of rounds are performed by a CH. As a result, CH duty is exchanged by the other nodes leading to equal energy consumption.

Adaptive Low-Power Reservation-Based MAC

The adaptive low-power reservation-based MAC protocol [19] also assumes a clustered hierarchical organization, where the CH is chosen as a result of contention similar to the BMA-MAC protocol.

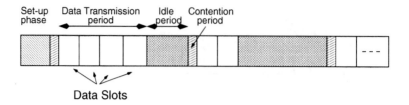

Figure 5.29   Frame structure of BMA-MAC.

The reservation period is composed of three parts. In the *control slot*, the CH broadcasts control information such as frame length and the end of cluster information. Since the role of the CH is energy consuming, the protocol aims to assign the CH duty to several nodes. After a specific duration, the CH announces the end of its duty and the CH is chosen again. The second slot in the frame is the *reservation request window*. In this window, nodes in the cluster transmit their requests to the CH. The transmissions in this window are performed through contention, where a basic CSMA scheme is used. During the *reservation confirmation slot*, slot allocations are communicated to the cluster nodes by the CH. Finally, a *data slot* is used for the nodes to transmit their data. The frame size is adjusted according to the probability of transmission failures of reservation request packets. If this probability is high, the frame size is increased. On the other hand, if there are no errors for reservation request packets, the frame size is decreased.

5.4.3   Summary

TDMA-based protocols provide collision-free communication in WSNs, achieving improved energy efficiency. However, TDMA-based protocols require an infrastructure consisting of CHs, which coordinate the time slots assigned to each node. Although many clustering algorithms have been proposed with these protocols, the optimality and energy efficiency of these algorithms still need to be investigated. While TDMA schedules within a cluster can be easily devised, the problem is more involved when individual CHs are not in direct range of the sink, thus necessitating intercluster multi-hop communication. An acceptable, non-overlapping slot assignment for all neighboring clusters needs to be derived in a distributed manner requiring coordination between different clusters. This problem has been shown to be NP-complete [11] by reduction to an instance of graph coloring, and the development of efficient heuristics is an open issue.

Network scalability is another important area of research and the TDMA schedules must be able to accommodate high node densities that are characteristic of sensor networks. As the channel capacity in TDMA is fixed, only slot durations or a number of slots in a frame may be changed keeping in mind the number of users and their respective traffic types. In addition, TDMA-based protocols result in high latency due to the frame structure. Hence, TDMA-based MAC protocols may not be suitable for some WSN applications where delay is important in estimating event features and the traffic has a bursty nature. Moreover, since time-slotted communication is performed in the clusters, intercluster interference has to be minimized such that nodes with overlapping schedules in different clusters do not collide which each other. Finally, time synchronization is an important part of the TDMA-based protocols and synchronization algorithms as explained in Chapter 11 are required.

5.5   Hybrid Medium Access

Contention-based and reservation-based protocols provide different advantages and disadvantages in the medium access performance. Since contention-based protocols require significantly less overhead, these protocols result in high utilization in cases where there is low contention. However, when the number of nodes contending for the channel increases, channel utilization decreases since these nodes are not coordinated. On the other hand, reservation-based protocols provide scheduled access to each node and decrease collisions. This results in high utilization when the competition is high but also in latency and overhead. Such a contrast produces a tradeoff between access capacity and energy efficiency. Hybrid schemes in MAC protocols aim to leverage the tradeoff introduced in channel allocation by combining random access schemes with reservation-based access TDMA approaches.

Hybrid solutions provide performance enhancements in terms of collision avoidance and energy efficiency due to improved channel organization and adaptivity to dynamic traffic load. Next, we describe two hybrid solutions developed for WSNs.

5.5.1   Zebra-MAC

In order to provide an adaptive operation based on the level of contention, Zebra-MAC (Z-MAC) combines the advantages of each scheme in a hybrid MAC solution [25]. The communication structure of Z-MAC still relies on time slots similar to TDMA-based solutions. Each slot is *tentatively* assigned to a node. However, the difference between TDMA-based solutions is that each slot can be *stolen* by other nodes if it is not used by its *owner*. Consequently, Z-MAC behaves like CSMA under low contention and like TDMA under high contention.

Similar to many reservation-based protocols, Z-MAC consists of a setup phase and a communication phase. The setup phase has four main components:

- Neighbor discovery
- Slot assignment
- Local frame exchange
- Global time synchronization.

Neighbor discovery is performed once by each node to gather information about its *two-hop neighborhood*. During this phase, each node broadcasts its one-hop neighborhood information to its neighbors. At the end of multiple message exchanges, each node is informed about its two-hop neighborhood information. Collisions in the wireless channel affect the two-hop neighborhood of each node because of the hidden terminal problem.

Slot assignment is performed by the DRAND protocol [25], which ensures a broadcast schedule such that each node is assigned a slot that will not coincide with the slots of its two-hop neighbors. DRAND first creates a radio interference map of the network. Nodes that can interfere with each other are connected by bidirectional links in the interference map. Slot assignment is performed iteratively according to this map.

■ **EXAMPLE 5.15**

A sample topology of six nodes is shown in Figure 5.30(a), where the connectivity of each node is also illustrated by ellipses. According to this topology, DRAND creates the radio interference map as shown in Figure 5.30(b).

The case where node C needs to assign a slot to itself is depicted in Figure 5.30(c). Node C first broadcasts a request message to its immediate neighbors, A, B, and D. If there is no conflict, each node responds with grant messages (Figure 5.30(d)). Accordingly, node C indicates that the particular slot is reserved by broadcasting a release message as shown in Figure 5.30(e). Each immediate neighbor also communicates this slot selection to each two-hop neighbor of node C, i.e., E and F, as shown in Figure 5.30(f).

In case there is a conflict with the slot selected by node C, this is indicated by a reject message after the request broadcast as shown in Figure 5.30(g), where node D sends a reject message to

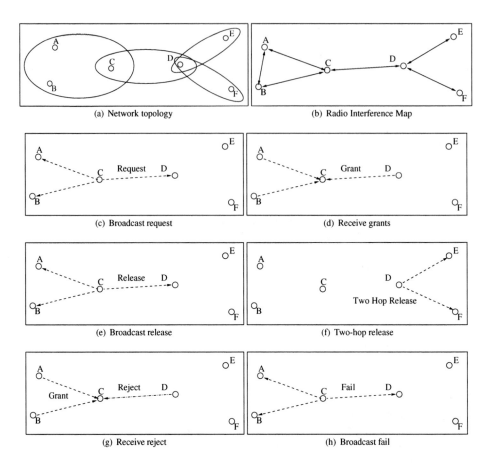

Figure 5.30   Operation of the DRAND protocol.

node C. If node C does not receive a grant message from all of its neighbors, it broadcasts a fail message as shown in Figure 5.30(h) to indicate that it cannot allocate the slot. As a result, collisions are prevented.

Z-MAC also introduces a local frame structure. In many TDMA-based solutions, the frame size is fixed for the whole network. While a large enough frame size is essential to accommodate all the contending nodes in a local neighborhood, the required size may not be the same for different parts of the network. Depending on the density at each location of the network, a node may have fewer neighbors to compete with. For these cases, smaller frame sizes are more efficient since the slots can be reused more frequently. Based on this observation, Z-MAC allows each node to specify its own local frame size.

The local frame size is selected according to the time frame rule. Accordingly, if a node i is assigned a slot s_i and the maximum slot number in its neighborhood is F_i, then node i's time frame size is selected as 2^a. The integer a is chosen such that $2^{a-1} \leq F_i < 2^a - 1$. This time frame rule ensures that if the node uses the slot s_i for each 2^a slots, then it does not collide with any of its two-hop neighbors. After the local frame sizes are selected, this information is exchanged between neighbors to constitute a stable operation.

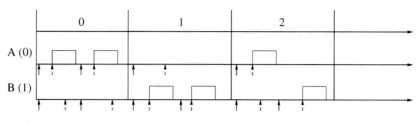

† Start backoff

‡ End backoff

Figure 5.31 Adaptive channel access mechanism of Z-MAC.

The transmission structure of Z-MAC follows the frame structure described above. For each slot, a node is denoted as the *owner* and other nodes are *non-owners*. Owners have higher priority over non-owners in accessing the channel during their assigned slots. This prioritization is performed similarly to CSMA. If an owner has a packet to transmit during its slot, it first performs CCA. If the channel is idle, it waits for a random amount of time within a time period T_o and transmits its packet. If the owner does not have a packet to send, non-owners can utilize its time slot. This is performed through a different backoff mechanism. The non-owners first wait for a time period T_o and then back off for a random amount of time within a time period T_{no}. This provides prioritization for owners and enable non-owners to *steal* the slot if it is not used.

■ **EXAMPLE 5.16**

The adaptive transmission mechanism of Z-MAC is illustrated in Figure 5.31, where two nodes A and B own slots 0 and 1, respectively. During slot 0, both nodes have packets to send and back off at the beginning of the slot. Since node A is the owner of this slot, its backoff timer expires first, it allocates the channel, and transmits its packet. On the other hand, since node B's backoff time expires later, it senses the channel as busy and waits for the packet transmission. If node A has packets to send, it can use slot 0. In the next slot, 1, node B is the owner, and can send its packets during this slot. The dynamic operation of Z-MAC is shown during slot 2, where node A is the owner. Node A first transmits its packet. When the transmission is over, node B starts its backoff timer and initiates transmission when the channel is idle. As a result, when node A does not have any packet to send during its assigned slot, node B can utilize the available channel.

Z-MAC also employs a mechanism to relieve the contention around a specific node. If an owner experiences a high contention during its time slot, it broadcasts an explicit contention notification (ECN) message to its neighbors. The ECN message is propagated toward the two-hop neighbors on the path to destination. The nodes receiving the ECN message switch to a *high contention level* (HCL) mode. Nodes in HCL mode for a specific node do not compete for the slots owned by that node. As a result, contention decreases on the path toward destination.

■ **EXAMPLE 5.17**

The ECN operation is illustrated in Figure 5.32, where node C tries to send packets to the sink through nodes D and F. If node C experiences high contention, it broadcasts a one-hop ECN message to A, B, and D. Since nodes A and B are not on the routing path, they discard the ECN message. However, since node D is on the routing path, it forwards the ECN message as a two-hop

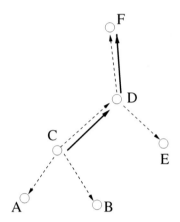

Figure 5.32   Operation of the ECN mechanism in Z-MAC.

ECN message to nodes E and F. As a result E and F do not compete during C's slot as non-owners. However, nodes A, B, and D are eligible to compete during C's slot.

Qualitative Evaluation

The hybrid medium access scheme of Z-MAC provides adaptive communication if a number of sensor nodes or the traffic load change. Compared to contention-based protocols, Z-MAC improves throughput when the traffic load is high and still maintains acceptable latency for low traffic compared to TDMA-based protocols.

Z-MAC has been developed for a specific scenario, where the sensor nodes are mobile and have intermittent communication capabilities. Hence, it may not be applicable for a large set of WSN applications, where the network is static.

IEEE 802.15.4

IEEE 802.15.4 was developed for low-data-rate wireless networks, and combines reservation-based and contention-based approaches [1]. The protocol, also supported by the ZigBee consortium, introduces a superframe structure with two disjoint periods, i.e. *contention access period* and *contention-free period*. The network is assumed to be clustered and each cluster head, i.e. PAN coordinator, broadcasts the frame structure and allocates slots to prioritized traffic in the contention-free period. In the contention period, nodes contend using CSMA/CA or slotted CSMA/CA to access the channel. The winners can allocate the channel for their transmissions for a specific amount of time. This provides a flexible access method for nodes with infrequent traffic. During the contention-free period, nodes with higher priority traffic are served by the PAN coordinator. Based on the traffic requirements, each node is assigned slots during the contention-free period. These slots are allocated to only one pair and channel contention is prevented to provide priority. As a result, the IEEE 802.15.4 protocol provides a hybrid operation through a CSMA-based and a TDMA-based operation. Although this protocol aims for prioritization and energy efficiency up to 1% duty cycles, it requires a cluster-based topology, which may not be applicable to some WSN scenarios.

References

[1] IEEE standard for information technology – telecommunications and information exchange between systems – local and metropolitan area networks – specific requirement part 15.4: Wireless medium access control (MAC) and physical layer (PHY) specifications for low-rate wireless personal area networks (WPANS). *IEEE Std. 802.15.4a-2007 (Amendment to IEEE Std. 802.15.4-2006)*, pp. 1–203, 2007.

[2] I. F. Akyildiz, W. Su, Y. Sankarasubramaniam, and E. Cayirci. Wireless sensor networks: a survey. *Computer Networks*, 38(4):393–422, March 2002.

[3] K. A. Arisha, M. A. Youssef, and M. Y. Younis. Energy-aware TDMA-based MAC for sensor networks. *Computer Networks*, 43(5):539–694, December 2003.

[4] M. Buettner, G. Yee, E. Anderson, and R. Han. X-MAC: a short preamble MAC protocol for duty-cycled wireless sensor networks. In *Proceedings of ACM SenSys'06*, pp. 307–320, Boulder, CO, USA, October 2006.

[5] I. Demirkol, C. Ersoy, and F. Alagoz. MAC protocols for wireless sensor networks: a survey. *IEEE Communications Magazine*, 44(4):115–121, April 2006.

[6] A. El-Hoiydi. Aloha with preamble sampling for sporadic traffic in ad hoc wireless sensor networks. In *Proceedings of IEEE ICC'02*, volume 5, pp. 3418–3423, New York, USA, April 2002.

[7] A. El-Hoiydi and J.-D. Decotignie. WiseMAC: an ultra low power MAC protocol for the downlink of infrastructure wireless sensor networks. In *Proceedings of the International Symposium on Computers and Communications (ISCC'04)*, volume 1, pp. 244–251, Alexandria, Egypt, July 2004.

[8] S. C. Ergen and P. Varaiya. PEDAMACS: power efficient and delay aware medium access protocol in sensor networks. *IEEE Transactions on Mobile Computing*, 5(7):920–930, July 2006.

[9] G. P. Halkes and K. G. Langendoen. Crankshaft: an energy-efficient MAC-protocol for dense wireless sensor networks. In *Proceedings of EWSN'07*, pp. 228–224, Delft, The Netherlands, January 2007.

[10] M. Hayajneh, I. Khalil, and Y. Gadallah. An OFDMA-based MAC protocol for under water acoustic wireless sensor networks. In *Proceedings of IWCMC'09*, pp. 810–814, Leipzig, Germany, June 2009.

[11] T. Herman and S. Tixeuil. A distributed TDMA slot assignment algorithm for wireless sensor networks. In *Proceedings of the Workshop on Algorithmic Aspects of Wireless Sensor Networks (ALGOSENSORS'04)*, pp. 45–58, Turku, Finland, July 2004.

[12] K. Kredo II and P. Mohapatra. Medium access control in wireless sensor networks. *Computer Networks*, 51(4):961–994, 2007.

[13] K. Jamieson, H. Balakrishnan, and Y. C. Tay. Sift: a MAC protocol for event-driven wireless sensor networks. In *Proceedings of EWSN'06*, pp. 260–275, Zurich, Switzerland, February 2006.

[14] J. Li and G. Y. Lazarou. A bit-map-assisted energy-efficient MAC scheme for wireless sensor networks. In *Proceedings of ACM IPSN'04*, pp. 55–60, Berkeley, CA, USA, April 2004.

[15] P. Lin, C. Qiao, and X. Wang. Medium access control with a dynamic duty cycle for sensor networks. In *Proceedings of IEEE WCNC'04*, pp. 1534–1539, Atlanta, GA, USA, March 2004.

[16] S. Liu, K.-W. Fan, and P. Sinha. CMAC: an energy efficient MAC layer protocol using convergent packet forwarding for wireless sensor networks. In *Proceedings of IEEE SECON'07*, pp. 11–20, San Diego, CA, USA, June 2007.

[17] S. Mahlknecht and M. Bock. CSMA-MPS: a minimum preamble sampling MAC protocol for low power wireless sensor networks. In *Proceedings of the IEEE International Workshop on Factory Communication Systems*, pp. 73–80, Vienna, Austria, September 2004.

[18] S. Meguerdichian, F. Koushanfar, M. Potkonjak, and M. B. Srivastava. Coverage problems in wireless ad-hoc sensor networks. In *Proceedings of IEEE INFOCOM 2001*, volume 3, pp. 1380–1387, Anchorage, AK, USA, April 2001.

[19] S. Mishra and A. Nasipuri. An adaptive low power reservation based MAC protocol for wireless sensor networks. In *Proceedings of the IEEE International Conference on Performance, Computing, and Communications*, pp. 731–736, Phoenix, AZ, USA, April 2004.

[20] A. Muqattash and M. Krunz. CDMA-based MAC protocol for wireless ad hoc networks. In *Proceedings of ACM MobiHoc*, volume 1, pp. 153–164, Annapolis, MD, USA, June 2003.

[21] J. Polastre, J. Hill, and D. Culler. Versatile low power media access for wireless sensor networks. In *Proceedings of SenSys'04*, pp. 95–107, Baltimore, MD, USA, 2004.

[22] D. Pompili, T. Melodia, and I. F. Akyildiz. A distributed CDMA medium access control for underwater acoustic sensor networks. In *Proceedings of Mediterranean Ad Hoc Networking Workshop (Med-Hoc-Net)*, Corfu, Greece, June 2007.

[23] V. Rajendran, J. J. Garcia-Luna-Aveces, and K. Obraczka. Energy-efficient, application-aware medium access for sensor networks. In *Proceedings of IEEE MASS'05*, Washington, DC, USA, November 2005.

[24] V. Rajendran, K. Obraczka, and J. J. Garcia-Luna-Aceves. Energy-efficient, collision-free medium access control for wireless sensor networks. In *Proceedings of ACM SenSys'03*, Los Angeles, CA, USA, November 2003.

[25] I. Rhee, A. Warrier, M. Aia, and J. Min. Z-MAC: a hybrid MAC for wireless sensor networks. In *Proceedings of ACM SenSys'05*, pp. 90–101, San Diego, CA, USA, 2005.

[26] M. Salajegheh, H. Soroush, and A. Kalis. HYMAC: hybrid TDMA/FDMA medium access control protocol for wireless sensor networks. In *Proceedings of IEEE PIMRC'07*, pp. 1–5, Athens, Greece, 2007.

[27] C. Schurgers, V. Tsiatsis, and M. B. Srivastava. STEM: topology management for energy efficient sensor networks. In *Proceedings of the IEEE Aerospace Conference*, volume 3, pp. 1099–1108, Big Sky, MT, USA, 2002.

[28] Y. Sun, O. Gurewitz, and D. B. Johnson. RI-MAC: a receiver-initiated asynchronous duty cycle MAC protocol for dynamic traffic loads in wireless sensor networks. In *Proceedings of ACM SenSys'08*, pp. 1–14, 2008.

[29] H.-X. Tan and W. K. G. Seah. Distributed CDMA-based MAC protocol for underwater sensor networks. In *Proceedings of IEEE LCN'07*, pp. 26–36, Washington, DC, USA, 2007.

[30] T. van Dam and K. Langendoen. An adaptive energy-efficient MAC protocol for wireless sensor networks. In *Proceedings of ACM SENSYS'03*, pp. 171–180, Los Angeles, CA, USA, November 2003.

[31] M. C. Vuran, O. B. Akan, and I. F. Akyildiz. Spatio-temporal correlation: theory and applications for wireless sensor networks. *Computer Networks*, 45(3):245–261, June 2004.

[32] M. C. Vuran and I. F. Akyildiz. Spatial correlation-based collaborative medium access control in wireless sensor networks. *IEEE/ACM Transactions on Networking*, 14(2):316–329, April 2006.

[33] Y. Wang and J. J. Garcia-Luna-Aceves. Performance of collision avoidance protocols in single-channel ad hoc networks. In *Proceedings of the IEEE International Conference on Network Protocols*, pp. 68–77, Paris, France, 2002.

[34] W. Ye, J. Heidemann, and D. Estrin. An energy-efficient MAC protocol for wireless sensor networks. In *Proceedings of IEEE INFOCOM'02*, volume 3, pp. 1567–1576, New York, USA, June 2002.

[35] W. Ye, J. Heidemann, and D. Estrin. Medium access control with coordinated adaptive sleeping for wireless sensor networks. *IEEE/ACM Transactions on Networking*, 12(3):493–506, June 2004.

[36] W. Ye, F. Silva, and J. Heidemann. Ultra-low duty cycle MAC with scheduled channel polling. In *Proceedings of ACM SenSys'06*, Boulder, CO, USA, November 2006.

[37] T. Zheng, S. Radhakrishnan, and V. Sarangan. PMAC: an adaptive energy-efficient MAC protocol for wireless sensor networks. In *Proceedings of the IEEE International Parallel and Distributed Processing Symposium (IPDPS '05)*, Denver, CO, USA, 2005.

6

Error Control

WSN applications rely on multiple sensor feedback for interaction with the environment. In general, the main objectives of the data link layer in this interaction are *multiplexing/ demultiplexing of data, data frame detection, medium access,* and *error control.* In addition to the MAC layer functionalities discussed in Chapter 5, in this chapter we explain the effects of error control on communication reliability and efficiency in WSNs.

Contrary to traditional networks, WSNs are characterized by low energy requirements and the collaborative nature of sensors. Hence, the design of data link layer protocols encounters unique challenges as opposed to traditional networking protocols. While fulfilling the application's objectives, the data link layer should provide reliable and energy-efficient point-to-point and point-to-multipoint communication throughout the network. An overview of the data link components is shown in Figure 6.1 [16].

In WSNs, where the correlation between sensors can be exploited in terms of *aggregation, collaborative source coding,* or *correlation-based protocols,* error control is extremely important. Since the communication solutions generally aim to reduce the redundancy in the traffic by filtering correlated data or switching off redundant nodes, it is essential for each packet to be transmitted reliably. Moreover, the multi-hop features of WSNs require a unique definition of reliability other than the conventional reliability metrics which focus on point-to-point reliability. Hence, the effects of error control schemes are analyzed using cross-layer analysis techniques to consider the effects on wireless channel, medium access, and multi-hop routing.

The main purpose of the error control schemes is to provide reliable communication in the wireless channel. This channel has adverse effects on communication because of the fading, interference, and loss of bit synchronization as discussed in Chapter 4. This results in wireless channel errors that impact the integrity of packets that are sent by sensor nodes. The wireless channel effects are further increased in WSNs because of the low-power communication techniques. As a result, in addition to the physical layer techniques that provide reliability at the *bit* level, error control schemes are used at the link layer to provide reliability at the *packet* level. Accordingly, error control schemes maintain *error-free, in-sequence, duplicate-free,* and *loss-free* communication between sensor nodes.

In general, the error control mechanisms in WSNs can be categorized into four main approaches: *power control, automatic repeat request* (ARQ), *forward error correction* (FEC), and *hybrid ARQ* (HARQ) as described next.

6.1 Classification of Error Control Schemes

6.1.1 Power Control

State-of-the-art transceivers such as the CC2420 used in MicaZ, TelosB, and SunSPOT nodes support multiple transmit power levels that can be controlled on a per-packet basis. Accordingly, the amount of

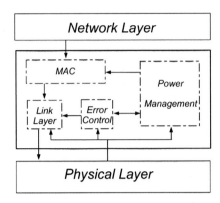

Figure 6.1   An overview of the data link layer providing reliability in WSNs [16].

power dissipated from the antenna can be adjusted to control the effective communication range of a node. Transmission power control can be used to achieve desired error rates in wireless communication. Higher transmit power reduces the packet error rate by improving the signal-to-noise ratio. However, energy consumption is therefore increased in addition to increased interference with other nodes. Power control requires sophisticated protocols to be implemented, which requires additional memory for the implementation. Transmit power control is generally used for long periods of time, such as topology control, as will be explained in Chapter 13. However, per-packet transmit power introduces significant overhead in the resource constraint sensor nodes. Further, as we will describe next, power control can be used in conjunction with forward error control schemes to improve energy efficiency.

6.1.2   Automatic Repeat Request (ARQ)

ARQ-based error control mainly relies on the retransmission of lost or corrupted packets to maintain reliability. For this purpose, a header is added to the payload of the packet to indicate the source node and a low-complexity error detection algorithm is used to calculate the checksum of the packet. This checksum is added to the end of the packet. Typically a cyclic redundancy check (CRC) mechanism, which is described in Chapter 4 is used. The receiver node computes the checksum of the received bits and compares this to the checksum of the received packet. Accordingly, the receiver can provide feedback to the source regarding the integrity of the received packet. This can be performed by sending either a *positive* acknowledgment (ACK) for each successfully received packet or a *negative* acknowledgment (NACK) for each corrupt packet. In the case of an ACK, the sender sets a timer to detect that ACKs have not arrived. If the sender infers that a packet is corrupted it retransmits this packet.

The main ARQ strategies can be summarized as *go-back-N*, *selective repeat*, and *stop-and-wait* [9, 5, 4]. Go-back-N lets the sender send up to N packets. If at least a packet has not been ACKed, then all the packets are retransmitted. For selective repeat, the sender retransmits only those packets that have not been ACKed. Finally, the stop-and-wait strategy lets only a single outstanding packet through and transmits the following packet after the outstanding packet is ACKed. Stop-and-wait is the most common ARQ strategy for WSNs.

ARQ-based error control mechanisms incur significant additional retransmission cost and overhead. Although these schemes are utilized at the data link layer for conventional wireless networks, the efficiency of ARQ in sensor network applications is limited due to the scarcity of the energy and processing resources of the sensor nodes.

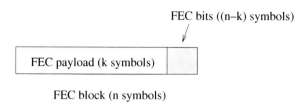

Figure 6.2   An illustration of FEC in WSNs.

6.1.3   Forward Error Correction (FEC)

As explained in Chapter 4, FEC codes (or channel codes) add redundancy to the transmitted packet such that it can be received error free at the receiver even if a limited number of bits are received in error. More specifically, in an (n,k,t) FEC code, as shown in Figure 6.2, $(n − k)$ redundant FEC symbols are added to the k-bit FEC payload to improve the error resilience of the wireless communication at the cost of increased bandwidth consumption. As a result, bit errors up to t bits can be recovered and the overall probability of error is decreased without any retransmissions. On the other hand, FEC codes incur overhead in terms of transmission and reception of additional redundant bits as well as decoding packets. In WSNs, where low-clock-rate CPUs are used, decoding energy should also be considered in assessing the performance of FEC protocols. There exist various FEC codes such as linear block codes (BCH and Reed–Solomon codes) and convolutional codes, which are optimized for specific packet sizes, channel conditions, and reliability notions. On the other hand, for the design of efficient FEC schemes, it is important to have a good knowledge of the channel characteristics and implementation techniques.

Although FEC codes provide flexible error control capabilities over a variety of ranges between nodes, such an advantage is limited in scenarios where limited error probabilities are acceptable. This impacts the type of FEC code that is suitable for WSNs. As an example, compared to convolutional codes, uncoded communication provides better energy efficiency for probability of error, $P_b > 10^{-4}$ [13]. This is due to the fact that the encoding/decoding energy is small at high P_b and the output power is limited. As a result, the transceiver energy dominates the overall energy consumption. Since the packet length is increased due to coding, the overall energy consumption increases. Consequently, if a lower P_b is not required by an application for individual packets from sensors, FEC coding can be inefficient.

6.1.4   Hybrid ARQ

ARQ and FEC schemes maintain the reliability in the channel through different techniques. ARQ protocols are generally efficient when the channel is good since no redundant bits are sent. However, these protocols become inefficient when the channel is bad since the whole packet is sent again, although only certain bits are corrupted. On the other hand, FEC schemes become inefficient when the channel is good since redundant bits are sent with the packet and are not exploited. However, FEC schemes become more efficient in bad channel conditions since retransmissions are avoided if a limited number of bits are received in error.

Hybrid ARQ (HARQ) schemes aim to exploit the advantages of both FEC and ARQ schemes by incrementally increasing the error resilience of a packet through retransmission. Firstly, an uncoded or a lightly coded packet is sent. In case of channel errors HARQ performs retransmission using a higher power FEC code. To this end, HARQ schemes can be classified into two based on the way the retransmissions are handled: Type I and Type II. With HARQ-I techniques, an uncoded packet or a packet that is coded with a lower error correction capability is sent first. If this packet is received in error, the receiver sends a NACK to the sender. This triggers a retransmission of the packet that is encoded

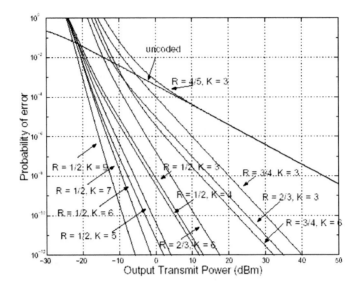

Figure 6.3   Probability of error versus transmission power for different convolutional codes.

with a more powerful FEC code. The difference in Type II is that, for retransmissions, only the redundant bits are sent. While Type II decreases the bandwidth usage of the protocol, Type I does not require the previously sent packets to be stored at the receiver side.

In the following, we describe the unique aspects of using error control schemes in WSNs.

6.2   Error Control in WSNs

In WSNs, energy consumption is the most important performance metric in the design of communication protocols. Since the sensor nodes have stringent energy capabilities, error control protocols should also consider energy efficiency as the first design goal. The sources of energy consumption in WSNs can be mainly classified into two types, namely computation power and transmission/receiving power. However, since transmission of a bit is more costly than processing power, protocols that exploit the on-board processing capabilities of sensor nodes are more favorable. Consequently, the use of FEC results in higher energy efficiency in certain situations given the constraints of the sensor nodes.

The use of FEC codes can decrease the transmit power due to the increased redundancy in the constructed packets [13]. The performance of convolutional codes in terms of probability of error and output transmit power is shown in Figure 6.3 [13]. As shown in this figure, lower transmit power is possible for a specific probability of error using FEC codes. However, the required processing power due to encoding/decoding of the packet increases the overall energy consumption. The required energy during the encoding and decoding of BCH codes is shown in Figure 6.4 along with the incurred encoding/decoding latency in Figure 6.5 [10]. Moreover, the increase in packet length also incurs additional energy cost. This additional cost is due to the longer packet transmission times and, hence, the increased packet collision rate.

Although the FEC can achieve significant reduction in the *bit error rate* (BER) for any given value of the transmit power, the additional processing power that is consumed during encoding and decoding must be considered when designing an FEC scheme. FEC is a valuable asset for the sensor networks if the additional processing power is less than the transmission power savings. Thus, the tradeoff between

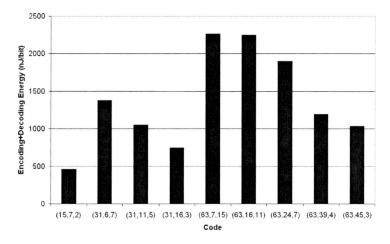

Figure 6.4   Total energy consumption for encoding and decoding using BCH codes [13].

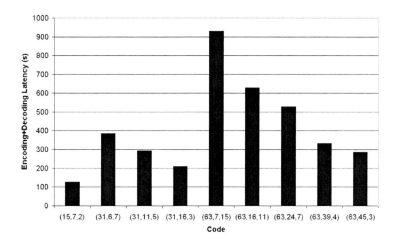

Figure 6.5   Total encoding and decoding latency using BCH codes [13].

this additional processing power and the associated coding gain needs to be optimized in order to have powerful, energy-efficient, and low-complexity FEC schemes for error control in sensor networks. Furthermore, powerful FEC codes incur additional decoding latency which should also be considered in the choice of error control schemes.

In addition to single-hop effects, the overall impact of error control schemes on multi-hop communication should also be considered. FEC coding and HARQ schemes improve the *error resilience* compared to ARQ schemes by sending redundant bits through the wireless channel. Therefore, lower SNR values can be supported to achieve the same error rate as an uncoded transmission. However, this advantage comes at the cost of increased packet sizes due to the redundant bits and encoding/decoding packets. To overcome the overhead because of increased communication and computational cost, the error resilience of the FEC and HARQ codes can be exploited in two ways in WSNs:

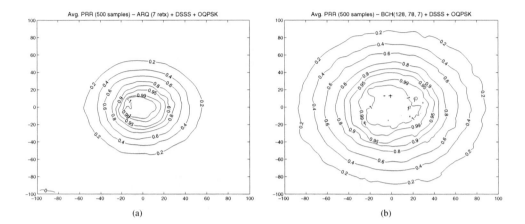

Figure 6.6 Two-dimensional average packet reception rate graphs for (a) ARQ ($N = 7$) and (b) FEC (BCH(128, 78, 7)) for MicaZ [14].

- **Hop length extension:** In multi-hop networks, the error resilience of FEC and HARQ schemes can be exploited to improve the effective transmission range of a node. This is illustrated in Figure 6.6. In these figures, the packet error rate (PER) contours of the ARQ and FEC codes are shown around a transmitter node. The contours show the locations where the corresponding PER is at a specific value. If we compare Figure 6.6(a) and Figure 6.6(b) for a target packet reception rate of 90%, the average transmission range of a node is 29 m for ARQ and 42 m for FEC with a transmit power of -5 dBm. This 45% increase in effective range clearly illustrates that FEC codes can increase the effective transmission range of a node compared to ARQ by using the same transmit power. While this property is generally not useful for single-hop communication, it is beneficial for WSNs where longer hops can be constructed in a multi-hop network such as WSNs. The *hop length extension* can be achieved through channel-aware cross-layer routing protocols.

- **Transmit power control:** The improved error resilience provided through FEC codes can also be exploited by reducing the transmit power. This technique is referred to as *transmit power control* and improves the capacity of the network by reducing the interference to other users.

■ EXAMPLE 6.1

The concepts of hop length extension and transmit power control for exploiting the error resilience of FEC codes are depicted in Figure 6.7 compared to an ARQ communication. In Figure 6.7(a), a typical communication link between nodes A and B is shown for a typical ARQ communication. In this case, node A uses a transmit power of $P_{t.ARQ}$ to send a packet to node B, which is at a distance d_{ARQ} away. As a result of the attenuation in the channel, this results in a SNR of SNR_{ARQ} at node B. Accordingly, a target PER value of PER^*_{ARQ} can be achieved.

Now consider the case where an FEC code is used for communication instead of ARQ. Based on the error resilience of FEC codes, the same target PER^*_{ARQ} can be reached through a lower SNR value $SNR_{FEC} < SNR_{ARQ}$. This advantage can be exploited in two ways in multi-hop networks. In the first case, node A selects another node C that is farther from node B as the next hop as shown in Figure 6.7(b). Since the distance is increased, i.e., $d_{FEC} > d_{ARQ}$, with the same transmit power,

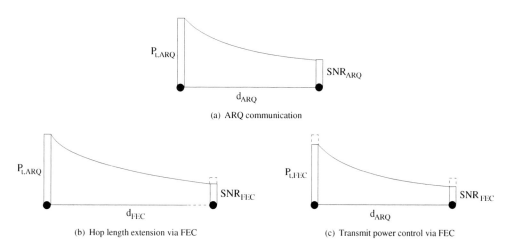

Figure 6.7   Error resilience of FEC vs. ARQ.

the received SNR value at node C is less than that at B. However, lower SNR values are adequate for FEC codes. Hence, the same PER can still be achieved. This is called *hop length extension*.

The second way is depicted in Figure 6.7(c). In this case, node A uses node B as the next hop by fixing the hop distance, i.e., $d_{FEC} = d_{ARQ}$. However, it uses a smaller transmit power, i.e., $P_{t,FEC} < P_{t,ARQ}$, than the ARQ case to reach the node. Since attenuation is constant at the same distance, the same target PER can be achieved with a smaller SNR. This is called *transmit power control*.

Both hop length extension and transmit power control improve the energy efficiency of communication. In hop length extension, fewer hops can be used to reach a destination, while transmit power control reduces the transmit power level at each hop without affecting the hop distance.

In addition to the advantages, FEC and HARQ schemes improve the error resilience at the cost of encoding/decoding energy and latency as well as communication overhead due to the transmission and reception of longer packets. As a result, the overall comparison of error control schemes requires a cross-layer analysis to weigh the advantages and disadvantages of these schemes. In the following, we present such an analysis model and describe the cases where FEC or HARQ is more efficient than ARQ.

6.3   Cross-layer Analysis Model

In this section, we describe a cross-layer analysis model that will be used to investigate the tradeoffs between ARQ, FEC, and HARQ schemes in terms of energy consumption, latency, and end-to-end PER considering the hop length extension and the transmit power control techniques [14, 15]. The objective of this model is to determine the following metrics for the analysis:

- Expected hop distance
- End-to-end energy consumption
- End-to-end latency
- End-to-end PER.

Next, we explain the system model and describe how these metrics are calculated.

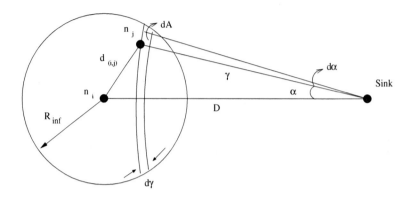

Figure 6.8   Reference model for the derivations [14].

6.3.1   Network Model

For the cross-layer analysis, consider a network composed of sensor nodes that are distributed according to a two-dimensional (2-D) Poisson distribution with density ρ. Duty cycle operation is deployed such that each node is active for a fraction δ of the time and is in sleep mode otherwise [14, 15]. The model is based on a monitoring application such that the reporting rates of sensors are low but the messages should be transmitted reliably subject to a certain end-to-end PER target.

The hop length extension can be realized through a channel-aware routing algorithm, where the next hop is determined according to the received SNR of a packet sent from a specific node i at a distance D from the sink. Among the neighbors of i, the neighbor j that is closest to the sink and with SNR value higher than a threshold, i.e., $\psi_j > \psi_{Th}$, is selected as the next hop. The medium access is performed through RTS–CTS–DATA exchange in addition to ACK and retransmissions for ARQ and NACK and retransmissions for HARQ.

The model used for derivations is shown in Figure 6.8, where a sender node n_i transmits a packet to the sink through a path that involves node n_j as the next hop. The log-normal shadow fading model explained in Chapter 4 is used for the wireless channel. In this case, the transmission range of a node is essentially infinite due to the shadow fading component. Hence, the transmission range of a node is R_{inf}, which is the distance at which the probability that a packet can be successfully received is negligible. Moreover, the distribution of hop distance and the associated parameters at each hop can be considered independent since duty cycle operation is performed. As a result, the state of the network will change at each hop since different nodes will be awake at different times. The following cross-layer analysis is based on a packet that is transmitted to the sink and captures the effects of neighbor nodes and wireless channel effects.

The cross-layer analysis framework enables a comprehensive comparison of ARQ, FEC, and HARQ schemes. To illustrate specific results, block codes are considered, which are represented by (n, k, t), where n is the block length, k is the payload length, and t is the error correction capability in bits. In the comparisons a series of extended BCH (n, k, t) codes with $n = 128$ and $t = \{2, 3, 5, 7, 9, 13, 21\}$ are used. Moreover, the $(7, 3, 2)$, $(15, 9, 3)$, and $(31, 19, 6)$ RS codes are considered. For the HARQ schemes, three different configurations are important. For Type I HARQ (HARQ-I), first an uncoded packet is sent followed by a packet encoded by BCH codes. Furthermore, a combination of two BCH codes is also considered for the HARQ-I. The Type II HARQ (HARQ-II) schemes necessitate incremental error control coding since only the difference in redundant bits is sent through retransmissions. Accordingly, the packet is encoded by a BCH code and then the payload is sent first. The receiver uses the CRC bits in the payload for decoding. In case of errors, the receiver sends a NACK packet and the transmitter sends the redundant bits for BCH decoding. The HARQ schemes are indicated by a tuple (t_1, t_2), where the first parameter, t_1, indicates the error correction capability of the first packet and t_2 is that of the second packet. As an example, HARQ-II $(0, 3)$ refers to the Type II HARQ, where first an uncoded packet is sent, i.e., $t_1 = 0$, and in case of channel errors, the redundant bits of the BCH $(128, 106, 3)$ encoded packet are sent. Hence, only single retransmission is considered.

The most important performance metrics in WSNs is the energy consumption. The total energy consumed as a result of a single flow from a source node at distance D from the sink is as follows:

$$E_{flow}(D) = E[E_h] E[n_h(D)] \tag{6.1}$$

where $E[E_h]$ is the expected energy consumption per hop and $E[n_h(D)]$ is the expected hop count from a source at distance D from the sink. Similarly, the end-to-end latency of a flow is given by

$$T_{flow}(D) = E[T_h] E[n_h(D)] \tag{6.2}$$

where $E[T_h]$ is the expected delay per hop.

The expected hop count $E[n_h(D)]$ is represented as [15]

$$E[n_h(D)] \simeq \frac{D - R_{inf}}{E[d_h]} + 1 \tag{6.3}$$

where D is the end-to-end distance, R_{inf} is the approximated transmission range, and $E[d_h]$ is the expected hop distance. In Sections 6.3.2, 6.3.3, and 6.3.4, we derive the expressions for the expected hop length, $E[d_h]$, the expected energy consumption per hop, $E[E_h]$, and the expected latency per hop, $E[T_h]$, respectively.

6.3.2   Expected Hop Distance

Let us consider the reference model in Figure 6.8, where a node j resides in the infinitesimal area $dA = d\gamma\, d\alpha$ at coordinates (γ, α) with respect to the sink. The distance from node j to node i is, hence, given by

$$d_{(i,j)} = d(D, \gamma, \alpha) = \sqrt{\gamma^2 + D^2 - 2\gamma D \cos \alpha} \tag{6.4}$$

where D is the distance between the source node i and the sink, γ is the distance between the sink and the relay node j, and the angle α is as shown in Figure 6.8.

Accordingly, the expected hop distance, $E[d_h]$, is represented as

$$E[d_h] = \int_{\gamma_{min}}^{D} \int_{-\alpha_\gamma}^{\alpha_\gamma} d_{(i,j)}\, dP\{\mathcal{N}_i = j\} \tag{6.5}$$

where $\gamma_{min} = D - R_{inf}$ is the minimum distance between a potential next hop node j and the sink. Moreover, $d_{(i,j)}$ is the distance between nodes i and j as given by (6.4) and $dP\{\mathcal{N}_i = j\}$ is the probability that the next hop $\mathcal{N}_i$ is selected as node j. Finally, $\alpha_\gamma = a\cos[(\gamma^2 + D^2 - R_{inf}^2)/(2\gamma D)]$ is the limits of the distance α_γ.

In order for node j to be selected as the next hop, three conditions should hold true:

- A node exists in the infinitesimal area dA.
- The received SNR of node j satisfies $\psi_j > \psi_{Th}$.
- The received SNR, ψ_k, at each node, k, which is closer to the sink than node j, should be lower than the SNR threshold, i.e., $\psi_k \leq \psi_{Th}$.

Consequently, the probability that node j is selected as the next hop is given by

$$dP\{\mathcal{N}_i = j\} = P\{N_{A(d\gamma)} = 1\}P\{\psi_j > \psi_{Th}\}P\{d_{(j,s)} \geq \gamma\} \tag{6.6}$$

where $N_{A(d\gamma)}$ is the number of nodes in the area, dA, at distance γ from the sink. The first component in (6.6) denotes the probability that there is a node inside the area $A(d\gamma)$. The second component, $P\{\psi_j > \psi_{Th}\}$, in (6.6) is the probability that the received SNR of the node j is above ψ_{Th}. Finally, the third component, $P\{d_{(j,s)} \geq \gamma\}$, is the probability that the next hop j is at least at a distance γ from the sink, s. By integrating over the feasible area as indicated by the limits in the integral in (6.5), the expected hop distance can be found.

The first component, $P\{N_{A(d\gamma)} = 1\}$, in (6.6) can be approximated by

$$P\{N_{A(d\gamma)} = 1\} \simeq 1 - e^{-\rho\delta\gamma\,d\gamma\,d\alpha} \quad \text{as } d\gamma \to 0$$

$$\simeq \rho\delta\gamma\,d\gamma\,d\alpha \tag{6.7}$$

where ρ is the node density in the network and δ is the duty cycle parameter. In the last step, the approximation $e^{-x} \simeq 1 - x$ is used since $(\rho\delta\gamma\,d\gamma\,d\alpha) \to 0$ as $d\gamma \to 0$, $d\alpha \to 0$.

For the calculation of $P\{\psi_j > \psi_{Th}\}$ and $P\{d_{(j,s)} \geq \gamma\}$ in (6.6), the log-normal channel model introduced in Chapter 4 is used. Accordingly, the received power at a receiver at distance d from a transmitter is given by

$$P_r(d) = P_t - PL(d_0) - 10\eta \log_{10}\left(\frac{d}{d_0}\right) + X_\sigma \tag{6.8}$$

where P_t is the transmit power in dBm, $PL(d_0)$ is the path loss at a reference distance d_0 in dB, η is the path loss exponent, and X_σ is the shadow fading component, with $X_\sigma \sim \mathcal{N}(0, \sigma)$. Moreover, the SNR at the receiver is given by $\psi(d) = P_r(d) - P_n$ in dB, where P_n is the noise power in dBm.

The second term in (6.5) is the probability that ψ_j is above the SNR threshold, ψ_{Th}, and is found by

$$P\{\psi_j > \psi_{Th}\} = P\{X_\sigma > \beta(d_{(i,j)}, \psi_{Th})\}$$

$$= Q\left(\frac{\beta(d_{(i,j)}, \psi_{Th})}{\sigma}\right) \tag{6.9}$$

where

$$\beta(d_{(i,j)}, \psi_{Th}) = \psi_{Th} + P_n - P_t + PL(d_0) + 10\eta \log_{10}\left(\frac{d_{(i,j)}}{d_0}\right) \tag{6.10}$$

and $Q(x) = 1/\sqrt{2\pi}(\int_x^\infty e^{-(t^2/2)})\,dt$. Since the shadow fading component is a log-normal random variable, this probability is based on the cdf of the normal random variable, which is represented by the $Q(x)$ function.

According to the channel model above, the probability $P\{d_{(j,s)} \geq \gamma\}$ that the nodes closer to the sink than node j have an SNR value below the threshold can be found. Let us denote the area that consists of nodes that are closer to the sink than node j as $A(\gamma)$. Then,

$$P\{d_{(j,s)} \geq \gamma\} = \sum_{i=0}^{\infty} P\{N_{A(\gamma)} = i\}p_k(\gamma)^i$$

$$= \sum_{i=0}^{\infty} \frac{e^{-M(\gamma)}M(\gamma)^i}{i!}p_k(\gamma)^i$$

$$= e^{-M(\gamma)(1-p_k(\gamma))} \tag{6.11}$$

where $A(\gamma)$ is the area of intersection of two circles with centers separated by D and with radii R_{inf} and γ, respectively. Consequently, $N_{A(\gamma)}$ is the number of nodes in $A(\gamma)$ and $M(\gamma) = \rho\delta A(\gamma)$ is the average number of nodes in this area. Moreover, $p_k(\gamma) = P\{\psi_k \leq \psi_{Th}, k \in A(\gamma)\}$ is the probability that, for a node k in $A(\gamma)$, the received SNR is less than the SNR threshold, i.e., $\psi_k \leq \psi_{Th}$. This can found as

$$p_k(\gamma) = \int_{\gamma_{min}}^{\gamma} \int_{-\alpha_\gamma}^{\alpha_\gamma} \left[1 - Q\left(\frac{\beta}{\sigma}\right)\right] \frac{1}{A(\gamma)} \, d\alpha \, d\gamma \tag{6.12}$$

where $\gamma_{min} = D - R_{inf}$.

Using (6.6), (6.7), (6.9), (6.11), and (6.12) in (6.5), the expected hop distance can be calculated as follows:

$$E[d_h] = \rho\delta \int_{\gamma_{min}}^{D} \int_{-\alpha_\gamma}^{\alpha_\gamma} \gamma d_{(i,j)} Q\left(\frac{\beta}{\sigma}\right) e^{-M(\gamma)(1-p_k(\gamma))} \, d\alpha \, d\gamma, \tag{6.13}$$

which will be used for energy consumption and latency analyses of the FEC, ARQ, and HARQ schemes according to (6.1), (6.2), and (6.3).

6.3.3   Energy Consumption Analysis

The expected energy consumption and latency per hop are also calculated by considering a node j as shown in Figure 6.8. The expected energy consumption between node i and node j is denoted by $E[E_j]$, which is a function of γ and α. Following the same derivations as in Section 6.3.2 and replacing $d_{(i,j)}$ by $E[E_j]$ in (6.13), the expected energy consumption per hop is found as

$$E[E_h] = \rho\delta \int_{\gamma_{min}}^{D} \int_{-\alpha_\gamma}^{\alpha_\gamma} \gamma E[E_j] Q\left(\frac{\beta}{\sigma}\right) e^{-M(\gamma)(1-p_k(\gamma))} \, d\alpha \, d\gamma. \tag{6.14}$$

Since a node can become a next hop if its received SNR value is above a certain threshold, the expected energy consumption, $E[E_j]$, in (6.14) can be found as

$$E[E_j] = \int_{\psi_{Th}}^{\infty} E_{comm}(d_{(i,j)}, \psi) f_\Psi(d_{(i,j)}, \psi) \, d\psi \tag{6.15}$$

where $E_{comm}(d_{(i,j)}, \psi)$ is the energy consumption for communication between nodes i and j given that they are at a distance $d_{(i,j)}$ and the SNR value at node j is ψ. Moreover, $f_\Psi(\cdot)$ is the pdf of the SNR, which is a function of $d_{(i,j)}$ as well. Since $P(\Psi \leq \psi) = P(X_\sigma \leq \beta(d_{(i,j)}, \psi))$, $f_\Psi(\cdot)$ is found as

$$f_\Psi(d_{(i,j)}, \psi) = f_{X_\sigma}(\beta(d_{(i,j)}, \psi)) = \frac{1}{\sigma\sqrt{2\pi}} e^{-(\beta(d_{(i,j)}, \psi))^2/2\sigma^2} \tag{6.16}$$

where $\beta(d_{(i,j)}, \psi)$ is given in (6.10).

The first component, $E_{comm}(d_{(i,j)}, \psi)$, in (6.15) is the energy consumption in transmitting a packet between two nodes at a distance $d_{(i,j)}$ with received SNR ψ. The energy consumption in a typical link in WSNs has three components: the energy consumption of the sender node, the receiver node, and the neighbor nodes. Accordingly, it is given as[1]

$$E_{comm} = E_{TX} + E_{RX} + E_{neigh} \tag{6.17}$$

where E_{TX} is the energy consumed by the sender node (node i), E_{RX} is the energy consumed by the receiver node (node j), and E_{neigh} is the energy consumed by the neighbor nodes.

The three energy consumption values depend on the underlying MAC and error control scheme deployed on the nodes. Accordingly, in order to successfully transmit the packet, a node needs to complete the four-way RTS–CTS–DATA–ACK handshake for ARQ, three-way RTS–CTS–DATA handshake for FEC codes, and RTS–CTS–DATA–NACK exchange for HARQ. Before transmitting an

---

[1] We drop the indices ψ and $d_{(i,j)}$ for ease of illustration.

RTS packet, for medium access, a node performs the carrier sense mechanism to assess the availability of the channel and transmits a packet thereafter.

A successful allocation of the channel depends on both successful carrier sense and the fact that the transmission encounters no collisions. The probability of successful carrier sense, p_{cs}, can be denoted as follows [12]:

$$p_{cs} = 1 - (1 - p_{cf})^{K+1} \tag{6.18}$$

where K is the number of re-sensings allowed for one transmission and p_{cf} is the probability of sensing the channel free, which is given by

$$p_{cf} = e^{-\lambda_{net}(\tau_{cs}+T_{comm})} \tag{6.19}$$

where τ_{cs} is the carrier sense period and T_{comm} is the duration of a packet transmission. After a successful carrier sense, a collision can only occur if another node transmits during the vulnerable period of τ_{cs}. As a result, the probability of no collisions, p_{noColl}, is given by

$$p_{noColl} = e^{-\lambda_{net}\tau_{cs}}. \tag{6.20}$$

The term λ_{net} that appears in both (6.19) and (6.20) refers to the overall traffic that is generated by all the nodes inside the transmission range of a node, which is given by

$$\lambda_{net} = \lambda \frac{p_{cs}}{p_{comm}}(1 - (1 - p_{comm})^{L+1}) \tag{6.21}$$

where λ is the total generated traffic in the transmission range of a node and p_{comm} is the probability of successful transmission. Accordingly, the probability that a node can successfully acquire the channel is given by $p_{cs}p_{noColl}$, which can be found by solving the system of equations (6.18)–(6.21).

The total generated packet rate, λ, depends on both the generated traffic rate and the size of the packet. Let us assume that the sensor node has an average sampling rate of b bps. Denoting the length of the packet payload as l_D, on average, the packet generation rate of a node i is $\lambda_{ii} = b/l_D$ pkts/s. Since a node will also relay packets from other nodes to the sink, the packet transmission rate of a node is higher than this value. If a routing scheme that equally shares the network load among nodes is considered, on the average, the packet transmission rate of a node is $\lambda_i = c_i\lambda_{ii}$, where $c_i > 1$. Consequently, λ in (6.21) is given by $\lambda = \sum_{i=1}^{M_n} \lambda_i$, where the number of nodes that are in the transmission range of a node is given by $M_n - 1$.

Upon accessing the channel, the energy consumption depends on the probability that a data and a control packet are successfully received at distance $d_{(i,j)}$ with SNR ψ, which are denoted as $p_s^D(\psi)$ and $p_s^C(\psi)$, respectively.[2] The derivations of $p_s^D(\psi)$ and $p_s^C(\psi)$ are explained in Section 6.3.6. In addition, an RTS packet transmission is only successful if a node can successfully acquire the channel (given by $p_{cs}p_{noColl}$). Accordingly, E_{TX} is given by

$$E_{TX}^{ARQ} = n_{ret}^{ARQ}\{E_{sense} + p_{cs}(E_{tx}^C + p_{noColl}p_s^C E_{rx}^C + p_{noColl}(p_s^C)^2 E_{tx}^D + p_{noColl}(p_s^C)^2 p_s^D E_{rx}^C$$
$$+ (1 - p_{noColl}p_s^C)E_{t/o}^C + p_{noColl}(p_s^C)^2(1 - p_s^D)E_{t/o}^D)\} \tag{6.22}$$

$$E_{TX}^{FEC} = n_{ret}^{FEC}\{E_{sense} + p_{cs}(E_{tx}^C + p_{coll}p_s^C (E_{rx}^C + E_{dec}^C) + (1 - p_{coll}p_s^C)E_{t/o}^C)\}$$
$$+ p_{cs}p_{coll}(p_s^C)^2 E_{tx}^D \tag{6.23}$$

$$E_{TX}^{HARQ} = n_{ret}^{HARQ}\{E_{sense} + p_{cs}(E_{tx}^R + E_{rx}^C + E_{dec}^C + E_{tx}^{D1})\}$$
$$+ (p_s^C)^2(1 - p_s^{D1})(E_{rx}^N + E_{dec}^C + E_{tx}^{D2}) \tag{6.24}$$

---

[2] The length of RTS, CTS, ACK, and NACK packets can be considered to be the same and the index ψ is dropped in the remainder for clarity.

for ARQ, FEC, and HARQ, respectively, where

$$n_{ret}^{ARQ} = (1 - p_{cs} + p_{cs} p_{noColl} (p_s^C)^3 p_s^D)^{-1} \tag{6.25}$$

$$n_{ret}^{FEC} = n_{ret}^{HARQ} = (1 - p_{cs} + p_{cs} p_{coll} (p_s^C)^2)^{-1} \tag{6.26}$$

are the expected number of retransmissions for ARQ, FEC, and HARQ, which are further limited by the allowed maximum number of retransmissions.

In (6.22), (6.23), and (6.24), E_{tx}^x and E_{rx}^x are the transmission and receiving energies for packets, where the superscripts R, C, D, and A refer to RTS, CTS, DATA, and ACK packets, respectively.

The first term in (6.22) is the expected number of retransmissions. This value is a function of probability of successful carrier sense, p_{cs}, the probability of no collisions, p_{noColl}, and the successful transmission probability of control and data packets. The first term in parentheses in (6.22), E_{sense}, is the energy consumption for sensing the region. If the channel sensing is successful, the node transmits an RTS packet and receives a CTS packet if there is no collision and the RTS packet is successfully received at the destination. Similarly, a data packet is sent if the CTS packet is received successfully, followed by an ACK packet reception. The last two terms in the parentheses are the energy consumption for the timeout for CTS and ACK packets in case of packet errors.

Similarly, E_{TX} for FEC and HARQ can be found as in (6.23) and (6.24), respectively, where E_{dec}^x is the decoding energy. Moreover, the superscripts $D1$ and $D2$ in (6.24) refer to the transmitted packets for the first and second transmission in HARQ. Using the same approach, the energy consumption of the receiver node, E_{RX} in (6.17), is given as follows:

$$E_{RX}^{ARQ} = n_{ret}^{ARQ} \{ E_{rx}^R + E_{tx}^C + E_{rx}^D + E_{tx}^A \} \tag{6.27}$$

$$E_{RX}^{FEC} = n_{ret}^{FEC} \{ E_{rx}^R + E_{dec}^R + E_{tx}^C + E_{rx}^D + E_{dec}^D \} \tag{6.28}$$

$$E_{RX}^{HARQ} = n_{ret}^{HARQ} \{ E_{rx}^R + E_{dec}^R + E_{tx}^C \} + E_{rx}^{D1} + E_{dec}^{D1} + (p_s^C)^2 (1 - p_s^{D1})(E_{tx}^N + E_{rx}^{D2} + E_{dec}^{D2}) \tag{6.29}$$

for ARQ, FEC, and HARQ, respectively.

In wireless networks, the neighbors of the sender and receiver are also affected by the communication due to the broadcast nature of the wireless channel. Hence, the last term in (6.17), E_{neigh}, is the energy consumed by the neighbors of the transmitter and receiver nodes. This term is as follows:

$$E_{neigh}^{ARQ} = n_{ret}^{ARQ} \{ (\rho \delta \pi R_{inf}^2 - 2) E_{rx}^R + \lfloor \rho \delta (\pi R_{inf}^2 - A(D, R_{inf}, D)) - 2 \rfloor E_{rx}^C \} \tag{6.30}$$

$$E_{neigh}^{FEC} = E_{neigh}^{HARQ} = n_{ret}^{FEC} (\rho \delta \pi R_{inf}^2 - 2) E_{rx}^R + \lfloor \rho \delta (\pi R_{inf}^2 - A(D, R_{inf}, D)) - 2 \rfloor E_{rx}^C \tag{6.31}$$

for ARQ, HARQ, and FEC codes, respectively.

Using these derivations in (6.1), the end-to-end energy consumption can be calculated for each error control technique.

6.3.4   Latency Analysis

The expression for end-to-end latency of a flow is derived using a similar approach to the above. The delay per hop is given by

$$E[T_h] = \rho \delta \int_{\gamma_{min}}^{D} \int_{-\alpha_\gamma}^{\alpha_\gamma} \gamma E[T_j] Q\left(\frac{\beta}{\sigma}\right) e^{-M(1-p_k)} \, d\alpha \, d\gamma \tag{6.32}$$

where

$$E[T_j] = \int_{\psi_{Th}}^{\infty} T_{comm}(\psi, d_{(i,j)}) f_\Psi(\psi, d_{(i,j)}) \, d\psi \tag{6.33}$$

and T_{comm} is the latency for communication given by

$$T_{comm}^{ARQ} = n_{ret}^{ARQ}\{T_{sense} + 2p_{cs}p_{noColl}(p_s^C)^2 T^{Ctrl}$$

$$+ p_{cs}(1 - p_{noColl}(p_s^C)^2)T_{t/o}^C + p_{cs}p_{noColl}(p_s^C)^3 p_s^D(T^D + T^{Ctrl})$$

$$+ p_{cs}p_{noColl}(p_s^C)^3(1 - p_s^C p_s^D)T_{t/o}^A\} \tag{6.34}$$

$$T_{comm}^{FEC} = n_{ret}^{FEC}\{T_{sense} + 2T^{Ctrl} + 2T_{dec}^C + T^D + T_{dec}^D\} \tag{6.35}$$

$$T_{comm}^{HARQ} = n_{ret}^{HARQ}\{T_{sense} + 2T^{Ctrl} + 2T_{dec}^C + T^{D1} + T_{dec}^{D1}\}$$

$$+ (p_s^C)^2(1 - p_s^{D1})(T^{Ctrl} + T_{dec}^C + T^{D2} + T_{dec}^{D2}) \tag{6.36}$$

for ARQ, FEC, and HARQ, respectively. In these expressions, T_{sense} is the time spent for sensing, T^{Ctrl} and T^D are the control and data packet transmission times, respectively, $T_{t/o}$ is the timeout value, and T_{dec}^{Ctrl} and T_{dec}^D are the decoding latency for control and data packets, respectively. The derivations of (6.34), (6.35), and (6.36) follow a similar approach as in (6.22).

6.3.5 Decoding Latency and Energy

One of the major overheads of FEC codes in addition to the transmission and reception of redundant bits is the energy consumption for encoding and decoding packets as well as the delay associated with it. It is well known that the encoding energy for block codes is negligible [8]. Hence, only the decoding energy and latency are considered in the analysis. The latency of decoding for a block code (n, k, t) is given as [8]

$$T_{dec}^{BL} = (2nt + 2t^2)(T_{add} + T_{mult}) \tag{6.37}$$

where T_{add} and T_{mult} are the latency for addition and multiplication, respectively, of field elements in $GF(2^m)$, $m = \lfloor \log_2 n + 1 \rfloor$ [11]. Both Mica2 and MicaZ nodes are implemented with 8-bit microcontrollers [6], which can perform addition and multiplication of 8 bits in one and two cycles, respectively. As a result,

$$T_{add} + T_{mult} = 3\left\lceil \frac{m}{8} \right\rceil t_{cycle} \tag{6.38}$$

where t_{cycle} is one cycle duration, which is 250 ns [6]. Consequently, the decoding energy consumption is $E_{dec}^{BL} = I_{proc} V T_{dec}^{BL}$, where I_{proc} is the current for processor, V is the supply voltage, and T_{dec}^{BL} is given in (6.37).

6.3.6 BER and PER

The energy consumption and the latency depend on the PER during the communication. In this section, we derive the expressions for the BER and PER for Mica2 and MicaZ nodes. Since the modulation schemes used in these nodes are significantly different, it is necessary to investigate the effects of FEC and HARQ on these nodes separately. Mica2 nodes are implemented with a non-coherent FSK modulation scheme. The BER of this scheme is given by [8]

$$p_b^{FSK} = \frac{1}{2}e^{-(E_b/N_0)/2}, \quad E_b/N_0 = \psi \frac{B_N}{R} \tag{6.39}$$

where ψ is the received SNR, B_N is the noise bandwidth, and R is the data rate. The modulation scheme used in MicaZ nodes is offset quadrature phase shift keying (O-QPSK) with direct sequence spread spectrum (DSSS). The BER of this scheme is given by [7]

$$p_b^{OQPSK} = Q(\sqrt{(E_b/N_0)_{DS}}) \tag{6.40}$$

where

$$(E_b/N_0)_{DS} = \frac{2N \times E_b/N_0}{N + 4(E_b/N_0)(K - 1)/3}$$

where N is the number of chips per bit, and K is the number of simultaneously transmitting users.

Based on the BER p_b, the PER for the error control schemes can be calculated as follows. For ARQ, the CRC-16 error detection mechanism is deployed in both Mica nodes. Assuming all possible errors in a packet can be detected, the PER of a single transmission of a packet with payload l bits is given by

$$PER^{CRC}(l) = 1 - (1 - p_b)^l. \tag{6.41}$$

For the BCH codes, assuming perfect interleaving at the transceiver, the block error rate (BLER) is given by

$$BLER(n, k, t) = \sum_{i=t+1}^{n} \binom{n}{i} p_b^i (1 - p_b)^{n-i}. \tag{6.42}$$

The BLER for RS codes is found through simulations. More specifically, the Berlekamp–Massey algorithm [3] is implemented and simulated to find the relationship between the BLER and the BER.

Since a packet can be larger than the block length n, especially where small block lengths are used, the PER for FEC is given by

$$PER^{FEC}(l, n, k, t) = 1 - (1 - BLER(n, k, t))^{\lceil l/k \rceil} \tag{6.43}$$

where $\lceil l/k \rceil$ is the number of blocks required to send l bits and $\lceil \cdot \rceil$ is the ceiling function. Using (6.41), (6.42), and (6.43), the PERs for the HARQ schemes are also found. The packet success probabilities p_s^C and p_s^D used in Section 6.3.3 and Section 6.3.4 for control and data packets can then be found by using $l = l_C$ and $l = l_D$, respectively.

6.4 Comparison of Error Control Schemes

In this section, we illustrate the effects of the FEC and HARQ schemes in terms of PER, energy consumption, and end-to-end latency in a multi-hop network via numerical evaluations in MATLAB and simulations. For this comparison, two sensor node architectures are considered, i.e., Mica2 [1] and MicaZ [2]. Unless otherwise noted, the parameters in Table 6.1 are used for the numerical results.

The expected hop distance d_{hop}, which is found in (6.5), is shown in Figure 6.9(a) as a function of the received SNR threshold, ψ_{Th}, for different transmit power values, P_t. It can be observed that for small values of the received SNR threshold, ψ_{Th}, the average hop distance increases. Since lower ψ_{Th} allows nodes with lower channel quality to be chosen as the next hop, further nodes may become the next hop. Therefore, the number of hops from a node to a sink decreases for smaller ψ_{Th} values. Moreover, when the transmit power of a node is decreased, the expected hop distance decreases as expected. The simulation results, also shown in Figure 6.9(a), follow the theoretical results closely.

In the following, we present the effects of hop length extension and transmit power control defined in Section 6.2.

6.4.1 Hop Length Extension

In Figure 6.9(b), the end-to-end energy consumption per useful bit is shown as a function of the SNR threshold, ψ_{Th}. The energy consumption is shown for ARQ with the maximum number of retransmissions ($N = 7$) and BCH(128, 78, 7) and RS(15, 9, 3) codes. The simulation results are also shown in Figure 6.9(b).

As shown in Figure 6.9(b), the energy consumption of a flow decreases when the ψ_{Th} value is decreased from 15 dB. This is mainly because of the increase in expected hop distance as shown in

Table 6.1 Parameters [14].

D	300 m	l_C	8 bytes
P_t	0, −5, −15 dBm	l_D	38 bytes
PL_{d0}	55 dB	t_{cycle}	250 ns
P_n	−105 dBm	I_{proc}	8 mA
η	3	V	3 V
σ	3.8		

	Mica2	MicaZ
e_{rx}	21 mJ	59.1 mJ
$e_{tx}\ (P_t = 0)$	24 mJ	52.2 mJ
$e_{tx}\ (P_t = -5)$	21.3 mJ	42 mJ
$e_{tx}\ (P_t = -15)$	16.2 mJ	29.7 mJ
$t_{bit} = 1/R$	62.4 μs	4 μs
N	N/A	16 chips
K	N/A	2

Figure 6.9(a). However, the energy consumption for ARQ significantly increases as ψ_{Th} is decreased below a specific value, e.g., 5 dB. A lower ψ_{Th} results in nodes with lower channel quality being selected as the next hop. As a result, retransmissions occur, which increase the energy consumption per hop. Although the expected number of hops decreases, the increase in energy consumption per hop dominates the total energy consumption for ARQ.

Note that, for ARQ, the energy consumption curve reaches a peak and decreases as ψ_{Th} is decreased. This point corresponds to the case where the maximum number of retransmissions is no longer sufficient for reliable communication.

When the FEC codes are considered, the energy consumption can be decreased at lower SNR threshold values. When ARQ and FEC codes are compared, for large ψ_{Th} values, ARQ clearly outperforms the FEC codes. However, the BCH(128, 78, 7) and RS(15, 9, 3) codes are more energy efficient for smaller ψ_{Th} ($\sim$ 2 dB). This figure clearly shows the energy consumption of the two schemes as a function of ψ_{Th}; the operating points of ψ_{Th} for ARQ and FEC are not the same and should be determined according to the target PER.

The PER for ARQ and FEC codes is given in (6.41) and (6.43), respectively. Since these equations show the PER for a single hop, they should be extended for the multi-hop case. The relation between ψ_{Th} and the end-to-end PER bound can be used to determine the optimal point for ψ_{Th} in Figure 6.9(b). Denoting the PER of a hop i by PER_i, there exists a π such that

$$PER_i \leq \pi, \quad \text{for } \psi_i \geq \psi_{Th},$$

where ψ_i is the received SNR at the hop i and $\pi = f(\psi_{Th})$, which can be calculated using (6.39)–(6.43) by replacing ψ with ψ_{Th} in (6.39). Since the end-to-end PER is

$$PER_{e2e} = 1 - \prod_{i=1}^{n_h}(1 - PER_i),$$

where n_h is the number of hops, PER_{e2e} is bounded by

$$PER_{e2e} \leq 1 - (1 - \pi)^{n_h}, \quad \text{for } \psi_i \geq \psi_{Th}, \ \forall i. \tag{6.44}$$

If the end-to-end PER needs to be bounded by a certain threshold, PER^*_{e2e}, according to the application requirements, the route selection needs to be performed such that

$$\psi_{Th} = f^{-1}(1 - [1 - PER^*_{e2e}]^{1/n_h}). \tag{6.45}$$

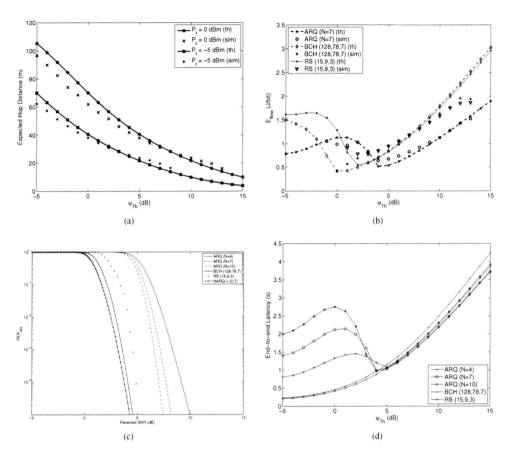

Figure 6.9 (a) Average hop distance and (b) average energy consumption of a flow vs. ψ_{Th}. (c) End-to-end PER vs. ψ_{Th} and (d) average end-to-end latency vs. ψ_{Th} (MicaZ) [15].

The relationship between the end-to-end PER, PER_{e2e}, and received SNR is shown in Figure 6.9(c) for the ARQ, BCH, RS, and HARQ schemes for MicaZ nodes. For RS codes, the results of the simulations are used as explained in Section 6.3.6, whereas for ARQ, BCH, and HARQ, (6.41) and (6.43) are used. According to Figure 6.9(c), the operating point for ψ_{Th} corresponding to a target end-to-end PER can be found.

As an example, if the target PER of an application is 10^{-2}, the minimum value for ψ_{Th} corresponds to 6.1 dB for ARQ, 3 dB for BCH(128, 78, 7), 4.8 dB for RS(15, 9, 3), and 2.5 dB for HARQ-I. As a result, it can be observed from Figure 6.9(b) that BCH(128, 78, 7) is slightly more energy efficient than ARQ. On the other hand, the RS(15, 9, 3) code results in higher energy consumption compared to the ARQ scheme. It is clear that more energy is consumed per hop for FEC codes due to both the transmission of redundant bits and decoding. However, since the error resilience is improved with FEC codes, lower SNR values can be supported. Through a channel-aware routing protocol that exploits this property, longer hop distances can be achieved leading to lower end-to-end energy consumption.

Exploiting FEC schemes with channel-aware routing not only improves energy consumption performance, but also decreases the end-to-end latency significantly as shown in Figure 6.9(d). Both FEC schemes outperform ARQ since their optimal ψ_{Th} value is lower than that of ARQ. This is due to both

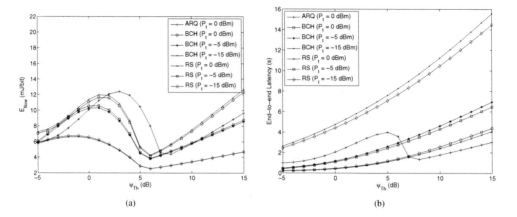

Figure 6.10 (a) Average energy consumption of a flow and (b) average end-to-end latency vs. ψ_{Th} for different values of transmit power (Mica2) [15].

the longer hops for FEC codes and the additional retransmissions of ARQ. Since the decoding delay of the FEC codes is lower than the time consumed for retransmission of a packet, FEC schemes improve the latency performance of the WSN. Furthermore, RS codes provide slightly lower end-to-end latency when compared to the BCH codes. This is related to the better error correction capability of RS codes when the same number of redundant bits is sent. Consequently, the end-to-end latency is slightly decreased.

6.4.2   Transmit Power Control

Another technique for exploiting the error resilience of FEC codes is to decrease the transmit power, P_t. In order to investigate the effect of transmit power, P_t, consider three power levels: 0, −5, and −15 dBm. Decreasing the transmit power can improve the energy efficiency of the FEC schemes since less power is consumed for the transmission of longer encoded packets. Although the receive power is fixed, since the interference range of a node decreases, the number of neighbors that consume idle energy also decreases. On the other hand, decreasing the transmit power increases the number of hops. In Figure 6.10(a), the energy consumption of BCH(128, 50, 13) and RS(15, 9, 3) are shown for three different transmit power levels and ARQ at $P_t = 0$ dBm. Note that the decrease in transmit power decreases the end-to-end energy consumption for both BCH and RS codes, which outperform the ARQ scheme for lower transmit power levels.

While transmit power control provides energy efficiency for particular FEC codes, its drawback is shown in Figure 6.10(b), where the end-to-end latency is shown. Contrary to the hop distance extension, since controlling transmit power has no effect on the time required for transmitting a packet, the end-to-end latency depends on the number of hops. Since transmit power control increases the number of hops, this technique introduces a significant increase in latency, which is a tradeoff for the BCH codes.

6.4.3   Hybrid Error Control

HARQ schemes exploit the advantages of both ARQ and FEC techniques. In this section, we present a comparison of the end-to-end energy consumption and the latency characteristics of these schemes with the ARQ scheme and BCH(128, 78, 7), which is found to be the most energy-efficient FEC scheme in the previous discussions.

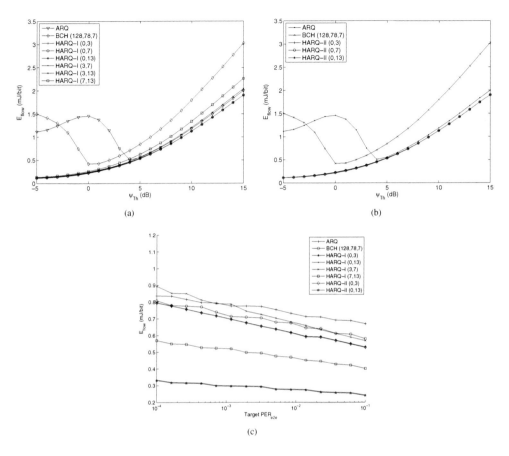

Figure 6.11 Average energy consumption vs. ψ_{Th} for HARQ (a) Type I and (b) Type II, and (c) average energy consumption vs. target end-to-end PER (MicaZ) [15].

The energy consumption of the Type I and Type II HARQ schemes is shown in Figure 6.11(a) and (b), respectively, for MicaZ. An important result is that Type II HARQ schemes are more energy efficient than both ARQ and FEC schemes. As shown in Figure 6.11(b), the energy consumptions of different HARQ-II schemes are similar for a given ψ_{Th} value. However, since the error resilience of these protocols depends on the BCH code used, the operating points of these schemes differ based on the target PER. It is shown in Figure 6.11(a) that the energy consumption of the HARQ-I scheme is dependent on the error correction code used in the first transmission. Consequently, the HARQ-I (7,13) scheme results in slightly higher energy consumption for a given ψ_{Th} value.

The energy consumption curves in Figure 6.11(a) and (b) as a function of the SNR threshold, ψ_{Th}, illustrate the effect of this threshold on the performance of these error control schemes. However, for a given end-to-end reliability requirement, the operating point of ψ_{Th} is different for each scheme because of the different error resilience. In Figure 6.11(c) the end-to-end energy consumption is shown as a function of the target PER for MicaZ nodes. It can be observed that HARQ-II schemes outperform ARQ, FEC, as well as HARQ-I schemes. This is particularly appealing since the HARQ-II scheme is implemented through only a single BCH code. Hence, the implementation cost of the HARQ-II scheme under consideration is also low compared to the HARQ-I case, where two different encoding schemes

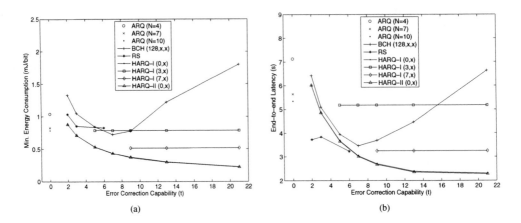

(a)   (b)

Figure 6.12 (a) Minimum energy consumption vs. error correction capability and (b) minimum end-to-end latency vs. error correction capability for MicaZ [15].

Table 6.2 Overview of results [15].

	Hop length extension		Transmit power control	
	Energy	**Latency**	**Energy**	**Latency**
Mica2	HARQ-I, II	HARQ-I, II	BCH ($t \geq 1$)	ARQ
MicaZ	HARQ-II (RS)	HARQ-II	BCH ($t \geq 1$)	ARQ

need to be implemented. Furthermore, the energy efficiency of HARQ-II schemes improves when more powerful FEC schemes are used.

6.4.4   Overview of Results

An overview of the energy and latency performance of the error control schemes is shown in Figure 6.12. In the figure, the minimum end-to-end energy consumption and latency of the ARQ, BCH, RS, HARQ-I, and HARQ-II schemes subject to an end-to-end PER target of 10^{-2} are shown as a function of their error correction capability. In particular, Figure 6.12(a) and (b), the minimum energy consumption and latency of these schemes are shown. Accordingly, both HARQ-I and HARQ-II schemes outperform other error control schemes. Moreover, it can be observed that for both BCH and RS codes, an optimum error correction capability, t, value can be found to minimize energy consumption and latency. The results reveal that HARQ-I codes are slightly inefficient in terms of energy consumption and latency. The HARQ-II scheme is more energy efficient compared to ARQ, BCH, and RS codes. Furthermore, it can be observed from Figure 6.12(b) that RS(31, 19, 6) performs very close to the HARQ-II scheme in terms of end-to-end latency, which makes both of these schemes suitable candidates for real-time traffic.

The results of the comparison of error control schemes are summarized in Table 6.2, where the efficient schemes are identified. Accordingly, hop length extension decreases both energy consumption and end-to-end latency for certain FEC codes when compared to ARQ. On the other hand, transmit power control can be exploited in situations where energy consumption is of paramount importance and can be traded off for end-to-end latency.

Error control constitutes an integral component of the link layer in addition to MAC protocol operations, which are explained in Chapter 5. As we have described in this chapter, these two main functionalities of the link layer are interdependent and should be designed and analyzed in a coherent manner.

References

[1] Crossbow MICA2 mote specifications. http://www.xbow.com.

[2] Crossbow MICAz mote specifications. http://www.xbow.com.

[3] E. R. Berlekamp. *Algebraic Coding Theory*. McGraw-Hill, 1968.

[4] I. Chlamtac, C. Petrioli, and J. Redi. Energy-conserving go-back-N ARQ protocols for wireless data networks. In *Proceedings of IEEE ICUPC'98*, volume 2, pp. 1259–1263, Piscataway, NJ, USA, October 1998.

[5] I. Chlamtac, C. Petrioli, and J. Redi. Energy-conserving selective repeat ARQ protocols for wireless data networks. In *Proceedings of the 9th IEEE International Symposium on Personal, Indoor and Mobile Radio Communications*, volume 2, pp. 836–840, Boston, MA, USA, September 1998.

[6] Atmel Corp. Atmega128 datasheet. Available at http://www.atmel.com.

[7] M. A. Landolsi and W. E. Stark. On the accuracy of Gaussian approximations in the error analysis of DS-CDMA with OQPSK modulation. *IEEE Transactions on Communications*, 50(12):2064– 2071, December 2002.

[8] S. Lin and D. J. Costello Jr. *Error Control Coding: Fundamentals and Applications*. Prentice Hall, 1983.

[9] H. Liu, H. Ma, M. El Zarki, and S. Gupta. Error control schemes for networks: an overview. *Mobile Networks and Applications*, 2(2):167–182, September 1997.

[10] R. Min, M. Bhardwaj, S.-H. Cho, E. Shih, A. Sinha, A. Wang, and A. Chandrakasan. Low-power wireless sensor networks. In *Proceedings of the 14th International Conference on VLSI Design*, pp. 205–210, Bangalore, India, January 2001.

[11] Y. Sankarasubramaniam, I. F. Akyildiz, and S. W. McLaughlin. Energy efficiency based packet size optimization in wireless sensor networks. In *Proceedings of the IEEE Internal Workshop on Sensor Network Protocols and Applications*, pp. 1–8, Anchorage, AK, USA, 2003.

[12] K. Schwieger, A. Kumar, and G. Fettweis. On the impact of the physical layer on energy consumption in sensor networks. In *Proceedings of EWSN'05*, pp. 13–24, Istanbul, Turkey, February 2005.

[13] E. Shih, S.H. Cho, N. Ickes, R. Min, A. Sinha, A. Wang, and A. Chandrakasan. Physical layer driven protocol and algorithm design for energy-efficient wireless sensor networks. In *Proceedings of ACM MobiCom'01*, pp. 272–287, Rome, Italy, 2001.

[14] M. C. Vuran and I. F. Akyildiz. Cross-layer analysis of error control in wireless sensor networks. In *Proceedings of IEEE SECON'06*, Reston, VA, USA, September 2006.

[15] M. C. Vuran and I. F. Akyildiz. Error control in wireless sensor networks: a cross layer analysis. *IEEE/ACM Transactions on Networking*, 17(4):1186–1199, 2009.

[16] L. C. Zhong and J. M. Rabaey. An integrated data-link energy model for wireless sensor networks. In *Proceedings of IEEE ICC'04*, Paris, France, June 2004.

7

Network Layer

The network layer is one of the most investigated research topics in WSNs [1, 2]. Especially, many routing algorithms and protocols have been proposed through the development of WSNs. In this chapter, the challenges for the network layer in WSNs are described. Moreover, the solutions for routing are described in detail. To provide a consistent discussion on these solutions, the routing protocols are broken down into four groups: (1) data-centric and flat architecture, (2) hierarchical, (3) location based, and (4) QoS based. The classification of routing protocols and the particular solutions are shown in Figure 7.1.

7.1 Challenges for Routing

Routing is one of the main problems in WSNs and many solutions have been developed to address this problem. Ensuring efficient routing faces many challenges due to both wireless communication effects and the peculiarities of sensor networks. These challenges preclude existing routing protocols developed for wireless ad hoc networks from being used in WSNs. Instead, novel routing protocols are required. Next, we describe these main challenges facing routing in WSNs.

7.1.1 Energy Consumption

The main objective of the routing protocols is efficient delivery of information between sensors and the sink. To this end, energy consumption is the main concern in the development of any routing protocol for WSNs. Because of the limited energy resources of sensor nodes, data need to be delivered in the most energy-efficient manner without compromising the accuracy of the information content. Hence, many conventional routing metrics such as the shortest path algorithm may not be suitable. Instead, the reasons for energy consumption should be carefully investigated and new energy-efficient routing metrics developed for WSNs. The major reasons of energy consumption for routing in WSNs can be classified as follows:

- **Neighborhood discovery:** Many routing protocols require each node to exchange information between its neighbors. The information to be exchanged can vary according to the routing techniques. While most geographical routing protocols require knowledge of the locations of the neighbor nodes, a data-centric protocol may require the information content of the observed values of each sensor in its surrounding. In each case, nodes consume energy in exchanging this information through the wireless medium, which increases the overhead of the protocol. In order to improve the energy efficiency of the routing protocols, local information exchange should be minimized without hampering the routing accuracy.

- **Communication vs. computation:** It is well known that computation is much cheaper than communication in terms of energy consumption. Moreover, in WSNs, the goal is to deliver

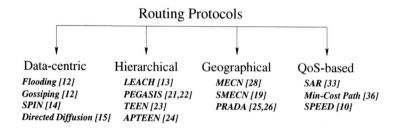

Figure 7.1   Overview of routing protocols.

information instead of individual packets. Consequently, in addition to the conventional packet switching techniques, computation should also be integrated with routing to improve energy consumption. As an example, data from multiple nodes can be aggregated into a single packet to decrease the *traffic volume* without hampering the *information content*. Similarly, computation at each relay node can be used to suppress redundant routing information.

7.1.2   Scalability

WSNs usually consist of a large number of nodes. The need to observe physical phenomena in detail may also require a high-density deployment of these nodes. The large number of nodes prevents global knowledge of the network topology from being obtained at each node. Hence, fully distributed protocols, which operate with limited knowledge of the topology, need to be developed to provide scalability. In addition, since the density is high in the network, local information exchange should also be limited to improve the energy efficiency of the network. Furthermore, since high-level information is more important than individual pieces of information from each sensor node, the routing protocol should support in-network combination of the information from a large number of nodes without hampering energy consumption.

7.1.3   Addressing

The large number of sensor nodes in a network prevents unique addresses from being assigned to each node. While local addressing mechanisms can still be used to facilitate communication between neighbors, address-based routing protocols are not feasible because of the large overhead required to use unique addresses for each communication. Consequently, the majority of the ad hoc routing protocols cannot be adopted for WSNs since these solutions require unique addresses for each node in the network. Furthermore, users are interested in collective information from multiple sensors regarding a physical phenomenon instead of information from individual sensors. Consequently, new addressing mechanisms or novel routing techniques that do not require unique IDs for each node are required.

7.1.4   Robustness

WSNs rely on the nodes inside the network to deliver data in a multi-hop manner. Hence, routing protocols operate on these sensor nodes instead of dedicated routers such as in the Internet. The low-cost components used in sensor nodes, however, may result in unexpected failures to such an extent that the sensor node may be non-operational. As a result, routing protocols should provide robustness to node failures and prevent single point-of-failure situations, where the information is lost if a sensor dies. Moreover, the wireless channel results in packets being lost during communication. In addition to robustness against node failures, the routing protocol should ensure that the effectiveness of the protocol does not rely on a single packet that can be lost. Even under very harsh conditions with frequent channel errors, the routing protocol should provide efficient delivery between the sensor and the sink.

7.1.5 *Topology*

The deployment of a WSN can be either predetermined or through a random strategy. While predetermined topology can be exploited to design more efficient routing protocols, this is usually not the case for WSNs. Consequently, individual nodes are usually unaware of the initial topology of the network. However, the relative locations of the neighbors of a node and the relative location of the nodes in the network significantly affect the routing performance. Therefore, routing protocols should provide topology-awareness such that the neighborhood of each node is discovered and the routing decisions are made accordingly. Furthermore, the network topology can change dynamically during the lifetime of the network. Since energy efficiency is crucial, nodes may switch off the transceiver circuitry, which, in effect, results in removing a node from the topology. Whenever the node is active again, it joins the network. These changes between active and sleep states of nodes dynamically affect the neighborhood topology of a sensor node.

WSN topology is usually assumed to be static. However, dynamic changes due to sink mobility or target mobility can affect the communication structure and, hence, the routes. Consequently, the routing protocol should also be adaptive to these changes in the network topology.

7.1.6 *Application*

The type of application is also important for the design of routing protocols. In monitoring applications, usually nodes communicate their observations to the sink in a periodic manner. As a result, static routes can be used to maintain efficient delivery of the observations throughout the lifetime of the network. In event-based applications, however, the sensor network is in sleep state most of the time. However, whenever an event occurs, routes should be generated to deliver the event information in a timely manner. Moreover, event location is not fixed since it is directly related to the event and, hence, new routes should be generated for each event. It can be seen that the routing technique is directly related to the application and significantly different techniques may be required for different kinds of applications.

7.2   Data-centric and Flat-Architecture Protocols

As explained in Section 7.1, one of the major differences between WSNs and ad hoc networks is that, because of the large number of sensor nodes, it is hard to assign specific IDs to each of the sensor nodes. Hence, address-based routing protocols are not preferred for WSNs. To overcome this, data-centric routing protocols have been proposed. As an example, "the areas where the temperature is over 70 °F (21 °C)" is a more common query than "the temperature read by a certain node." Attribute-based naming is used to carry out queries by using the attributes of the phenomenon.

■ **EXAMPLE 7.1**

An example for data-centric routing is illustrated in Figure 7.2. In this case, the sink is interested in areas where the temperature is higher than 70 °F. Consequently, the nodes with sensor readings matching this request are addressed. Note that data-centric routing provides routes according to the query content and, hence, the nodes that send information change for each query. Moreover,

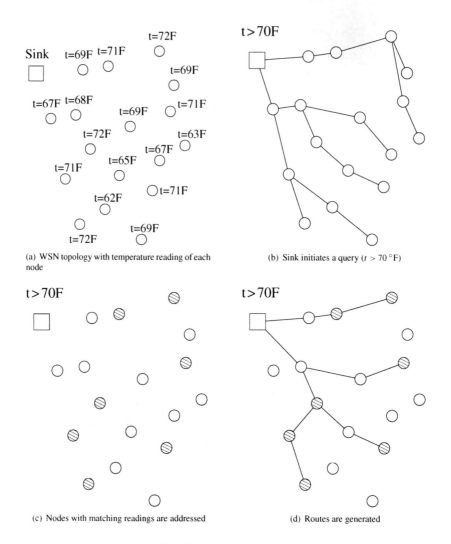

(a) WSN topology with temperature reading of each node

(b) Sink initiates a query ($t > 70 °$F)

(c) Nodes with matching readings are addressed

(d) Routes are generated

Figure 7.2   Illustration of data-centric routing.

using a single data-centric query, multiple nodes in distant locations can be addressed as also shown in Figure 7.2.

Data-centric routing refers to the type of query message initiated through the sink as described in Chapter 9. Instead of node IDs, data-centric routing requires attribute-based naming [7, 8, 26, 32]. For attribute-based naming, users are more interested in querying an attribute of the phenomenon, rather than querying an individual node.

Some of the protocols that may apply data-centric principles are flooding [11], gossiping [11], SPIN [13], directed diffusion [9, 14], energy-aware routing protocol [31], rumor routing [4], gradient-based routing [29], CADR [6], COUGAR [35], ACQUIRE [28], Shortest Path Minded SPIN (SPMS) [17], and solar-aware routing [34].

To provide insight into these solutions, we discuss some of the data-centric routing mechanisms.

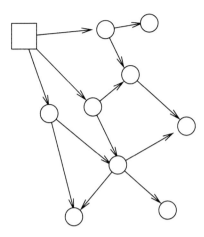

Figure 7.3   Flooding.

7.2.1   *Flooding*

The simplest routing algorithm, which has been developed for multi-hop networks, is the flooding technique. Accordingly, whenever a node receives a packet, it broadcasts this packet to all of its neighbors. This continues until all the nodes in the network receive the packet as shown in Figure 7.3. As a result, a packet can be *flooded* through the whole network. The flooding can be controlled by limiting the rebroadcast until the packet reaches the destination or the maximum number of hops is reached.

Flooding is a reactive protocol and its implementation is fairly straightforward. The advantages of flooding are in its simplicity since a node does not require neighborhood information, and flooding does not require costly topology maintenance and complex route discovery algorithms. However, flooding has several deficiencies as described below [13]:

- **Implosion:** The flooding mechanism does not restrict multiple nodes from broadcasting the same packet to the same destination. As a result, duplicated messages are usually sent to the same node. As shown in Figure 7.4(a), if node A shares N neighbor nodes with another node B, the sensor node B receives N copies of the message sent by node A.
- **Overlap:** The information sent by the sensor nodes is closely related to their sensing regions. As shown in Figure 7.4(b), if two nodes have overlapping sensing regions, then both of them may sense the same stimuli at the same time. As a result, neighbor nodes receive duplicated messages.
- **Resource blindness:** The most important resource in WSNs is the available energy, which should be efficiently consumed by networking protocols. However, the flooding protocol does not take into account the available energy resources. An energy resource-aware protocol must take into account the amount of energy available to it at all times.

7.2.2   *Gossiping*

One of the main inefficiencies of flooding is the implosion problem, where multiple copies of the same packet traverse the network. This can be avoided through gossiping, which is a derivation of flooding [11]. Gossiping avoids implosion by selecting a single node for packet relaying. As a result, whenever a node receives a packet it does not broadcast the packet but selects a random node among its neighbors and forwards the packet to that particular node. Once the neighbor node receives the packet, it randomly selects another sensor node.

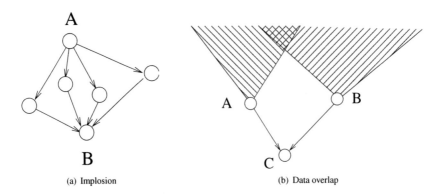

(a) Implosion   (b) Data overlap

Figure 7.4   Main problems with flooding.

Although gossiping avoids the implosion problem by just having one copy of a message at any node, it increases the latency in propagating the message to all sensor nodes. Since a single node at a time is informed about the packet, the information is distributed slowly. On the other hand, since multiple copies of a packet are prevented, the energy consumption of gossiping is lower than that of flooding.

Although simple and inefficient, flooding and/or gossiping techniques can still be used by recent routing protocols for specific functions. As an example, during the deployment phase, the sink can use flooding or gossiping protocols to determine the active nodes. Similarly, during sensor network initialization, limited flooding can be used to gather information from neighbors in close proximity.

7.2.3   Sensor Protocols for Information via Negotiation (SPIN)

SPIN is a family of routing protocols designed to address the deficiencies of flooding by negotiation and resource adaptation [13]. For this purpose, two main approaches are followed. Firstly, instead of sending all the data, sensor nodes negotiate with each other through packets that describe the data. Consequently, the observed information is only sent to interested sensor nodes as a result of this negotiation. Secondly, each node monitors its energy resource, which is used to perform *energy-aware* decisions.

■ **EXAMPLE 7.2**

The negotiation mechanism of SPIN is performed through three types of messages, namely advertisement (ADV), request (REQ), and DATA, which are illustrated in Figure 7.5. Before sending a DATA packet, a node advertises its intent by broadcasting an ADV packet (Step 1). The ADV packet contains a description of the DATA packet to be sent, which is much smaller in size than the DATA packet. Then, if a neighbor is interested in the ADV packet, it replies back with a REQ message (Step 2). Finally, the DATA packet is sent to the node that requests it (Step 3). Data propagation in WSNs is coordinated through this mechanism at each hop. As shown in Steps 4, 5, and 6 of Figure 7.5, multiple nodes can send REQ messages back to a node, which sends DATA to each node until all the nodes get a copy. As a result of the SPIN protocol, the sensor nodes in the entire sensor network which are interested in the data will get a copy.

The basic operation explained in Example 7.2 is referred to as the point-to-point SPIN protocol (SPIN-PP). In addition to SPIN-PP, several variations have been proposed to address some of the disadvantages of SPIN-PP. We explain these variations of SPIN next, i.e., SPIN with energy consumption awareness (SPIN-EC), SPIN for broadcast networks (SPIN-BC), and SPIN with reliability (SPIN-RL).

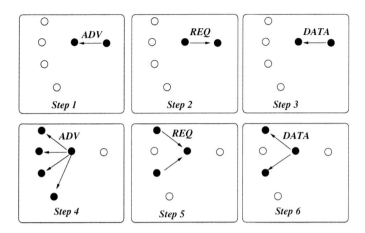

Figure 7.5 The SPIN protocol [13].

SPIN-PP does not address the resource-blindness problem of conventional flooding or gossiping protocols. Although the DATA packet transmission is limited to nodes that provide interest, energy consumption is still a concern. SPIN-EC addresses this through a simple energy conservation heuristic such that whenever the residual energy of a node is lower than a threshold, the node does not participate in the protocol operation, i.e., it does not send a REQ packet if it does not have enough energy to transmit the REQ packet and receive a DATA packet. Since node participation is dependent on the residual energy, if a node has plenty of energy SPIN-EC behaves like SPIN-PP.

Another disadvantage of SPIN-PP is shown in Steps 5 and 6 of Figure 7.5. Whenever there is more than one node that sends REQ packets, the DATA packet is sent to each node individually. Considering the broadcast nature of the wireless channel, this approach is a waste of resources since each neighbor of a node can receive the packet in each unicast. Furthermore, SPIN-PP does not provide any mechanism to prevent collisions when multiple REQ packets are send. This is addressed through SPIN-BC, which is developed for broadcast networks. In contrast to SPIN-PP, SPIN-BC introduces a randomized backoff mechanism for the nodes before transmitting a REQ packet. As a result, if a node has an interest in a packet but hears a REQ packet related to that particular packet, it drops its REQ packet and waits for the DATA packet. Upon receiving a REQ packet, a transmitter node broadcasts a single DATA packet which can be received by all the interested neighbors. As a result, SPIN-BC decreases the energy consumption and overhead caused by multiple interested neighbors.

SPIN-RL provides a reliability mechanism to the SPIN-BC protocol such that if a node receives an ADV packet but does not receive a DATA packet followed by it (due to wireless channel errors), it requests the DATA packet from the neighbors that may have received the DATA packet. Moreover, SPIN-RL limits the retransmission period of the nodes such that they do not retransmit a DATA packet before a specified period.

SPIN is based on data-centric routing [13] where the sensor nodes broadcast an advertisement for the available data and wait for a request from interested sinks. Compared to flooding, SPIN-PP reduces energy consumption by 70% since redundant transmissions are prevented. SPIN-EC provides a further 10% increase in energy consumption, through energy-aware operation. Moreover, since local interactions are required for routing, SPIN is also scalable. However, compared to flooding, the latency in data dissemination is higher because of the overhead in the handshake mechanism.

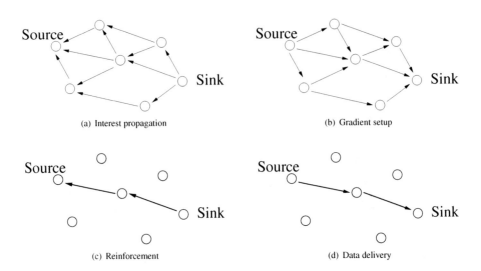

Figure 7.6   Operation of the directed diffusion protocol [14].

7.2.4   Directed Diffusion

SPIN provides efficient mechanisms for a sensor node to disseminate its observations to interested nodes. As a result, the traffic flow in SPIN is initiated from the sensors and usually ends up at the sink. However, this type of traffic may not always be preferable when the user (i.e., sink) requests specific information from the sensors. The *directed diffusion* data dissemination paradigm has been developed to address this requirement [14]. Directed diffusion consists of four stages to construct routes between the sink and the sensors of interest to the sink's request. The four stages are: (1) interest propagation, (2) gradient setup, (3) reinforcement, and (4) data delivery, as described next.

The information request is provided through *interest* messages initiated by the sink. Directed diffusion is initiated when the sink sends out interest messages to all sensors as shown in Figure 7.6(a). This phase is called *interest propagation*, where the interest messages flood through the network. The interest messages act as *exploratory* messages to indicate the nodes with matching data for the particular task. During the task, the sink continues to periodically broadcast the interest message.

Upon receiving the interest message, each sensor node stores it in an interest cache. The interest cache has several fields including *timestamp, gradient, interval*, and *duration*. The timestamp field indicates the local time when the interest is received. The gradient indicates the node from which the interest has been received. This gradient field is used to form reverse paths towards the sink. Each interest is stored at the cache for a specific time indicated by the duration field. After receiving the interest, the sensor node forwards the message to its downstream neighbors. This forwarding can be similar to flooding or more limited according to the task description. Local rules can be specified to define different gradient setup techniques. Accordingly, the gradient can be sent to the node which first sends the interest. Similarly, the gradient can be set up such that nodes with the highest remaining energy are chosen. As the interest is propagated throughout the sensor network, the gradients from the source back to the sink are set up as shown in Figure 7.6(b).

The interest messages indicate the required data at a given time from the sensor networks. Each node checks its sensor observations and becomes a source node if it has data matching the interest. When the source has data for the interest, the source sends the data along the interest's gradient path as shown in Figure 7.6(b).

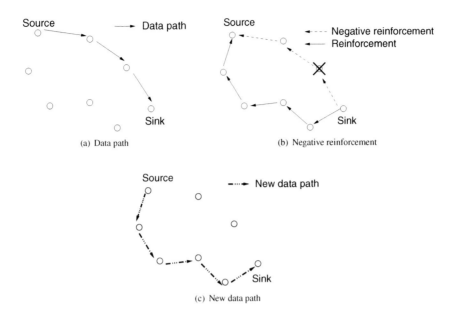

Figure 7.7 Negative reinforcement in directed diffusion.

Since the gradient setup phase does not limit the number of gradients that a node may have, the source node may have several gradients for the same interest. As a result, the data can be sent through multiple routes to the sink. In this case, the sink can *reinforce* a particular path by resending the interest through the specified node in that path as shown in Figure 7.6(c). This path can be selected according to several rules such as best link quality, number of packets received from a neighbor, or lowest delay. Once a specified node is selected, the interest is sent only to that node to reinforce the path associated with that node. Each node along this path forwards the reinforcement to its next hop. Finally, a route between the source and sink can be established as shown in Figure 7.6(d).

Moreover, using reinforcement, the data routes can be dynamically changed according to the changes in the WSN. In this case, the sink sends reinforcement messages through a new path other than the current path. Furthermore, negative reinforcement messages are sent through the current path to suppress data transfer through that path as shown in Figure 7.7.

The interest propagation, gradient setup, reinforcement, and data propagation are performed according to local rules. Consequently, different local rules can result in different transmission techniques. While the basic operation of directed diffusion as described above results in single path delivery, the sink may select multiple paths during the reinforcement stage to provide multi-path delivery, where the traffic on each path is proportional to the gradient. Similarly, sensor information can be delivered to multiple sinks according to the interest propagation and gradient setup techniques.

An important overhead of the directed diffusion protocol is the flooding operation during interest propagation. This operation is also referred to as *two-phase pull* since the sink initiates the communication. For applications where data need to be initiated from the sensors, an extension of the directed diffusion, i.e., *push diffusion*, has also been developed. Push diffusion omits the interest propagation phase and sensors advertise their data to the sink much like the operation of SPIN. Upon receiving the advertisements, the sink sends reinforcement packets to establish routes between the sources and the sink.

7.2.5   Qualitative Evaluation

Data-centric routing protocols provide application-dependent routes based on the interests of the user. A data-centric mechanism is used to determine end-points in the network, which results in a dynamic operation. Whenever the sensor readings change, the network routes are adapted to these changes to satisfy the user's requests. Moreover, since a data-centric mechanism is used, a global node addressing mechanism is not required. This provides energy efficiency since the routes are constructed only when there is an interest and there is no need to maintain a global network topology. Data-centric solutions also decrease energy consumption in the network by specifying selected routes for a specific interest from the sink. Accordingly, only those nodes that have matching information are involved in information generation.

A major disadvantage of the data-centric routing protocols is that they are generally based on a flat topology. This causes scalability problems as well as increased congestion among the nodes closer to the sink. Distributed aggregation mechanisms are necessary to decrease the information content flowing in each part of the network. Moreover, protocols such as directed diffusion are applicable to a subset of applications in WSNs, since the communication is initiated by queries generated from the sink. This also makes directed diffusion not a good choice for dynamic applications, where continuous data delivery is important. Moreover, the query types as well as the interest matching procedures need to be defined for each application. Furthermore, the data-centric approach results in application-dependent naming schemes. Accordingly, for each change in the application, these schemes should be defined a priori. Finally, the matching process for data and queries causes some overhead at the sensors.

7.3   Hierarchical Protocols

The data-centric and flat-architecture protocols result in the majority of the information generated at the sensors being concentrated near the sink. As a result, flat-architecture protocols suffer from data overload close to the sink as the density increases. The nodes which are located near the sink route more information than nodes in other parts of the network. As a result, these nodes die faster and produce a disconnection between the sink and the WSN. Consequently, flat-architecture protocols result in uneven energy consumption throughout the network and limit the scalability of the protocols.

The disadvantages of the flat-architecture protocols can be addressed by forming a hierarchical architecture, where the nodes are grouped in *clusters* and the local interactions between cluster members are controlled through a *cluster head* as shown in Figure 7.8. Based on this architecture, several hierarchical routing protocols have been developed to address the scalability and energy consumption challenges of WSNs. Sensor nodes form clusters where the cluster heads aggregate and fuse data to conserve energy. The cluster heads can also form another layer of clusters among themselves before reaching the sink. Some of the hierarchical protocols proposed for sensor networks are the low-energy adaptive clustering hierarchy (LEACH) [12], power-efficient gathering in sensor information systems (PEGASIS) [20, 21], threshold-sensitive energy-efficient sensor network (TEEN) [22], and adaptive threshold-sensitive energy-efficient sensor network (APTEEN) [23].

7.3.1   LEACH

The LEACH protocol aims to minimize energy consumption in WSNs through a cluster-based operation [12]. The goal of LEACH is to dynamically select sensor nodes as cluster heads and form clusters in the network. The communications inside the clusters are directed to the cluster head, which performs aggregation. Cluster heads then directly communicate with the sink to relay the collected information from each cluster. LEACH also changes the cluster head role dynamically such that the high-energy consumption in communicating with the sink is spread to all sensor nodes in the network.

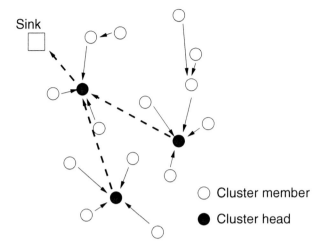

Sink

Cluster member

Cluster head

Figure 7.8   Hierarchical cluster-based architecture in WSNs.

The operation of LEACH is controlled through *rounds*, which consist of several phases. During each round, each cluster formation stays the same, and the cluster heads are selected at the beginning of each round. A round is separated into two phases, the *setup phase* and *steady state phase*. During the setup phase, cluster heads are selected, clusters are formed, and the cluster communication schedule is determined. During the steady state phase, data communication between the cluster members and the cluster head is performed. The duration of the steady state phase is longer than the duration of the setup phase in order to minimize the overhead.

The setup phase of LEACH consists of three phases: advertisement, cluster setup, and schedule creation. LEACH aims to randomly select sensors as cluster heads during the beginning of each round. The cluster head selection is performed through the advertisement phase, where the sensor nodes broadcast a cluster head advertisement message. Firstly, a sensor node chooses a random number between 0 and 1. If this random number is less than a threshold $T(n)$, the sensor node becomes a cluster head. $T(n)$ is calculated as

$$T(n) = \begin{cases} \dfrac{P}{1 - P[r \bmod(1/P)]} & \text{if } n \in G \\ 0 & \text{otherwise} \end{cases} \qquad (7.1)$$

where P is the desired percentage to become a cluster head, r is the current round, and G is the set of nodes that have not been selected as a cluster head in the last $1/P$ rounds. The selected cluster heads then advertise to their neighbors in the network that they are the new cluster heads. For this operation, LEACH relies on a CSMA-based random access scheme to avoid advertisement collisions from multiple cluster heads.

Once the sensor nodes receive the advertisement, they determine the cluster that they belong to. If a node receives an advertisement from a single cluster head, then it automatically becomes a member of that cluster. However, if a sensor node receives advertisements from multiple cluster heads, the cluster selection is performed based on the signal strength of the advertisement from the cluster heads to the sensor nodes. The cluster head with the highest signal strength is selected. Consequently, the channel quality between the cluster members and the cluster head is aimed to be high.

After the advertisement phase, the sensor nodes inform the associate cluster head that they will be members of the cluster, which is called the cluster setup phase. Again, LEACH relies on a CSMA-based

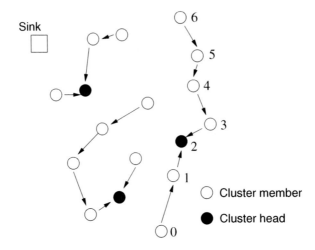

Figure 7.9 Chain structure of PEGASIS.

random access scheme to prevent collisions between packets sent by each node. Finally, the schedule creation phase is performed, where the cluster heads assign the time during which the sensor nodes can send data to the cluster heads. This selection is based on a time division multiple access (TDMA) approach, which is followed throughout the steady state phase.

Once the cluster formation is completed in the setup phase, LEACH switches to the steady state phase. During this phase, the sensor nodes can begin sensing and transmitting data to the cluster heads. The cluster heads also aggregate data from the nodes in their cluster before sending these data to the sink. At the end of the steady state phase, the network goes into the setup phase again to enter into another round of selecting the cluster heads. As a result, energy consumption due to the cluster head duty is equally distributed among sensor nodes.

The cluster-based operation of LEACH improves the energy efficiency of WSNs. During the steady state phase, only the cluster heads are active all the time. A cluster member in a cluster is active only during its allocated time slot and the setup phase. Consequently, the energy consumption of a regular node is minimized significantly. Since LEACH performs periodic cluster head selection, the energy consumption burden of the cluster head nodes is also shared. Accordingly, LEACH provides a factor of 4–8 reduction in energy consumption compared to a flat-architecture routing protocol.

7.3.2   PEGASIS

The PEGASIS protocol aims to provide improvements to the LEACH protocol [20, 21]. PEGASIS aims to address the overhead caused by the cluster formation in LEACH by constructing chains of nodes instead of clusters as shown in Figure 7.9. The chain construction is performed according to a greedy algorithm, where nodes select their closest neighbors as next hops in the chain. It is assumed that the nodes have a global knowledge of the network and the chain construction starts from the nodes that are farthest from the sink. As a result of chain operation, instead of maintaining cluster formation and membership, each node only keeps track of its previous and next neighbor in the chain.

Communication in the chain is performed sequentially such that each node within a chain aggregates data from its neighbor until all the data are aggregated at one of the sensor nodes, i.e., chain leader. The chain leader controls the communication order by passing a token among the nodes.

■ **EXAMPLE 7.3**

An example of chain communication is shown in Figure 7.9. The chain leader in this example is node 2. Node 2 first passes the token to node 0 to initiate communication. Node 0 transmits its data to node 1, which aggregates these data with its own to create a packet of the same length. This packet is transmitted to node 2. Once node 2 receives the packet from node 1, it passes the token to the other end of the chain, i.e., node 6. Information from nodes 6, 5, 4, and 3 is also aggregated and sent to node 2 in the same fashion. Upon receiving the aggregated information in the chain, node 2 uses a single hop communication to transmit the data to the sink.

PEGASIS provides performance enhancement of 100–300% over LEACH in energy consumption. This improvement is due to the limited overhead in chain communication compared to cluster formation. However, PEGASIS results in significant delays since the data have to be sequentially transmitted in the chain and the chain leader waits until all the messages are received before communicating with the sink. Moreover, PEGASIS requires all the information in the chain to be aggregated into a single packet, which may cause inaccuracy in the information sent to the sink.

7.3.3   TEEN and APTEEN

The LEACH and PEGASIS protocols support applications where information from sensor nodes is periodically transmitted to the sink. Consequently, the information content from multiple nodes is decreased through aggregation techniques. However, these protocols may not be responsive to event-based applications, where information is generated only when certain events occur. The TEEN protocol aims to provide event-based delivery in the network [22]. As the name implies, multi-hop routes are generated according to a threshold related to sensory data, which is set by the application.

The TEEN protocol organizes the sensor nodes into multiple levels of hierarchy as shown in Figure 7.10. In this hierarchical architecture, data are transmitted first by sensor nodes to cluster heads, which collect, aggregate, and transmit these data to a higher level cluster head until the base station is reached. In order to evenly distribute the energy consumption, the cluster heads are periodically changed inside the cluster.

Based on this hierarchical network structure, TEEN provides event-based communication through two thresholds: *hard threshold (H_T)* and *soft threshold (S_T)*. The sensor nodes are programmed to respond to sensed-attribute changes, e.g., temperature or magnetic flux, by comparing the measured value to the hard threshold. If the hard threshold H_T is exceeded, the sensor node sends its observed data to the cluster head. Consequently, data are collected only if an event of interest occurs. It is clear that events can last for a long time, which requires frequent data transmission. In order to reduce the redundancy in this transmission, the soft threshold S_T is used. Whenever the hard threshold is exceeded, the sensor node also checks the soft threshold for consequent observations. If the difference between consecutive observations does not exceed the soft threshold, the sensor node does not transmit this information. This informs the cluster head that similar values are observed. New observations are only transmitted if the soft threshold is exceeded. Consequently, the hard threshold limits the transmissions to those observations that match the sink's interests (exceed the hard threshold) and the soft threshold further limits the transmitted information when there is no or little change in the sensed value.

Since TEEN is based on fixed threshold limits, it is not suitable for periodic reports required by some applications. In order to provide periodic information retrieval, the adaptive threshold-sensitive energy-efficient sensor network (APTEEN) protocol has been developed as an advancement of TEEN [23]. APTEEN provides a TDMA-based structure for information transmission in each cluster. Consequently, each node transmits its information periodically to the cluster head. Moreover, the hard and soft threshold values control when and how frequently to send the data. As a result, both event-based and monitoring applications can be served.

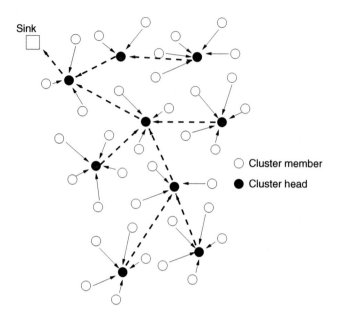

Figure 7.10   Hierarchical architecture of TEEN and APTEEN.

7.3.4   *Qualitative Evaluation*

Hierarchical protocols, in general, provide scalability in the network by limiting most of the communication inside the clusters. Consequently, the traffic generated in the network can be limited by the cluster heads. This enables large-scale networks to be deployed without traffic overload in certain parts of the network. Moreover, dynamic clustering mechanisms result in better energy efficiency compared to flat-topology protocols. In event-based WSNs, where the nodes are mostly dormant, most of the sensors can be put to sleep with the help of the cluster heads. Furthermore, most of the intelligence is passed to a small number of cluster heads and the rest of the nodes perform simple tasks. This improves the overall network lifetime.

Despite their advantages, cluster-based protocols significantly rely on cluster heads and face robustness issues such as failure of the cluster heads. Moreover, cluster formation requires additional signaling, which increases the overhead in case of frequent cluster head changes as will be discussed in more detail in Chapter 13. As a result, the tradeoff between increased energy consumption of the cluster heads and the overhead in cluster formation needs to be considered for efficient operation. Furthermore, intercluster communication is still a major challenge for many hierarchical routing protocols. Generally, cluster heads are assumed to directly communicate with the sink using higher transmit power. This limits the applicability of these protocols to large-scale networks, where single hop communication with the sink is infeasible. Hierarchical clustering mechanisms are generally required to provide multi-hop intercluster communication.

7.4   **Geographical Routing Protocols**

Location information is essential in many applications of WSNs. Since the main interest in such applications is the physical phenomenon in the environment, it is important to associate the sensor observations with the locations of the wireless sensor nodes. To provide location information to each

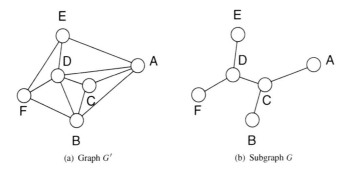

(a) Graph G' (b) Subgraph G

Figure 7.11   Subgraph formation in MECN.

sensor node, GPS devices can be integrated in the embedded board. Although GPS is not envisioned for all types of WSNs, it can still be used if stationary nodes with a large amount of energy are used. In addition, the localization protocols that will be explained in Chapter 12 provide accurate location information even if not all sensor nodes are equipped with GPS.

Since location information is necessary (and usually available) for most applications, it is natural to use this information for routing as well. Location-based protocols or *geographical routing protocols* exploit the location information of each node to provide efficient and scalable routing. Since the destination in WSNs is usually fixed, the location of the destination can be used to develop localized rules to route the data closer to that location. Geographical routing protocols provide several techniques to route that data to improve energy efficiency or latency.

7.4.1   MECN and SMECN

The MECN (Minimum Energy Communication Network) solution computes an energy-efficient sub-network when a communication network is given [27]. The resulting subnetwork minimizes the communication energy consumption between any pair of nodes in the network. The solution starts with a network graph G', where each vertex represents a node and the edges connect two nodes that can communicate with each other. From this network graph, G', MECN derives a subgraph, G, with the same number of vertices but a smaller number of edges. Consequently, the number of nodes that a particular node can communicate with is decreased. However, if two nodes are connected in G', they may be connected in G through other nodes. In other words, subgraph G contains all the vertices in graph G with fewer edges.

■ **EXAMPLE 7.4**

An example of subgraph formation is shown in Figure 7.11. In Figure 7.11(a), a portion of the graph G', which models the neighbors of a node A, is shown. In this graph, node A can directly communicate with node B. However, the power required to transmit between a pair of nodes increases as the nth power of the distance, where $n \geq 2$. Consequently, it might be much more energy efficient for node A to use node C as a relay to reach node B. Considering the energy efficiency in communication, then, the subgraph G can be computed as shown in Figure 7.11(b). As a result, the energy required to transmit data from node A to all its neighbors in subgraph G is less than the energy required to transmit to all its neighbors in graph G'.

The power required to transmit data between node A and B is modeled as $p(A, B) = td(A, B)^n$, where t is a constant, $d(A, B)$ is the distance between node A and B, and $n \geq 2$ is

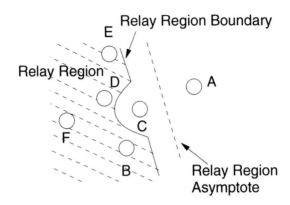

Figure 7.12   Relay region concept in MECN.

the path-loss exponent experienced by radio transmission. Also, the power needed to receive data is c. Since $p(A, B)$ increases by the nth power of the distance between node A and B, it may take less power to relay data than directly transmit data between node A and B. The path between node A (i.e., A_0) and B (i.e., A_k) is represented by r, where $r = (A_0, A_1, \ldots, A_k)$ in the subgraph $G = (V, E)$ is an ordered list of nodes such that the pair $(A_i, A_{i+1}) \in E$. Also, the length of r is k. The total power consumption between node A_0 and A_k is

$$C(r) = \sum_{i=0}^{k-1}(p(A_i, A_{i+1}) + c) \tag{7.2}$$

where $p(A_i, A_{i+1})$ is the power required to transmit data between node A_i and A_{i+1}, and c is the power required to receive data. A path r is a *minimum energy path* from A_0 to A_k if $C(r) \leq C(r')$ for all paths r' between node A_0 and A_k in G'. As a result, a subgraph G has the minimum energy property if for all $(A,B) \in V$, there exists a path r in G which is a minimum energy path in G' between node A and B.

The operation of MECN relies on a localized search for each node through a *relay region* concept as shown in Figure 7.12. According to the subgraph formalization as explained in Example 7.4, the relay region consists of nodes in a surrounding area where transmitting through those nodes is more energy efficient than direct transmission. The concept of relay region is explained in the following example.

■ **EXAMPLE 7.5**

The relay region between nodes A and C is shown in Figure 7.12. In this case, if node A wants to communicate with node B, which is in its relay region with node C, then it is more energy efficient to use node C as the intermediate node. The same is true for any node inside the relay region, e.g., node D or node E. The union of the relay regions with all the neighbors of a node constitutes the *enclosure* of that node.

Based on MECN, a new algorithm called small MECN (SMECN) has been proposed to improve the channel modeling in MECN [18]. Instead of a circular transmission range, where all the nodes in this range are connected all the time, SMECN also considers the obstacles between nodes, which results in a graph with fewer edges than that of MECN. SMECN also follows the minimum energy path, which MECN uses to construct the subnetwork.

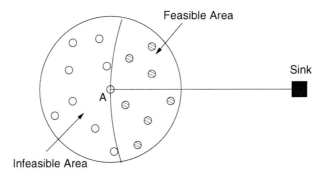

Figure 7.13   Geographical routing.

7.4.2   *Geographical Forwarding Schemes for Lossy Links*

As explained above, the wireless channel is prone to failures and packet errors. Hence, the network graph formed by MECN or SMECN can change frequently, which necessitates frequent message exchanges between nodes to reconstruct the graph. Moreover, instead of constructing a network graph, which is costly due to the overhead, it may be much more efficient to define localized rules, which all nodes in the network follow to forward packets. In addition, energy consumption may not be the only important metric for establishing the route.

Based on the information used, the localized forwarding schemes can be classified mainly into two classes: namely, distance-based forwarding, where only the distance information is used; and reception-based forwarding, where the state of the channel is also incorporated into route decisions. Next, we explain several variations of these techniques [30, 37]. The discussion is based on the following example.

■ **EXAMPLE 7.6**

We consider the scenario shown in Figure 7.13, where the source node A tries to forward its packet to destination "Sink" through one of the neighbors in its transmission range, which is represented by the circle around node A. In any geographical routing algorithm, the area around node A defined by its transmission range is divided into two regions as shown in Figure 7.13. The area where the neighbors that are farther away from the destination "Sink" than the source node A reside is denoted as the *infeasible region*. Consequently, the nodes in this area are denoted as *infeasible nodes*. Similarly, the area where the neighbors that are closer to the destination than the source node reside is the *feasible region* and the neighbors in this area are *feasible nodes*.

Geographical routing algorithms aim to select one of the feasible nodes as the next hop to advance the packet toward the destination. As a result, routing loops are prevented. The selection of the next hop inside the feasible region depends on the forwarding algorithm as explained next.

Greedy forwarding

The simplest geographical routing protocol is the greedy forwarding protocol, which is explained next.

■ **EXAMPLE 7.7**

According to the scenario in Example 7.6, greedy forwarding is shown in Figure 7.14(a). Node A selects the feasible node that is closest to the destination. There exists several metrics that

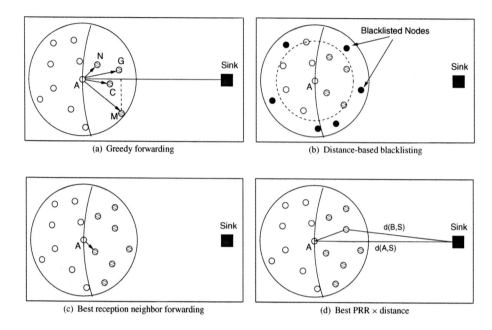

Figure 7.14   Geographical forwarding techniques for lossy links.

define the *closest* node to the destination as shown in Figure 7.14(a) [24, 25]. Node M is selected according to the *most forward within radius* (MFR) metric, where the node with the largest advancement on the line connecting the source and destination is chosen. The *nearest forward progress* metric chooses node N, which is the closest node to the source that is closer to the destination. Node G is selected according to the *greedy routing scheme* (GRS), which selects the node that is closest to the destination. Finally, the *compass* metric chooses node C, which is closest to the straight line between the source node and the destination.

Each greedy forwarding metric has certain advantages for establishing the overall route. In an ideal setting, where the transmission range of a node can be defined as a circle around that node, these techniques provide fast delivery since a packet traverses through fewer nodes. However, in addition to the location of the nodes, the channel quality between two nodes is also important for wireless multi-hop routing. As an example, although geographically selecting the closest neighbor to the destination may lead to fewer hops, since the channel quality degrades as the distance increases, choosing the farthest node usually results in retransmissions.

Distance-Based Blacklisting

In order to alleviate the effects of distance between nodes, some of the nodes that are at the boundary of the transmission range of a node can be blacklisted as shown in Figure 7.14(b). As an example, if a node's transmission range is 100 m with a blacklisting threshold of 20%, the source node selects nodes that are only closer than 80 m.

Reception-Based Blacklisting

The distance between two nodes does not directly relate to the channel quality. Hence, distance-based blacklisting does not always prevent the selection of nodes with lower channel quality. Instead, the absolute reception-based blacklisting protocol blacklists neighbors that have a packet reception rate below some threshold. For this purpose, each node keeps track of its packet reception rate (PRR) with each of its neighbors. This information is then exchanged with these neighbors. As a result, when a node has a packet to send, it chooses the next hop among the feasible nodes with a PRR higher than the threshold.

Absolute reception-based blacklisting results may lead to disconnections in the network if none of the neighbors of a node satisfy the threshold constraint. Consequently, relative reception-based blacklisting is used, which orders the neighbors of a node according to the PRR values. Then, according to a blacklisting threshold value, the percentage of nodes with the worst PRR value is blacklisted. This technique provides adaptive operation according to the neighbors of each node. A special case for the absolute reception-based blacklisting is the *best reception neighbor* algorithm, where the next hop is the neighbor with the highest PRR, as shown in Figure 7.14(c). By controlling the threshold value, the absolute reception-based blacklisting algorithm can operate between two extremes: greedy forwarding and best reception neighbor.

Best PRR × distance

Forwarding mechanisms that rely only on the PRR suffer from increased latency. As an example, as shown in Figure 7.14(c), the best reception neighbor algorithm tends to select the node closest to the source node. As a result, a packet traverses through more hops to reach the destination. The PRR×distance algorithm aims to exploit the tradeoff between PRR and distance.

■ **EXAMPLE 7.8**

An illustration of the algorithm is shown in Figure 7.14(d). The next hop, B, is selected such that the product of PRR and the distance improvement achieved by forwarding to a node is maximized. The distance improvement is found as follows:

$$1 - \frac{d(B, S)}{d(A, S)} \tag{7.3}$$

where $d(B, S)$ is the distance between the neighbor, node B, and the destination "Sink" and $d(A, S)$ is the distance between the source node, A, and the destination, "Sink."

A comparison of these schemes shows that the PRR × distance algorithm provides the highest delivery rate followed by the best reception, relative reception-based, absolute reception-based, distance-based, and greedy forwarding ones. Consequently, considering channel quality in geographical routing decisions improves the performance compared to algorithms which only consider geographical location. In terms of energy efficiency, the absolute reception-based algorithm outperforms the other algorithms followed by the PRR × distance, relative reception-based, best reception, distance-based, and greedy forwarding ones. As a result, considering both the channel quality and advances in geographical routing protocols provides the most effective operation.

7.4.3   PRADA

Geographical routing protocols require neighborhood information to select the next hop according to distance, PRR, etc. However, collecting this neighborhood information is costly. Moreover, neighborhood information provides only a limited view of the network upon which the node has to make decisions.

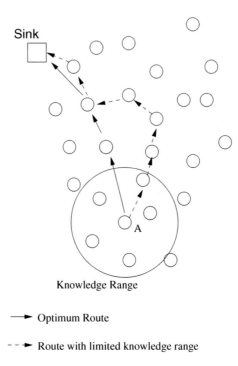

Knowledge Range

⟶ Optimum Route

- - ➤ Route with limited knowledge range

Figure 7.15   The impact of knowledge range on geographical routing.

It can be argued that if a node has global knowledge of the network, the optimum path to the destination
can be found. Accordingly, the next hop in this optimum route can be selected. However, in practice,
global network knowledge is impossible to gather in WSNs due to the high density and associated high
cost. Hence, geographical routing protocols aim to provide decision-making mechanisms using a limited
view of the network, i.e., information about neighbors of a node. In Figure 7.15, the optimum route
between a source node A and destination S is shown with a solid line when the global knowledge of the
network is known. However, if each node in the network only knows the locations of their immediate
one-hop neighbors, the resulting route is as shown by the dashed line. It is clear that with limited topology
knowledge, the established route deviates from the optimum route. However, the difference decreases
when the topology knowledge of a node increases since more informed decisions can be made. As
a result, the accuracy of the forwarding decisions is related to the topology knowledge of the node.
However, to improve the topology knowledge, a node has to consume more energy.

   PRADA, the probe-based distributed protocol for knowledge range adjustment [24, 25], exploits the
tradeoff between the cost associated with larger topology knowledge and the accuracy of the forwarding
decision. In order to obtain larger topology knowledge, a node has to use a larger transmit power
and increase its transmission range. As a result, the cost of obtaining topology knowledge increases.
However, with increased topology knowledge, routes closer to the optimum can be created which can
decrease the energy consumption for data delivery.

   PRADA is based on a centralized forwarding scheme called *partial topology knowledge forwarding*
(PTKF), which aims to minimize the energy consumption for data communication as well as the cost
for topology information. Accordingly, each node constructs a route based on a weighted shortest path
algorithm with the link cost given as the energy consumption. According to this route, the next hop in it
is selected and this hop calculates the route according to *its* topology knowledge. PTKF aims to find the

optimum set of knowledge range values, **R**, of each node to minimize the overall energy consumption. In other words, the objective is to minimize

$$C^{TOT} = \sum_{i \in V} (C_i^{COM} + C_i^{INF}) \tag{7.4}$$

where V is the set of nodes in the network, C_i^{COM} is the communication cost for node i, and C_i^{INF} is the topology information cost for node i. Note that the values of C_i^{COM} and C_i^{INF} depend on the knowledge range of each node and PTKF finds the optimum set of these knowledge ranges.

PRADA is a distributed version of PTKF, where each node adjusts its knowledge range according to the feedback information it receives from neighbor nodes. PRADA first selects a random knowledge range (KR), $r_{current}$, and constructs a route to the sink according to its current KR. Then, each node periodically selects a certain KR, r_{probe}, to be probed and calculates

$$C_i^{TOT}(r_{probe}) = C_i^{INF}(r_{probe}) + \sum_{p \in P_i} c_i^p(r_{probe}) \tag{7.5}$$

where $c_i^p(r_{probe})$ is the cost of communication for each connection p with the KR r_{probe}. If $C_i^{TOT}(r_{probe}) < C_i^{TOT}(r_{current})$, then the knowledge range is updated as $r_{current} = r_{probe}$.

PRADA converges quickly and, according to the simulations, achieves a performance very close to the global optimum.

7.4.4   Qualitative Evaluation

Geographical routing protocols exploit local position information for routing decisions. Instead of routing tables that are explicitly constructed, neighborhood information is implicitly inferred from the physical placement of nodes. This results in scalable routing protocols. Moreover, geographical protocols have low complexity since the next hops can be selected based on local information.

On the other hand, the performance of geographical routing depends on accurate knowledge of the location. The error in location detection can cause an error in routing. For example, sensor nodes are prone to failure and may be deployed in a hostile environment. If the GPS or location device is damaged, sensor nodes that depend on the device are rendered useless. In addition, the monetary cost for the GPS may be expensive for some applications. Also, the power consumption and size of the GPS may not be appropriate if sensor nodes are operated by batteries and deployed in their thousands.

7.5   QoS-Based Protocols

Many of the routing protocols explained above focus only on energy consumption in the network. Hence, route generation is performed to minimize energy consumption in the network. While energy consumption is one of the most important performance metrics in WSNs, it is not the only one. In particular, the development of novel applications including multimedia communication necessitates other performance metrics such as throughput, delay, and jitter also being considered in protocol design. Moreover, in some cases delay may be even more important than energy consumption. Hence, the quality of service (QoS) requirements of these applications needs to be guaranteed in addition to the energy efficiency in WSNs.

Some of the routing protocols [1, 3, 5, 6, 10, 15, 16, 18, 19, 33] aim to minimize the energy consumption of the network by using the remaining energy of the sensor nodes as a metric of optimization.

7.5.1 SAR

The SAR (Sequential Assignment Routing) protocol provides a table-driven multi-path approach and was one of the first routing protocols developed for WSNs which considers QoS requirements [33]. The main goal of SAR is to create multiple trees originating from a root node which is one of the single hop neighbors of the sink. Each tree grows outward from the sink while avoiding nodes with very low QoS (i.e., low throughput/high delay) and energy reserves. Consequently, multiple paths that connect any node in the network to the sink can be created.

Each node specifies two parameters for each path to the sink:

- **Energy resources:** The energy resource of a path is estimated as the maximum number of packets that can be sent by the node, if the node has exclusive use of the path.

- **Additive QoS metric:** Each path is also associated with an additive QoS metric, which is related to the energy and delay at each link. A high QoS metric means low QoS.

Using the multiple paths, a node can select the routing path according to the QoS of each path and the packet priority. As explained above, the QoS of a path is determined as an additive metric, which is a function of energy and delay. As the QoS metric increases, the QoS of the route deceases. Whenever a node has a packet to send, it calculates a *weighted QoS metric* for the packet, which is a product of the QoS metric of the path and the priority level of the packet. As a result, paths with higher QoS are used for higher priority packets.

7.5.2 Minimum Cost Path Forwarding

The minimum cost path forwarding protocol combines the delay, throughput, and energy consumption characteristics to establish routes between nodes in the network [36]. The protocol assigns a cost function, which captures the delay, throughput, and energy consumption of the node, to each link. According to the cost function, the minimum cost forwarding algorithm specifies a *cost field* at each node. As a result, packets flow through the cost field, which specifies the next hop with the lowest cost.

The minimum cost path algorithm has two phases: cost field establishment and cost path forwarding. The cost field establishment phase aims to determine the minimum cost between any node and the sink. The sink broadcasts an advertisement (ADV) message with an initial cost of 0. The ADV message is then forwarded by updating the cost. Each node j, which receives an ADV message from node i, calculates its cost

$$L_i + C_{j,i} \qquad\qquad (7.6)$$

where L_i is the cost of node i ($= 0$ for the sink) and $C_{j,i}$ is the cost from node j to node i. Each node then sets up a backoff timer proportional to its cost to node i, $C_{j,i}$, and broadcasts the ADV message. The backoff time helps the node to update its cost to the sink by selecting the node with the minimum cost to the sink.

■ **EXAMPLE 7.9**

The operation of the cost field establishment phase is shown in Figure 7.16 for nodes A, B, and C. Assume that node A has a minimum cost of $L_A = 0.5$ to the sink and the link costs are as follows: $L_{B,A} = 1.5$, $L_{C,A} = 4$, and $L_{C,B} = 1$, as shown in Figure 7.16(a). Node A first broadcasts an ADV message as shown in Figure 7.16(b) and, upon receiving the ADV message, node B updates its cost from inf to $L_A + L_{B,A} = 2$ and sets up its backoff timer to $\gamma L_{B,A}$, where γ is a constant. Node C also updates its cost to $L_A + L_{C,A} = 4$, 5 and sets its backoff timer to $\gamma L_{C,A}$. Since node B's backoff timer expires first, node B broadcasts an ADV with cost 2 as shown in Figure 7.16(c). Upon receiving the ADV, node C determines that its cost to the sink is lower through node B than through A ($L_B + L_{C,B} = 3 < 4$, 5) and hence updates its minimum

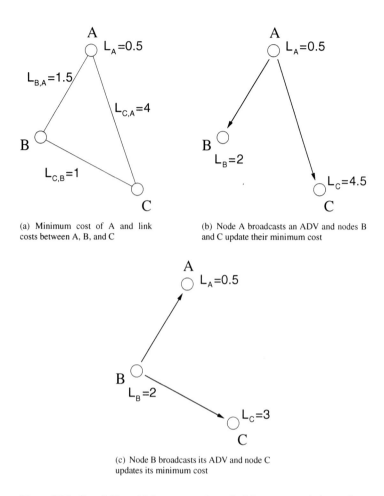

(a) Minimum cost of A and link costs between A, B, and C

(b) Node A broadcasts an ADV and nodes B and C update their minimum cost

(c) Node B broadcasts its ADV and node C updates its minimum cost

Figure 7.16   Cost field establishment procedure of minimum cost path forwarding.

cost with an updated backoff timer $\gamma L_{C,B}$, after which it also broadcasts its ADV message. As a result of the backoff-based cost field establishment phase, each node determines its minimum cost to the sink by broadcasting a single ADV message.

In the second phase, the source node broadcasts the data message to its neighboring nodes. The message is routed through the cost field of the nodes. When a source node broadcasts a message, a budget for the message is also included. The budget is equal to the minimum cost of the source to the sink. As an example, assume that node C in Figure 7.16(c) has a packet to send to the sink. It broadcasts its message with a budget of 3. When node A receives this message, it compares its minimum cost (0.5) plus the link cost to node C (4) to the message budget and determines that the total cost exceeds the budget. Consequently, it does not forward the packet. On the other hand, node B forwards the packet since it is in the optimum route from node C to the sink.

Based on the cost field concept, the messages can be routes without specific route information, neighborhood information, or node IDs. Each packet is broadcast without specifying the next hop and the next hop is selected as a result of the budget and (minimum cost + link cost) comparison. If the cost is not sufficient to reach the sink, the message is dropped; otherwise, the message is forwarded until it reaches the sink. As a result, the minimum cost path forwarding mechanism delivers the messages according to the minimum costs at each node.

7.5.3 SPEED

The QoS requirements in traditional networks are usually related to end-to-end delay and throughput. However, in WSNs, where the relative location of a node significantly affects the end-to-end performance, distance is another factor for guaranteeing QoS requirements. The SPEED protocol exploits this fact and aims to provide a packet guaranteed advancement at a given time (i.e., speed) [10]. Consequently, the end-to-end delay experienced by the packets is proportional to the distance between the source and the destination.

The SPEED protocol has several components to provide speed guarantees to the packets in the network. The *neighbor beacon exchange* protocol is periodically run to exchange location information between neighbors. As a result, each node constructs a neighbor table and stores the information about its neighbors with the following fields: NeighborID, Position, SendToDelay, and ExpireTime, where SendToDelay is the estimated delay to the neighbor node and ExpireTime controls the expiration time when the particular entry is deleted from the table if no updates are received.

As explained above, in addition to the location, each node stores the estimated delay to its neighbors in its neighbor table. Delay estimation is performed at the sender whenever a packet is sent to the particular neighbor. The sender stores the time it takes between placing the packet in its queue and receiving the ACK from its neighbors. Moreover, the receiver neighbor communicates the duration of processing for the ACK using the ACK packet, which is subtracted from the delay estimate. In case multiple packets are sent to the same neighbor, the moving average of the delay estimates is stored in the neighbor table.

The SendToDelay value stored for each neighbor is used for forwarding using the stateless non-deterministic geographic forwarding (SNGF) algorithm of SPEED. SNGF aims to forward the packets to neighbors that can provide a minimum delivery speed of $S_{setpoint}$.

■ **EXAMPLE 7.10**

The operation of SNGF is illustrated in Figure 7.17, where a sender node A aims to forward its packet toward a destination D. Firstly, nodes that are inside the transmission range of node A and closer to the destination D than node A are selected as forwarding candidates. In Figure 7.17, these nodes are E, F, and G. Then, the *relay speed* between node A and a node $j \in \{E, F, G\}$ for destination D is calculated for each forwarding candidate as follows:

$$Speed_A^j(D) = \frac{d_{A,D} - d_{j,D}}{HopDelay_{A,j}} \ \forall j \in \{E, F, G\} \tag{7.7}$$

where $d_{A,D}$ and $d_{j,D}$ are distances to destination D from nodes A and j, respectively, and $HopDelay_{A,j}$ is the estimated hop delay between nodes A and j.

SNGF chooses the next hop among the forwarding candidates with estimated speed higher than $S_{setpoint}$. As a result, forwarded packets are guaranteed to reach a minimum speed. However, if no node is found to satisfy this requirement, the packet is randomly dropped according to the neighborhood feedback loop (NFL).

The NFL component of SPEED controls the packet drop procedure in case no neighbor satisfying the minimum speed exists. For this purpose, the miss ratio of each neighbor, which is the rate at which the neighbor node does not satisfy the speed requirement, is calculated. Based on the miss ratio of the neighbor, the NFL determines either to drop or forward the packet.

In some cases, the packets can be routed toward *hot spots*, where there exists a high contention. To prevent further packets from being forwarded to the same region, SPEED also employs a backpressure

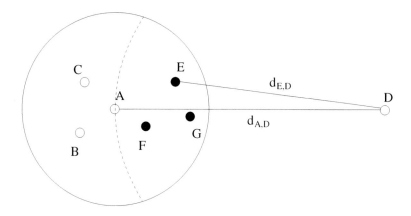

Figure 7.17   SNGF operation.

mechanism. Nodes which experience high miss rates send backpressure beacons to the upstream nodes. The backpressure beacons are used to remove these neighbors from the neighbor list and, as a result, packets are rerouted around the hot spots to relieve congestion.

SPEED is shown to provide lower delay and miss rates compared to the DSR and AODV protocols, which are developed for ad hoc networks. Moreover, SPEED is energy efficient because of the small overhead for establishing the route. One of the drawbacks of SPEED, however, is that only a single speed guarantee can be supported in the network. This prevents heterogeneous applications, where different classes of traffic reside in the network. SPEED is enhanced to provide multiple speed guarantees for different traffic types through the MMSPEED protocol, which is described in Chapter 15.

7.5.4   Qualitative Evaluation

QoS-based routing protocols consider additional metrics in addition to energy consumption for constructing routes. This provides additional capabilities to WSNs, where more sophisticated applications can be developed. However, providing additional guarantees increases the cost in terms of energy consumption and, hence, network lifetime. The tradeoff between these additional capabilities and their costs should be carefully tailored to the requirements of the applications.

References

[1] K. Akkaya and M. Younis. A survey on routing protocols for wireless sensor networks. *Ad Hoc Networks*, 3(3):325–349, May 2005.

[2] J. N. Al-Karaki and A. E. Kamal. Routing techniques in wireless sensor networks: a survey. *IEEE Wireless Communications*, 11(6):6–28, December 2004.

[3] J. Aslam, Q. Li, and D. Rus. Three power-aware routing algorithms for sensor networks. *Wireless Communications and Mobile Computing*, 3(2):187–208, 2003.

[4] D. Braginsky and D. Estrin. Rumor routing algorithm for sensor networks. In *Proceedings of the First Workshop on Sensor Networks and Applications (WSNA'02)*, pp. 22–31, Atlanta, GA, USA, September 2002.

[5] J. H. Chang and L. Tassiulas. Maximum lifetime routing in wireless sensor networks. In *Proceedings of the Advanced Telecommunications and Information Distribution Research Program (ATIRP'2000)*, pp. 609–619, College Park, MD, USA, March 2000.

[6] M. Chu, H. Haussecker, and F. Zhao. Scalable information-driven sensor querying and routing for ad hoc hetero-geneous sensor networks. *International Journal of High Performance Computing Applications*, 16(3):293–313, 2002.

[7] J. Elson and D. Estrin. Random, ephemeral transaction identifiers in dynamic sensor networks. In *Proceedings of the 21st International Conference on Distributed Computing Systems*, pp. 459–468, Mesa, AZ, USA, April 2001.

[8] D. Estrin, L. Girod, G. Pottie, and M. Srivastava. Instrumenting the world with wireless sensor networks. In *Proceedings of ICASSP'01*, volume 4, pp. 2033–2036, Salt Lake City, UT, USA, May 2001.

[9] D. Estrin, R. Govindan, J. Heidemann, and S. Kumar. Next century challenges: scalable coordination in sensor networks. In *Proceedings of ACM/IEEE MobiCom'99*, pp. 263–270, Seattle, WA, USA, August 1999.

[10] T. He, J. A. Stankovic, C. Lu, and T. Abdelzaher. SPEED: a stateless protocol for real-time communication in sensor networks. In *Proceedings of the 23rd International Conference on Distributed Computing Systems*, pp. 46–55, Providence, RI, USA, May 2003.

[11] S. Hedetniemi and A. Liestman. A survey of gossiping and broadcasting in communication networks. *Networks*, 18(4):319–349, 1988.

[12] W. R. Heinzelman, A. Chandrakasan, and H. Balakrishnan. Energy-efficient communication protocol for wire-less microsensor networks. In *Proceedings of the IEEE Hawaii International Conference on System Sciences*, pp. 1–10, Maui, HI, USA, January 2000.

[13] W. R. Heinzelman, J. Kulik, and H. Balakrishnan. Adaptive protocols for information dissemination in wireless sensor networks. In *Proceedings of MobiCom'99*, pp. 174–185, Seattle, WA, USA, August 1999.

[14] C. Intanagonwiwat, R. Govindan, and D. Estrin. Directed diffusion: a scalable and robust communication paradigm for sensor networks. In *Proceedings of MobiCom'00*, pp. 56–67, Boston, MA, USA, August 2000.

[15] K. Kalpakis, K. Dasgupta, and P. Namjoshi. Maximum lifetime data gathering and aggregation in wireless sensor networks. Technical report, University of Maryland, 2002.

[16] R. Kannan, L. Ray, R. Kalidindi, and S. S. Iyengar. Max-min length-energy-constrained routing in wireless sensor networks. In *Proceedings of the 1st European Workshop on Wireless Sensor Networks*, pp. 234–249, Berlin, Germany, January 2004.

[17] G. Khanna, S. Bagchi, and Y. Wu. Fault tolerant energy aware data dissemination protocol in sensor networks. In *Proceedings of the International Conference on Dependable Systems and Networks*, pp. 795–804, Florence, Italy, June 2004.

[18] L. Li and J. Y. Halpern. Minimum energy mobile wireless networks revisited. In *Proceedings of the IEEE International Conference on Communications (ICC'01)*, Helsinki, Finland, June 2001.

[19] Q. Li, J. Aslam, and D. Rus. Distributed energy-conserving routing protocols. In *Proceedings of the 36th Hawaii International Conference on System Science*, January 2003.

[20] S. Lindsey and C. S. Raghavendra. PEGASIS: power efficient gathering in sensor information systems. In *Proceedings of the IEEE Aerospace Conference*, Big Sky, MT, USA, March 2002.

[21] S. Lindsey, C. S. Raghavendra, and K. Sivalingam. Data gathering in sensor networks using the energy*delay metric. In *Proceedings of the IPDPS Workshop on Issues in Wireless Networks and Mobile Computing*, San Francisco, USA, April 2001.

[22] A. Manjeshwar and D. P. Agrawal. TEEN: a protocol for enhanced efficiency in wireless sensor networks. In *Proceedings of the 1st International Workshop on Parallel and Distributed Computing Issues in Wireless Networks and Mobile Computing*, San Francisco, USA, April 2001.

[23] A. Manjeshwar and D. P. Agrawal. APTEEN: a hybrid protocol for efficient routing and comprehensive information retrieval in wireless sensor networks. In *Proceedings of the 2nd International Workshop on Parallel and Distributed Computing Issues in Wireless Networks and Mobile Computing*, Ft. Lauderdale, FL, USA, April 2002.

[24] T. Melodia, D. Pompili, and I. F. Akyildiz. Optimal local topology knowledge for energy efficient geographical routing in sensor networks. In *Proceedings of IEEE INFOCOM 2004*, Hong Kong, China, March 2004.

[25] T. Melodia, D. Pompili, and I. F. Akyildiz. On the interdependence of distributed topology control and geographical routing in ad hoc and sensor networks. *Journal of Selected Areas in Communications*, 23(3):520–532, March 2005.

[26] J. Mirkovic, G. P. Venkataramani, S. Lu, and L. Zhang. A self-organizing approach to data forwarding in large-scale sensor networks. In *Proceedings of the IEEE International Conference on Communications (ICC'01)*, Helsinki, Finland, June 2001.

[27] V. Rodoplu and T. H. Meng. Minimum energy mobile wireless networks. *IEEE Journal of Selected Areas in Communications*, 17(8):1333–1344, 1999.

[28] N. Sadagopan, B. Krishnamachari, and A. Hemly. The ACQUIRE mechanism for efficient querying in sensor networks. In *Proceedings of the First International Workshop on Sensor Network Protocol and Applications*, Anchorage, AK, USA, May 2003.

[29] C. Schurgers and M. B. Srivastava. Energy efficient routing in wireless sensor networks. In *Proceedings of the MILCOM on Communications for Network-Centric Operations: Creating the Information Force*, McLean, VA, USA, 2001.

[30] K. Seada, M. Zuniga, A. Helmy, and B. Krishnamachari. Energy-efficient forwarding strategies for geographic routing in lossy wireless sensor networks. In *Proceedings of SenSys'04*, pp. 108–121, Baltimore, MD, USA, 2004.

[31] R. Shah and J. Rabaey. Energy aware routing for low energy ad hoc sensor networks. In *Proceedings of the IEEE Wireless Communications and Networking Conference (WCNC)*, Orlando, FL, USA, March 2002.

[32] C. Shen, C. Srisathapornphat, and C. Jaikaeo. Sensor information networking architecture and applications. *IEEE Personal Communications*, 8(4):52–59, August, 2001.

[33] K. Sohrabi, J. Gao, V. Ailawadhi, and G.J. Pottie. Protocols for self-organization of a wireless sensor network. *IEEE Personal Communications*, 7(5):16–27, October, 2000.

[34] T. Voigt, H. Ritter, and J. Schiller. Solar-aware routing in wireless sensor networks. In *Proceedings of Personal Wireless Communication 2003*, Venice, Italy, September 2003.

[35] Y. Yao and J. Gehrke. The Cougar approach to in-network query processing in sensor networks. *SIGMOD Record*, 31(3):9–18, September 2002.

[36] W. Ye, J. Heidemann, and D. Estrin. An energy-efficient MAC protocol for wireless sensor networks. In *Proceedings of IEEE INFOCOM'02*, volume 3, pp. 1567–1576, New York, USA, June 2002.

[37] M. Z. Zamalloa, K. Seada, B. Krishnamachari, and A. Helmy. Efficient geographic routing over lossy links in wireless sensor networks. *ACM Transactions on Sensor Networks*, 4(3):1–33, 2008.

8

Transport Layer

The success and efficiency of WSNs directly depend on reliable communication between the sensor
nodes and the sink. In a multi-hop, multi-sensor environment, to accomplish this a reliable transport
mechanism is imperative in addition to robust modulation and media access, link error control, and
fault-tolerant routing. In this chapter, the functionalities and design of transport layer solutions for WSNs
are described.

The main objectives of a transport layer protocol for WSNs are as follows:

- **Congestion control:** Packet losses due to congestion can impair reliability at the sink even
  when enough information is sent out by the sources. Hence, congestion control is an important
  component of the transport layer to achieve the required reliability. Furthermore, congestion
  control not only increases the network efficiency, but also helps conserve scarce sensor resources.
- **Reliable transport:** Based on the application requirements, the extracted event features should
  be reliably transferred to the sink. Similarly, the programming/retasking data for sensor operation,
  command, and queries should be reliably delivered to the target sensor nodes to assure the proper
  functioning of the WSNs.
- **(De)multiplexing:** Different applications can be served on sensor nodes through the same
  network. The transport layer should bridge the application and network layers by using
  multiplexing and demultiplexing.

There exist several transport layer solutions developed for conventional wireless networks [3].
Although these solutions may be relevant, they are not suitable for WSNs, In particular, to satisfy the
above objectives and to accommodate the unique characteristics of WSNs, significant modifications
to the existing transport layer protocols are needed. The majority of existing solutions focuses on
reliable data transport following the end-to-end TCP semantics and addresses the challenges posed by
wireless link errors and mobility [4]. Moreover, the notion of end-to-end reliability, which is based on
acknowledgments and end-to-end retransmissions, imposes significant overhead for the implementation
of these solutions in WSNs. The inherent correlation in the data flows generated by the sensor nodes
makes these mechanisms for strict end-to-end reliability significantly energy inefficient. Furthermore,
all these protocols bring considerable memory requirements to buffer transmitted packets until they
are ACKed by the receiver. In contrast, sensor nodes have limited buffering space (<4 kB in MICA
motes [8]) and processing capabilities.

8.1 Challenges for Transport Layer

The energy, processing, and hardware limitations of the wireless sensor nodes bring further challenges
for the design of the transport layer protocol. The development of transport layer protocols is a
considerable effort because the limitations of the sensor nodes and the specific application requirements
primarily determine the design principles. In this respect, the main objectives and challenges of the

transport layer are stated next. Moreover, the essential features to address the unique questions posed by the characteristics of the WSNs are also indicated.

8.1.1   End-to-End Measures

Conventional transport layer solutions such as TCP adopt end-to-end retransmission-based error control and window-based additive-increase multiplicative-decrease (AIMD) congestion control mechanisms. These mechanisms provide end-to-end and point-to-point reliability and congestion control solutions. More specifically, the packet losses and congestion mitigation are performed through communication between a source and a destination without any involvement from the intermediate parties. Essentially, the transport control mechanisms reside only on the source and destination. Furthermore, each flow is considered independently to provide a point-to-point communication solution.

These end-to-end control mechanisms used for conventional transport layer protocols usually lead to resource wastage in WSNs, where collective information from a group of sensors is much more important than the individual information from each sensor node. Therefore, conventional end-to-end, point-to-point transport layer techniques may lead to waste of scarce wireless sensor resources. Instead, local measures for reliability and congestion control are usually exploited to improve the energy efficiency of the transport layer protocols. Similarly, reliability of the collective information from a group of sensors is controlled instead of the reliability of information from each individual sensor node.

8.1.2   Application-Dependent Operation

Another difference in the operation of WSNs is that they are deployed with a specific sensing application objective. For example, sensor nodes can be used within a certain deployment scenario to perform continuous sensing of a specific phenomenon such as temperature monitoring. On the other hand, another application may require multiple sensors to carry out event detection and event identification tasks. Similarly, another application may require location sensing and local control of actuators for a wide range of application areas. Consequently, while for a monitoring application reliability is the most important metric, for event detection applications timeliness is crucial. Thus the focus of the transport layer solutions should be tailored to the application. Moreover, the application can be developed for a wide range of purposes in the military, environment, health, space exploration, and disaster relief areas. The importance of these metrics can also vary according to the application area. Consequently, the specific objective of WSNs influences the design requirements of the transport layer protocols.

8.1.3   Energy Consumption

Energy efficiency is the most important concern in the design of WSNs, and also affects the design of transport layer solutions. Scarce energy resources affect the design of transport layer solutions. Consequently, the transport layer functionalities should be energy aware, i.e., the error and congestion control objectives must be achieved with the minimum possible energy expenditure. For instance, if reliability levels at the sink are found to be in excess of that required for event detection, the source nodes can conserve energy by reducing the amount of information sent out or temporarily powering down. Similarly, end-to-end measures, which are successfully used for conventional networking for reliability, usually require significant energy consumption in a multi-hop network. Hence, these solutions may not be scalable for WSNs. Consequently, a transport layer protocol can be designed such that the reliability level can be traded off for decreased energy consumption through local reliability measures.

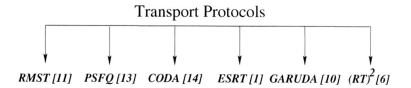

Transport Protocols

RMST [11]   PSFQ [13]   CODA [14]   ESRT [1]   GARUDA [10]   (RT)^2 [6]

Figure 8.1   Overview of transport protocols.

8.1.4   Biased Implementation

WSNs are usually deployed with a large number of resource-constrained sensor nodes that are connected to a resource-rich sink. The limited processing power and memory capacity of the sensor nodes prevent sophisticated algorithms from being run locally. Hence, the transport layer algorithms should be designed such that most of the functionalities are performed at the sink with minimum functionalities required at the sensor nodes. More specifically, the intelligence should be passed to the sink instead of the sensors. This helps to conserve limited sensor resources and shifts the burden to the high-powered sink.

Moreover, the traffic in the WSNs exhibits significantly different characteristics depending on the flow direction. While the flow in the sensors-to-sink direction may require timely delivery with loss-tolerant operation, the sink-to-sensors direction usually requires a high delivery ratio. Consequently, transport protocols should be designed also by considering these biases in traffic.

8.1.5   Constrained Routing/Addressing

As explained in Chapter 7, wireless sensor nodes may not be assigned unique addresses. Therefore, unlike protocols such as TCP, in the design of transport layer protocols for WSNs the existence of an end-to-end global addressing should not be assumed. It is more likely to have attribute-based naming and data-centric routing, which call for different transport layer approaches.

Several transport layer protocols have been developed for WSNs to address these challenges [15]. An overview of the transport layer solutions that will be discussed next is shown in Figure 8.1. Next, we describe the following protocols: reliable multi-segment transport (RMST), pump slowly, fetch quickly (PSFQ), congestion detection and avoidance (CODA), event-to-sink reliability (ESRT), GARUDA, and real-time and reliable transport (RT)^2.

8.2   Reliable Multi-Segment Transport (RMST) Protocol

The RMST protocol is one of the first transport layer protocols developed for WSNs [11]. The main goal of RMST is to provide end-to-end reliability. Accordingly, RMST is built on top of the directed diffusion protocol [7], which is explained in Chapter 7, and uses some of the functionalities of this protocol. More specifically, RMST is designed as a filter that could be attached to the directed diffusion protocol.

RMST provides two of the three functionalities required for a transport layer protocol: reliable transport and multiplexing/demultiplexing. Multiplexing and demultiplexing are carried at the source nodes and the sink, respectively. In the meantime, RMST provides mechanisms to handle errors throughout the routes in the network. To this end, RMST utilizes in-network caching and provides guaranteed delivery of the data packets generated by the event flows.

RMST relies on the directed diffusion routing mechanism for a certain path between a source and destination. Hence, an implicit assumption is that the packets of a flow follow the same path unless there is a node failure. In case of node failures, directed diffusion is assumed to reroute packets. Based on this assumption, RMST has two modes of operation:

- **Non-caching mode:** This mode of operation is very similar to conventional transport layer protocols, where only the source and destination play a role in providing reliability. Consequently, the packet losses are detected at the sink and requested from the source node in an end-to-end fashion through a NACK packet. The advantage of this mode is that it requires no involvement – and, hence, additional processing, storage, and energy consumption – from the intermediate nodes in the multi-hop network.

- **Caching mode:** In this mode, the intermediate nodes on the reinforced path cache the transmitted packets to decrease the overhead in end-to-end retransmissions.

In RMST, each packet of a flow is labeled by a unique sequence number. Accordingly, the packet errors are detected whenever there is a hole in the sequence numbers received. In case of packet errors, nodes request retransmission by sending a NACK packet toward the reverse route from the sink to the sensor, i.e., the *reverse path* as explained in the next examples for the non-caching and caching modes.

■ **EXAMPLE 8.1**

Error recovery in non-caching mode is illustrated in Figure 8.2, where a sensor node is trying to transmit a series of packets to the sink through the multi-hop route. The sequence number of the last received packet at the sink is also shown in each case. In non-caching mode, end-to-end retransmissions are performed to provide reliability. In Figure 8.2(a), packet 4 is lost before reaching the sink. The sink can only recognize the packet loss after it receives packet 5 (Figure 8.2(b)). Then, the sink transmits a NACK packet back to the source node as shown in Figure 8.2(c). The lost packet is then retransmitted to the sink through the multi-hop route as shown in Figure 8.2(d).

■ **EXAMPLE 8.2**

In caching mode, as shown in Figure 8.3, certain sensor nodes are assigned as *caching nodes* (denoted as black nodes in Figure 8.3) on the reinforced path from the sensor node to the sink. In addition to the sink, loss packet detection is also performed at the caching nodes. Similar to the non-caching case, the loss of packet 3 in Figure 8.3(a) can only be detected by the caching node after receiving packet 4 as shown in Figure 8.3(b). Then, the caching node transmits a NACK packet back to the source node as shown in Figure 8.3(c). In this case, as shown in Figure 8.3(d), the first caching node with the missing packet on the reverse path replies and the packets are transmitted to the sink in order. If the packet cannot be found in one of the caching nodes, the NACK packet is propagated until it reaches the source node.

8.2.1   Qualitative Evaluation

In caching mode, RMST essentially creates reliable segments between two consecutive caching nodes and the retransmissions are performed inside these segments instead of through the end-to-end route. As a result, the cost associated with end-to-end retransmissions is minimized. Moreover, RMST aims to provide guaranteed delivery for each flow in the WSN. This is helpful for applications where individual node information is important, such as in network management solutions as discussed in Chapter 9.

RMST may incur additional overhead since caching requires additional processing and memory at the caching nodes. This may increase the overall complexity and energy consumption of the network. Most applications related to event detection/tracking may not require 100% reliability since the individual data flows are correlated and loss tolerant. However, RMST treats each flow separately, which may lead to overutilization of the resources in WSNs. Moreover, guaranteed reliability via in-network caching may bring significant overhead for WSNs with power and processing limitations. Finally, RMST focuses

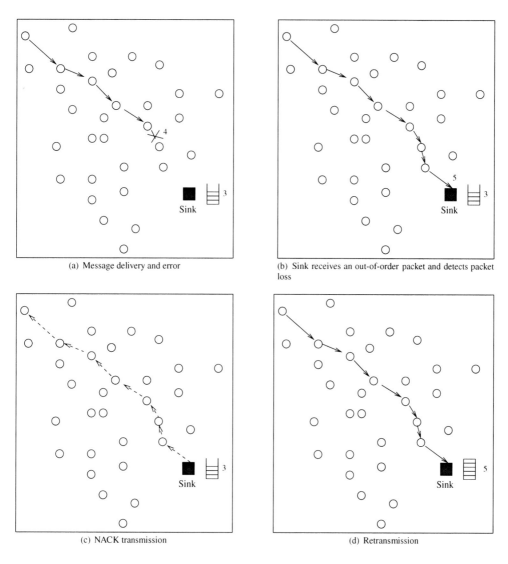

(a) Message delivery and error

(b) Sink receives an out-of-order packet and detects packet loss

(c) NACK transmission

(d) Retransmission

Figure 8.2   Error recovery with RMST in non-caching mode.

only on the reliability aspects of communication. Since a packet-by-packet reliability notion is followed, a large amount of information may flow inside the network. This will result in congestion and associated packet drops, which is not addressed in RMST.

8.3   Pump Slowly, Fetch Quickly (PSFQ) Protocol

The PSFQ protocol [13] has been developed to address the path from *sink to sensors*. Contrary to many transport layer approaches that focus on the sensors-to-sink path, the reverse path is generally used for network management tasks and retasking of the sensor nodes. Hence, reliability is of major concern.

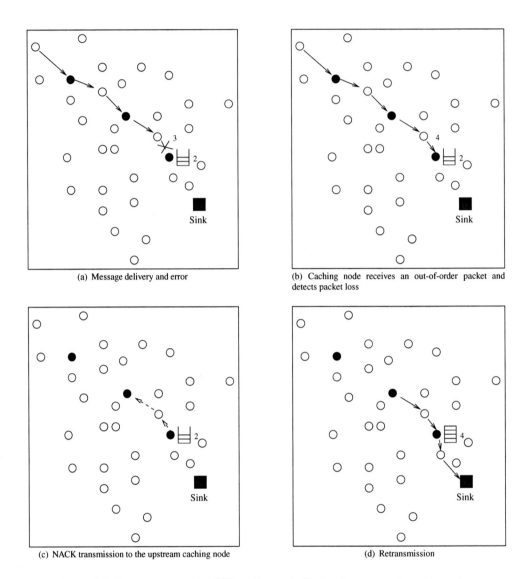

(a) Message delivery and error

(b) Caching node receives an out-of-order packet and detects packet loss

(c) NACK transmission to the upstream caching node

(d) Retransmission

Figure 8.3   Error recovery with RMST in caching mode. Black nodes represent the caching nodes.

While the sensors-to-sink path may tolerate information loss due to the highly correlated data generated by sensors, the reverse path requires *one-to-one* communication support to reliably disseminate control information that is sent by the sink to each sensor. Accordingly, the PSFQ protocol provides three main functions:

- **Pump operation:** Since reliability is generally more important than timeliness, PSFQ employs a slow pump mechanism that is based on slowly injecting packets into the network. In this case, each node on the route to the destinations waits for a specific amount of time before relaying the messages.

- **Fetch operation:** In case of packet errors, each node performs aggressive hop-by-hop recovery to fetch the lost packets from neighbor nodes.
- **Status reporting:** PSFQ also provides a reporting functionality that creates closed-loop communication between sensors and sink. Through this functionality, the sink can collect information related to operation of the network.

Pump operation is the default strategy in PSFQ for information dissemination from sink to sensors unless there are any errors. This is illustrated next.

■ **EXAMPLE 8.3**

The PSFQ pump operation is illustrated in Figure 8.4(a), where a sensor is transmitting packets to its neighbor node A. Node A relays this information to node B. Two timers, T_{min} and T_{max}, are used to schedule the transmission times of nodes along a path from a sensor to sink. A sensor broadcasts packets every T_{min} to its immediate neighbors. Upon receiving a packet, the neighbor nodes relay the packets after a random wait time, which is selected between T_{min} and T_{max} as shown in Figure 8.4(a). Hence, a node has to wait at least T_{min} between packet transmissions. This duration between packet transmissions is integrated to allow nodes to recover the missing packets. Moreover, the random delay allows a reduction in the number of redundant broadcasts of the same packet by the neighbors. If the packet is forwarded by one of the nodes, other neighbors suppress their transmissions.

A sensor sends packets related to a particular message with consecutive sequence numbers. If a node on the path from the sink to sensors detects a sequence number gap in this sequence of packets, the fetch operation is initiated. Through the fetch operation, a node aggressively sends out NACK packets to quickly recover the loss packets from its immediate neighbors as explained in the following example.

■ **EXAMPLE 8.4**

The fetch operation is illustrated in Figure 8.4(b). When node A detects a packet loss, it broadcasts a NACK packet to its immediate neighbors. It may be the case that no reply is heard or only a fraction of the lost packets are received. In this case, if node A does not receive any reply within a period of T_r, where $T_r < T_{max}$, it persistently continues to send NACK packets for every T_r. If one of node A's neighbors has the packet in its cache, it sends the packet in an interval between $(1/4)T_r$ and $(1/2)T_r$ as shown in Figure 8.4(b).

The fetch operation results in local error recovery through persistent NACK messages between two packet transmissions. However, to prevent message implosion, PSFQ limits NACK message transmission to the one-hop neighborhood. In other words, NACK messages are not propagated through multi-hop routes.

The loss detection mechanism in PSFQ relies on the sequence numbers of packets in a flow. In case of packet losses in the middle of a flow, this mechanism can effectively detect those packet losses. However, the mechanism is not suitable for cases where the last packet of a flow is lost or all the packets related to a flow are lost. For this case, PSFQ employs a *proactive fetch operation*, where the receivers follow a timer-based fetch operation.

■ **EXAMPLE 8.5**

As shown in Figure 8.4(c), if a node does not receive a packet for T_{pro}, it sends a NACK packet to its neighbors. The wait time, T_{pro}, is determined so that it is proportional to the difference between the last highest sequence number, S_{last}, of the packet received and the largest sequence number, S_{max}, of the flow, i.e., $T_{pro} = \alpha(S_{max} - S_{last})T_{max}$, where $\alpha \geq 1$. Consequently, the node

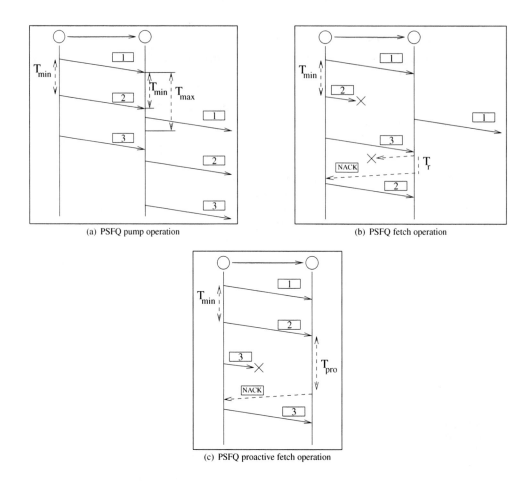

Figure 8.4   Pump and fetch operations of PSFQ protocol.

proactively sends a NACK packet earlier if it is closer to the end of the message. In the case where buffer length is limited, the waiting time, T_{pro}, is determined as follows: $T_{pro} = \alpha n T_{max}$, where n is the buffer length.

The third component of the PSFQ protocol is the *report operation*, which allows the sink to request feedback from the sensor nodes in a scalable manner. Report operation is initiated by the sink by setting a *report bit* in the packet header. This packet is sent through the network to the intended nodes. Upon receiving the report request, a sensor node responds immediately by transmitting a report message. Each node along the path to the sink adds its status information to this report by piggybacking. If a node in the upstream path does not receive the report response in T_{report}, it creates its own report packet and sends it to the sink.

The report operation is also used for single packet delivery. If a message that the sink will send can fit into one packet, then the report operation is initiated by setting the report bit. Whenever the intended sensor node receives this packet, it responds with a report packet. As a result, an end-to-end error control mechanism is integrated into PSFQ for single packet flows.

8.3.1   Qualitative Evaluation

PSFQ employs a simple protocol operation that does not require an end-to-end reliability mechanism. As a result, the protocol scales well with network size. Moreover, the communication requirements of the sink-to-sensors direction are significantly different than that of the sensors-to-sink direction. PSFQ aims to provide reliable and orderly transfer of packets in a distributed manner in this direction.

PSFQ relies on the pump slowly operation to limit congestion. However, as the number of sources in the network increases, congestion will occur. Moreover, PSFQ handles reliability only by considering possible wireless channel problems. Hence, there is no mechanism in PSFQ which captures the congestion problems in WSNs. In addition, PSFQ is rather a hop-by-hop reliability protocol than an end-to-end solution. Hence, end-to-end reliability cannot be maintained in all cases even though hop-by-hop reliability is ensured. The pump slowly operation introduces an artificial delay in the network at each hop. As a result, networks of large size may suffer from high latency. Whenever a packet is lost, the receiver node stores the out-of-order packets until the lost packet(s) are received. This increases the memory required for the PSFQ protocol, which may be limited in the sensor nodes.

8.4   Congestion Detection and Avoidance (CODA) Protocol

The aim of CODA is to detect and avoid congestion in WSNs [14]. To this end, three main congestion scenarios are considered. In WSN applications, where the source nodes frequently generate traffic, congestion builds up close to the source nodes because of the contention in the wireless channel. Moreover, for low-rate traffic, congestion can occur temporarily in *hot spots*, where multiple flows are served. For these two types of congestion, CODA provides local congestion control mechanisms. Furthermore, because of the topology and the capacity of the network, certain persistent hot spots may exist. This requires end-to-end mechanisms to regulate the data rate of source nodes.

To address congestion and the different scenarios that cause it, CODA provides three mechanisms via *receiver-based congestion detection*, *open-loop hop-by-hop backpressure* signaling to inform the source about the congestion, and *closed-loop multi-source regulation* for persistent and larger scale congestion conditions.

Accurate congestion detection is important since congestion control mechanisms usually impose additional processing and communication requirements on sensor nodes. It is clear that congestion occurs because of an increase in the buffer occupancy of a node and the consequent packet drops. Hence, the buffer occupancy level has been mostly used as a congestion detection metric. However, in a multi-node environment, because of packet errors and collisions in the wireless channel, the buffer occupancy level does not always accurately reflect the congestion. More specifically, while the congestion in a certain area of the network can increase, the buffer occupancy level of the nodes in that area may not be affected. Consequently, CODA uses a combination of buffer occupancy level and channel load condition as an indication of congestion. Moreover, since congestion usually occurs in receiver nodes, a *receiver-based congestion detection mechanism* is designed.

The receiver-based congestion detection mechanism relies on both buffer occupancy and channel load. The channel load is estimated by listening to the channel whenever a node has a packet to send. CODA decides on congestion if the channel load is higher than a fraction of the maximum channel utilization. Consequently, congestion is detected at the receivers.

■ **EXAMPLE 8.6**

The congestion mitigation mechanism of CODA is illustrated in Figure 8.5, where a sensor node is trying to transmit a packet to the sink through a congested area. When the congestion is detected by one of the nodes inside the congested area (Figure 8.5(a)), the receiver broadcasts a *suppression message* toward the source node in the reverse path as shown in Figure 8.5(b). The suppression

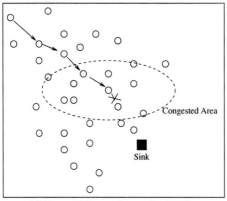

(a) Message delivery and congestion detection

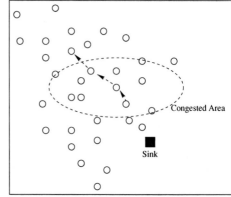

(b) Suppression message transmission

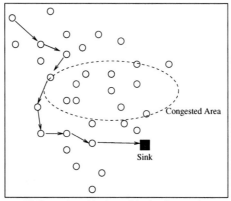

(c) Congestion mitigation and (optional) re-routing

Figure 8.5 Open-loop hop-by-hop backpressure mechanism of CODA.

message is used to inform the upstream nodes about congestion. Upon receiving the suppression, the upstream nodes decrease their sending rates or drop packets to relieve the congestion in the forward path. Moreover, the suppression message is rebroadcast until a non-congested node receives the message as shown in Figure 8.5(b). This type of congestion control is referred to as *open-loop hop-by-hop backpressure*. While the suppression message may not reach all the way to the source nodes, they serve to relieve local congestion in a hot spot area. Moreover, as shown in Figure 8.5(c), using cross-layer routing techniques the forward path can be rerouted to avoid these hot spots.

In addition to local congestion that may occur as a result of the dynamics in the network, the traffic generated by the source nodes can create network-wide congestion. More specifically, if a source node injects a traffic load larger than the network can handle, the local congestion control mechanism cannot relieve the congestion. For this case, CODA uses a *closed-loop multi-source regulation* mechanism as shown in Figure 8.6. This mechanism is similar to conventional end-to-end congestion control mechanisms. Each source node monitors the source rate, r. If the source rate exceeds a threshold, $r \geq \nu S_{max}$, which is a fraction of the maximum theoretical throughput of the channel, the source node

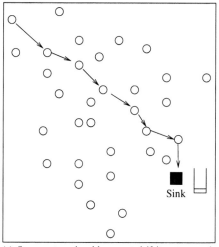
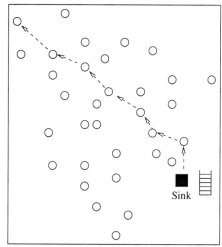

(a) Sensor enters closed-loop control if its rate exceeds $r \geq v S_{max}$

(b) Sink replies with ACK for each n packets received

Figure 8.6   Closed-loop multi-source regulation mechanism of CODA.

enters closed-loop control. In this case, a regulation bit is set in the header of the packets to inform the sink. Consequently, the sink begins sending periodic ACK messages for each n number of packets received. If the source cannot receive the ACK messages, it determines network-wide congestion and decreases its source rate.

8.4.1   Qualitative Evaluation

Identifying congestion in a WSN is essential for accurate congestion control. CODA uses a combination of buffer occupancy level and channel load condition to provide efficient congestion detection. Similarly, CODA addresses both local and end-to-end congestion in WSNs. As a result, a complete congestion control mechanism is provided for downstream traffic.

CODA can increase the network performance by avoiding congestion. However, the CODA protocol does not address reliability. The operation of CODA reveals that the congestion control performed at the sensor nodes without considering reliability impairs the end-to-end transport reliability [14]. CODA only provides congestion avoidance and does not provide a reliable operation. Hence, it should be coupled with a reliability mechanism. Furthermore, the closed-loop multi-source regulation mechanism incurs additional latency for cases where the network traffic is high.

8.5   Event-to-Sink Reliable Transport (ESRT) Protocol

In contrast to the transport layer protocols developed based on conventional end-to-end reliability, the ESRT protocol [1] is based on the *event-to-sink reliability* notion and provides reliable event detection without any intermediate caching requirements. ESRT aims to address both the reliability and congestion problems in WSN.

The main characteristic of information delivery in WSNs is the data-centric communication paradigm. More specifically, to the user, *collective* information from multiple sensors regarding an *event* is much more important than the *individual* information sent from each node. Consequently, the notion of flow,

Table 8.1   Network operating regions based on congestion and reliability levels.

State	Decision boundaries	Definition
(NC, LR)	$f < f_{max}$ and $\eta > 1 + \epsilon$	(No Congestion, Low Reliability)
(NC, HR)	$f \leq f_{max}$ and $\eta < 1 - \epsilon$	(No Congestion, High Reliability)
(C, HR)	$f > f_{max}$ and $\eta < 1$	(Congestion, High Reliability)
(C, LR)	$f > f_{max}$ and $\eta \geq 1$	(Congestion, Low Reliability)
OOR	$f < f_{max}$ and $1 - \epsilon \leq \eta \leq 1 + \epsilon$	(Optimal Operating Region)

which is defined as the messages that are sent from a source to a destination, no longer applies. Instead an *event-to-sink* information flow exists, which is the collection of "flows" from a group of sensors associated with a single event. ESRT aims to provide reliability of the event information flow through congestion control mechanisms. Moreover, the algorithms of ESRT run mainly on the sink, with minimal functionality required at resource-constrained sensor nodes. It mainly exploits the fact that the sheer amount of data flows generated by the sensor nodes toward the sink is correlated due to spatial and temporal correlation among the individual sensor readings [12].

ESRT relies on the sink to measure the reliability in the WSN every τ time units, which is also referred to as the *decision interval*. The reliability is measured in terms of the number of data packets from all the sensor nodes associated with an event. Consequently, the following definitions are used:

- *Observed event reliability*, r_i, is the number of received data packets in decision interval i at the sink.
- *Desired event reliability*, R, is the number of data packets required for reliable event detection. This is determined by the application.

Accordingly, ESRT aims to configure the reporting rate, f, of source nodes so as to achieve the required event detection reliability, R, at the sink with minimum resource utilization.

■ **EXAMPLE 8.7**

In Figure 8.7, the reliability of a WSN as a function of the reporting rate, f, of the sensor nodes is shown. It can be observed that the reliability, r, shows a linear increase (note the log scale) with source reporting rate, f, until a certain $f = f_{max}$, beyond which the reliability drops. This is because the network is unable to handle the increased injection of data packets and packets are dropped due to congestion. Moreover, for higher reporting rates, $f > f_{max}$, the behavior is rather wavy and not smooth, and the reliability always stays well below the maximum reliability at $f = f_{max}$.

The relation between congestion and reliability in a WSN is clearly visible in Example 8.7. Accordingly, ESRT aims to control congestion to provide reliable event delivery by defining five operating regions as shown in Figure 8.7. The network operating regions based on congestion and reliability levels are given in Table 8.1.

To control congestion in the network, the sink needs to determine the operating region of the network. This operating region depends on two factors: whether congestion exists in the network and whether the required reliability is achieved. Hence, two mechanisms are required to estimate the congestion and the reliability state of the network. In each decision interval, τ, the sink determines whether the network is operating in the low- or high-reliability region according to the number of packets received in that interval. Moreover, a congestion detection algorithm is employed at each sensor node in the network. Consequently, the sink can determine the operating region of the network. According to the operating region, the sink controls the reporting rate of the sensor nodes to operate at the optimal operating region (OOR).

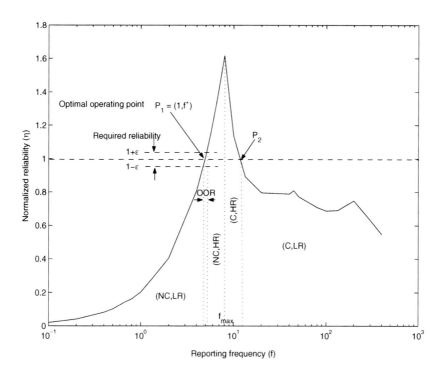

Figure 8.7   Relationship between reliability and reporting rate and the operating regions defined for ESRT.

The congestion detection algorithm is based on monitoring the local buffer-level monitoring at each sensor node. Assuming that network conditions such as reporting rate and the number of source nodes do not change significantly between two consecutive decision intervals, the increment in the buffer fullness level at the end of each reporting interval is expected to be constant.

Accordingly, as shown in Figure 8.8, let b_k and b_{k-1} be the buffer fullness levels at the end of the kth and $(k-1)$th reporting intervals respectively and B be the buffer size. Then, Δb is the buffer length increment observed at the end of last reporting period, i.e., $\Delta b = b_k - b_{k-1}$. Hence, if the sum of the current buffer level at the end of the kth reporting interval and the last experienced buffer length increment exceeds the buffer size, i.e., $b_k + \Delta b > B$, the sensor node infers that it is going to experience congestion in the next reporting interval. If a sensor node detects congestion in the network, it notifies the sink by piggybacking the packets that are sent. This is notified the sink for the upcoming congestion condition to be experienced in the next reporting interval.

The state of the network is determined at each decision interval through the reliability measurement and congestion notifications. Accordingly, the sink updates the reporting rates of the sensor nodes by broadcasting a packet to each sensor node. Denoting the reporting rate at decision interval i as f_i and the reliability level as η_i, the sink updates the reporting rate, f_{i+1}, at the next interval as follows: if the network is in the (NC, LR) (No Congestion, Low Reliability) region, the reporting rate needs to be increased to achieve a higher reliability level at the sink. Hence, the reporting rate is aggressively increased as $f_{i+1} = f_i/\eta_i$. In the (NC, HR) (No Congestion, High Reliability) region, the network is above the required reliability level and the network resources can be saved by decreasing the reporting rate such that OOR is achieved. In this case, the reporting rate is decreased as $f_{i+1} = f_i/2(1 + 1/\eta_i)$ to conserve energy. In the (C, HR) (Congestion, High Reliability) region, the network is congested and the

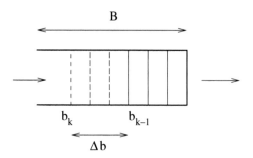

Figure 8.8   Buffer occupancy analysis in ESRT.

reporting rate needs to decreased to relieve the congestion and improve the energy efficiency. As a result, a multiplicative decrease policy is used for the reporting rate as $f_{i+1} = f_i/\eta_i$. Finally, in the (C, LR) (Congestion, Low Reliability) region, the reporting rate is exponentially decreased to improve reliability and relieve congestion in the network as $f_{i+1} = f_i^{\eta_i/k}$. If the optimal operating point is reached, the reporting rate is kept constant.

8.5.1   Qualitative Evaluation

ESRT introduces a novel reliability notion called the *event-to-sink* reliability. The notion of the event-to-sink reliability is necessary for reliable transport of event features in WSNs. Based on such a collective reliability notion, ESRT introduces a new reliable transport scheme for WSNs. Moreover, efficient congestion control and reliability mechanisms are provided by shifting the decision mechanism to the power-rich sink node. ESRT has a congestion control component that serves the dual purpose of achieving reliability and conserving energy. The algorithms of ESRT run mainly on the sink and require minimal functionality at resource-constrained sensor nodes.

The primary objective of ESRT is to configure the network as close as possible to the optimal operating point, where the required reliability is achieved with minimum energy consumption and without network congestion. Through distributed update mechanisms, ESRT converges to the optimum operating state regardless of the initial network state.

ESRT relies on the assumption that the base station is one hop away from all the sensor nodes, i.e., a sink broadcast for frequency update will be received by all sensor nodes. Although this assumption may hold true for a certain class of application, as the network size increases, multi-hop strategies are required for information delivery from the sink to the sensors. Besides, ESRT maintains the *same* value of operating frequency, f, at the end of each decision interval for each sensor node. In case a large group of sensors sense an event, the required operating frequency may not be the same for all these sensors because of their local interactions in the network.

The event-to-sink reliability notion applies to cases where the information at each sensor is correlated. Hence, ESRT cannot be applied for applications where individual sensor information is necessary.

8.6   GARUDA

While the data flows in the forward path carry correlated sensed/detected event features, the flows in the reverse path mainly contain data transmitted by the sink for operational or application-specific purposes. This may include the operating system binaries, programming/retasking configuration files, application-specific queries and commands. Dissemination of these types of data mostly requires 100%

reliable delivery. Therefore, the event-to-sink reliability approach, which is suitable for sensors-to-sink communication, would not be suitable for addressing such tighter reliability requirements of the flows in the reverse path.

Such a strict reliability requirement for the sink-to-sensors communication involves a certain level of retransmission as well as acknowledgment mechanisms. However, these mechanisms should be incorporated into the transport layer protocols cautiously in order not to totally compromise scarce wireless sensor resources. In this respect, local retransmissions and negative acknowledgment approaches would be preferable over the end-to-end retransmissions and acknowledgments to maintain minimum energy expenditure.

On the other hand, the sink is involved more in the sink-to-sensor data transport on the reverse path. Hence, the sink, with plentiful energy and communication resources, can broadcast the data with its powerful antenna. Consequently, the amount of traffic forwarded in the multi-hop WSN infrastructure can be reduced and help sensor nodes conserve energy. As a result, data flows in the reverse path may experience less congestion in contrast to the forward path. Therefore, reliability is significantly important in the reverse path compared to aggressive congestion control mechanisms.

The reverse path consists of multi-hop and one-to-many flows, which resemble conventional wireless networks. There exist many transport layer schemes that address reliable transport and congestion control for the case of a single sender and multiple receivers [5]. Although the communication structure of the reverse path is an example of multicast, these schemes do not stand as directly applicable solutions; rather they need significant modifications/improvements to address the unique requirements of the WSN paradigm. Next, we will describe the GARUDA protocol, which provides reliable delivery between a resource-rich sink and resource-constrained sensor nodes.

Reliable sink-to-sensors delivery is addressed through the GARUDA protocol [9, 10]. The name GARUDA is based on a mythological bird that reliably transported gods. GARUDA incorporates an efficient pulsing-based solution for successful delivery of single packets. Moreover, as in the RMST protocol, certain nodes in the network are selected to perform caching and manage the loss recovery process. In GARUDA, these nodes are referred to as *core* nodes. To this end, a two-stage loss recovery mechanism is incorporated for each node in the network to reliably receive the packets sent from the sink. GARUDA consists of three mechanisms that accomplish reliable data delivery from sink to sensors for different scenarios:

- **Single/first packet delivery:** GARUDA mainly follows a NACK-based recovery policy for the lost packets. However, NACK-based recovery cannot address packet losses in single packet delivery or the loss of all the packets in a message. Consequently, a pulse-based delivery mechanism is incorporated so that the sink informs the sensors about the subsequent delivery of a single packet flow or the first packet of a multi-packet flow through short-duration pulses.

- **Core construction:** GARUDA relies on specific nodes in the network, the core nodes, for caching. Consequently, a simple and efficient core construction mechanism is designed to select these nodes without significant overhead.

- **Two-phase loss recovery:** Once the core is constructed, information delivery is performed through control of the core nodes. In the case of packet losses, a two-phase loss recovery mechanism is utilized, where a core node first recovers the lost packets from upstream core nodes and then delivers the packet to its neighboring non-core nodes.

GARUDA supports the delivery of single packet flows or the first packet of a flow through a pulse-based mechanism as explained in Example 8.8.

■ **EXAMPLE 8.8**

The sink informs the sensor nodes about an impending reliable short-message delivery by transmitting a specific series of *wait-for-first-packet* (WFP) pulses at a certain amplitude and period as shown in Figure 8.9(a). The WFP pulses are of very small duration and only used to

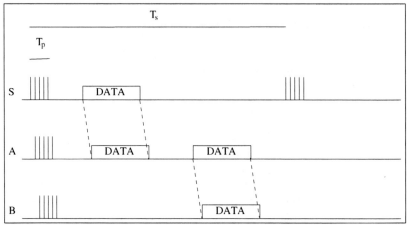

(a) First packet delivery with wait-for-first-packet (WFP) pulse transmission

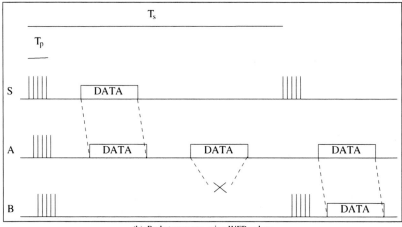

(b) Packet recovery using WFP pulses

Figure 8.9   First packet delivery mechanism in GARUDA.

indicate the subsequent transfer of a packet. When the sink has a packet to send, it transmits a
series of WFP pulses of duration T_p as shown in Figure 8.9(a). The duration of the WFP pulse
is significantly smaller than the duration of a packet transmission and incurs a small overhead.
Moreover, the sink continues to transmit WFP pulses with a period of T_s which is significantly
larger than the packet transmission duration. Each node on the path from the sink to the intended
sensor(s) broadcasts the WFP pulses on reception. Consequently, each node in the network is
informed about the packet to be received.

   After waiting a sufficient time for the WFP pulses to propagate in the network, the sink
transmits the packet as shown in Figure 8.9(a). Upon receiving this packet, each node stops
transmitting WFP pulses, switches to delivery mode, and relays the packet as shown in
Figure 8.9(a). The error recovery phase is initiated if a node does not receive the packet. As
shown in Figure 8.9(b), if node B does not receive the data packet from node A, it will continue
to transmit WFP pulses every T_s. These WFP pulses also serve as a NACK indication for node A

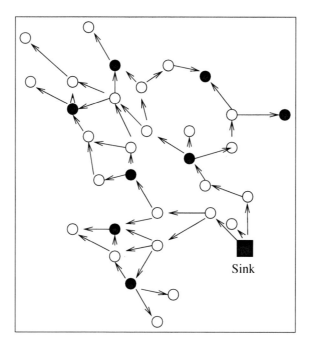

Figure 8.10 First packet transmission and core construction in GARUDA. The black nodes represent the core nodes.

to perform retransmission. As shown in Figure 8.9(b), if node A receives a WFP pulse after transmitting a packet, it retransmits the packet to node B. Consequently, GARUDA ensures that a single packet or the first packet of a series of packets is reliably received at the intended recipients.

During the transmission of the first packet, GARUDA also implicitly employs the core construction mechanism. The core approximates a near-optimal assignment of local designated servers for loss recovery. The assignment is performed according to the number of hops between the sink and the sensor nodes. During the delivery of the first packet, the sensor network is partitioned into *bands*. Consequently, a node is selected as a core node if (1) its hop count, i.e., *band-id*, is a multiple of 3 and (2) it does not hear that any of its neighbors have been selected as the core node as shown in Figure 8.10. Upon receiving the first packet from the sink, each node determines its band-id and forwards the packet. The core nodes also indicate their duty in the forwarded packets. Other non-core nodes associate themselves with one of the core nodes for loss recovery purposes.

Loss recovery in the GARUDA protocol is mainly performed between core nodes. An important distinction of the GARUDA protocol compared to other reliable transport layer solutions such as PSFQ is that it allows out-of-order forwarding. More specifically, if a node detects a packet loss, it continues to forward the next packets in the sequence before recovering the loss packet. This minimizes the impact of the loss recovery mechanism on packet delivery delay. Moreover, each core node exchanges an *A-map* (availability map) to indicate the available packets in its cache. Using the A-map information, a two-stage loss recovery process is conducted. This loss recovery process is described next.

■ **EXAMPLE 8.9**

The recovery process in GARUDA consists of two stages as shown in Figure 8.11. Each core node tries to recover lost packets by exchanging unicast messages among other core nodes as quickly as

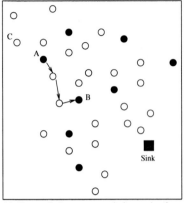

(a) Core node A requests a missing packet from another core node B

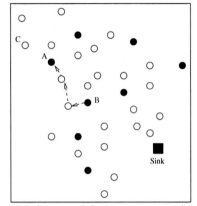

(b) Packet retransmission between two core nodes

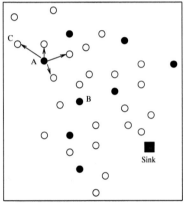

(c) Node A broadcasts its A-map when it has all the packets

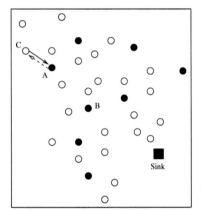

(d) Non-core node C requests loss recovery from core node A

Figure 8.11   Two-phase loss recovery mechanism of GARUDA.

possible. This process is performed in parallel with the out-of-sequence forwarding mechanism. As shown in Figure 8.11(a), whenever, a core node, A, has a missing packet, it checks the A-map of node B, which is the first node on the path to the sink. If node B's A-map indicates that it has the missing packet, node A sends a unicast message to node B. As shown in Figure 8.11(b), node B replies with the missing packet. Once node A receives all the missing packets, it indicates this by broadcasting its own A-map as shown in Figure 8.11(c). The non-core nodes associated with node A, e.g., node C, then request the missing packets as shown in Figure 8.11(d). Using this two-phase loss recovery mechanism, all the packets sent by the sink are reliably delivered to the receivers.

GARUDA also supports other reliability semantics that might be required for sink-to-sensors communication, such as: (1) reliable delivery to all nodes within a subregion of the sensor network; (2) reliable delivery to a minimal number of sensors required to cover the entire sensing area; and (3) reliable delivery to a probabilistic subset of the sensor nodes in the network. For each of these cases, the first packet is delivered to each node in the network and the core is constructed. Then, depending

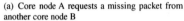

on the set of destinations, GARUDA operations are performed to reliably deliver the messages to only these sets of nodes.

8.6.1   Qualitative Evaluation

GARUDA provides a wide variety of reliability mechanisms for downstream data flow. Since high reliability is required for the sink-to-sensors path, these mechanisms provide flexibility in protocol operation. Compared to PSFQ, GARUDA supports out-of-order packet delivery. This decreases the packet delivery delay and places less burden on the memory of sensor nodes.

GARUDA relies on a core construction mechanism, during which the core nodes are chosen. For large WSNs, this mechanism may impact the overall delay in communication. Moreover, the reliability mechanism in GARUDA is partitioned between core nodes. Hence, end-to-end reliability of the packets may not be guaranteed. Similar to PSFQ, GARUDA focuses on reliability rather than congestion control. Hence, packet losses due to congestion may not be mitigated.

8.7   Real-Time and Reliable Transport $(RT)^2$ Protocol

The $(RT)^2$ protocol has been developed to reliably and collaboratively transport event features from the sensor field with minimum energy dissipation in a timely manner [6]. Accordingly, the $(RT)^2$ protocol simultaneously addresses congestion control and timely event transport reliability objectives. It is one of the first protocols to focus on real-time guarantees in addition to reliability in WSNs.

The $(RT)^2$ protocol addresses the following challenges. Certain applications need to immediately react to sensor data based on the application-specific requirements. Therefore, real-time communication within certain delay bounds is a crucial concern, which is addressed through a delay-based reliability notion in $(RT)^2$. The wireless channel errors and node failures lead to bursts of packet loss [2]. Although the effects of these losses can be alleviated by the channel coding schemes to some extent, packet-level transport layer reliability mechanisms are still required. This is addressed by the reliability mechanism of $(RT)^2$. Finally, energy efficiency is sought for real-time reliable data delivery.

As we have discussed in Section 8.5, WSNs are characterized by the dense deployment of sensors and the sensor observations are highly correlated in the space domain. Furthermore, the physical phenomenon exhibits temporal correlation. Accordingly, event transport from a group of sensor nodes to the sink does not require 100% reliability. Thus, the $(RT)^2$ protocol follows the collective *event transport reliability* notion similar to ESRT rather than the traditional end-to-end reliability notions. To this end, a new notion of *event delay bound* is considered to meet the application-specific deadlines. Based on both event transport reliability and event delay bound notions, the following definitions are used in protocol operation:

- **Observed delay-constrained event reliability (DR_i):** The number of received data packets within a certain delay bound at the sink in a decision interval i. In other words, DR_i accounts for the number of correctly received packets that are received within the application-specific delay bound. The value of DR_i is measured in each decision interval i.
- **Desired delay-constrained event reliability (DR^*):** The minimum number of data packets required for reliable event detection within a certain application-specific delay bound. This lower bound for the reliability level is determined by the application and based on the physical characteristics of the event signal being tracked.
- **Delay-constrained reliability indicator (δ_i):** The ratio of the observed and the desired delay-constrained event reliabilities, i.e., $\delta_i = DR_i/DR^*$.

The operation of $(RT)^2$ is based on these definitions. At each decision interval i, the sink observes DR_i to determine the necessary actions. If the observed delay-constrained event reliability is higher than the reliability bound, i.e., $DR_i > DR^*$, then the event is reliably detected within a certain delay bound.

Otherwise, appropriate action needs to be taken to assure the desired reliability level. This can be accomplished by increasing the amount of information transported from the sensors to the actor. Accordingly, the reporting frequency of the sensors is increased properly while avoiding congestion in the network.

The transport reliability problem of $(RT)^2$ is to *configure the reporting rate, f, of source nodes so as to achieve the required event detection reliability, DR^*, at the actor node within the application-specific delay bound.*

The notion of delay-constrained reliability is based on the end-to-end delay bound specified by the application. This is denoted by Δ_{e2a} and is specific to application requirements. The event delay can be decomposed into three main components:

1. *Event transport delay (Γ^{tran}):* The overall time between the event occurrence time and the time when the event is reliably transported to the sink. The event transport delay can further be decomposed into the following delay components: *buffering delay, channel access delay, transmission delay,* and *propagation delay.*

2. *Event processing delay (Γ^{proc}):* The processing delay experienced at the sink when the desired features of the event are estimated using the data packets received from the sensor field. This may include a certain decision interval [1] during which the sink waits to receive adequate samples from the sensor nodes.

3. *Action delay (Γ^{act}):* The action delay is the time it takes from the instant when an event is reliably detected at the sink to the instant when a particular action is performed.

For a given event delay bound, Δ_{e2a}, the following condition should be met:

$$\Delta_{e2a} \geq \Gamma^{tran} + \Gamma^{proc} + \Gamma^{act} \qquad (8.1)$$

where Γ^{tran} is directly affected by the current network load and the congestion level in the network. In addition, the network load depends on the event reporting frequency, f, which is used by the sensor nodes to send their readings of the event. Moreover, the end-to-end delay depends on the queuing delay at each relay node. $(RT)^2$ employs a *time-critical event first* (TCEF) scheduling policy so that events with smaller remaining deadline are given priority. The remaining deadline of a packet is updated at each node by measuring the elapsed time and piggybacking the elapsed time to the event packet so that the following sensor can determine the remaining time to the deadline without a globally synchronized clock. In addition to the TCEF scheduling policy, $(RT)^2$ also updates the reporting rates so that congestion is mitigated in the network. The effect of reporting frequency can be observed with the following example.

■ **EXAMPLE 8.10**

In a sensor network, 200 sensor nodes were randomly positioned in a 200 m × 200 m sensor field. Several events were generated at different event centers denoted as (X_{ev}, Y_{ev}) and all sensor nodes within the event radius behave as sources for that event.

The impact of the event reporting frequency on the *on-time event delivery ratio* is shown in Figure 8.12(a) for different number of source nodes, i.e, $n = 41, 62, 81, 102$. Here, the on-time event delivery ratio is the fraction of data packets received within the sensor–actor delay bound over all data packets received in a decision interval. When the reporting frequency is small enough, the on-time event delivery is ensured. However, above a certain $f = f_{max}$, network congestion is experienced. After this point, the average transport delay starts to increase and, accordingly, the on-time event delivery ratio decreases below 1. Since the network load is increased as a result of higher reporting frequency, the buffer occupancy and network channel contention increases.

To further elaborate the relationship between observed delay-constrained event reliability, DR_i, and the event reporting frequency, f, in Figure 8.12(b), the number of packets received at the

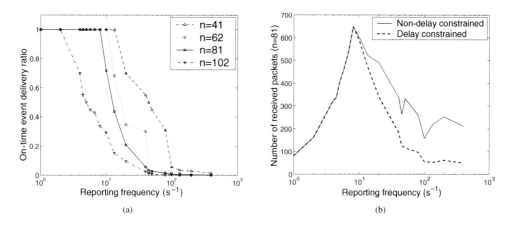

Figure 8.12 Effect of reporting frequency on (a) on-time delivery and (b) reliability.

sink in a decision interval, τ, is shown. In the figure, two values are shown. The solid line is the number of received packets irrespective of the delay associated with the packets. The dashed line shows the number of packets received within the required delay.

Accordingly, until a certain $f = f_{max}$, the delay-constrained event reliability and non-delay-constrained event reliability are the same. However, beyond this value, the former significantly deviates from the latter. Further, the observed delay-constrained event reliability, DR_i, shows a linear increase (note the log scale) with source reporting rate, f, until a certain $f = f_{max}$, beyond which the observed delay-constrained event reliability drops. The decrease is because of the congestion that builds up in the network. For $f > f_{max}$, delay-constrained event reliability drops significantly due to network congestion. $(RT)^2$ aims to control the reporting frequency so that the delay-constrained event reliability is above the required value.

The delay-constrained event reliability depends on the congestion in the network. If the congestion is detected accurately, the reporting rate can be adjusted so that maximum reliability can be achieved without causing congestion in the network. To this end, the $(RT)^2$ protocol uses a *combined congestion detection* mechanism based on both average node delay calculation and local buffer level monitoring of the sensor nodes to accurately detect congestion in the network.

The average node delay at the sensor node gives an idea about the contention around the sensor node, i.e., how busy the surrounding vicinity of the sensor node is. To compute the average node delay at the sensor node i, the sensor node takes the exponential weighted moving average of the elapsed time. In addition to the node delay, the sensor nodes also monitor their buffer. If a sensor node experiences buffer overflow due to excessive incoming packets or the average node delay is above a certain delay threshold value, this node is regarded as congested. Accordingly, the sink is notified of the upcoming congestion condition in the network by utilizing the *congestion notification* (CN) bit in the header of the event packet transmitted from the sensors to the actor node. In conjunction with the delay-constrained reliability indicator, δ_i, the sink determines the current network condition and dynamically adjusts the reporting frequency of the sensor nodes.

The $(RT)^2$ protocol decisions are made at the sink node according to the network state. The sink determines the network state based on the delay-constrained reliability indicator $\delta_i = DR_i/DR^*$, and T_i and T_{sa}, which are the amount of time needed to provide delay-constrained event reliability for a decision interval i and the application-specific communication delay bound, respectively. In conjunction with the congestion notification information (CN bit) and the values of f_i, δ_i, T_i, and T_{sa}, the sink updates the reporting frequency, f_{i+1}. This information is broadcast to the source nodes in each decision interval. This updating process is repeated until the optimal operating point is found, i.e., adequate reliability and no congestion condition is obtained. The protocol operation is explained based on the network states next:

- **Early reliability and no congestion:** This state represents the case where the required reliability level is reached, the communication delay is within the bound, i.e., $T_i \leq T_{sa}$, and no congestion is observed in the network, i.e., $CN = 0$. However, the observed delay-constrained event reliability, DR_i, is larger than the desired delay-constrained event reliability, DR^*. This is because source nodes transmit event data more frequently than required. To save energy, the reporting frequency is updated as follows:

$$f_{i+1} = f_i \frac{T_i}{T_{sa}}. \tag{8.2}$$

- **Early reliability and congestion:** In this case, the required reliability level is reached, the communication delay bound, i.e., $T_i < T_{sa}$, but congestion is observed in the network, i.e., $CN = 1$. Moreover, the observed delay-constrained event reliability, DR_i, is larger than the desired delay-constrained event reliability, DR^*. In this situation, the $(RT)^2$ protocol decreases the reporting frequency to avoid congestion and save the limited energy of sensors. More specifically, the updated reporting frequency can be expressed as follows:

$$f_{i+1} = \min\left(f_i \frac{T_i}{T_{sa}}, \ f_i^{(T_i/T_{sa})} \right). \tag{8.3}$$

- **Low reliability and no congestion:** This state represents the case where the required reliability level is not reached before the communication delay bound, i.e., $T_i > T_{sa}$, and no congestion is observed in the network, i.e., $CN = 0$. However, the observed delay-constrained event reliability, DR_i, is lower than the desired delay-constrained event reliability, DR^*. This state can be reached by (1) packet loss due to wireless link errors, (2) failure of intermediate relaying nodes, or (3) inadequate data packets transmitted by source nodes. To achieve the required event reliability, the data reporting frequencies of the source nodes are increased such that

$$f_{i+1} = f_i \frac{DR^*}{DR_i}. \tag{8.4}$$

- **Low reliability and congestion:** In this condition, the required reliability level is not reached before the communication delay bound, i.e., $T_i > T_{sa}$, and congestion is observed in the network, i.e., $CN = 1$. However, the observed delay-constrained event reliability, DR_i, is lower than the desired delay-constrained event reliability, DR^*. This situation is the worst possible case, since the desired delay-constrained event reliability is not reached, network congestion is observed, and, thus, the limited energy of sensors is wasted. Hence, $(RT)^2$ aggressively reduces the reporting frequency to reach the optimal reporting frequency as follows:

$$f_{i+1} = f_i^{DR_i/(DR^* * k)} \tag{8.5}$$

where k is the number of successive decision intervals for which the network has remained in the same situation, including the current decision interval, i.e., $k \geq 1$. Here, the purpose is to decrease the reporting frequency with greater aggression, if a network condition does not improve.

- **Adequate reliability and no congestion:** Finally, this state represents the desired network state where the network is within tolerance β of the optimal operating point, i.e., $f < f_{\max}$ and $1 - \beta \leq \delta_i \leq 1 + \beta$, and no congestion is observed in the network. Hence, the reporting frequency of source nodes is left constant for the next decision interval:

$$f_{i+1} = f_i. \tag{8.6}$$

The main goal of $(RT)^2$ is to operate as close to $\delta_i = 1$ as possible, while utilizing minimum network resources and meeting event delay bounds.

8.7.1   Qualitative Evaluation

$(RT)^2$ is self-configuring and can perform efficiently under the random, dynamic topology frequently encountered in WSN applications. In addition to real-time and reliable delivery, the average energy consumed per packet during communication is also decreased as the (no congestion, adequate reliability) state is approached through reporting rate control. Similar to ESRT, the $(RT)^2$ protocol simultaneously addresses congestion control and timely event transport reliability objectives in WSNs. In addition, real-time delivery is also supported, which is important for applications where delay is a major concern.

Despite its advantages, the $(RT)^2$ protocol operation is based on feedback from the sink, where a broadcast medium is assumed for the downstream traffic. However, this may not be applicable to deployments where multi-hop communication is necessary. Similar to ESRT, the same reporting frequency value is used for all sensors in a decision interval. This approach may not be suitable for networks where local interactions vary the network conditions significantly.

References

[1] Ö. B. Akan and I. F. Akyildiz. Event-to-sink reliable transport in wireless sensor networks. *IEEE/ACM Transactions on Networking*, 13(5):1003–1016, October 2005.

[2] I. F. Akyildiz and I. H. Kasimoglu. Wireless sensor and actor networks: research challenges. *Ad Hoc Networks Journal*, 2(4):351–367, October 2004.

[3] A. Al Hanbali and E. Altman. A survey of TCP over ad hoc networks. *IEEE Communications Surveys & Tutorials*, 7(3):22–36, 2005.

[4] H. Balakrishnan, V. N. Padmanabhan, S. Seshan, and R. H. Katz. A comparison of mechanisms for improving TCP performance over wireless links. *IEEE/ACM Transactions on Networking*, 5(6):756–769, December 1997.

[5] S. Floyd, V. Jacobson, C. Liu, S. Macanne, and L. Zhang. A reliable multicast framework for lightweight sessions and application level framing. *IEEE/ACM Transactions on Networking*, 5(6):784–803, December 1997.

[6] V. C. Gungor, O. B. Akan, and I. F. Akyildiz. A real-time and reliable transport protocol for wireless sensor and actor networks. *IEEE/ACM Transactions on Networking*, 16(2):359–370, April 2008.

[7] C. Intanagonwiwat, R. Govindan, and D. Estrin. Directed diffusion: a scalable and robust communication paradigm for sensor networks. In *Proceedings of MobiCom'00*, pp. 56–67, Boston, MA, USA, August 2000.

[8] Mica motes and sensors. Available at http://www.xbow.com/Products/Wireless_Sensor_Networks.htm.

[9] S.-J. Park, R. Vedantham, R. Sivakumar, and I. F. Akyildiz. A scalable approach for reliable downstream data delivery in wireless sensor networks. In *Proceedings of ACM MobiHoc'04*, pp. 78–89, Tokyo, Japan, May 2004.

[10] S.-J. Park, R. Vedantham, R. Sivakumar, and I. F. Akyildiz. GARUDA: achieving effective reliability for downstream communication in wireless sensor networks. *IEEE Transactions on Mobile Computing*, 7(2):214–230, 2008.

[11] F. Stann and J. Heidemann. RMST: reliable data transport in sensor networks. In *Proceedings of the 1st IEEE International Workshop on Sensor Network Protocols and Applications*, pp. 102–112, Anchorage, AK, USA, May 2003.

[12] M. C. Vuran, O. B. Akan, and I. F. Akyildiz. Spatio-temporal correlation: theory and applications for wireless sensor networks. *Computer Networks*, 45(3):245–261, June 2004.

[13] C.-Y. Wan, A. T. Campbell, and L. Krishnamurthy. Pump-slowly, fetch-quickly (PSFQ): a reliable transport protocol for sensor networks. *IEEE Journal on Selected Areas in Communications*, 23(4):862–872, April 2005.

[14] C.-Y. Wan, S. B. Eisenman, and A. T. Campbell. CODA: congestion detection and avoidance in sensor networks. In *SenSys'03: Proceedings of the 1st International Conference on Embedded Networked Sensor Systems*, pp. 266–279, Los Angeles, USA, 2003.

[15] C. Wang, K. Sohraby, B. Li, M. Daneshmand, and Y. Hu. A survey of transport protocols for wireless sensor networks. *IEEE Network*, 20(3):34–40, May/June 2006.

9

Application Layer

The role of the application layer is to abstract the physical topology of the WSN for the applications. Moreover, the application layer provides necessary interfaces to the user to interact with the physical world through the WSN. We will describe the application layer solutions in three categories: source coding, query processing, and network management. An overview of the application layer protocols is shown in Figure 9.1.

9.1 Source Coding (Data Compression)

As explained in Chapter 4, source coding is the first step in information delivery through wireless communication. Whenever a sensor node has information to transmit, the information source is first encoded with a source encoder. In effect, source coding exploits the information statistics to represent the source with fewer bits, i.e., source codeword. Therefore, source coding is also referred to as *data compression*. Accordingly, redundant information is compressed to decrease the data volume while preserving some or all of the information content.

The basic compression solutions can be classified into two based on the preservation of information after coding: *lossless compression* and *lossy compression*. Lossless compression reduces the data content, i.e., packet size, without hampering the information integrity. Much better compression efficiency can be achieved by lossy compression, which allows information loss for a higher reduction in data. Due to the higher complexity required for lossy compression, lossless compression techniques are generally adopted in WSNs.

Several lossless compression algorithms have been developed for both digital communication and computation. However, the applicability of these algorithms to WSNs is limited. Since the programming memory of many sensor nodes is limited, higher complexity algorithms such as LZW or gzip [13] cannot be directly implemented. Furthermore, several compression algorithms require a static dictionary to be stored for compression/decompression and frequent memory access is required. Since read/write operations result in high energy consumption for wireless sensor nodes, these solutions are not energy efficient [6].

In addition to the inefficiency of the compression algorithms in terms of energy consumption, the redundancy of information collected at each individual sensor may also not allow efficient compression at each source. However, the high-density deployment of WSNs results in highly correlated information being collected by multiple, closely located nodes. Consequently, instead of compressing on an individual basis, redundant information collected by a group of sensor nodes can be compressed. This leads to significant reductions in data to be delivered to the sink and, hence, reduced energy consumption. As a result, distributed compression techniques that exploit correlation at different sources for compression have been investigated for WSNs. Next, we describe two major approaches for compression in WSNs. Firstly, a node-centric source coding techniques, Sensor LZW, is described. Then, we explain distributed source coding, which distributes compression to multiple sensor nodes.

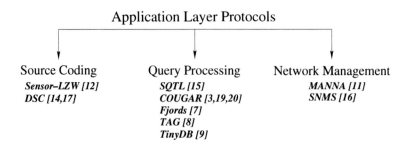

Figure 9.1   Overview of application layer protocols.

9.1.1   Sensor LZW

There exist several general purpose compression algorithms, where the main design goal is to increase the *compression ratio* [13]. However, energy and memory efficiency are not major concerns since they are generally designed for resource-rich devices. Furthermore, the code space of existing algorithms is too high to be implemented in memory-constrained sensor nodes [2, 12]. As a result, these existing algorithms are not suitable for WSNs and energy- and memory-efficient compression algorithms are required so that node-centric compression can be deployed in WSNs. To this end, it is important to consider that, in a typical sensor node platform, the energy consumed for computation is significantly less than that for memory access and communication. Hence, computation-intensive solutions with a small memory footprint are required.

Sensor LZW (S-LZW) [12] is a variant of the Lempel–Ziv–Welch (LZW) compression algorithm that is tailored to the resource constraints of wireless sensor nodes. LZW is generally used as part of the Unix *compress* function and the GIF standard.[1]

LZW is a dictionary-based method, which encodes strings of symbols as a *token* in a dictionary. The dictionary is partly initialized before encoding and is populated based on the sequence of symbols encountered during encoding. The main challenge for exploiting the LZW algorithm for WSNs is the adverse effects of wireless channel errors. LZW decoding cannot be performed if the encoder output stream is not completely received at the decoder. Consequently, the data to be compressed should be limited to small blocks so that retransmissions can be performed without large data loss and energy consumption. Furthermore, due to memory constraints, the dictionary size should be limited. S-LZW was designed to address these issues, and uses a dictionary size of 512 entries to compress data in blocks of 528 bytes (two flash pages).

The compression of observational data can be further improved by exploiting their repetitive nature. Since each sensor node observes a physical phenomenon, the data to be compressed are highly correlated in time. This means that there will be several repetitions in the data to be compressed. This can be addressed by adding a mini-cache structure that stores the most recent entries. S-LZW with mini-cache (S-LZW-MC) improves the compression ratio in cases where the data contain many repetitions.

In addition to relying on the internal interdependencies of data, compression efficiency can be further improved by transforming the data to a form with several patterns before compression. This is generally referred to as the *precondition step* and the S-LZW algorithm is improved by using two different methods.

The first precondition method is the Burrows–Wheeler transform (BWT), which is generally used for compression of images, sound, and text. An example transformation is discussed next.

---

[1] For details of LZW, refer to [13].

Table 9.1 Burrows–Wheeler transform [13].

swiss_miss	_missswiss	s
wiss_misss	iss_misssw	w
iss_misssw	issswiss_m	m
ss_missswi	missswiss_	_
s_missswis	s_missswis	s
_missswiss	ss_missswi	i
missswiss_	ssswiss_mi	i
issswiss_m	sswiss_mis	s
ssswiss_mi	swiss_miss	s
sswiss_mis	wiss_misss	s

■ **EXAMPLE 9.1**

Sample information of string "swiss_miss" is shown in Table 9.1. For the transformation, a string of n symbols is used to create an $n \times n$ matrix (first column of Table 9.1), where the first row is the original string of symbols. Each row of the matrix is constructed as the shifted copy of the previous row. Then, the rows are ordered according to their first character (second column of Table 9.1). Finally, the last column is used as the output from the BWT (last column of Table 9.1). The BWT reorders the symbols such that occurrences of the same symbol are located close by, which significantly improves the compression ratio. This flavor is denoted as S-LZW-MC-BWT.

The second precondition method used with S-LZW is based on the internal data structure of monitoring data. Generally, in monitoring applications, the data length and the contents are well known and fixed. Moreover, for a set of observation values, the entire observations or at least the most significant bits are the same. Accordingly, a *structured transpose* (ST) can be performed to improve the compression ratio further. More specifically, the data for each observation are used to fill the rows of a matrix until a maximum length is reached. Then, the matrix is transposed and the data are compressed using the transposed matrix. Since the transposed matrix has several repetitions, the compression ratio can be further decreased. This version of the protocol is denoted as S-LZW-MC-ST.

S-LZW affects the energy consumption of source nodes depending on the mote architecture. Since the compression is performed at the source nodes, the major overhead is found at those nodes. In cases where transceiver energy consumption dominates memory access, a reduction up to 2.6× is possible. However, for a platform such as MicaZ, where the energy consumptions for communication and memory access are comparable, energy savings depend on the type of information to be compressed [12]. However, the main advantage of compression techniques in WSNs is on end-to-end energy consumption. The amount of information to be transmitted is decreased through compression. As a result, less amounts of data are transmitted throughout the network. Therefore, even if compression leads to slightly increased energy consumption at the source nodes, overall energy savings are substantial for transmitting the information from the source node to the sink in a multi-hop path. This leads to reductions of up to 2.9× in energy consumption.

Precondition steps also improve the compression ratio as well as energy efficiency despite the penalty in computing the transformations and the additional storage required. For S-LZW-MC-BWT, decreases up to 1.96× and 2.1× are possible for local and end-to-end energy consumption, respectively. Similarly, for monitoring applications, S-LZW-MC-ST saves energy by 2.8× both locally and overall.

Source coding (compression) also helps save energy under wireless channel errors. As discussed in Chapter 6, reliability under wireless channel errors can be provided by retransmitting the corrupted packets. Since compression decreases the amount of information to be sent, retransmission cost is also decreased. This results in significant energy savings for multi-hop WSNs.

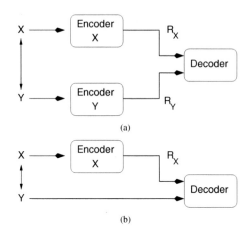

Figure 9.2 Slepian–Wolf coding: (a) the general case and (b) coding with side information.

9.1.2   Distributed Source Coding

Node-centric compression schemes such as S-LZW improve energy efficiency by decreasing the data content at each sensor node locally. The temporal relations between consecutive observations can be utilized for local compression. However, most applications may not require very high sampling rates that will lead to high temporal correlation. Moreover, because of the high density in sensor deployment, the information content observed by closely located sensors is highly correlated. This correlation can be exploited to significantly decrease the data content.

Distributed source coding (DSC) that dates back to the work of Slepian and Wolf [14] and Wyner and Ziv [17] in the 1970s stands as a promising solution for this type of application. Distributed source coding refers to the compression of multiple correlated sensor outputs that do not communicate with each other [18]. More specifically, the information at a sensor node is compressed assuming correlation with another node without exchanging any information. This information is sent to the sink and joint decoding is performed by the sink that receives data independently compressed by different sensors.

The lossless DSC, also known as *Slepian–Wolf coding* , focuses on distributed coding of two correlated sources, X and Y, as shown in Figure 9.2(a). Each source is encoded separately without any information from the other. Information theory states that without any information from other sources, the discrete random variable X can be encoded by at least $H(X)$ bits without any information loss, where $H(X)$ is the entropy of X. Accordingly, with separate entropy encoders and decoders, the following rates can be achieved: $R_X \geq H(X)$ and $R_Y \geq H(Y)$, where $H(X)$ and $H(Y)$ are entropies of X and Y, respectively. The Slepian–Wolf theorem states that by exploiting the statistical properties of X and Y at each encoder, the performance of joint encoding of X and Y can be achieved. In other words,

$$R_X + R_Y \geq H(X, Y) \tag{9.1}$$

$$R_X \geq H(X|Y), \quad R_Y \geq H(Y|X) \tag{9.2}$$

where $H(X, Y)$ is the joint entropy of X and Y and $H(X|Y)$ and $H(Y|X)$ are the conditional entropies.

A special case of Slepian–Wolf coding is where the source X is encoded while considering the statistics of source Y and source Y is encoded separately. Consequently, $R_Y = H(Y)$ and $R_X \geq H(X|Y)$, where $R_X + R_Y \geq H(X, Y)$. As shown in Figure 9.2(b), in this case, source Y is regarded as *side information*. The joint decoding of X and Y is performed at a single decoder such that the information of Y is used to jointly decode the information from both sources. This type of encoding/decoding reverses the traditional balance of encoder and decoder complexity leading to simple encoders and complex decoders, which are suitable for WSNs.

Next, we will discuss Wyner–Ziv coding, which constitutes the basis for lossy compression as well as special cases where DSC is used for WMSNs.

Although Slepian–Wolf coding constructs the foundations of DSC, it does not consider the existence of a channel and is purely a source coding solution. Wyner–Ziv coding considers lossy compression with side information as an extension to Slepian–Wolf coding [17]. Similarly, a source X is encoded with the statistical information of source Y. The decoder uses this information as well as the source Y as side information to decode $\hat{X}$. Given an acceptable distortion, $D = E[d(X, \hat{X})]$, Wyner–Ziv theory states that a *rate loss* is observed for lossy coding such that $R_{X|Y}^{WZ}(D) - R_{X|Y} \geq 0$, where $R_{X|Y}^{WZ}(D)$ and $R_{X|Y}$ are the achieved rates for Wyner–Ziv and joint encoding of X, respectively.

9.2   Query Processing

WSNs consist of many sensor nodes that monitor the physical phenomenon according to the requirements of the application. The sink ensures the delivery of the data of interest through queries sent to the nodes containing information about the requested information. The query replies can be made simply by sending the requested raw data immediately to the sink. However, the processing capabilities of sensor nodes provide alternative ways of processing these queries inside the network leading to significant energy conservation [1]. This phenomenon is referred to as *query processing*.

The WSN can be viewed as a distributed database where nodes continuously deliver streams of data to the sink [7]. Although there exists many database management systems (DBMSs) developed for traditional distributed databases, the unique characteristics of WSNs make these solutions not applicable. The unique characteristics of the WSN can be listed as follows:

- **Streaming data:** Sensor nodes produce data continuously, usually at well-defined time intervals, without having been explicitly asked for those data [7].
- **Real-time processing:** Sensor data usually represent real-time events. Moreover, it is often expensive to save raw sensor streams to disk at the sink. Hence, queries over streams need to be processed in real time [7].
- **Communication errors:** Since sensors deliver data through multi-hop wireless communication, wireless errors affect the reliability and delay of the distributed information reaching the sink.
- **Uncertainty:** The information gathered by the sensors contains noise from the environment. Moreover, factors such as sensor malfunction and sensor placement might bias individual readings [20].
- **Limited disk space:** Sensor nodes have strictly limited disk space. Hence, the information sent by the sensors may not be queried later.
- **Processing vs. communication:** As explained in Chapter 3, energy expenditure in data processing in WSN is much less than that in data communication. Hence, the data processing capabilities of sensor nodes should be exploited in query processing.

The processing power available in sensor nodes provides potential solutions for the challenges encountered in WSN query processing. It is clear that the queries sent by the sink can be easily replied to by sending the raw sensor observation to the sink. This approach is referred to as *warehousing* [3], where the processing of sensor queries and access to the sensor network are separated. As a result, a centralized DBMS can be developed to provide access to the collected data through classical database management techniques.

The warehousing approach, however, leads to both overutilization of communication resources in the WSN and accumulation of highly redundant data at the sink. As an example, if an application is only interested in the average value of specific information at a specific location, it would be more

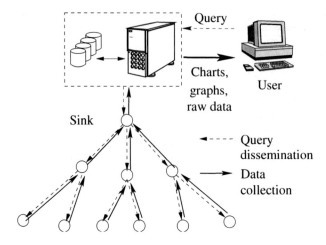

Figure 9.3   Query processing architecture.

efficient for nodes at this location to calculate the average locally and send this information as a single packet to the sink instead of sending all the individual information. Hence, exploiting existing techniques for query processing in WSNs leads to inefficient solutions. Instead, novel approaches tailored to the WSN paradigm are required. In this section, we describe these solutions, which have been developed for WSNs.

The overall architecture of a query processing framework is illustrated in Figure 9.3, where a user interacts with the sensor network. The query processing solutions provide the necessary tools to realize this interaction between a user and a distributed set of sensor nodes. Hence, the services provided by these solutions can be divided into two: server-side services and WSN-side services. In each query processing approach, first, the user interests are represented through a well-established syntax that can be easily converted into *queries* to the rest of the network. This constitutes the server-side part of the query processing solutions. The WSN-side services can be broadly divided into two steps: (1) query dissemination and (2) data collection. *Query dissemination* is performed by transmitting the queries, which are generated according to the user interests, to all or a subset of nodes in the network. In the second step, *data collection*, sensor nodes respond to these queries if their locally observed sensor data match the queries. Energy-efficient *representation* and *dissemination* of queries as well as the collection of interested data are of major concern in the design of query processing algorithms.

In the following, we describe the query representation approaches and aggregation issues in WSNs in general. Existing query processing techniques are described in detail.

9.2.1   Query Representation

Representation of the user requirements or interests is a key factor in the design of efficient query processing algorithms. The high-level interests of the user should be effectively represented as distributed queries that will be sent to each sensor in the WSN. To this end, an interest can be viewed as a task descriptor, which is named by assigning a list of attribute–value pairs that describe the task. This is illustrated through the following example.

■ **EXAMPLE 9.2**

Consider an animal tracking application, where a group of sensors are deployed to detect particular animals and communicate their locations. Let us assume that the user is interested in receiving information if a four-legged animal is detected. This is represented as a query using following attribute–value pairs:

```
type = four-legged animal           // detect animal location
interval = 20 ms              // send back events every 20 ms
duration = 10 seconds               // for the next 10 seconds
rect = [-100, 100, 200, 400]     // from sensors within rectangle
```

In this case, the sink is interested in receiving packets with 20 ms intervals through the next 10 seconds from the location specified by the *rect* attribute. A sensor can reply to this interest through the attribute–value pair list shown below:

```
type = four-legged animal           // detect animal location
instance = elephant                 // instance of this type
location = [125,220]                      // mote location
intensity = 0.6               // signal amplitude measure
confidence = 0.85                // confidence in the match
timestamp = 01:20:40
```

The response indicates that an elephant was detected by the sensor at location [125, 220] at time 01:20:40. In addition, more detailed information about the confidence of the detection as well as the signal amplitude measure, which is used for this detection, can be provided. This information is then used at the sink with the other responses to estimate the location of the animal and track its movements.

The representation of the user interest through efficient syntax to be used by sensors is a major challenge in query representation. The Sensor Query and Tasking Language (SQTL) has been developed to address this challenge through a procedural scripting language [15]. Accordingly, SQTL defines two primitives for the user to interact with the WSN as follows:

- **Querying:** The user generates formal statements to gather information related to a certain interest. These statements are called *queries* and are of the form: "What is the temperature in the southeast corner of the sensor field?" Querying is regarded as a synchronous operation, where the user waits for the answer until further processing is performed, i.e., it is blocked.

- **Tasking:** Certain long-term operations cannot be performed through queries. For interests related to the long-term changes regarding a certain phenomenon, instead, the user assigns *tasks* to the sensor nodes. Consequently, the WSN can respond efficiently to an interest of the form "Track any moving object inside the sensor field." The user is not blocked during tasking but rather can perform certain operations on the already collected data.

WSNs can be regarded as distributed databases, where the data are continuously generated based on the interests of the user. Accordingly, traditional database solutions may seem applicable to data processing in WSNs. In other words, the WSN can be regarded as a distributed database and the data related to the user interest can be sent to the sink from each sensor that observes the events of interest. According to the collected information, the sink can perform high-level operations to provide a global view of the data of interest to the user. As an example, if the user is interested in the average temperature of a given area, temperature information from each sensor in the particular area can be collected at the sink. Then, the sink finds the average of these values to provide a response to the user.

The traditional distributed database approaches, however, do not consider the cost of information delivery. In WSNs, information delivery is the dominant cost because of the required communication power and frequent losses. Instead, the local processing capabilities of sensors need to be exploited to distribute the intelligence and processing tasks throughout the network. Accordingly, the amount of data, which will be transmitted through the network, can be decreased while preserving the information content of interest to the user. Considering the same example, instead of sending their raw temperature data, closely located sensors can exchange their readings and locally calculate the average value to be sent to the sink. As a result, instead of sending the raw temperature data from each sensor in an event area, only a single packet representing the average in this area is transmitted. Although the information content is kept intact, significant savings are possible in terms of communication cost.

In-network processing, however, cannot be accomplished through traditional queries. Instead, through tasking, the user can ask the sensor nodes to perform certain long-term tasks that include gathering data, local processing, and collaboration among neighbors. SQTL provides the required syntax to define these interactions.

Each SQTL program is developed at the sink and encapsulated within an *SQTL wrapper*. Upon receiving the SQTL packet, each sensor node performs the actions specified by the SQTL wrapper. These actions include storing the SQTL program for further use, forwarding the program to a particular set of nodes, or executing the program through a specific application. To this end, a *sensor execution environment (SEE)* is used at each sensor to dispatch incoming messages.

The SEE provides specific low-level primitives that can be used by the user according to the SQTL commands. More specifically, three groups of primitives are provided:

- **Sensor access:** `getTemperatureSensor()`, `turnOn()`, `turnOFF()`
- **Communication:** `tell()`, `execute()`, `send()`
- **Location-aware:** `isNorthOf()`, `isNear()`, `isNeighbor()`

Using these primitives, sensors can be remotely programmed to execute tasks that include collaboration among local neighbors. As a result, the SQTL framework is performed at each node. The incoming messages are filtered based on the SQTL wrappers by the SEE. Certain messages can be immediately discarded or forwarded to certain neighbor nodes. If a certain criterion is met, the SQTL programs are then executed in the node as running applications. A sample SQTL program for temperature reading is provided next.

■ **EXAMPLE 9.3**

Consider a temperature reading application, where the sink is interested in the maximum temperature of the temperature readings in the sensor network. Accordingly, a simple SQTL application is sent by the sink to the sensor nodes including the following SQTL wrapper [15]:

```
(execute
    :sender          SINK
    :receiver (:group NODE(1) :criteria TRUE)
    :application-id 123
    :language SQL
    :content ( SELECT Max(getTemperature()) FROM ALL_NODES ))
```

The SQTL wrapper includes the SQTL function, which requires the node to get the maximum temperature from its children. The corresponding architecture is shown in Figure 9.4. Firstly, the sink sends the SQTL application with the wrapper to node A as shown in Figure 9.4(a). Upon receiving the message, node A executes the SQTL wrapper and sends a confirmation message to the sink. Then, node A sends the SQTL message to its children, i.e., nodes B, C, and D (Figure 9.4(b)). Each node responds with a confirmation message to node A and reiterates the same procedure for its children. Finally, the maximum temperature information is returned to node A as shown in Figure 9.4(c). By using its own temperature information, node A returns the maximum temperature information to the sink.

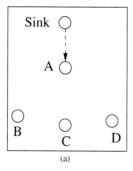 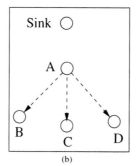 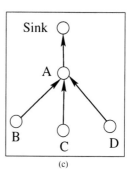

Figure 9.4   Query dissemination and data collection using SQTL.

SQTL provides an efficient method to disseminate user interests to the sensors in a WSN. Through the use of SQTL wrappers, higher level operations are supported without executing the main program. Accordingly, user queries and tasks are executed at the sensor nodes specified by the SQTL wrapper. Since different application types are supported, SQTL can be used for any type of query or task. This allows even the sensor software to be updated based on SQTL messages.

Query representation techniques enable flexible operation for information delivery in WSNs. Thus, information delivery not only depends on the requests generated by the sink, but can also be initiated by each sensor in the network. Consider the SQTL wrapper code used in Example 9.3. The same SQTL message can be generated by a sensor in the field by identifying the sender as SENSOR. This is particularly important for event-based WSNs, where the user is not aware of the events that may occur in the network. If a certain activity is detected by a sensor, this sensor can *publish* the availability of such data and inform the user. If the user is interested, further queries can then be sent to improve the fidelity of the event information by assigning more sensors to gather data. According to the way the query messages are generated, query processing approaches can then be classified into three as follows:

- **Push-based query processing:** Also referred to as *sensor-initiated information delivery*, in this case sensors publish/advertise their information readings, which may be locally distributed or sent to the sink for further processing.

- **Pull-based query processing:** Also referred to as *sink-initiated information delivery*, here the sink disseminates the interests of the user to a set of sensor nodes in the network. In this case, the interest may be related to an attribute of the physical field (query) or an event of interest to be observed (task). Based on the user interest, the sensor nodes respond with the requested information.

- **Push–pull query processing:** Both push- and pull-based query approaches can be used in a network, in which case both the sensors and the sink are actively involved in query processing.

The type of query processing approach dictates the suitable architecture to be used. In addition to the high-level architecture to be used with query processing, different query types can be used. Queries can be classified based on two different criteria: (1) temporal and (2) spatial content.

According to temporal content, queries are classified into three as follows:

- **Continuous query:** This type of query is generally related to monitoring applications, where sensors continuously send their observations at the specified frequency. The duration of the query can also be specified in the query description. Continuous queries are generally associated with tasking.

- **Snapshot query:** This type of query relates to collecting a specific attribute at the current instant or at some specific point in time. In the case of queries related to future instances, snapshot queries are associated with tasking.
- **Historical query:** In addition to performing sensing based on continuous or historical queries, the sensor can also store information related to previously sensed information. Historical queries collect information (or a summary thereof) about the past. They are generally regarded as closest to traditional distributed database approaches.

In addition to their temporal content, queries are also classified according to the way they address sensors in the WSNs, i.e., the spatial content. A general query can address *all* the sensors in the network, which is generally related to system-level tasks. However, most queries in sensor networks result in a group of sensors responding instead of all the sensors in the network. According to the way these sensors are addressed, queries can further be classified into three as follows:

- **Data-centric query:** This is the most used query type, where a particular event of interest is defined in the query. Fire detection according to a sudden decrease in ambient temperature is an example of a data-centric query. The query does not directly address any nodes to send their temperature readings. Instead, nodes respond to the query only if their data content matches the query requirements.
- **Geographical query:** Instead of detecting an event, the user may be interested in monitoring a certain area of the network. In this case, geographical queries are used, which specify sensor nodes according to their location.
- **Real-time detection query:** For applications such as intruder detection, the data originators are not known in advance. Instead, a sensor responds to the query only if the object of interest moves close to it.

The type of query also impacts the routes created in the network to deliver the query messages. It is important to note that the set of sensors that respond to a query is not always the same as the recipients of the query message. As an example, data-centric queries may be delivered to all the sensors in the network. However, nodes respond only if the event of interest occurs. On the other hand, for a geographical query, the query messages can be specifically sent to the area of interest. Therefore, routing decisions are taken based on the type of query that is used in the network.

9.2.2   Data Aggregation

The main peculiarity of WSN traffic is that the sensed information is usually collected at a single point, i.e., the sink. Hence, information flow can usually be perceived as a reverse multicast tree as shown in Figure 9.5, where the sink queries the sensor nodes to report the ambient conditions of a physical phenomenon. The *many-to-one* nature of the information flow results in high contention and congestion at nodes which are located closer to the sink, since these nodes are exposed to a larger volume of traffic compared to nodes that are located away from the sink. This causes an increase in contention and congestion, which may lead to packet losses. Moreover, nodes closer to the sink consume higher energy, which may result in these nodes depleting their residual energy earlier. As a result, the sink may be disconnected from the rest of the network.

The adverse effects of many-to-one flows can be mitigated by exploiting the content of information that is sent throughout the network. WSNs observe physical phenomena which are correlated in space and time. As a result, information sent by each sensor network is correlated with another to some extent. For nodes that are located very close to each other, the correlation is so high that these nodes may send identical packets. On the other hand, for distant nodes, although the information content may differ considerably, the type of information is still the same. As an example, for a query related to temperature sensing, all nodes send temperature information. As a result, the packets that are being sent contain similar header information even if the reported temperature value may be different.

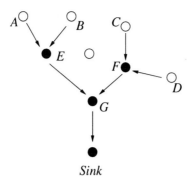

Figure 9.5   Example of data aggregation.

The similarity in both the information content and packet content in WSNs can be exploited to minimize data flow from several sensor nodes to the sink. As an example, for closely located nodes that observe the same information, instead of sending packets from each node, a single packet can be sent to represent these groups of sensors. Similarly, packets from different locations of the sensor field can be merged in a single packet by maintaining the temperature information as an array. In addition, traffic load can further be reduced by performing *in-network processing* based on the query sent by the user. According to Example 9.3, if the maximum temperature value is required, a sensor can merge the information that it receives from its children with its own reading to report only the maximum of the received samples. As a result, several packets injected into the network can be merged into fewer packets, which relieve the traffic at the bottleneck nodes as the information progresses through the network. These constitute the basics of *aggregation* techniques in WSNs, which are generally used in conjunction with query processing methods. This is illustrated by the following example.

■ **EXAMPLE 9.4**

Consider the network topology shown in Figure 9.5, where a tree rooted at the sink is used for information delivery from the sensors to the sink. Data aggregation can be performed in this tree, where the data received from multiple sensor nodes are aggregated as if they are about the same attribute of the phenomenon. The nodes that receive information from multiple nodes in this tree are usually used for aggregation and are called *aggregation points*. As an example, the sensor node E aggregates the data from sensor nodes A and B while sensor node F aggregates the data from sensor nodes C and D.

Data aggregation can be perceived as a set of automated methods of combining the data that come from many sensor nodes into a set of meaningful information [4]. In this respect, data aggregation is also referred to as *data fusion* [5]. Data aggregation solutions generally consist of three main components: data storage, aggregation function, and aggregation path. We describe these components next.

Data aggregation takes place in *aggregation points*, where information from multiple sensor nodes is collected. Due to the unsynchronized nature of sampling and communication, data to be aggregated do not arrive simultaneously at the aggregation points. Hence, these data should be stored until sufficient information is available for aggregation. Since on-board memory is limited, sensor data should be stored in a memory-efficient manner. On the other hand, the accuracy of information should be preserved. Depending on the type of query, data can be stored with different degrees of accuracy. If individual information is important for aggregation functions, gathered information is stored as lists or histograms. As a result, information content is highly preserved at the aggregation points. This, however, results in higher memory allocation for the data and limits the number of data points that can be stored. On the other hand, if the query is related to the statistical properties of the data, such as mean, variance, or maximum, then only statistical representations of the collected data are stored. This results in more efficient memory allocation.

The second main component of aggregation solutions is the selection of the aggregation functions. As explained above, the type of aggregation function affects the data storage approach. This is directly related to the efficiency of the aggregation solutions as well. Aggregation functions can be as simple as first-order statistics such as mean, maximum, minimum, or very simple operations such as suppression of duplication. This simplifies the data storage approach as well as the processing that needs to be performed at the aggregation points. On the other hand, more sophisticated functions, which require the calculation of spatial or temporal correlation, signal processing, or algorithmic approaches, can be selected as aggregation functions. As a result, the complexity of aggregation increases, resulting in higher load for aggregation points.

Finally, the construction of aggregation paths is a major challenge for aggregation solutions. The aggregation points should be selected so that information is collected at these points efficiently and rapidly. This is, however, difficult due to the dynamic nature of WSN communication. The information to be gathered at aggregation points may not be available due to uncertainties in the sensor hardware, physical phenomenon, and wireless channel. Accordingly, sensors may not gather the requested data due to intermittent problems with the sensory hardware. Moreover, data delivery is generally based on data-centric queries. Hence, the number of nodes that respond to the query depends on the properties of the physical phenomenon, which are non-deterministic. Similarly, packets sent from certain nodes may be lost before reaching the aggregation points due to wireless channel errors.

Because of these uncertainties, it is not possible to determine the amounts of data that will be available at the aggregation points in advance. More importantly, the amounts of data vary throughout the lifetime of the network. Moreover, the multi-hop communication structure and the broadcast nature of the wireless channel result in information trickling at one message a time. Hence, a node may not have complete information about the neighborhood and the information to be aggregated. This results in a tradeoff between accuracy in aggregation and the associated delay. Thus, finding the most efficient aggregation points for each aggregation task in the network is not trivial.

Aggregation has several benefits for communication in WSNs. The high asymmetry between the energy consumption for computation and that for communication motivates the use of in-network processing solutions. Aggregation is a typical example, where the in-network processing solutions help reduce the traffic in the network. As a result, energy efficiency of the network can be improved. Since the traffic load is minimized through aggregation, the network can be made scalable for large numbers of sensors and sinks.

Despite its advantages, aggregation incurs additional storage requirements in resource-constrained sensor nodes. Therefore, aggregation solutions should be designed to minimize the impact on memory, while preserving the accuracy. Furthermore, in-network processing may result in information loss due to imperfect aggregation approaches. As an example, outliers may distort the results for simple max/min aggregates if care is not taken. However, providing robust aggregation solutions with resource-constrained sensor nodes is a major challenge. Several solutions have been developed for aggregation in WSNs [10]. Most of these solutions are implemented as part of the query processing protocols, which will be explained next.

9.2.3   COUGAR

The COUGAR sensor database system provides a distributed data gathering framework for individual sensor data through user-defined queries [3, 19, 20]. The information in the WSN is classified into two as: *stored data* and *sensor data*. Accordingly, the WSN is regarded as a database system with two classes of available data.

The stored data refer to the information about the sensors in the field including their type, their locations (if available), as well as the properties of the physical phenomena. The stored data are available at the sink and modeled as relations.

The sensor data refer to the information gathered by the sensors according to the queries. In COUGAR, the sensor data are represented as time series. More specifically, the sensors are assumed to be synchronized and sensor data are regarded as outputs of a signal processing function at a specific location, at the time of the record. Each sensor represents the data based on a sequence model, which is defined as a 3-tuple that includes the set of records, an ordered domain, and the ordering of the records by the ordering domain.

According to the representation of data in two classes, two types of queries are used based on relational and sequence operators. The relational operators are applied to stored data, while the sequence operators apply to sensor data. Moreover, three operators are defined to combine relations and sequences. Accordingly, (1) sequences can be used as input to project out the position information to obtain a relation, (2) a sequence and a relation can be used in a product operation to obtain a sequence, and (3) a sequence can be aggregated by excluding the position attribute.

The sensor nodes observe data that are gathered based on certain signal processing functions, e.g., sampling, comparison, integration. Accordingly, in COUGAR, the outputs of individual sensors are represented as abstract data type (ADT) functions. As a result, an ADT object in the database corresponds to a physical sensor in the real world. Using this abstraction, the COUGAR database system uses an SQL-like query language to issue queries to the sensors. This is illustrated next.

■ EXAMPLE 9.5

Let us consider an indoor WSN that is deployed on different floors of a building for a temperature monitoring application. In this case, the sensor database contains a single type of relation *R(loc point, floor int, s sensorNode)*, which defines the location, floor, and the sensor ADT that provide temperature data. Accordingly, the query: "Return the temperature on the third floor every minute" is represented as

```
SELECT R.s.getTemp()
FROM R
WHERE R.floor = 3 AND $every(60);
```

using slightly modified SQL expressions. In the representation, the function `getTemp()` is a method which returns the observed temperature. With the definition of `floor` and `every()`, the sensors on the third floor are instructed to return this value every 60 s.

In addition to data representation, COUGAR also provides a query layer that is regarded as a layer between the application and network layers. Accordingly, cross-layer interactions between the routing protocol and distributed query protocol are provided. The main goal of the design of the query layer is to abstract the functionality of a large class of applications into a common interface of declarative queries. However, since the query layer has different requirements in terms of routing, some modifications to existing routing protocols are needed. Hence, COUGAR follows a specific topology for query processing and aggregation in WSNs.

A typical architecture for the COUGAR protocol is shown in Figure 9.6, where the WSN is composed of *sensor* nodes and *leader* nodes. Sensor nodes send their responses to the leader node, which performs data aggregation optimally. Based on the query, a specific sensor node is designated as the leader node to perform aggregation. The leader selection can be performed before the deployment by specifically deploying nodes with rich processing power or leader selection can be distributively performed in the network based on a distributed leader election algorithm. Based on the leader and non-leader sensor nodes, the query processing and algorithm are performed as follows.

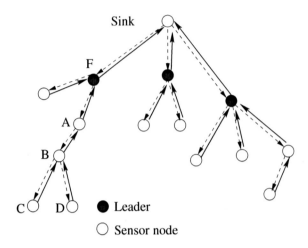

Figure 9.6   COUGAR architecture [3, 19, 20].

While the aggregation is done optimally at the leader node, the non-leader sensor node can also perform *partial* aggregation based on the information it receives from neighboring sensors. This is illustrated in the following example.

■ **EXAMPLE 9.6**

Consider the network architecture shown in Figure 9.6, where the information generated by nodes C and D is forwarded to node B, which uses node A to reach the leader node F. Accordingly, a partial aggregation of the data from nodes B, C, and D can be performed at node B and this aggregation can be further refined at node A later. In this way, the processing burden at node F is partly relieved by pushing the aggregation operations inside the network. More importantly, instead of sending four individual packets from nodes A, B, C, and D, a single packet traverses through the network, which significantly reduces the cost for communication. Partial aggregation, as a result, provides scalability to the distributed aggregation operation.

Based on the observation in Example 9.6, the query plan at the sensor nodes is as shown in Figure 9.7. Each sensor performs partial aggregation based on two sources of data. Firstly, a node locally generates information based on the collected data from its on-board sensors. Furthermore, if the node is on the path from a sensor node to the leader node, it receives these data from the radio. These two types of data can be partially aggregated for distributive and algebraic aggregate operators such as SUM, AVG, MAX, MIN [20]. Each sensor node then sends the partially aggregated data toward the leader.

The query plan at the leader nodes is also shown in Figure 9.7, where an aggregate and a selection operator exist. The partially aggregated results from multiple sensor nodes are first aggregated optimally. The results are then sent to the sink only if they match the query specification sent by the user. Hence, in addition to aggregation, the leader nodes serve as a first level of query processing so that the information is filtered inside the network if it does not meet the requirements of the user.

The COUGAR protocol requires several modifications at the underlying layers. Firstly, the intermediate sensor nodes need to access the data in the packets that they receive from their neighbors. However, generally, relay packets are handled at the network layer through the header information but the payload is not accessed by the relay nodes. Therefore, COUGAR requires the relay nodes to *intercept* the packets.

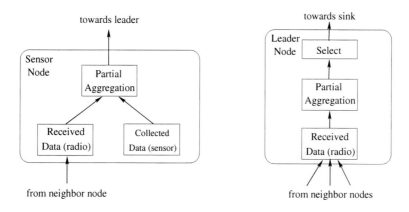

Figure 9.7   Query plans at the sensor and leader nodes.

This is performed using *filters*, which lets the network layer pass the received packet through a set of predefined functions. Accordingly, data aggregation can be performed at the query layer.

Secondly, COUGAR requires paths to be constructed from each sensor to the leader node associated with a query. Hence, modifications are required at the network layer for route initialization and maintenance. Based on the selected leader, the routing tree needs to be built according to a query tree originated at the leader of aggregation. Moreover, since packets containing aggregated results are more important than individual readings of sensors, a route maintenance technique that considers the depth of the node in a query tree is required.

Qualitative Evaluation

COUGAR provides efficient operators to combine stored data with streams of sensor data. Accordingly, the efficiency of the query processing is improved since static information related to sensors can be stored at the sink without requiring delivery of frequency information. Furthermore, the cluster-based approach for aggregation results in a distributed operation, where certain query processing tasks are distributed to leader nodes in the network. More efficient aggregation results can be obtained through this architecture.

COUGAR focuses on long-term queries in a push-based architecture. Hence, it is applicable to monitoring data. However, pull-based queries and event-based applications are not supported. Moreover, the cluster-based architecture increases the load on leader nodes since most of the query processing, aggregation, and filtering is performed at these nodes. Hence, these nodes should be frequently changed in the network to distribute the load among other sensors, which increases the communication overhead due to the leader selection algorithm.

9.2.4   Fjords Architecture

The Framework in Java for Operators on Remote Data Streams (Fjords) has been developed to handle *streams* of data, which are generated by sensors, and perform in-network operations on these data [7]. Accordingly, certain potions of the streams can be combined for efficient data delivery and the network can support both push- and pull-based queries. In addition to defining operators for query processing, the architecture consists of *sensor proxies* that reside at the sensor nodes and provide an interface between the query operations and the physical sensors.

The Fjords architecture has been developed for data streams that originate at each sensor. A stream is defined as a continuous set of data that is *pushed* into the network by the sensors. Accordingly, each sensor transmits sensed information at a certain rate, which is carried over the network. Streams are considered to have infinite length. Hence, operations over entire streams are not possible. On the other hand, query operations can be performed on individual data on-the-fly or for small groups of data that are stored locally. Fjords supports query processing on these limited blocks of streams.

The streams are managed through a *queue* structure, which defines a pipe between any two components. The pipes may be defined inside a single node between two components or between remote components that are connected wirelessly. Based on the queue structure, query processing in Fjords is performed through operators. An operator O is characterized by a set of input queues, Q_i, and a set of output queues, Q_o. Based on the particular operation, the operator O can select a subset of input queues in any order and provide an output to some of the output queues. Manipulation of the data is performed only at the operators, which are connected by queues. Thus, when information is output to a queue, it is routed to the next operator connected to the queue. Therefore, data integrity is preserved in the queues.

Since queues act as pipes between operators in the network, both push- and pull-based queries can be supported. For a push query, the input operator places data on the push queue continuously, which can be received by the output operator. Similarly, for a pull query, the output operator requests data from the input operator, which responds by placing the requested data on a pull queue. Fjords has been specifically developed to handle flows. Consequently, some of the traditional operators that operate on groups of data in a database cannot be directly used. To accommodate these operators, several non-blocking operators are defined. As an example, aggregate operators such as average, count, and sort are modified to operate incrementally, without requiring the entire stream to be available.

Fjord operators and queues provide a distributed architecture to communicate and process queries on a WSN. The same architecture can be used by multiple users, which may send similar queries to the network. However, trying to handle each query independently consumes significant resources at each sensor node. Instead, Fjords relies on *sensor proxies*, which serve as a sensor interface with the rest of the query processor. Sensor proxies are responsible for a wide range of operators, including adjusting sensor sampling rate, directing sensors to aggregate samples, packaging samples as tuples, routing these tuples to user queries as needed, and downloading new programs to sensors.

A sensor is limited in the services that it can provide. However, several users may be interested in the same information from a particular sensor. Sensor proxies ensure that sensors do not consume unnecessary resources for each query. Instead, queries are aggregated so that a single copy of an observation is generated at the sensor but this copy is sent as a response to all the queries associated with it.

■ **EXAMPLE 9.7**

Let us consider a real-life scenario, where a group of sensor nodes are deployed alongside a freeway to detect the speed of passing vehicles. The network sensors collect speed information and relay this information to a sink, i.e., data server. For this deployment, the following query can be sent to the network:

```
SELECT AVG(s.speed, w)
FROM sensorReadings AS s
WHERE s.segment in knownSegments
```

where the average speed in road segments indicated as *knownSegments* is sought. The query specifies an averaging window of size w.

Since the query defines a continuous task, each sensor generates a push queue between the sensor and the sink. The received stream at the sink is forwarded to Fjords. Firstly, the sensor proxy uses the received streams to find the average over a time interval w. The aggregates are then sent to a multiplex operator, which forwards these data to two operators. The save-to-disk operator saves the information for future use and the filter operator filters the aggregates related to only the known segments according to the user query. The response to the user query is then provided through an output stream. Using this architecture, if multiple queries related to different segments of the road are generated by different users, the same Fjords can be used by adding multiple filter operators according to each query. In this way, the same sensor proxy and multiplex operator can be used.

Qualitative Evaluation

Fjords provides efficient methods for query processing on streams of data. Accordingly, continuous streams can be aggregated over a time interval. Moreover, sensor proxies significantly reduce the load on sensors by aggregating similar queries and adjusting operations on the individual sensors accordingly. As a result, a sensor is not required to provide services for each query sent to the network.

Since the Fjords architecture collects most of the data at the sink and performs query processing operations at the sink, the raw data can also be saved. This provides additional capabilities for postprocessing data if further information regarding an event is required. Although this decreases the complexity at the sensor nodes, the communication cost is significantly high. Sensor information should be collected at the query operator for further processing.

The concepts of input and output queues are mostly focused on the sink architecture rather than the rest of the network. Issues such as packet losses and energy efficiency at the sensor nodes have not been addressed. Moreover, similar to COUGAR, Fjords does not support event-based queries.

9.2.5 Tiny Aggregation (TAG) Service

The TAG service combines distributed query processing techniques with in-network aggregation [8]. TAG provides a simple and declarative interface for data collection and aggregation, while distributively executing aggregation queries in the WSN. To this end, a tree-based query dissemination approach is employed. Queries from the users are distributed into the network, which results in a routing tree rooted at the sink. Using this resulting tree, data transmissions are scheduled to collect the responses to the queries from the leaves to the root. The protocol operation is divided into *epochs*, which define the unit time each operation is based on.

TAG relies on a SQL-like query representation, where the information is stored in a table called sensors. Accordingly, a typical TAG query is represented as follows:

```
SELECT {agg(expr), attrs} FROM sensors
   WHERE {selPreds}
   GROUP BY {attrs}
   HAVING {havingPreds}
   EPOCH DURATION i
```

which follows the SQL syntax closely except for the EPOCH DURATION clause, which defines the unit time in terms of local ticks for each node to use. The TAG query can be used to specify certain attributes for gathering by selecting certain sensors using the WHERE clause. The sensor data can further be grouped based on one or more attributes through the GROUP BY clause. Moreover, certain data can be further filtered out through the HAVING clause if certain predicates are not satisfied. The main difference from the traditional SQL applications is that the output of a TAG query is a stream of values rather than a single aggregate value.

The agg function in the TAG query syntax constitutes the core of the TAG protocol. Similar to aggregation computation in traditional large-scale networks, the aggregation is performed via three functions, i.e., a merging function, an initializer, and an evaluator. As an example, for an average

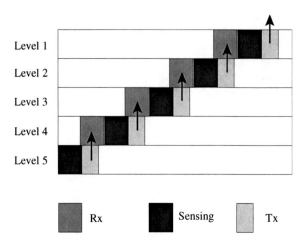

Level 1

Level 2

Level 3

Level 4

Level 5

Rx   Sensing   Tx

Figure 9.8   Data delivery during one epoch in TAG [8].

operator, the merging function computes the sum and count of the aggregated data, the initializer instantiates the state records, and the evaluator uses the sum and count to calculate the average by a simple division.

In addition, as in a SQL-like query representation scheme, the TAG protocol provides distributed protocols to disseminate queries and collect data associated with these queries. For this purpose, TAG relies on the routing tree built by broadcast messages sent by the sink. More specifically, the TAG protocol consists of two phases, the distribution phase and collection phase, as explained below:

- **Distribution phase:** In this phase, the aggregate queries are disseminated into the network through the sink. This results in a routing tree rooted at the sink. Each node is associated with a parent node, which is a node closer to the sink and used to forward the local and relay the aggregated data toward the sink.

- **Collection phase:** In this phase, responses to the queries are sent from children to parents in the routing tree toward the sink. The communication is scheduled based on epochs and each level at the routing tree performs communication according to this schedule.

When a query is generated, an epoch is also associated with this query. The epoch is defined as the amount of time the network is expected to take in collecting the responses to the query at the sink. Since the network consists of many levels as part of the routing tree, the epoch is used by the parents to assign a deadline to the children according to their level at the routing tree. The dissemination of the query during the distribution phase also serves as a routing tree construction mechanism.

When a node i receives a query r, it first chooses the sender of this message as the parent. This parent is used to send the aggregates back to the sink. Moreover, r indicates the time when node i is expected to report back to the parent. Accordingly, node i estimates the time it takes for the local computations, and sets an earlier time as a deadline for *its* children. The updated query r is forwarded to the children. Since a broadcast medium is used, nodes that receive this query become part of the routing tree if they are not already and select node i as their parent. If a node receives query messages from multiple nodes, it chooses one of them as a parent based on the channel quality or other criteria defined by the protocol. The query is disseminated into the network until all nodes receive the message.

In the collection phase, the responses to the query are collected from the leaves of the network in an epoch. To this end, the deadline, which is communicated by the parents during the distribution phase, is used to synchronize the children and parents. During the specific deadline, the parent node activates its

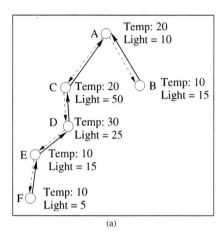
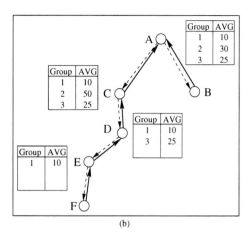

Figure 9.9   Grouping concept for Example 9.8.

transceiver to receive the aggregate information from its parents. Then, local sensing and processing are performed to aggregate the received data with the local observations. Finally, the aggregate information is passed to a higher level during the associated deadline. This is illustrated in Figure 9.8, where an epoch is shown for a network with a routing tree depth of 5. At the beginning of the epoch, the nodes at the edges of the network (level 5) perform sensing and transmit this information to the higher level (level 4). The parent node at level 4 starts to listen to the channel at a time earlier than the associated deadline to address any clock drifts that may occur. Once all the information is collected from all the children, the parent node at level 4 performs sensing and processing and transmits its aggregate data to its parent at level 3. Accordingly, the information trickles in this manner toward the sink.

The synchronized collection approach allows the parent nodes to receive information from all of their children and perform aggregation efficiently. Accordingly, if the aggregation function allows, each node transmits only one packet during each epoch. This significantly reduces the traffic load on nodes closer to the sink compared to a scheme, where tree-based aggregation is not performed. Moreover, the aggregated data can be classified into groups to preserve higher granular information during collection. This is performed through the GROUP BY clause. Grouping is illustrated in the following example.

■ **EXAMPLE 9.8**

Consider the following query

```
SELECT AVG(light), temp/10
    FROM sensors
    GROUP BY temp/10
```

and the topology shown in Figure 9.9(a). The query requests an average of the light intensity sensed by each sensor in the network but groups the aggregates with respect to temperature in groups of 10 °F. The light and temperature values observed by each node in a given epoch are also shown in Figure 9.9(a). According to the query description above, each node computes the aggregate of light intensity in each group based on the values it receives from its children and its locally generated value. The resulting aggregate results are shown in Figure 9.9(b).

Since the temperature values observed by nodes E and F are within the same group $(0 < \text{temp} \leq 10)$, only a single aggregate entry is sent to node D. At node D, the locally generated value is in another group (20 < temp ≤ 30), so a new entry is added for the locally generated value and transmitted to node C. At node C, a similar operation is performed and finally node A receives the aggregates from nodes B and C. As a result, the average light values corresponding to a particular temperature range are collected through the GROUP BY clause.

TAG provides several additional services for distributed query processing and aggregation. Since aggregation does not depend on entire streams, aggregation results tolerate losses in the network. In case of disconnections, the routing tree is updated so that nodes select parents that provide better channel quality. If a node experiences frequent channel errors with its parent, it listens to the broadcasts from other nodes in its neighborhood to select another node with a better channel quality and the same or better tree depth as its parent. Moreover, the routing tree is constructed based on each query, which minimizes the need for continuous route maintenance.

Qualitative Evaluation

Since aggregation does not depend on entire streams, aggregation results tolerate losses in the network. Moreover, the traffic load on each node is decreased through the synchronized communication scheme. If the packet sizes and the aggregation function allows, each node needs to transmit only a single packet during an epoch. TAG synchronizes nodes at different levels so that aggregation operations can be performed more efficiently. This synchronization also provides opportunities for energy management by turning off nodes during idle durations. Since the times when a node should sense and communicate are clearly identified, the nodes can be completely turned off without affecting the connectivity of the network.

The energy consumption for receiving packets from multiple children nodes is not clearly maintained. During the specified communication, multiple children nodes may contend to send their aggregates. In addition, the synchronized communication scheme incurs additional delay for event delivery, which is proportional to the tree depth. Hence, the scalability of the protocol is limited. Although TAG supports distributed aggregation to reduce data traffic, queries are still sent to the entire network even if some nodes will not respond. This increases the communication overhead.

9.2.6 TinyDB

COUGAR, Fjords, and TAG query processing solutions employ several traditional distributed query processing approaches with appropriate modifications for the WSN paradigm. Both solutions provide a query processing architecture that is tailored to monitoring traffic, where nodes periodically sample a physical phenomenon and send their readings to the sink. This architecture is applicable to a large class of WSN applications. However, for event-based applications, information delivery is bursty and may not result in continuous streams. Hence, these solutions may not be applicable to event-based applications. Moreover, these solutions consider that the sensor data are available and address issues related to information delivery and in-network processing. Except for TAG, where sampling times are adjusted according to the collection tree, the cost for sensing, i.e., generating database entries, has not been considered. Furthermore, none of these solutions address the energy consumption due to sampling as a result of query processing.

The TinyDB protocol has been developed based on an acquisitional query processing (ACQP) approach, where the cost for data generation is also considered as part of the query processing solutions [9]. By controlling the sensing operations based on the requirements of the user, significant energy savings are promised.

Similar to COUGAR and TAG, TinyDB also views sensor data as a single table with one column per sensor type. In addition to periodic monitoring data, TinyDB also supports event-based data using an *event* mechanism for initiating data collection. As a result, significant energy savings are possible for event-based applications. Moreover, as with earlier solutions, TinyDB relies on a routing tree that is built during the query dissemination process. Similarly, the network operation is managed in terms of *epochs*, which also define the sample intervals that are used to append information to the sensor table periodically. In-network processing, including query processing and aggregation, are performed at each node in this tree during data delivery to the sink. Furthermore, similar extensions to SQL are considered for controlling data acquisition using the sensor table. In summary, in TinyDB, users send their queries to the sink, and the sink performs a simple query optimization for the correct ordering of sampling, selections, and joins.

The TinyDB protocol provides efficient support for event-based traffic through several modifications of SQL. The events are defined by the user, i.e., the sink, and disseminated to the network. However, the responses to these queries are only generated if the event of interest occurs at a particular node. This results in considerable energy savings since the network can be kept idle during periods when an event of interest does not occur. The following convention can be used to define an event query:

```
ON EVENT bird-detect(loc):
    SELECT AVG(light), AVG(temp), event.loc
    FROM sensors as s
    WHERE dist(s.loc, event.loc) < 10m
    SAMPLE INTERVAL 2s FOR 30s
```

where a bird detection event is defined. Upon detecting a bird, nodes located at a distance less than 10 m to this bird are issued to report the average light and temperature information at an interval of 2 s for the next 30 s. As a result, a query is *generated* as a result of an event, where certain properties of the event can be used to define the query. In this case, the event location is used to identify the data generator sensors for this query such that nodes within the 10 m radius of the event are addressed. In addition to event-based queries, TinyDB supports periodic queries in a similar way as in COUGAR or TAG.

Energy consumption is a major concern for WSNs. Hence, query processing solutions should also be tailored to energy-efficient mechanisms. TinyDB addresses this concern through a third class of queries: *lifetime-based queries*. Similar to periodic queries, lifetime-based queries are used to define monitoring tasks. However, in lifetime-based queries, the query is defined using a LIFETIME clause, which defines the expected lifetime of the query, instead of using a fixed rate as defined by the periodic queries. To this end, each node monitors its energy consumption rate to estimate the lifetime of the node. According to the remaining energy, the reporting rate is adjusted to meet the lifetime requirement. In addition, the user can also define a MIN SAMPLE RATE to prevent the nodes from reporting with too low rates.

TinyDB uses very similar techniques as in COUGAR and TAG for query dissemination and data collection. Query dissemination is performed through controlled flooding, as a result of which a routing tree is formed. In addition, TinyDB ensures that a query does not propagate below a certain parent, if this parent and all of its neighbors would not respond to the query. The resulting tree is called the *semantic routing tree* (SRT), since query semantics are also considered for tree formation. As a result, unnecessary queries are prevented from being sent to the network.

To efficiently build a SRT, a parent node needs to have information about the capabilities of its children so that it can prune queries not related to them. This requires an initial tree formation and is performed in two phases. In the first phase, an *SRT build request* is disseminated to the whole network, which includes the type of information requested by the following query. During this phase, a node selects its parent and forwards its own and its children's observed values about the requested information. As a result, each parent has a range of values that its children observes. In the second phase, the actual query is disseminated to nodes that have overlapping values with the requested values in the query. In case of disconnections between a parent and a child, the child chooses a different parent following techniques similar to those in TAG.

Qualitative Evaluation

Information dissemination is performed through the routing tree, which is similar to most commonly used query processing algorithms. Through the SRT concept, routing trees can be constructed according to the contents of the query instead of a network-wide approach. This semantic approach significantly improves energy efficiency since routing trees are not formed at locations where a response is not accepted. On the other hand, SRT formation relies on network-wide flooding. Hence, the applicability of this scheme is limited in networks where the observed phenomenon dynamically changes. As a result, the information at the parents becomes obsolete with time, which requires frequent updates of SRT formation and, hence, network flooding.

9.3   Network Management

The dynamic nature of WSNs requires efficient schemes to monitor and manage the components [21]. Similar to any network architecture, WSNs require efficient administrator tools so that network administrators or system users can easily interact with the sensor nodes in the network. However, information delivery in WSNs is costly and unreliable because of the low-power wireless links. Hence, reliable management with high energy efficiency is a major challenge. Similarly, sensor nodes are constrained in terms of processing and memory. Hence, network management overhead at the sensor nodes should be as small as possible.

Network management tasks are generally performed in two steps: *network monitoring* and *management control*. As a part of network monitoring, information from the WSN should be efficiently collected to estimate the current state of the network. The collected information includes network-wide data such as connectivity, coverage, topology, and mobility, as well as node-based data such as node state, residual energy, memory usage, and transmit power. Considering the impacts of wireless communication, collecting these distributed sets of information is highly challenging. Moreover, the dynamic changes in these properties require frequent network monitoring for accurate state representation.

Based on the information received as a result of network monitoring, several management control tasks can be performed to maintain the desired state of operation. These tasks include network-wide operations such as route management, protocol update, and traffic management, as well as node-based operations such as turning node components on or off, transmit power management, sampling rate control, and moving nodes. Further network monitoring is required to ensure that the desired network state is reached. Therefore, network monitoring and management control tasks should operate in harmony for network management solutions.

Network management encounters several difficulties due to the unique properties of WSNs. These major challenges are as follows:

- **Data overload:** The high density of WSNs increases the amounts of data to be collected from the network. Moreover, different types of data, which are required by network management solutions, dramatically increase the information from each sensor.

- **Unreliable communication:** Environmental conditions, energy constraints, wireless channel errors, and network congestion make accurate monitoring of nodes difficult. The large number of packet losses that may occur because of these factors should be considered for network management decisions.

- **Information visualization:** The sheer amounts of data continuously received from a large number of sensor nodes impose many challenges in visualization of these data.

To overcome all of these, several network management solutions have been developed [21]. The solutions generally share several common characteristics that are specifically used for WSNs. Due to the resource constraints of sensor nodes, the network management components that run on sensor nodes follow a *lightweight operation* to minimize the energy, processing, and memory overhead. Moreover, most solutions rely on event-based approaches so that sensor nodes act autonomously to changes in their state so that valuable resources are not consumed for end-to-end polling. To combat the adverse effects of the wireless channel, network management solutions are developed to be *robust* and *fault tolerant*. Accordingly, network monitoring and management control tasks do not rely on certain nodes, but rather a distributed operation is sought. The dynamic changes due to channel errors, environmental factors, as well as node movement require *adaptive* solutions that dynamically change operation. Moreover, as in other networking solutions, network management protocols are designed to be *scalable*. Finally, network management solutions sit on top of the WSN architecture. Hence, appropriate interfaces are required to integrate the WSN with the rest of the communication architecture, e.g., Internet, cellular networks, etc.

In addition to these common characteristics, the network management solutions can be classified according two different criteria: management type and architecture. According to management type, generally three different approaches exist:

- **Passive:** Network monitoring is continuously performed and the received data are processed at the sink. This type of management scheme relies on postprocessing to detect any state changes. In case of state changes, management control tasks are dispatched.

- **Reactive (event triggered):** Reactive solutions rely on on-board processing capabilities of sensor nodes for network management. Certain events of interest related to the various states of the node are defined similar to event-based queries. Accordingly, network monitoring is initiated if an event of interest occurs. Management control tasks are dispatched as a response to these events. The network management solutions rely on real-time processing of the collected data.

- **Proactive:** Network monitoring is performed actively and the collected data are processed in real time. The network management protocol reacts to any changes to predict and control any future events.

In addition to network management type, three different network architectures exist depending on where in the network the management operations reside, as shown in Figure 9.10:

- **Centralized:** As shown in Figure 9.10(a), the network management functionalities reside at the sink. In this case, the WSN is used for data delivery purposes for both network monitoring and management control. The sink gathers global information about the network and can perform complex management decisions based on this information. This architecture reduces the processing load on the resource-constrained sensor nodes. However, monitoring data should be transmitted from each node in the network to the sink, which increases the communication load in the network. Since individual node information is important in this case, aggregation solutions may not be applicable either.

- **Distributed:** Instead of relying on a single network manager, i.e., the sink, distributed solutions employ several network managers distributed in the network as shown in Figure 9.10(b). Certain sensor nodes or specific devices are used as network managers, which distribute the management tasks. Each network manager is associated with a cluster of sensor nodes. This improves the robustness of the solutions and decreases the communication cost, since nodes are controlled by local network managers instead of a single centralized manager. However, the coordination and the associated communication between distributed network managers is a major challenge. Moreover, processing load is high at the network managers, which either require specific hardware or result in limited network management functionalities.

- **Hierarchical:** Distributed network management solutions face problems in terms of gathering a global view of the network since managers are distributed in the network. Instead, hierarchical network management solutions employ network managers in the network in a hierarchical manner as shown in Figure 9.10(c). While the network management tasks are distributed among network managers, each manager reports to a higher level manager, where the sink is at the top level. Accordingly, local management tasks can be handled at the lower level managers without increasing the communication cost in the network. Besides, a global view of the network can still

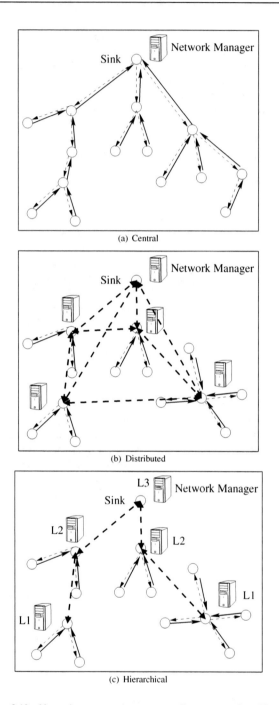

(a) Central

(b) Distributed

(c) Hierarchical

Figure 9.10   Network management types according to network architecture.

be maintained by reporting necessary information to the higher level managers and the sink. This enables network-wide management control decisions to be taken at the sink.

Several network management solutions have been developed for WSNs. In the following, we describe two representative solutions: MANNA [11] and SNMS [16].

9.3.1   Management Architecture for Wireless Sensor Networks (MANNA)

The MANNA architecture supports dynamic information collection from wireless sensor nodes and maps this information in *WSN models* or *WSN maps* [11]. The WSN maps provide a global view of certain parameters in the network and its state. Moreover, the collected information and its relations are defined in a management information base (MIB).

The MANNA architecture relies on a distributed management architecture, where managers are deployed in the network in a purely distributed or hierarchical manner. Centralized network management is also supported by assigning the management tasks to the sink node. These entities maintain two kinds of management information, namely static and dynamic information. Static information refers to the configuration parameters for the services, the network, and the network components. Dynamic information, on the other hand, is represented by the WSN maps. Due to the dynamic nature, the managers are required to collect this information from the network. Based on the management architecture, different manager entities may be involved in creating the WSN maps. Some examples of WSN maps are as follows:

- **Sensing coverage area map:** This map provides a global view of the areas covered by existing sensors based on the location and coverage information received from each sensor.

- **Communication coverage area map:** Similar to the sensing coverage area map, the communication ranges and the available neighbors of each node are represented. This global view helps identify connectivity problems in the network.

- **Residual energy map:** Based on the residual energy information received from the nodes in the network, a topological view of the remaining energy in the network is presented. Such information is essential in identifying the energy consumption profile of the network. Accordingly, the application parameters such as information delivery rate can be modified. Moreover, additional nodes can be added to the network in places where the residual energy is low.

MANNA provides a structural view of the management activities in WSNs. The management functionalities are abstracted in three dimensions. Traditional network management is generally organized into *management functional areas* (fault detection, configuration, performance, security, and accounting) and *management levels* (business, service, network, network element management, as well as network element), which constitute the first two dimensions. In addition to these dimensions, *WSN functionalities* are considered as the third dimension in terms of configuration, maintenance, sensing, processing, and communication. The third dimension is necessary since network management is closely coupled with the communication architecture in WSNs.

In addition to the abstract view of the management functionalities, MANNA provides three main management architectures: functional, information, and physical. The functional architecture is used to identify the functionalities of each component. More specifically, MANNA uses two different components to support distributed management. Managers perform management control decisions based on the information received from the network. In addition to managers, certain nodes are defined as *agents*, which collect information from a certain group of sensor nodes. The definition of agents helps decouple the network monitoring and management control functionalities and assign these tasks to different components. The network managers can be deployed in a centralized, distributed, or hierarchical manner. Similarly, agents can be distributed in the network to form clusters or an agent can be used at the sink to collect network-wide information. Accordingly, MANNA supports flexible network management without relying on a particular management architecture. In addition, the functional architecture identifies the relationships between the managers, agents, and the management information base (MIB), which stores the definitions of the objects and the relationships among them given the current network state. The physical architecture closely follows the functional architecture. Similarly, the information architecture defines the object relations based on manager and agent duties.

The network management operations in MANNA are based on the MANNA network management protocol (MNMP). Sensor nodes are organized into clusters and they communicate with their cluster heads to communicate any state changes. The cluster heads may be agents that collect state information or managers that directly execute management tasks. They are responsible for aggregating the monitoring data from the cluster nodes and sending this information to the managers or the sink. For all network architectures, a central network manager exists at the sink. This manager is generally responsible for complex management tasks that require a global view of the network.

Qualitative Evaluation

MANNA allows distributed management in WSNs through a flexible architecture. Accordingly, the network management tasks can be performed locally through the distributed managers. This improves system flexibility, making it more adaptive to frequent network state changes. Moreover, the WSN maps provide a global view of the network, which can be used for visualization purposes as well. Accordingly, network administrators are informed about the global changes in the network and management tasks can be determined accordingly.

Due to its distributed architecture, MANNA requires higher processing at intermediate nodes that are assigned manager or agent duties. This increases the processing and memory burden at these nodes. To alleviate this problem, additional higher power devices may be need in the network topology.

9.3.2   Sensor Network Management System (SNMS)

Most network management systems rely on underlying networking services such as routing and reliability to interact with the WSN. This approach, however, results in a strict interdependency between the network management solutions and the network that these systems are trying to monitor. As a result, any failure in the underlying application or the networking protocols affects the network management system. Thus, these failures cannot be detected. SNMS addresses this issue by providing a set of query services that are independent of the underlying network protocols [16]. Furthermore, some of the management information is stored on-board for later retrieval in case of network disconnections.

Instead of relying on underlying network protocols, SNMS uses a stand-alone networking stack that is tailored to reliable query dissemination and distributed data collection. The compiled binary can be loaded based on a failure in the application or on a command. Accordingly, SNMS operates independently from the network operation.

The network architecture of SNMS is illustrated in Figure 9.11. Similar to the query processing mechanisms explained in Section 9.2, SNMS relies on a routing tree that is rooted at the node and initiates the query. In most cases, the network manager resides at the sink and, hence, the routing tree is rooted at the sink. With this architecture SNMS supports centralized network management. The SNMS architecture supports networking for the two major operations in network management: network monitoring (collection) and management control (dissemination).

For data collection, a routing tree is generated based on a message sent from the sink. This constitutes the *tree construction and refinement* phase. The sink sends a tree construction message to its neighbors. Each node estimates the link quality based on the received signal strength (RSS) of the tree construction message and selects a parent. In the meantime, the RSS value is added to the message as a cumulative RSS value. Accordingly, a channel-aware routing tree is constructed as the construction message propagates through the network. Further messages from the root can be used to refine the tree structure by consecutive RSS measurements. Accordingly, if a node receives another message, it adds the RSS

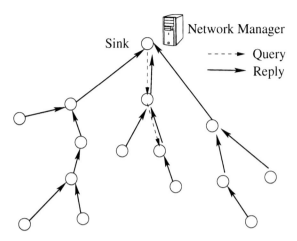

Figure 9.11 SNMS network architecture [16].

value of the received message to the cumulative RSS value in the message. If the total is less than the current parent's cost, the node selects the sender as the new parent.

Since the routing tree is constructed according to the messages sent from the parents, the resulting links may exhibit asymmetry. More specifically, although the parents are selected so that parent–child link quality is high, the child–parent link is not considered for this selection. To maintain high communication quality for both directions, the nodes keep track of the acknowledgment messages they receive from their parents when they forward messages. If the success rate is low, another parent is selected.

SNMS also supports routing tree construction such that only a subset of the nodes in the network is involved. This is accomplished by including three fields in the messages: destination address, destination group, and time to live. Based on the destination address or the destination group, a node determines whether the message is destined for itself or any of its children. Accordingly, the packet can be processed, forwarded, or dropped. Moreover, if the time to live of a packet expires, it is not retransmitted.

In a WSN, a wide range of queries can be sent to the network. To provide flexibility, these queries are not defined in SNMS. Instead, the network administrator can define any query and develop the corresponding components that will reside in the sensor nodes. Accordingly, for each query, an associated component can be generated, which makes the network management system customizable. The administrator can use a string to define each query. At the network level, though, query names with arbitrary sizes are not used. Based on the administrator's definitions, each query is mapped to an integer key and the sensors programmed accordingly. The query definitions can further be extended by assigning new keys and programming nodes accordingly.

Based on these definitions, an SNMS query consists of a list of attribute keys and a sample period. Based on the destination requirements, the query is disseminated to the network using the SNMS network stack through the generated routing tree. After a node receives the query, the sample period is used to periodically observe the attributes requested by the query. The results are included in a response and sent back to the sink using the same routing tree. Since predefined queries are not used but representative keys are, the data structure is not self-describing. Therefore, SNMS limits the aggregation capabilities of the sensor nodes and operations are mainly performed at the network manager, i.e., the sink.

At the sink, the received responses are collected to represent the query results to the user. Since strings are used for the definition of queries and attributes, a human manager is required by the SNMS architecture for management task decisions. In addition to data delivery, SNMS also supports local caching for certain events of interest to the network manager. In case of any failures in the SNMS networking stack, the locally cached event information can be retrieved, which provides a safeguarding mechanism.

Qualitative Evaluation

SNMS provides a stand-alone networking stack for query dissemination and data collection phases that are essential for network management in WSNs. Accordingly, the network management tool is not affected by failures in the application or the network stack but can also identify these failures. In other words, the stand-alone networking stack of SNMS can be viewed as an *off-band* management approach, which is not employed in many network management solutions. In addition, the networking stack does not require a neighborhood table for data delivery and relies on a routing tree. This minimizes the overhead for storage and network traffic. Furthermore, query definitions and the associated components can easily be defined by the network administrators. This provides flexibility and ease of use for the network management tools.

SNMS relies on a centralized approach, where the network management is performed at a single entity. Moreover, since user-defined queries are supported, aggregation cannot be performed efficiently in the network. This approach decreases the energy efficiency of the network, since the information from each individual sensor should be carried to the sink for processing. More importantly, SNMS network management tasks rely on human intervention. Hence, an automated operation cannot easily be provided.

References

[1] G. Amato, A. Caruso, S. Chessa, V. Masi, and A. Urpi. State of the art and future directions in wireless sensor network's data management. Technical report 2004-TR-16, ISTI, May 2004.

[2] K. Barr and K. Asanovic. Energy aware lossless data compression. In *Proceedings of ACM MobiSys'03*, pp. 231–244, San Francisco, USA, May 2003.

[3] P. Bonnet, J. E. Gehrke, and P. Seshadri. Towards sensor database systems. In *Proceedings of Second International Conference on Mobile Data Management*, pp. 3–14, Hong Kong, China, January 2001.

[4] W. R. Heinzelman, A. Chandrakasan, and H. Balakrishnan. Energy-efficient communication protocol for wireless microsensor networks. In *Proceedings of IEEE Hawaii International Conference on System Sciences*, pp. 1–10, January 2000.

[5] W. R. Heinzelman, J. Kulik, and H. Balakrishnan. Adaptive protocols for information dissemination in wireless sensor networks. In *Proceedings of MobiCom'99*, pp. 174–185, Seattle, WA, USA, August 1999.

[6] N. Kimura and S. Latifi. A survey on data compression in wireless sensor networks. In *Proceedings of ITCC'05*, volume 2, pp. 8–13, Las Vegas, NV, USA, April 2005.

[7] S. Madden and M. J. Franklin. Fjording the stream: an architecture for queries over streaming sensor data. In *Proceedings of the 18th International Conference on Data Engineering*, pp. 555–566, San Jose, CA, USA, 2002.

[8] S. R. Madden, M. J. Franklin, J. M. Hellerstein, and W. Hong. TAG: a tiny aggregation service for ad-hoc sensor networks. In *Proceedings of OSDI*, Boston, MA, USA, December 2002.

[9] S. R. Madden, M. J. Franklin, J. M. Hellerstein, and W. Hong. The design of an acquisitional query processor for sensor networks. In *Proceedings of SIGMOD*, San Diego, CA, USA, June 2003.

[10] R. Rajagopalan and P. K. Varshney. Data aggregation techniques in sensor networks: a survey. *IEEE Communication Surveys & Tutorials*, 8:48–63, 2006.

[11] L. B. Ruiz, J. M. Nogueira, and A. A. F. Loureiro. MANNA: a management architecture for wireless sensor networks. *IEEE Communications Magazine*, 41(2):116–125, February 2003.

[12] C. M. Sadler and M. Martonosi. Data compression algorithms for energy-constrained devices in delay tolerant networks. In *Proceedings of ACM SenSys'06*, Boulder, CO, USA, November 2006.

[13] D. Salomon. *Data Compression: The Complete Reference*. Springer, 1997.

[14] D. Slepian and J. K. Wolf. Noiseless coding of correlated information sources. *IEEE Transactions on Information Theory*, 19(4):471–480, July 1973.

[15] C. Srisathapornphat, C. Jaikaeo, and C.-C. Shen. Sensor information networking architecture. In *Proceedings of the International Workshop on Parallel Processing (ICPP'00)*, p. 23, Washington, DC, USA, August 2000.

[16] G. Tolle and D. Culler. Design of an application-cooperative management system for wireless sensor networks. In *Proceedings of EWSN'05*, pp. 121–132, Istanbul, Turkey, January 2005.

[17] A. Wyner and J. Ziv. The rate-distortion function for source coding with side information at the decoder. *IEEE Transactions on Information Theory*, 22:1–10, January 1976.

[18] Z. Xiong, A. D. Liveris, and S. Cheng. Distributed source coding for sensor networks. *IEEE Signal Processing Magazine*, 21:80–94, September 2004.

[19] Y. Yao and J. Gehrke. The COUGAR approach to in-network query processing in sensor networks. *SIGMOD Record*, 31(3):9–18, September 2002.

[20] Y. Yao and J. Gehrke. Query processing in sensor networks. In *Proceedings of the 1st Biennial Conference on Innovative Data Systems Research*, Asilomar, CA, USA, January 2003.

[21] B. Zhang and G. Li. Analysis of network management protocols in wireless sensor network. In *Proceedings of the International Conference on Multimedia and Information Technology*, pp. 546–549, Los Alamitos, CA, USA, 2008.

10

Cross-layer Solutions

The communication solutions described so far follow the layered protocol stack as explained in Chapter 1. More specifically, the communication protocols are classified according to the layered architecture, which includes the application, transport, network, link, and physical layers. Historically, this layered architecture has been motivated by the extensive success of Internet protocols, which adopt the layered protocol stack approach. The key to success with this architecture is the modularized design, where the design of each protocol layer is decoupled from other layers. This has enabled the rapid improvement of Internet protocols, which have been designed and implemented by independent entities. The use of standardized interfaces between layers has enabled compatibility between these independently designed protocol layers.

In WSNs, however, efficiency is the major design parameter. Wireless sensor nodes are constrained in terms of memory and processing power. Hence, protocol stacks that have large memory footprints are not desirable. Similarly, interfaces that require high processing between layers cannot be supported. In addition, the broadcast and non-deterministic nature of the wireless channel creates interdependencies between each layer, especially with the low-power communication techniques. Recent empirical studies motivate that the properties of low-power radio transceivers and the wireless channel conditions should be considered in protocol design. Finally, the event-centric approach of WSNs requires application-aware communication protocols. These unique characteristics of WSNs have motivated cross-layer protocols that include the functionalities of two or more layers in a single coherent framework. Recent studies on WSNs reveal that cross-layer integration and design techniques result in significant improvement in terms of energy conservation [9, 22, 25].

The existing layered communication protocols improve the energy efficiency to a certain extent by exploiting the collaborative nature of WSNs and its correlation characteristics as we have seen in previous chapters. However, the main drawback of these protocols is that they follow the traditional layered protocol architecture. While they may achieve very high performance in terms of the metrics related to each of the individual layers, they are not jointly optimized to maximize the overall network performance while minimizing energy expenditure. Considering the scarce energy and processing resources of WSNs, joint optimization and design of networking layers, i.e., cross-layer design, stands as the most promising alternative to inefficient traditional layered protocol architectures.

In this chapter, we overview the motivations for cross-layer design by listing the interlayer effects and the corresponding challenges for layer protocols in multi-hop wireless networking. Moreover, several approaches that combine two traditional protocol layers, i.e., *pairwise cross-layering*, are described. Finally, we explain the *cross-layer protocol* (XLP) that merges several communication functionalities into a single coherent protocol module.

Wireless Sensor Networks   Ian F. Akyildiz and Mehmet Can Vuran
© 2010 John Wiley & Sons, Ltd

10.1   Interlayer Effects

The performance of the layered protocol stack in realistic settings reveals several important interactions between different layers. Both analytical studies and experimental work in WSNs highlight these important interactions between different layers of the network stack. These interactions are especially important for the design of cross-layer communication protocols for WSNs.

The dominant factor in the performance of communication protocols in WSNs is the wireless channel. The low-power communication capabilities and the rather limited capabilities of low-cost transceivers result in a significant impact on higher layers. Consequently, the effects of the wireless channel cannot be confined to only the physical layer.

As we explained in Chapter 4, wireless communication is non-deterministic because of the random effects of multi-path and shadow fading. This diminishes the effect of distance on the communication quality in low-power wireless links. More specifically, given a particular distance between two nodes, the communication quality can fluctuate significantly with time. In addition, the wireless channel is not symmetric, i.e., the channel quality from a node A to node B may not be same as that from node B to node A.

A typical interlayer effect in WSNs is that between the physical and the routing layers. Many characteristics of the wireless channel violate the assumptions of the unit disk graph (UDG) model that is used for the design of several routing protocols.[1] As an example, the asymmetry in the wireless link signifies the importance of communication direction for the routing protocols. More specifically, if a node uses a particular path for communication with the sink, the same path may not be available for the reverse direction if the sink needs to communicate with the node. As a result, solutions that rely on flooding messages from the sink to construct a routing tree do not perform well under realistic scenarios. Although a node receives a flooding packet from one of its neighbors, it may not send packets to that neighbor successfully because of the asymmetric channel.

Similarly, the relation between the MAC and routing layers also impacts communication success. Since the wireless channel is essentially a broadcast medium, only a single transmission is allowed in a transmission area by the MAC protocol. As a result, simultaneous transfers are not possible. Moreover, the MAC layer introduces a non-deterministic delay for channel access because of the activities of other nodes. If a neighbor of a node is transmitting a packet, the MAC protocol delays the transmission for a random amount of time to prevent collisions with the ongoing transmission as well as other neighbors that are trying to access the channel. This may significantly impact the performance of routing protocols, especially when distance or hop length measures are closely tied to the time in which a packet is received from a particular node. A fundamental example is the operation of a flooding protocol under CSMA-based MAC operation [11].

■ EXAMPLE 10.1

Testbed experiments on a flooding protocol with a CSMA-based MAC protocol reveal several interdependencies between the MAC and routing layers [11]. A typical flooding protocol disseminates information to each node in the network by initiating the flood at a specific point. Each node that receives this information forwards the packet to its neighbors. Accordingly, the information propagates throughout the network as shown in Figure 10.1(a). MAC protocols, on the other hand, limit the transmission of a packet in a broadcast region so that only a single communication occurs at a given time to prevent collisions. As a result of the random access mechanism of CSMA protocols, a flooding protocol may produce a completely different tree as shown in Figure 10.1(b). There exist nodes that are farther from the sink geographically but closer in terms of number of hops to the sink (compare nodes A and B). In addition, certain nodes may never receive the flood packets due to wireless channel errors (node C). Testbed experiments show

---

[1] For details of the UDG model, see Chapter 4.

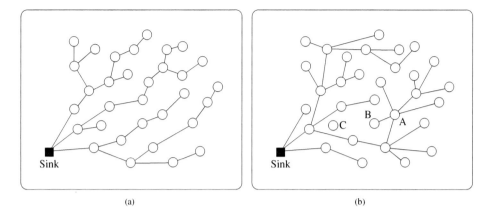

(a)   (b)

Figure 10.1   (a) A typical flooding tree and (b) a flooding tree with the MAC protocol.

that it may take almost the same amount of time to disseminate information to 90% of the network as it takes to disseminate the information to the remaining 10%. The close interaction between the physical, MAC, and routing layers results in a completely different performance than that assumed for a typical flooding protocol.

The experimental studies reveal that the UDG model can be misleading in protocol design and evaluation because of the existence of a *transitional region* in low-power links. More specifically, the wireless connectivity exhibits a probabilistic characteristic rather than a deterministic one. Therefore, a node is not *connected* to a particular node for all times. Instead, there exists some probability that two nodes will not be able to communicate with each other. Hence, the assumption that a node has certain *neighbors*, to which it is connected, is misleading. Instead of neighbor-based approaches, channel-aware solutions that consider the quality of the wireless channel are suitable for WSNs.

Similarly, there is a close interdependency between local contention and end-to-end congestion in multi-hop WSNs [25]. The broadcast nature of the wireless channel prevents simultaneous transmissions in close proximity. If multiple nodes have packets to send, they contend with each other to gain access to the channel. The contention results in increased medium access delay and affects the queuing delay of each packet. Considering that the traffic generation rate does not change, this leads to congestion at certain parts of the network, where local contention is high. In order to address this problem, the MAC and transport layer protocols should be designed in a cross-layer fashion so that local congestion is communicated to the data sources to regulate the input traffic rate. In addition, the multi-hop routes should be selected so that local contention among multiple paths is minimized.

In addition to the wireless channel impact and cross-layer interactions in the protocol stack, the data content also plays an important role in the design of cross-layer protocols. The data-centric nature of the WSN applications impacts the requirements of communication protocols as we discussed in Chapter 3. Accordingly, the event information is much more important than the individual readings of the sensor nodes. This leads to close interaction between the physical properties of the event and the traffic generated in the network. Dense deployment of sensor nodes results in highly correlated sensor observations for closely located nodes. Similarly, the nature of the energy-radiating physical phenomenon yields temporal correlation between each consecutive observation of a sensor node. Exploiting the spatial and temporal correlation further improves the energy efficiency of communication protocols without hampering the fidelity of event information in WSNs [23].

The interdependency between each different network layer calls for cross-layer solutions for efficient data delivery in WSNs. In the following, we describe cross-layer solutions for two main topics. Cross-layer solutions have so far been developed in two main contexts in WSNs. Many solutions employ cross-layer *interaction*, where the traditional layered structure is preserved, and each layer is designed by considering the interdependencies among other layers. In addition, the information from certain layers is effectively communicated to other layers for cross-layer decisions. In this case, the mechanisms of each layer still remain intact. On the other hand, there is still much to be gained by rethinking the mechanisms of network layers in a unified way so as to provide a single communication module for efficient communication in WSNs. In the following, we also focus on the cross-layer *module* design, where functionalities of the traditional layers are merged into a single functional module.

10.2 Cross-layer Interactions

The layered structure of the protocol architecture can be altered to address many challenges that arise because of the interactions between different layers. In particular, the direct influence of the wireless channel on higher layers requires cross-layer solutions at each layer. Here, we provide several cross-layer solutions that are relevant to the cross-layering philosophy. These solutions are classified in terms of interactions among the physical (PHY), MAC, network, and transport layers.

10.2.1 MAC and Network Layers

The cross-layer interactions between the MAC and network layers are mostly exploited in WSNs. Since the routes constructed by the network layer constitute the source of contention, the MAC layer has to be aware of the multi-hop nature of the communication. Moreover, the network layer relies on the MAC layer to provide bounded delay in forwarding packets through the multi-hop path. This requires the MAC protocol to be designed considering this multi-hop nature.

Joint Scheduling and Routing

Sensor nodes generally employ a duty cycle mechanism so that the transceiver is turned on and off in certain intervals. Since nodes are not active for the entire time, routing should be performed in coordination with the scheduling component of the MAC protocols. For periodic traffic, this is exploited to form on–off schedules for each flow in the network through a joint scheduling and routing scheme [19]. The schedules of the nodes depend on the routes that are established between the sources and the sink. Since the traffic is periodic, the schedules are then maintained to favor maximum efficiency. The on–off schedules in this case are also determined according to the connectivity of the network. The usage of on–off schedules in a cross-layer routing and MAC framework is also investigated in a TDMA-based MAC scheme [22]. The nodes distributively select their appropriate time slots based on local topology information. The routing protocol also exploits this information to establish a route. Compared to a strictly layered approach, the cross-layer scheme provides several advantages.

The interdependency between scheduling and routing can also be observed from the performance of the S-MAC [27] protocol in multi-hop networks. S-MAC schedules the activities of the nodes according to a duty cycle operation as explained in Chapter 5. As a result, a group of nodes are on for a certain duration of the time and off for the remainder of the frame. This duty cycle operation of S-MAC introduces a network delay in multi-hop networks. This is illustrated in Chapter 5 through the multi-hop awareness feature of S-MAC. Since a group of nodes are all off during the sleep duration, a packet cannot be forwarded. More specifically, the duty cycle operation of S-MAC increases the multi-hop latency linearly with the hop count. The multi-hop latency can be decreased by the *adaptive listening* feature of S-MAC, which lets potential receivers remain awake until the end of a packet transmission.

In this way, the node that receives the packet can find an active next hop node. As a result, the multi-hop latency can be decreased by as much as half if the adaptive listening algorithm is used [27]. However, the protocol still introduces significant latency compared to non-duty cycle, contention-based protocols.

To address the multi-hop latency of S-MAC, the route-aware contention-based MAC protocol for data gathering (DMAC) introduces a sleep schedule so that the nodes on a multi-hop path wake up sequentially as the packet traverses the path [14]. The sleep schedules are determined based on the data gathering tree that is constructed as rooted at the sink with branches spreading through the WSN. In this tree, the number of hops from a node to the sink is referred to as its *depth*. DMAC aims to schedule nodes that are on the same gathering tree to offset their sleep schedules according to their depths in the tree. Consequently, the data are collected through the unidirectional tree. DMAC incorporates a local synchronization protocol in order to perform local scheduling and uses data prediction in case a node requires a higher duty cycle for data transmission. This results in a wakeup schedule that resembles the flashing lights on a Christmas tree. As a packet traverses the data gathering tree, the nodes at respective depths wake up. DMAC does not deploy RTS/CTS exchanges. Instead, the simple CSMA structure is used with ACK packets. Based on these techniques, multi-hop effects on the delay performance are minimized specifically for data gathering applications where a unidirectional tree is used.

Receiver-Based Routing

Receiver-based routing is a novel cross-layer approach that exploits the traditional MAC functionalities in routing decisions. Traditional network and MAC protocols operate in a layered stack so that the routing decisions are made at the network layer and communicated to the MAC layer. At the MAC layer, the selected next hop is communicated through unicast messages. This layered approach is suitable for networks where the nodes are always active and any node can respond to a unicast message at any time. In WSNs, however, nodes employ duty cycle operation. As a result, a node is not active at all times. This introduces significant delays if the MAC protocol needs to communicate with a particular sensor node. Instead, the available sensor nodes can be used to relay the packet. This is referred to as *receiver-based routing* or *anycast routing*.

In this approach, the next hop is chosen as a result of the contention in the neighborhood [20], [29], and [30]. The receiver-based routing technique closely couples the functionalities related to determining the next hop in a network with MAC. In this approach, the next hop is chosen as a result of the contention in the neighborhood. Receiver-based routing techniques exploit the geographical routing and channel-aware routing techniques with medium access procedures. Generally, receiver-based routing can be described as shown in Figure 10.2 as follows. When a node A has a packet to send, the neighbors of this node are notified through a broadcast message. Based on geographical routing techniques, the nodes that are closer to the sink than node A are feasible nodes. Consequently, the nodes in this feasible region, denoted by $A(R, D)$, perform contention for routing. In order to provide a minimum number of hops in routing, the feasible region is further divided into multiple priority zones. The nodes closer to the sink, e.g., node B in D_1, perform backup for a smaller amount of time and contend for the packets. Moreover, each node sleeps periodically in order to save energy and contend for the relay role based on the priority-based backoff policy. Consequently, energy consumption and latency are decreased since a specific node does not wait for the next hop transmission. Similarly, MAC layer information is used as the basis for achieving energy-efficient routing.

In addition to the basic receiver-based routing, the routing decisions can also be performed as a result of successive *competitions* at the medium access level [10]. More specifically, the next hop is selected based on a weighted progress factor and the transmit power is increased successively until the most efficient node is found. This results in the use of on–off schedules in the protocol operation.

Incorporating physical layer and network layer information into the MAC layer design improves the performance in WSNs. Route-aware protocols provide lower delay bounds while physical layer coordination improves the energy efficiency of the overall system. Moreover, since sensor nodes are

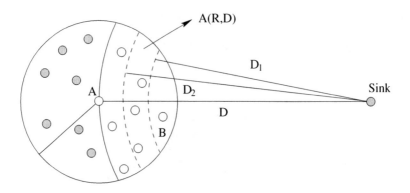

Figure 10.2   Receiver-based routing.

characterized by their limited energy capabilities and memory capacities, cross-layer solutions provide efficient solutions in terms of both performance and cost. However, care must be taken while designing cross-layer solutions since the interdependency of each parameter should be analyzed in detail.

10.2.2   MAC and Application Layers

The application layer, i.e., the content of the information that is carried by the sensor nodes, has a direct impact on the design of communication protocols. Contrary to the traffic model in ad hoc networks, where each node generates independent traffic, the information content of the sensor nodes is correlated. Since the main goal is to collect information regarding a physical phenomenon, i.e., event, the collective information from sensors is more important than individual readings. This fundamental difference in the application layer influences how communication protocols are designed.

An example of a cross-layer solution between the MAC and application layers for WSNs is the CC-MAC protocol, which is explained in Chapter 5 [24]. In fact, spatial correlation is a significant characteristic of WSNs. Dense deployment of sensor nodes results in the sensor observations being highly correlated in the space domain. Exploiting the spatial correlation in the design of communication protocols further improves energy efficiency in WSNs. The CC-MAC protocol exploits the spatial correlation in the observed physical phenomenon for MAC.

The theory of spatio-temporal correlation in WSNs states that the correlation between the observations of nodes can be modeled by a correlation function based on two different source models, i.e., point and field sources [23]. Based on this theory, the estimation error resulting from the collection of observations from multiple nodes can be calculated. This error is defined as distortion.

Figure 10.3 shows the effect of spatial correlation on the distortion in event reconstruction. In general, lower distortion results in more accurate estimation of the event features. Hence, using more nodes in locating an event results in lower distortion. However, Figure 10.3 reveals that by using a small subset of nodes for reporting an event, e.g., 15 out of 50, the same distortion in event reconstruction can be achieved. This helps to reduce the redundancy from multiple highly correlated sensor readings. From the communication perspective, these results also reveal that significant energy savings can be achieved when the correlation in the content of information is exploited.

CC-MAC performance evaluations show that exploiting spatial correlation for medium access results in high performance in terms of energy, packet drop rate, and latency.

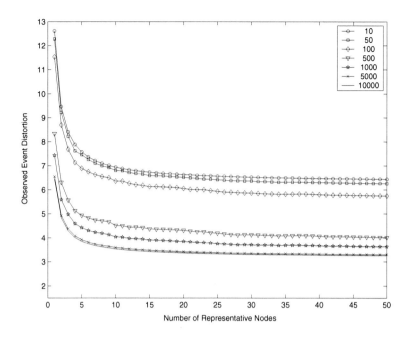

Figure 10.3   Effect of spatial correlation on the distortion [24].

10.2.3   Network and PHY Layers

The wireless channel dynamics impact the channel quality throughout the lifetime of the network. Although the distance between two nodes does not change, the random effects of the wireless channel result in significant fluctuations in communication quality. This affects the quality of the routes that are constructed based on static decisions.

In Chapter 7, geographical routing protocols have been explained as one of the scalable approaches for routing. These protocols have been improved for lossy links by considering channel quality information from the PHY layer in routing decisions [28]. Two examples of this cross-layer approach are the *reception-based blacklisting* and *best PRR × distance* algorithms.

Reception-based blacklisting requires the nodes to keep track of their packet reception rates with their neighbors. This information is periodically exchanged so that each node is aware of the packet reception rate (PRR) between its neighbors. Among these neighbors, the ones with a PRR lower than some threshold are blacklisted. As a result, the node does not communicate with the blacklisted nodes since the communication will lead to errors in most cases. Based on this blacklisting approach, other routing metrics can be used among the nodes that are not blacklisted for routing decisions. In this way, channel information is used as a filter for the routing decisions in a cross-layer fashion. In addition to blacklisting based on a fixed threshold, relative reception-based blacklisting can also be used to prevent those cases where there are no neighbors that satisfy the fixed threshold. In this case, a certain fraction of the nodes with low PRR values are blacklisted. The same notion can be used as a routing metric as a *best reception neighbor* algorithm, where the next hop is selected as the neighbor with this highest PRR.

The second forwarding mechanism is a combination of the progress based on the distance between the source, relay, and destination and the PRR as explained in Chapter 7. The PRR × distance algorithm selects the next hop so that the channel quality is high and the progress is long enough. The PRR information is exchanged between the nodes similar to the reception-based blacklisting schemes.

The distance improvement is found according to $1 - d(B, S)/d(A, S)$, where $d(B, S)$ is the distance between the neighbor, B, and the destination *Sink* and $d(A, S)$ is the distance between the source node, A, and the destination, *Sink*.

The cross-layer interactions between the network and PHY layers result in informed decisions for selecting the next hop. The channel quality information can be exploited to develop new metrics for routing. An important challenge with exploiting the channel information is that the wireless channel is generally asymmetric. As a result, an information exchange mechanism is generally used, such as in reception-based blacklisting. Nodes exchange PRR information with each neighbor to provide information for selecting routes. This requirement, however, may result in a high overhead in networks where the density is high. Since the PRR may fluctuate frequently, nodes need to exchange this information in small intervals. As the network density increases, the cost for exchanging this information increases.

10.2.4   Transport and PHY Layers

The transport layer regulates the transmission rate of traffic flows so that congestion is mitigated in the network. In wireless networks, the congestion is closely tied to the local contention. The contention as well as the channel rate can be controlled through transmit power control at the PHY layer. As a result, this power control can be tied to congestion control in multi-hop networks.

The cross-layer, jointly optimal congestion control and power control (JOCP) algorithm aims to reach an equilibrium among the nodes in the network through an iterative update policy at each node [7]. Accordingly, each node updates the queuing delay of each packet according to its input and output rates. The difference between the total ingress flow intensity and the egress link capacity, divided by the egress link capacity, gives the average time that a packet needs to wait before being sent out on the egress link. Moreover, each source uses the communication delay to update the traffic rate to mitigate congestion in the network. Furthermore, each node j calculates the following quantity:

$$m_j(t) = \frac{\lambda_j(t)SIR_j(t)}{P_j(t)G_{jj}}.$$

(10.1)

$\lambda_j(t)$ is called the *shadow price* for node j, $SIR_j(t)$ is the signal-to-interference ratio, $P_j(t)$ is the transmit power level, and G_{jj} incorporates propagation loss, spreading gain, and other normalization constants on link j. Finally, the transmit power of a node l is updated as follows:

$$P_l(t + 1) = P_l(t) + \frac{\kappa \lambda_l(t)}{P_l(t)} - \kappa \sum_{j \neq l} G_{lj} m_j(t).$$

(10.2)

Accordingly, the transmit power is increased according to the shadow price $\lambda_l(t)$ and decreased according to the weighted sum of the messages from each of the neighbors. Since the weights are based on the path loss, the transmit power is adjusted according to the status of the closer nodes. This results in an increase in transmit power, if the queuing delay of a node is high. On the other hand, if the queuing delay of the neighboring nodes is high, the node decreases its transmit power to decrease interference in the neighborhood. Accordingly, each node reaches an equilibrium such that congestion is minimized.

The JOCP protocol decomposes the cross-layer operation of multi-hop communication into congestion control and transmit power control components. Through four main update policies, a distributed and cross-layer operation of transport and physical layers is realized. The transmit power control mechanism is updated according to the queuing states of the nodes in a local vicinity, while the traffic rates are adjusted according to the transmit power levels. The distributed solution also performs very close to the global optimum without any centralized control.

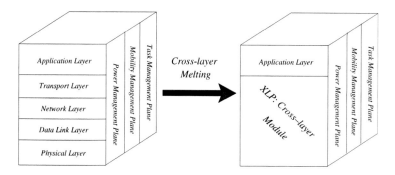

Figure 10.4 The cross-layer module concept for WSNs.

10.3 Cross-layer Module

The close coupling between protocol layers suggests that a novel perspective is necessary for networking in WSNs. More specifically, the design of a cross-layer protocol that addresses the communication functionalities required by these networks is important. These results have, recently, led to several solutions for the cross-layer interaction and design [3, 5, 4, 6, 7, 8, 10, 13, 15, 16, 17, 18, 20, 29, 30].

In this section, we describe the *cross-layer protocol (XLP)*, which unifies several networking functionalities into a single functional module. XLP is based on the *initiative determination* concept, which enables the nodes to make distributed decisions for communication based on various factors.

XLP employs a *cross-layer module* concept, such that in addition to information sharing, the functionalities of three fundamental communication paradigms (medium access, routing, and congestion control) are also combined in a single protocol operation. This is illustrated in Figure 10.4, where the traditional layered protocol stack is converted into a single module approach through XLP. This allows the nodes to incorporate the required functionalities into a single protocol by considering the channel effects. Accordingly, a cross-layer design can be implemented without impacting the physical layer architecture of sensor nodes. The nodes are equipped with a typical RF transceiver which has all necessary physical layer functionalities, e.g., modulation/demodulation, channel coding, RF power control, specified according to the specific deployment and application requirements.

XLP is based on the following major components:

- **Initiative concept:** The communication incentive is passed to the receiver. Each receiver participates in communication based on its local state, which is represented by the *initiative* function.

- **Receiver contention:** Similar to the receiver-based routing protocols, potential receivers contend for packets and become next hop.

- **Local cross-layer congestion control:** Highly congested nodes do not participate in communication through the local congestion control component.

- **Angle-based routing:** The local minimum is avoided through a distributed angle-based routing concept in the network.

- **Channel adaptive operation:** The receivers adapt communication parameters based on channel conditions.

- **Duty cycle operation:** The nodes employ a distributed duty cycle operation, which minimizes energy consumption as well as protocol overhead.

Details of these functionalities are explained in the following sections.

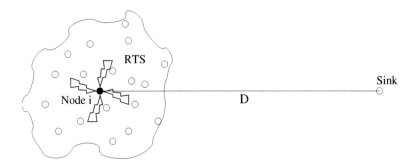

Figure 10.5   The communication is initiated through an RTS packet in XLP [3].

10.3.1   Initiative Determination

The *initiative determination* concept constitutes the core of the XLP. Through this concept, the intrinsic communication functionalities that are required for successful communication are implicitly incorporated. Coupled with the receiver-based contention mechanism, the initiative determination concept provides freedom for each node to participate in communication. Accordingly, multi-hop routes are created between the data sources and the sink by using the most appropriate nodes.

The protocol operation will be explained according to the scenario in Figure 10.5. Here, a sender node i is located at a distance D from the sink. The communication is initiated by broadcasting a RTS packet as shown in Figure 10.5. The neighbors of node i that receive this RTS message are also shown in this figure. The neighbors of node i decide to participate in the communication through the initiative determination procedure. The initiative, $\mathcal{I}$, is determined based on the state of each node's communication capabilities as follows:

$$
\mathcal{I} = \begin{cases} 1, & \text{if } \begin{cases} \xi_{RTS} \geq \xi_{Th} \\ \lambda_{relay} \leq \lambda_{relay}^{Th} \\ \beta \leq \beta^{\max} \\ E_{rem} \geq E_{rem}^{\min} \end{cases} \\ 0, & \text{otherwise.} \end{cases}
\tag{10.3}
$$

Accordingly, the initiative determination is a binary operation, where a node decides to participate in communication if its $\mathcal{I} = 1$. The initiative is set to 1 if all four conditions in (10.3) are satisfied. Each condition in (10.3) constitutes a certain communication functionality in XLP. These functionalities are listed as follows:

- **Channel quality ($\xi_{RTS} \geq \xi_{Th}$):** The first condition requires that the received signal-to-noise ratio (SNR) of the RTS packet, ξ_{RTS}, is above some threshold ξ_{Th} for a node to participate in communication. Accordingly, this condition ensures that reliable links are constructed for communication based on the current channel conditions. Only those nodes with high SNR participate in communication.
- **Local congestion ($\lambda_{relay} \leq \lambda_{relay}^{Th}$):** The second condition requires that the relay input rate, λ_{relay}, is below some threshold λ_{relay}^{Th}. This condition prevents congestion by limiting the traffic that a node can relay. More specifically, a node participates in the communication if it can relay the packet based on its communication activity.
- **Buffer length ($\beta \leq \beta^{\max}$):** The third condition ensures that the buffer occupancy level, β, of a node does not exceed a specific threshold, $\beta^{\max}$. This prevents congestion and ensures that the

node does not experience buffer overflow. The second and third conditions are a part of the local congestion control mechanism, which is explained in Section 10.3.5.

- **Residual energy** $(E_{rem} \geq E_{rem}^{min})$: The last condition ensures that the remaining energy of a node E_{rem} stays above a minimum value, E_{rem}^{min}. This constraint helps preserve a uniform distribution of energy consumption throughout the network.

The four conditions in the initiative function define the cross-layer functionalities of XLP. Accordingly, XLP performs receiver-based contention, initiative-based forwarding, local congestion control, hop-by-hop reliability, and distributed operation. Details of these functionalities will be explained in the following sections.

In addition to the initiative determination, each node employs a duty cycle operation such that it is active or in sleep mode based on an individual schedule. The active–sleep periods are based on a *duty cycle parameter*, δ. The duty cycle operation is performed according to a sleep frame of length T_S s. A node is active for $\delta \times T_S$ s and is in sleep state for $(1 - \delta) \times T_S$ s. The duty cycle operation is implemented independently at each node. Therefore, the start and end times of each node's sleep cycle are not synchronized. This results in a distributed duty cycle operation. Moreover, nodes do not wait for a specific node to wake up but, instead, relay packets using the neighbors that are active at a given time.

The initiative determination concept ensures that each node contributes to the transmission of the event information based on its local conditions. As a result, each node is assigned a duty in the network. The duties of a node in WSNs can be classified as follows:

- **Source duty:** Source nodes observe event information through on-board sensors and generate packets to be transmitted to the sink. These nodes are responsible for managing the medium access to inject these packets into the network and maintaining the injection rate to control congestion in the network.

- **Router duty:** Sensor nodes also forward the packets received from other nodes to the next destination in the multi-hop path to the sink. These nodes are responsible for creating efficient routes in the network by accepting appropriate flows through their path to the destination.

The initiative determination is performed as part of the router duty. Each node determines its initiative as shown in (10.3) to participate in the transmission of an event, i.e., relay the event information. Details of the protocol operation are explained in the following sections based on these definitions.

XLP operation is explained next. More specifically, the XLP consists of the following functionalities:

- **Transmission initiation:** For a successful communication, a node first initiates transmission as explained in Section 10.3.2.

- **Receiver contention:** The nodes that receive the transmission initiation perform initiative determination. The nodes that decide to participate in communication perform receiver-based contention as described in Section 10.3.3

- **Angle-based routing:** The voids in the network are handled based on angle-based routing as described in Section 10.3.4.

- **Local cross-layer congestion control:** The local congestion control component ensures energy efficient as well as reliable communication by two-step congestion control as explained in Section 10.3.5.

10.3.2   Transmission Initiation

The transmission initiation is performed whenever a node has a packet to send. This may be due to the source duty, when the node has a packet regarding the sensor measurements, or due to the router duty, when the node relays a packet from another node. Firstly, the node listens to the channel for a specific period of time. If the channel is occupied, backoff is performed based on the contention window size, CW_{RTS}. If the channel is idle, the node broadcasts an RTS packet as shown in Figure 10.5.

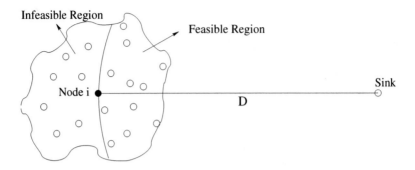

Figure 10.6   The feasible and infeasible regions in XLP.

The RTS packet contains the location information of the sensor node i and the sink and serves as a link quality indicator.

When a neighbor of node i receives the RTS packet, it first checks the source and destination locations. Accordingly, the node determines its region based on the source and destination locations. As shown in Figure 10.6, the transmission region of a node i can be divided into two regions: the *feasible region* contains the neighbors of a node that are closer to the sink; the remaining nodes reside in the *infeasible region*. The nodes inside the feasible region are used for data relay to prevent loops in the routes.

The nodes that receive the RTS packet and reside inside the feasible region of the transmitting node i evaluate their initiatives according to (10.3). If the initiative function results in 1, the node decides to participate in communication and performs *receiver contention* as explained in Section 10.3.3. On the other hand, if a node is inside the infeasible region, it switches to sleep to save energy.

10.3.3   Receiver Contention

Following the transmission initiation, the next hop is selected based on receiver contention in XLP as shown in Figure 10.7. This operation leverages the initiative determination concept with the receiver-based routing approach [20, 29]. After an RTS packet is received, if a node has an initiative, i.e., $\mathcal{I} = 1$, to participate in the communication, it performs receiver contention to forward the packet.

The receiver contention is based on the routing priority of each node. The routing priority of a node is determined from the progress a packet would make if the node were to relay the packet. As shown in Figure 10.7, the feasible region is divided into N_p priority regions, i.e., A_i, $i = 1, \ldots, N_p$. The priority region of a node is also associated with the priority in channel access. Accordingly, the nodes with the longer progress have higher priority over other nodes.

The prioritization is realized through the backoff mechanism. Each priority region, A_i, corresponds to a backoff window size, CW_i, as shown in the table in Figure 10.7. Based on the location, a node determines its region and performs backoff for $\sum_{j=1}^{i} CW_j + cw_i$, where cw_i is randomly chosen such that $cw_i \in [0, CW_i]$. This backoff scheme helps differentiate nodes of different progress into different prioritization groups. Only nodes inside the same group contend with each other and this contention is governed by the random cw_i value.

The node with the shortest contention window transmits a CTS packet to node i to indicate that it will forward the packet. According to Figure 10.7, a node in A_1 has the highest priority and can transmit its CTS when the contention window timer expires. A node in A_2 has to wait for any CTS messages from A_1. If a node CTS is transmitted, then a node in A_2 can respond. Once a node transmits a CTS packet, other potential receivers determine that another potential receiver with a longer progress has accepted to forward the packet. Accordingly, these nodes switch to sleep state.

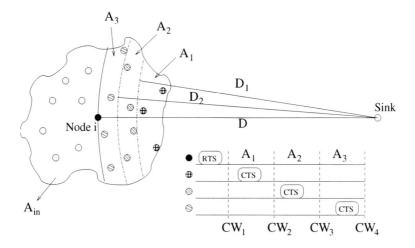

Figure 10.7 Priority regions and the prioritization mechanism.

When node i receives a CTS packet from a potential receiver, it determines that receiver contention has ended and sends a DATA packet indicating the position of the winner node in the header. The CTS and DATA packets both inform the other contending nodes about the transmitter–receiver pair. Hence, other nodes stop contending and switch to sleep. In the case when two nodes send CTS packets without hearing each other, the DATA packet sent by node i can resolve the contention. Finally, the receiver node sends an ACK packet to acknowledge receipt of the DATA packet. An example of successful communication using XLP is given next.

■ **EXAMPLE 10.2**

The scenario for $N_p = 3$ priority regions is shown in Figure 10.8. The transmission is initiated by node i through an RTS packet broadcast as shown in Figure 10.8(a). Following receipt of the RTS packet, each potential relay evaluates its position with respect to node i and the sink. Based on their potential advancement, each feasible node corresponds to one of the three priority regions (A_1, A_2, or A_3). The backoff scheme is also illustrated in Figure 10.8(a), where the possible times when a CTS packet can be sent from a priority region are shown. As an example, if a node in A_1 satisfies the initiative function, it first waits for a random cw_1 value and transmits a CTS packet as shown in Figure 10.8(b). The CTS packet resolves the contention among potential relay nodes and the nodes that receive the CTS packet stop their contention timers and do not contend for the packet. Node i receives the CTS packet and transmits a DATA packet as shown in Figure 10.8(c). Finally, the ACK packet is sent back to node i as shown in Figure 10.8(d).

The receiver contention is completed if a node sends a CTS packet and the DATA packet is sent by the sender node i. However, there may be cases when node i does not receive a CTS packet. This may occur for the following three reasons:

- multiple nodes inside the same priority region send CTS packets and these packets collide at the sender node;
- there are no potential neighbors with $\mathcal{I} = 1$; or
- there are no nodes in the feasible region.

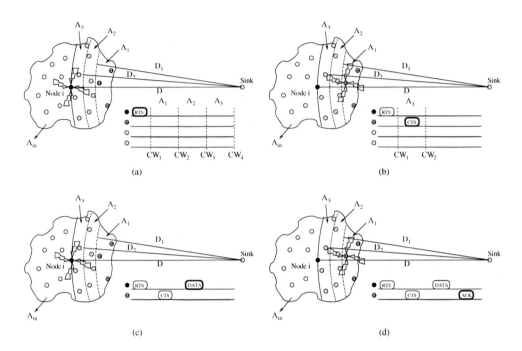

Figure 10.8 Receiver contention mechanism in XLP: (a) RTS, (b) CTS, (c) DATA, and (d) ACK exchange.

While the lack of a CTS packet may occur because of these reasons, node i cannot differentiate these cases. If the CTS packets collide, node i needs to retransmit the RTS packet to restart the receiver contention. If no potential neighbor satisfies the initiative function, the node may retransmit to find potential neighbors that wake up later. However, if there are no nodes in the feasible region, then the node should find an alternate route. Therefore, a node should differentiate between these three cases if there is a lack of CTS packets.

This is performed by a KEEP ALIVE packet. As shown in Figure 10.9, at the end of the contention phase, if a node in the feasible region does not receive any CTS or DATA packets, it sends a KEEP ALIVE packet. This is sent after $\sum_{j=1}^{N_p} CW_j + cw$ if no communication is overheard. In this case, cw is a random number, where $cw \in [0, CW]$ and N_p is the number of priority regions. The existence of a KEEP ALIVE packet notifies the sender that there exist nodes closer to the sink in the feasible region, but the initiative shown in (10.3) is not met for any of these nodes. On reception of this packet, the node continues retransmission. However, if a keep alive packet is not received, the node continues retransmission in case there is a CTS packet collision. If no response is received after k retries, node i determines that there is no node inside the feasible region. This notifies that a local minimum has been reached and the node switches to angle-based routing mode as explained next.

10.3.4 Angle-Based Routing

The receiver-based routing decisions depend, in part, on the locations of the receivers. However, there may be cases where the packets reach local minima. In other words, a node may not find any feasible nodes that are closer to the sink than itself as shown in Figure 10.10, where the packet reaches node i. In XLP, this case is addressed through a distributed solution, i.e., an *angle-based routing* technique.

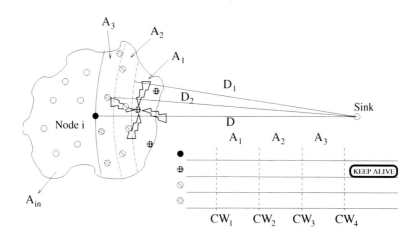

Figure 10.9   KEEP ALIVE packet transmission.

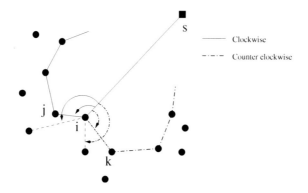

Figure 10.10   Illustration of angle-based routing.

Angle-based routing is depicted in Figure 10.10, where a packet reaches node i, which is a local minimum toward the sink. In this case, the packet should be routed *around the void* using the nodes that are farther away from the sink than node i. The packet can be routed either in a clockwise direction (through node j) or in a counterclockwise direction (through node k).

Since the sender node i is not aware of the locations of all its active neighbors, receiver selection is based on their relative angle. Assume that lines are drawn between node i and the sink, s, as well as between node i and its neighbors. If we compare the angles between the line $i–s$ and the other lines, the angle $\angle sij$ (respectively, angle $\angle sik$) has the smallest angle in the counterclockwise (clockwise) routing direction. Once a measuring direction is set (clockwise or counterclockwise), the packet traverses around the void using the same direction.

In angle-based routing, *traversal direction* is first set to indicate the direction in which the angle measurement should be performed. When a node switches to angle-based routing mode as explained in Section 10.3.3, it also sets the traversal direction to clockwise and sends an RTS packet, which indicates both the routing mode and the traversal direction. The nodes that receive this packet calculate their angle relative to the source–sink direction. Denoting the angle by θ_{ij}, node j sets its contention window to

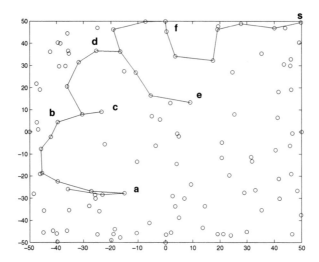

Figure 10.11   Sample route created by angle-based routing.

$c\theta_{ij} + cw_i$, where cw_i is a random number, and c is a constant. The node with the smallest contention window (hence, the smallest angle) sends a CTS packet and the data communication takes place.

This procedure is repeated until the packet reaches a local minimum. In this case, the traversal direction is set to counterclockwise and the procedure is repeated. Angle-based routing is terminated and default routing is performed when the packet reaches a node that is closer to the sink than the node that started the angle-based routing.

■ **EXAMPLE 10.3**

A sample route found by this algorithm is shown in Figure 10.11, where the sink is denoted by s. XLP switches to the angle-based routing mode in clockwise direction at node a. At nodes b and c, the traversal direction is changed until the packet reaches node d. Since node d is closer to the sink than node a, the angle-based routing mode is terminated and packet is forwarded until node e, where a local minimum is reached and angle-based routing mode is used again. Finally, at node f, this mode is terminated since this node is closer to the sink than node e.

10.3.5   Local Cross-layer Congestion Control

In addition to channel quality, the congestion level at a node is also important for relay selection. The congestion state at a node is incorporated into the initiative function as discussed in Section 10.3.1. Accordingly, non-congested nodes are selected as next hops. Further, the source nodes regulate their traffic rate according to the local congestion in their neighborhood. This is performed by the hop-by-hop local cross-layer congestion control component.

By performing hop-by-hop congestion control, this component exploits the local information in the receiver contention to avoid end-to-end congestion control. Since high-quality channels are selected as part of the initiative determination concept, the traditional end-to-end reliability measures are also not necessary. The hop-by-hop congestion control is also motivated by the fact that the sink is only interested in reliable detection of event features from the collective information provided by numerous

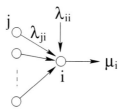

Figure 10.12   The input and output rates for the congestion control mechanism in XLP.

sensor nodes. Moreover, the correlated data flows from closely located source nodes result in loss-tolerant traffic. This diminishes the need for an end-to-end reliability mechanism and the local reliability measures of XLP suffice for an event in the sensor field to be tracked with a certain accuracy at the sink.

The congestion control mechanism ensures that the buffer length of each node does not exceed the maximum capacity. The sources of input to the buffer of each node are therefore important. According to the source and router duties of a sensor node, two sources of traffic are considered as shown in Figure 10.12:

1. *Generated packets:* The first source is the application layer, i.e., the sensing unit of a node. Whenever an event is sensed, the sensing unit generates information that is encapsulated in data packets. These packets should be transmitted by the sensor node due to its source duty. The rate of the generated packets is denoted by λ_{ii}.

2. *Relay packets:* In addition to generated packets, each node receives packets from its neighbors as part of its router duty. These relay packets should be forwarded to the sink through the multi-hop path. The rate at which node i receives relay packets from node j is denoted as λ_{ji}.

The local cross-layer congestion control mechanism aims to regulate the input rate to the buffer of a node according to its output rate so that the buffer does not overflow. Accordingly, the generated and relay packets contribute to the input rate to the buffer of node i. The output rate, on the other hand, depends on channel activity as well as the duty cycle operation of the node. If a node is a source node, the buffer occupancy builds up while it sleeps because of the generated packets, unless appropriate action is taken. Accordingly, the relay traffic that a node accepts should be regulated according to its buffer activity.

The local cross-layer congestion control component of XLP has two main congestion control measures. Firstly, the congestion is controlled in *router duty* by preventing the sensor node from participating in communication if the current load is high. Secondly, the rates of the generated packets are explicitly controlled in *source duty* so that the network is not overloaded.

The first measure is the upper bound for the relay packet rate according to which the node accepts relay packets. This measure is incorporated into the initiative function in (10.3) as discussed in Section 10.3. This bound is denoted by λ_{relay}^{Th} and is used in the XLP initiative determination.

According to the definitions of generated packet rate, λ_{ii}, and the relay packet rate, λ_{ji}, the overall input packet rate at the node i, λ_i, is represented as

$$\lambda_i = \lambda_{ii} + \lambda_{i,relay} = \lambda_{ii} + \sum_{j \in \mathcal{N}_i^{in}} \lambda_{ji} \qquad (10.4)$$

where $\mathcal{N}_i^{in}$ is the set of nodes from which node i receives relay packets. Node i aims to transmit all the packets in its buffer and, hence, the overall output rate of node i can be given by

$$\mu_i = (1 + e_i)(\lambda_{ii} + \lambda_{i,relay}) \qquad (10.5)$$

where e_i is the packet error rate. Note that since the node retransmits the packets that are not successfully sent, the output rate is higher than input rate.

According to (10.4) and (10.5), in a long enough interval, T_∞, the average time that node i spends in transmitting and receiving is given by

$$T_{rx} = \lambda_{i,relay} T_\infty T_{PKT}, \tag{10.6}$$

$$T_{tx} = (1 + e_i)(\lambda_{ii} + \lambda_{i,relay}) T_\infty T_{PKT} \tag{10.7}$$

respectively. The T_{PKT} in (10.6) and (10.7) is the average duration for a packet to be transmitted to another node. Moreover, λ_{ii} is the generated packet rate, $\lambda_{i,relay}$ is the relay packet rate of node i, and e_i is the packet error rate. In both equations, the first component in parentheses shows the average number of packets received and sent, respectively, in a long enough interval T_∞.

In order to prevent congestion at a node, the generated and received packets should be transmitted while the node is active. Because of the duty cycle operation, on average, a node is active for δT_∞ s. Therefore,

$$\delta T_\infty \geq [(1 + e_i)\lambda_{ii} + (2 + e_i)\lambda_{i,relay}] T_\infty T_{PKT}. \tag{10.8}$$

Consequently, the input relay packet rate, $\lambda_{i,relay}$, is bounded by

$$\lambda_{i,relay} \leq \lambda_{i,relay}^{Th} \tag{10.9}$$

where the relay rate threshold, $\lambda_{i,relay}^{Th}$, is given by

$$\lambda_{i,relay}^{Th} = \frac{\delta}{(2 + e_i)T_{PKT}} - \frac{1 + e_i}{2 + e_i}\lambda_{ii}. \tag{10.10}$$

By throttling the input relay rate according to the relay rate threshold, congestion at a node can be prevented. This is incorporated into XLP through a hop-by-hop congestion control mechanism, where nodes participate in routing packets as long as (10.9) is satisfied. For this inequality, a node calculates the parameters e_i, T_{PKT}, and λ_{ii}. The generated packet rate, λ_{ii}, is essentially the rate of injected packets from the sensing boards to the communication module. The packet error rate, e_i, is stored as a moving average of the packet loss rate encountered by the node. Similarly, T_{PKT} is determined by using the delay encountered in sending the previous packet by the node. Consequently, each node updates these values after a successful or unsuccessful transmission of a packet.

According to (10.10), the relay rate threshold, $\lambda_{i,relay}^{Th}$, is directly proportional to the duty cycle parameter, δ. This suggests that the capacity of the network will decrease as δ is reduced. Moreover, the inequality in (10.9) ensures that the input relay rate of source nodes, i.e., nodes with $\lambda_{ii} > 0$, is lower than that of the nodes that are only relays, i.e., $\lambda_{ii} = 0$. This provides homogeneous distribution of traffic load in the network, where source nodes relay less traffic.

The inequality given in (10.9) controls the congestion in the long term. However, in some cases the buffer of a node can still be full due to short-term changes in the traffic. In order to prevent buffer overflow in these cases, the third inequality, $\beta \leq \beta^{max}$, in (10.3) is used by the nodes to determine the initiative. This ensures that the buffer level, β, is lower than the threshold β^{max}, which is the maximum buffer length of a node. Consequently, if a node's buffer is full, it does not participate in communication.

The first measure of the local cross-layer congestion control component is the relay input rate regulation. As a second measure, the amount of traffic generated and injected into the network by the source nodes is directly regulated by the XLP. This is performed by local feedback from the neighborhood of a source node.

As described in Section 10.3.3, during receiver contention, node i may not receive any CTS packets but receive a KEEP ALIVE packet. This notifies that there is congestion in the neighborhood of the node. Accordingly, the source node reduces its transmission rate by decreasing the amount of traffic generated by itself. In other words, the congestion control of source nodes is performed by controlling the rate of generated packets λ_{ii} at node i.

The source generation rate is controlled using an additive-increase, multiplicative-decrease method. Whenever congestion is detected, the rate of generated packets, λ_{ii}, is decreased such that $\lambda_{ii} = \lambda_{ii} \cdot 1/\nu$, where ν is defined to be the transmission rate throttle factor. If the packet is successfully transmitted, then the packet generation rate is increased additively. This conservative increase is performed so that oscillation in the local traffic load is prevented. XLP increases the generated packet rate linearly for each ACK packet received, i.e., $\lambda_{ii} = \lambda_{ii} + \alpha$. XLP adopts a rather conservative rate control approach mainly because it has two functionalities to control the congestion for both source and router duties of a sensor node. As the node decides to participate in relaying based on its buffer occupancy level and relay rate, it already performs congestion control. Hence, XLP does not apply the active congestion control measures, i.e., linear increase and multiplicative decrease, to the overall transmission rate. Instead, only the generated packet rate, λ_{ii}, is updated.

> **Pseudocode of XLP**
> **if** *has packet to send* **then**   /*Source node i*/
>   Perform carrier sense and transmit RTS;
>   **if** *CTS received* **then**
>   Transmit DATA;
>   **if** *ACK received* **then**
>   Increase λ_{ii};
>   Update $\lambda_{i,relay}$, e_i, T_{PKT};
>   **else** Retransmit RTS, update e_i;
>   **elseif** *KEEP ALIVE packet received* **then**
>   Decrease λ_{ii};
>   Update $\lambda_{i,relay}$, e_i;
>   Retransmit RTS;
>   **else**
>   Switch to angle-based routing;
>   Retransmit RTS;
>   **end**
>   **end**
> **elseif** *packet received*   /* Neighbor node j*/
>   **switch** *packet type* **do**
>   **case** *RTS*
>   Calculate $\mathcal{I}$;
>   **if** $\mathcal{I}=1$ **then** Set backoff timer;
>   **else** Set backoff for keep alive;
>   **case** *CTS*
>   Reset backoff timers, switch to sleep;
>   **case** *DATA*
>   **if** *Destined for itself* **then**
>   Transmit ACK, update $\lambda_{j,relay}$;
>   **else** Reset backoff timers, switch to sleep;
>   **case** *ACK*
>   Reset backoff timers;
>   **end**
>   **end**

10.3.6   Recap: XLP Cross-layer Interactions and Performance

XLP merges several functionalities into a single protocol as explained in the previous sections. Here, we revise the cross-layer interactions provided by the XLP based on the traditional layered stack:

- **Network, MAC, and PHY:** The physical layer information such as the channel condition is incorporated into higher layer functionalities. A node monitors its channel quality using the received packet. Accordingly, it participates in communication based on the recent channel state. As a result, the receiver-based initiative concept provides accurate channel-aware operation.

- **Network and MAC:** The receiver-based contention and routing scheme merges the MAC and network layer functionalities. As a result, potential receivers contend for the transmitted packets and become the next hop. In addition, the feasible receivers are determined from the geographical information of source and sink and the voids are avoided by angle-based routing.

- **Transport, Network, and MAC:** The traditional transport layer functionalities are incorporated into the local cross-layer congestion control mechanism in XLP. The nodes monitor their buffer states as well as their input relay rates. Accordingly, highly congested nodes are prevented from participating in contention and routing. In addition, the relay rate is controlled to prevent congestion and the source rate is controlled based on local information at the source nodes.

The cross-layer module concept results in significant gains in performance in terms of energy consumption, throughput, and efficiency in WSNs. When compared to traditional layered protocols such as geographical routing [18], directed diffusion [12], ESRT [2], RMST [21], CC-MAC [24], and S-MAC [26], as well as the cross-layer ALBA-R protocol [5, 4], the throughput achieved by XLP provides up to 80% improvement. Especially, end-to-end reliability mechanisms such as the RMST protocol significantly decrease the throughput. Similarly, XLP also provides high reliability even through the duty cycle is decreased since the next hop nodes are selected among the nodes that are available. This is in contrast to layered protocols, where the routing layer directs the MAC layer to communicate with a specific node that may not be awake for a long time. While providing these improvements, XLP performs similar to the layered protocols in terms of latency, which may important for certain applications. The major advantage of XLP is the significant reductions in energy consumption due to the efficient use of resources through the cross-layer communication techniques. Nodes with high channel quality, low congestion, and ample residual energy are chosen in advance, which proactively prevents several problems that would otherwise need to be addressed reactively by traditional layered protocols.

In addition to the communication performance, implementation efficiency is also of importance in WSNs. In a traditional layered protocol architecture, each layer has clear boundaries. This layered structure leads to computational delays due to the sequential handling of a packet. For example, in TinyOS [1], each layer has to wait for the lower layers to process a packet since a single buffer is used for a packet for all layers. XLP, however, melts the functionalities of traditional medium access, routing, and congestion control into a unified cross-layer communication module by considering physical layer and channel effects. Hence, these functionalities are performed as a whole and overall protocol efficiency can be improved using XLP.

XLP does not require any tables or extra buffer space for routing and congestion control functionalities. The routing is performed based on receiver initiatives, which eliminates the need for a routing table at each node. The implementation of XLP is both simple and compact. On the other hand, as an example, the S-MAC protocol maintains a schedule table for each of the one-hop neighbors to provide synchronized sleeping cycles. Similarly, in directed diffusion, each node has to implement a reinforcement table for each source indicating the next hop in the reinforced path. In case a node is a source node, it also has to keep track of multiple neighbors, which have a path to the sink for exploratory messages. At the transport layer, RMST requires a separate queue to cache data locally to support loss recovery at all hops. These requirements, due to either layered operation of the protocol stack or the internal protocol structure at each layer, place a burden on memory space for communication in sensor nodes. This extra space required by the communication stack limits the available space for developing new applications for sensor networks. On the other hand, careful use of code space and cross-layer implementation of communication functionalities in XLP provides a more efficient operation in WSNs. When coupled with the noticeably better communication performance, XLP becomes a successful candidate for communication protocols in WSNs.

References

[1] TinyOS. Available at http://www.tinyos.net/.

[2] Ö. B. Akan and I. F. Akyildiz. Event-to-sink reliable transport in wireless sensor networks. *IEEE/ACM Transactions on Networking*, 13(5):1003–1016, October 2005.

[3] I. F. Akyildiz, M. C. Vuran, and Ö. B. Akan. A cross layer protocol for wireless sensor networks. In *Proceedings of the Conference on Information Sciences and Systems (CISS'06)*, pp. 1102–1107, Princeton, NJ, USA, March 2006.

[4] P. Casari, M. Nati, C. Petrioli, and M.Zorzi. Efficient non planar routing around dead ends in sparse topologies using random forwarding. In *Proceedings of the IEEE International Conference on Communications (ICC'07)*, pp. 3122–3129, Glasgow, UK, June 2007.

[5] P. Casari, M. Nati, C. Petrioli, and M. Zorzi. ALBA: an adaptive load-balanced algorithm for geographic forwarding in wireless sensor networks. In *Proceedings of the IEEE Military Communications Conference (MILCOM'06)*, Washington, DC, USA, October 2006.

[6] M. Chiang. To layer or not to layer: balancing transport and physical layers in wireless multihop networks. In *Proceedings of IEEE INFOCOM 2004*, volume 4, pp. 2525–2536, Hong Kong, China, March 2004.

[7] M. Chiang. Balancing transport and physical layers in wireless multihop networks: jointly optimal congestion control and power control. *IEEE Journal on Selected Areas in Communications*, 23(1):104–116, January 2005.

[8] S. Cui, R. Madan, A. Goldsmith, and S. Lall. Joint routing, MAC, and link layer optimization in sensor networks with energy constraints. In *Proceedings of IEEE ICC 2005*, volume 2, pp. 725–729, Seoul, South Korea, May 2005.

[9] Y. Fang and B. McDonald. Dynamic codeword routing (DCR): a cross-layer approach for performance enhancement of general multi-hop wireless routing. In *Proceedings of IEEE SECON'04*, pp. 255–263, Santa Clara, CA, USA, October 2004.

[10] D. Ferrara, L. Galluccio, A. Leonardi, G. Morabito, and S. Palazzo. MACRO: an integrated MAC/Routing protocol for geographical forwarding in wireless sensor networks. In *Proceedings of IEEE INFOCOM 2005*, volume 3, pp. 1770–1781, Miami, FL, USA, March 2005.

[11] D. Ganesan, B. Krishnamachari, A. Woo, D. Culler, D. Estrin, and S. Wicker. An empirical study of epidemic algorithms in large scale multihop wireless networks. Technical report IRB-TR-02-003, Intel Research, March 2002.

[12] C. Intanagonwiwat, R. Govindan, D. Estrin, J. Heidemann, and F. Silva. Directed diffusion for wireless sensor networking. *ACM/IEEE Transactions on Networking*, 11(1):2–16, February 2002.

[13] S. Liu, K.-W. Fan, and P. Sinha. CMAC: an energy efficient MAC layer protocol using convergent packet forwarding for wireless sensor networks. In *Proceedings of IEEE SECON'07*, pp. 11–20, San Diego, CA, USA, June 2007.

[14] G. Lu, B. Krishnamachari, and C. S. Raghavendra. An adaptive energy-efficient and low-latency MAC for data gathering in wireless sensor networks. In *Proceedings of the IEEE International Parallel and Distributed Symposium*, pp. 224–231, Santa Fe, NM, USA, April 2004.

[15] M. Mastrogiovanni, C. Petrioli, M. Rossi, A. Vitaletti, and M. Zorzi. Integrated data delivery and interest dissemination techniques for wireless sensor networks. In *Proceedings of IEEE Globecom'06*, pp. 1–6, San Francisco, CA, USA, November 2006.

[16] T. Melodia, M. C. Vuran, and D. Pompili. The state-of-the-art in cross-layer design for wireless sensor networks. In *Proceedings of EuroNGI Workshops on Wireless and Mobility* (Springer Lecture Notes in Computer Science, volume 3883), June 2006.

[17] L. Savidge, H. Lee, H. Aghajan, and A. Goldsmith. Event-driven geographic routing for wireless image sensor networks. In *Proceedings of COGIS'06*, Paris, France, March 2006.

[18] K. Seada, M. Zuniga, A. Helmy, and B. Krishnamachari. Energy-efficient forwarding strategies for geographic routing in lossy wireless sensor networks. In *Proceedings of SenSys'04*, pp. 108–121, Baltimore, MD, USA, 2004.

[19] M. L. Sichitiu. Cross-layer scheduling for power efficiency in wireless sensor networks. In *Proceedings of IEEE INFOCOM 2004*, volume 3, pp. 1740–1750, Hong Kong, China, 2004.

[20] P. Skraba, H. Aghajan, and A. Bahai. Cross-layer optimization for high density sensor networks: distributed passive routing decisions. In *Proceedings of Ad-Hoc Now'04*, Vancouver, Canada, July 2004.

[21] F. Stann and J. Heidemann. RMST: reliable data transport in sensor networks. In *Proceedings of the First IEEE International Workshop on Sensor Network Protocols and Applications*, pp. 102–112, Anchorage, AK, USA, May 2003.

[22] L. van Hoesel, T. Nieberg, J. Wu, and P. J. M. Havinga. Prolonging the lifetime of wireless sensor networks by cross-layer interaction. *IEEE Wireless Communications*, 11(6):78–86, December 2004.

[23] M. C. Vuran, Ö. B. Akan, and I. F. Akyildiz. Spatio-temporal correlation: theory and applications for wireless sensor networks. *Computer Networks*, 45(3):245–261, June 2004.

[24] M. C. Vuran and I. F. Akyildiz. Spatial correlation-based collaborative medium access control in wireless sensor networks. *IEEE/ACM Transactions on Networking*, 14(2):316 –329, April 2006.

[25] M. C. Vuran, V. B. Gungor, and Ö. B. Akan. On the interdependency of congestion and contention in wireless sensor networks. In *Proceedings of SENMETRICS'05*, San Diego, CA, USA, July 2005.

[26] W. Ye, J. Heidemann, and D. Estrin. An energy-efficient MAC protocol for wireless sensor networks. In *Proceedings of IEEE INFOCOM'02*, volume 3, pp. 1567–1576, New York, USA, June 2002.

[27] W. Ye, J. Heidemann, and D. Estrin. Medium access control with coordinated adaptive sleeping for wireless sensor networks. *IEEE/ACM Transactions on Networking*, 12(3):493–506, June 2004.

[28] M. Z. Zamalloa, K. Seada, B. Krishnamachari, and A. Helmy. Efficient geographic routing over lossy links in wireless sensor networks. *ACM Transactions on Sensor Networks*, 4(3):1–33, 2008.

[29] M. Zorzi and R. Rao. Geographic random forwarding (GeRaF) for ad hoc and sensor networks: multihop performance. *IEEE Transactions on Mobile Computing*, 2(4):337–348, October–December 2003.

[30] M. Zorzi and R. Rao. Geographic random forwarding (GeRaF) for ad hoc and sensor networks: energy and latency performance. *IEEE Transactions on Mobile Computing*, 2(4):349–365, October–December 2003.

11

Time Synchronization

WSNs are distributed systems where each sensor device is equipped with its own local clock for internal operations. Each event that is related to operation of the sensor device including sensing, processing, and communication is associated with timing information controlled through the local clock. Since users are interested in the collaborative information from multiple sensors, timing information associated with data at each sensor device needs to be consistent. Moreover, the WSN should be able to correctly order the events sensed by distributed sensors to accurately model the physical environment.

These timing requirements make time synchronization an important part of communication for WSNs. Distributed synchronization protocols are required to coordinate the nodes in the network so that they follow the same reference frame. As a result, the following capabilities are provided:

- **Temporal event ordering:** Network-wide synchronized operation enables correct ordering of events sensed by different devices. The multi-hop and packet-based information delivery in WSNs results in variations in the information delivery time. Moreover, the delay between a node and the sink is proportional to the distance between them. Consequently, the received time of the packets at the sink and the order in which they are received do not correctly represent the sensing time of the events. In a synchronized network, the sink can easily reorder the received packets according to their timestamps.

- **Synchronization to global time:** A WSN that is internally synchronized can also be synchronized to a global time, i.e., coordinated universal time (UTC), through a simple conversion between the local time of the network and UTC. As a result, the integration of multiple WSNs with the Internet will be trivial.

- **Synchronized network protocols:** The synchronization between sensor nodes also provides many capabilities in terms of protocol development. As an example, time division multiple access (TDMA) protocols require neighbor nodes to be synchronized so that they can follow a common time frame for medium access as explained in Chapter 5. Hence, synchronization protocols generally require MAC protocols to provide a common frame of reference in protocol operation.

While synchronization is essential for a number of applications in WSNs, there exist many challenges because of the distributed nature of these networks as well as hardware and software constraints on the sensor nodes. In the following we describe the design challenges and the factors that affect synchronization.

11.1 Challenges for Time Synchronization

The time synchronization protocols have to address design challenges and factors affecting time synchronization, including *low-cost clocks, effects of wireless communication, resource constraints, high density*, and *node failures*. In addition, the synchronization techniques have to address the mapping between the sensor network time and Internet time, e.g., UTC. The main design challenges and the factors affecting time synchronization are as follows.

11.1.1 Low-Cost Clocks

Wireless sensor nodes are typically implemented from low-cost materials, where low-end crystals are used for local clocks, which results in frequent clock drift and clock jitter throughout the lifetime of the node. These changes can be explained as follows.

The clock of each sensor i represents the time as a value $\tau_i(t)$ at each instant, which is a function of time, t, as shown below:

$$\tau_i(t) = \alpha_i(t)t + \theta_i(t) \tag{11.1}$$

where $\alpha_i(t)$ is the *clock drift* and $\theta_i(t)$ is the *clock offset*. In a perfect clock, $\alpha_i(t) = 1$ and $\theta_i(t) = 0$. In WSNs, however, the low-cost crystals introduce both clock drift, i.e., $\alpha_i(t) \neq 1$, and clock offset, i.e., $\theta_i(t) > 0$. Since the clock drift and the clock offset can be different for each sensor node, the nodes are unsynchronized. Moreover, the local clock of a node i can be represented relative to a node j as follows:

$$\tau_i(t) = \alpha_{ij}(t)\tau_j(t) + \theta_{ij}(t) \tag{11.2}$$

where $\alpha_{ij}(t)$ and $\theta_{ij}(t)$ are the *relative* clock drift and offset, respectively. A synchronization protocol does not aim to provide perfect operation of the clock, i.e., $\alpha_i(t) = 1$ and $\theta_i(t) = 0$. Instead, the relative differences between two clocks are corrected. Consequently, providing a common reference frame requires that the reference clock drift and offset between the nodes in the network must be minimized.

The synchronization between two nodes i and j can be performed according to the relation between their clocks as shown in (11.2). Through message exchanges, the clocks of two nodes can be corrected. However, a single exchange of messages can only correct the relative offset, θ_{ij}, between the clocks. Since the individual drifts of the nodes are usually different, i.e., $\alpha_{ij}(t) \neq 0$, the clocks tick at different rates and eventually become unsynchronized. Therefore, periodic message exchanges or relative clock drift estimation are required between nodes to correct for the relative clock drifts as well.

Another important factor in time synchronization is the temporal variations in a clock's behavior. Note that the clock drift and offset of a local clock shown in (11.1) are also a function of time. As a result, correcting for the relative clock drift at a single instant may not synchronize the clocks of two nodes for the rest of the network lifetime, which calls for adaptive synchronization methods.

11.1.2 Wireless Communication

Synchronization requires nodes to communicate with each other to exchange clock information through which the local clocks can be synchronized. While this communication is usually trivial in wired networks such as the Internet, WSNs require wireless communication between nodes, which creates additional challenges for synchronization as a result of the error-prone communication and non-deterministic delays.

Firstly, the wireless channel errors result in some of the synchronization messages being lost. Thus, some nodes in the network may be unsynchronized. More importantly, the synchronization messages sent by the unsynchronized nodes force other nodes to adapt to their local clocks. Consequently, the network is partitioned into different zones with different synchronized times, which prevents network-wide synchronization. Therefore, robust synchronization methods are required.

Secondly, the broadcast nature of the wireless channel necessitates MAC protocols being utilized for efficient channel access. These MAC protocols introduce a non-deterministic *access delay*, which is the time between the synchronization protocol issuing a synchronization packet to be sent and the time this packet is actually transmitted. As we will see next, the channel access operation introduces an important randomness in time synchronization and needs to be accounted for in calculations.

Finally, the wireless channel introduces an asymmetric delay between two nodes for message exchanges. Since the access times as well as transmission times can vary because of channel errors and retransmissions, the communication delay may be different for each direction. This is an important point for synchronization since most solutions rely on consecutive message exchanges where the round-trip time between two nodes is considered for calculations. While, for wired networks, the round-trip time is roughly equal to twice the delay in one direction, wireless communication results in asymmetric delays in each direction.

11.1.3   Resource Constraints

Wireless sensor nodes are constrained in terms of energy consumption, processing, and memory because of limited resources. Therefore, many traditional synchronization methods cannot be used in WSNs. Since the transmission and reception of packets constitute major energy consumption in a sensor node, frequent synchronization message exchanges should be avoided. On the other hand, the low-end clock crystals introduce significant amounts of drift, which may lead to frequent synchronization. Hence, this tradeoff needs to be carefully considered.

11.1.4   High Density

WSNs are characterized by high density in the network. As a result, solutions based on pairwise synchronization may not be suitable for WSNs. Considering the high cost of communication in terms of energy consumption, pairwise synchronization among each node in the neighborhood may not be the most efficient solution. Consequently, synchronization solutions that exploit the broadcast nature of the wireless channel, such as reference broadcast synchronization (RBS) [4], which is described in Section 11.5, will lead to scalable solutions.

11.1.5   Node Failures

The low-cost hardware and limited energy resources result in frequent node failures in WSNs. Consequently, synchronization solutions that rely on a single node as a master clock cannot be applied. Distributed and robust solutions that are not affected by node failures in WSNs are therefore required.

Next, we introduce the Network Time Protocol (NTP) [11], which is a traditional synchronization protocol used for the Internet. This protocol also contains several synchronization approaches that have been adopted by the synchronization protocols developed for WSNs [5, 15, 17]. Then, we discuss synchronization solutions developed for WSNs.

11.2   Network Time Protocol

In the Internet, the Network Time Protocol (NTP) is used to discipline the frequency of each host's oscillator. Synchronization among hosts is accomplished through a hierarchical structure of time servers as shown in Figure 11.1. In this hierarchical structure, the root is synchronized with UTC and at each level the time servers synchronize the clocks of their subnetwork peers [1].

NTP relies on a two-way handshake between two nodes to estimate the delay between these nodes and calculate the relative offset accordingly, as described next.

■ EXAMPLE 11.1

Consider a server i and peer j at different levels of the hierarchy as shown in Figure 11.2. The server i issues a SYNC message to be sent to node j at time t_1^i according to its local time. This

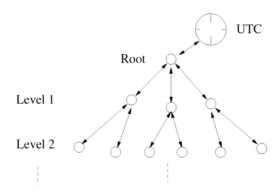

Figure 11.1   The hierarchical time server architecture of NTP [11].

message is received and timestamped at node j at t_2^j according to node j's local time. Then, node j issues a REPLY message at time t_3^j, which is received and timestamped at server i at time t_4^i.

Denoting two variables $a = t_2^j - t_1^i$ and $b = t_3^j - t_4^i$, the round-trip delay between two nodes, δ_{ij}, and the relative clock offset, θ_{ij}, are estimated as follows [11]:

$$\delta_{ij} = a - b, \quad \theta_{ij} = \frac{a+b}{2}. \tag{11.3}$$

Using the relative clock offset, the clocks of the server and the peers can be synchronized.

The accuracy of NTP synchronization is on the order of milliseconds [1]. However, in NTP, it is assumed that the transmission delay between two hosts is the same in both directions, which is reasonable for the Internet.

While NTP provides robust time synchronization in a large-scale network, many characteristics of WSNs make this protocol unsuitable. It may be useful to use NTP to discipline the oscillators of the sensor nodes, but connection to the time servers may not be possible because of frequent sensor node failures. In addition, synchronizing all sensor nodes into a single clock reference may be a problem due to interference from the environment and the large variations in delays between the different parts of the sensor field. The interference can temporarily disjoint the sensor field into multiple smaller fields leading to undisciplined clocks among these smaller fields.

Moreover, NTP is computationally intensive and requires a precise time server to synchronize the nodes in the network. In addition, it does not take into account the energy consumption required for time synchronization. Although NTP is robust, it may suffer from large propagation delays when sending timing messages to the time servers. In addition, the nodes are synchronized in a hierarchical manner and some of the time servers in the middle of the hierarchy may fail causing unsynchronized nodes in the network. If these nodes fail, it is hard to reconfigure the network. NTP also assumes symmetric link delays between two nodes. Because of the wireless link errors and contention in the broadcast channel, this may not be true for WSNs. Consequently, the delay measurements may lead to inaccurate offset estimation.

11.3   Definitions

The unique challenges posed by WSNs as explained in Section 11.1 call for synchronization solutions tailored to these networks. One of the main challenges in this respect is the effect of the broadcast

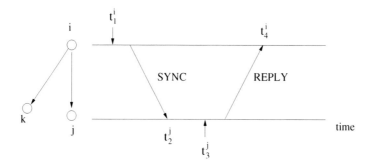

Figure 11.2   NTP two-way handshake mechanism [11].

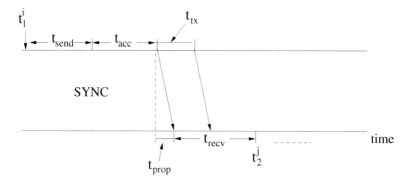

Figure 11.3   Synchronization delay between a pair of nodes.

wireless channel. Since synchronization can only be achieved through communication between nodes, the effects of the wireless channel need to be carefully considered in the design of synchronization protocols. However, wireless communication introduces randomness in the delay between two nodes. Considering the handshake scheme shown in Figure 11.3, which is redrawn from Figure 11.2, the delay between two nodes has four components:

- **Sending delay (t_{send}):** The handshake between two nodes is initiated when node i issues a SYNC packet transmission with the timestamp t_1^i. However, there is a non-zero delay between the time when the synchronization protocol issues this command and the time when the packet is prepared. This delay is a combination of operating system, kernel, and transceiver delays on the embedded system board. More importantly, the sending delay is non-deterministic because of the complex and time-varying interactions between each hardware and software component in the embedded system.

- **Access delay (t_{acc}):** The broadcast nature of the wireless channel also introduces an additional delay after the packet has been prepared and transferred to the transceiver. Depending on the MAC protocol, an additional delay is introduced in waiting for access to the channel. While this delay may be bounded in a TDMA protocol because of reserved slots, a CSMA-based protocol may introduce a significant amount of access delay if the channel is highly loaded. In either case, the access delay cannot be determined a priori.

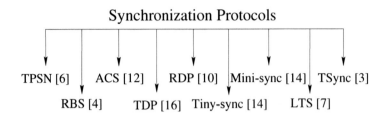

Figure 11.4   Overview of synchronization protocols.

- **Propagation time (t_{prop}):** Once a node accesses the channel, the SYNC packet is transmitted and it takes some amount of time for the packet to reach the intended receiver. While the propagation delay is negligible in communication through air, underground and underwater environments introduce significant propagation delay, which is important for synchronization. Moreover, this delay is directly proportional to the distance between the nodes.

- **Receiving delay (t_{recv}):** The last component for communication delay is the time required for the transceiver of the receiver node j to receive the packet, decode it, and notify the operating system. This includes the transmission and processing delay required for the antenna to receive the message from the channel, perform A/D conversion, and notify the operating system of its arrival. An important component of the receiving delay is the transmission delay, t_{tx}, which is the time needed for the SYNC packet to be completely received as shown in Figure 11.3. The transmission delay depends on the SYNC packet length as well as the transmission rate.

The four main components that contribute to the communication delay are referred to as the *critical path* in synchronization. The critical path is non-deterministic in nature and, hence, creates a major challenge in exploiting the traditional offset estimation methods used by NTP explained in Section 11.2. Many synchronization protocols for WSNs aim to minimize the effects of this random delay.

In the following, we explain the existing synchronization protocols in detail. These protocols are shown in Figure 11.4 and are the *Timing-sync Protocol for Sensor Networks* (TPSN) [6], *Reference-Broadcast Synchronization* (RBS) [4], *Adaptive Clock Synchronization* (ACS) [12], the *Time Diffusion Synchronization Protocol* (TDP) [16], and the *Rate-based Diffusion Protocol* (RDP) [10], as well as the *Tiny- and Mini-sync protocols* [14]. The representative protocols other than these protocols are also summarized.

11.4   Timing-Sync Protocol for Sensor Networks (TPSN)

TPSN [6] adopts some concepts from NTP. Similar to NTP, a hierarchical structure is used to synchronize the whole WSN to a single time server. TPSN requires the root node to synchronize all or parts of the nodes in the sensor field. It consists of two phases: (1) the level discovery phase, where the hierarchical structure is built in the network starting from the root node; and (2) the synchronization phase, where pairwise synchronization is performed throughout the network.

The first step of TPSN is the *level discovery phase*, which is described through Example 11.2.

■ **EXAMPLE 11.2**

In the level discovery phase, the root node is assigned to level 0, and the nodes in the network are assigned levels in a hierarchy according to their distance to the root node as shown in Figure 11.5. Firstly, the root node constructs the hierarchy by broadcasting a *level_discovery* packet. The first level of the hierarchy is level 0, which is where the root node resides. The nodes that receive

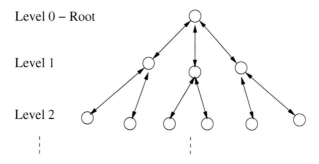

Figure 11.5 Hierarchical synchronization architecture of TPSN.

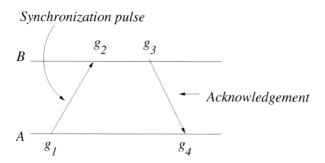

Figure 11.6 Two-way message handshake.

the *level_discovery* packet from the root node are the nodes that belong to level 1. Afterward, the nodes in level 1 broadcast their *level_discovery* packet, and neighbor nodes receiving the *level_discovery* packet for the first time are labeled as level 2 nodes. This process is repeated until all the nodes in the sensor field have a level number.

The second phase of TPSN is the *synchronization phase*, where each node in the hierarchy is synchronized with a node from a higher level as explained next.

■ EXAMPLE 11.3

Based on the hierarchy described in Example 11.2, the two-way handshake mechanism used in NTP is as follows. The root node sends a *time_sync* packet to initialize the time synchronization process. Then, the nodes in level 1 start synchronization with the root node by sending a *synchronization_pulse* to level 0 as shown in Figure 11.6. To prevent collisions with multiple nodes, each node in level 1 waits for a random amount of time before transmitting the *synchronization_pulse*. The root node sends an *acknowledgment* back to the node to finalize synchronization. As a result, level 1 nodes are synchronized to the root.

The *synchronization_pulse* from level 1 also serves as a *time_sync* packet to the neighbors of the level 1 nodes, i.e., level 2 nodes. Upon hearing this packet from a node in level 1, the nodes wait for a random amount of time that is long enough for the level 1 nodes to finish their synchronization. Then, they initialize the synchronization process by transmitting a *synchronization_pulse* as

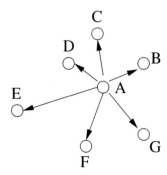

Figure 11.7 Reference broadcasting, where node A's broadcast message is used by the remaining nodes for synchronization.

shown in Figure 11.6. The level 1 node then sends an acknowledgment. The synchronization of the whole network is performed level by level in this way.

TPSN is based on a sender–receiver synchronization model, where the receiver synchronizes with the local clock of the sender according to the two-way message handshake as shown in Figure 11.6. Since the design of TPSN is based on a hierarchical methodology similar to NTP, nodes within the hierarchy may fail and may cause nodes to become unsynchronized. In addition, node movements may render the hierarchy useless, because nodes may move out of their levels. Hence, nodes at level i cannot synchronize with nodes at level $i - 1$. This may lead to inter-synchronized islands that are not in sync with each other. The advantages and disadvantages of TPSN are listed below.

11.4.1 Qualitative Evaluation

The hierarchical structure of TPSN provides scalability. The synchronization of a node depends on its parent in the hierarchical structure. Therefore, even if the number of nodes in the network increases, the high synchronization accuracy can still be achieved. Since the hierarchical structure covers the entire network based on a root node, the whole network can be synchronized to the same time reference. As a result, network-wide synchronization is possible. In addition, TPSN requires each node to exchange timing information with its parent in the hierarchical structure. Consequently, in TPSN, each node has to exchange synchronization information with only a single node and the protocol ensures that it is synchronized with all the remaining nodes in its neighborhood. Moreover, the synchronization cost is relatively low compared to NTP [1].

TPSN requires a hierarchical structure to exist for synchronization. The maintenance of this structure in the case of failed nodes increases the energy consumption. The hierarchical structure also prevents accurate synchronization of mobile nodes. Since the connectivity of different nodes changes as nodes move, the hierarchical structure needs to be formed accordingly. Moreover, the synchronization procedure of TPSN is based on adjusting the clocks according to the parent nodes in the hierarchy. This increases the cost of synchronization compared to other methods where the relative offsets of the neighbor nodes are stored and the time is translated without adjusting the physical clock. When multiple root nodes are used in large networks, each cluster can be synchronized to a different reference time. As a result, the protocol forms islands of times. To prevent this, each root node should be synchronized in advance. Of course, this increases the overall cost for synchronization if the root nodes are located far from each other. Furthermore, multi-hop synchronization is not supported since nodes synchronize only to their parent node.

11.5   Reference-Broadcast Synchronization (RBS)

The traditional sender–receiver handshake scheme shown in Figure 11.3 introduces a significant amount of non-deterministic delay from both sender and receiver. RBS aims to minimize the critical path in synchronization by eliminating the effect of the sender [4]. This is accomplished by exploiting the broadcast nature of the wireless channel.

Instead of synchronizing a sender with a receiver, RBS provides time synchronization among a set of receivers that are within the reference broadcast of a sender. Since the propagation times are negligible, once a packet is transmitted by a sender, it is received at its neighbors almost at the same instant. Consequently, the synchronization accuracy can be improved by synchronizing only the receivers. This is explained through the critical path concept next.

■ **EXAMPLE 11.4**

As shown in Figure 11.7, the transmitter broadcasts m reference packets. Each of the receivers, which are within the broadcast range, records the time of arrival of the reference packets. Then, the receivers communicate with each other to determine the offsets. Let us compare this method to traditional synchronization, where the transmitter and each receiver are synchronized.

As shown in Figure 11.8(a), the critical path for traditional synchronization includes the sender-side uncertainty as well. The sending delay as well as the access delay should be accurately estimated to improve the synchronization accuracy. Broadcast-based synchronization, however, does not involve the transmitter (node A) in the synchronization. Instead, only the receivers (nodes B, C, D, E, F, G) synchronize among themselves based on a broadcast message from the transmitter (node A). As shown in Figure 11.8(b), this reduces the critical path. The only uncertainty in RBS is the time between when a packet is received and when it is timestamped. Considering the hardware capabilities of the sensor node, this delay can also be estimated up to a certain accuracy.

The receiver–receiver synchronization method provides an efficient way for each node to determine its clock offset relative to its neighbors. Following message exchanges from each neighbor, each node populates a table consisting of relative offsets with its neighbors. Consequently, RBS does not aim to correct the clocks of the nodes. Rather, whenever a packet with a timestamp is received, this timestamp is converted to the local clock using the relative offset information.

The receiver–receiver synchronization method can provide synchronization only in a broadcast area. To provide multi-hop synchronization, RBS uses nodes that are receiving two or more reference broadcasts from different transmitters. As shown in Figure 11.9, these nodes are denoted as *translation nodes* and are used to translate the time between different broadcast domains.

As shown in Figure 11.9, nodes A, B, and C are the transmitter, receiver, and translation nodes, respectively. The transmitter nodes broadcast their timing messages, and the receiver nodes receive these messages. Then, the receiver nodes synchronize with each other. The sensor nodes that are within the broadcast regions of both transmitter nodes A and B are the translation nodes. When an event occurs, a message describing the event with a timestamp is translated by the translation nodes between each broadcast range. As a result, the message can be routed back to the sink.

11.5.1   Qualitative Evaluation

RBS eliminates the sender-side uncertainty from the critical path. This decreases the synchronization error and improves the efficiency. Each node stores the offset and skew of its neighbors and the time is translated between nodes according to this information. As a result, local clocks are not corrected for each synchronization attempt. Moreover, the time synchronization mechanism is tunable and lightweight since it relies on broadcast messages only. In addition, multi-hop synchronization is provided through

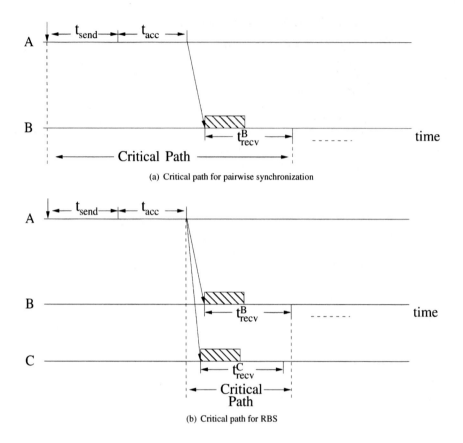

(a) Critical path for pairwise synchronization

(b) Critical path for RBS

Figure 11.8   Critical path for different synchronization approaches.

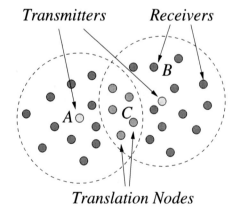

Figure 11.9   RBS multi-hop synchronization scheme.

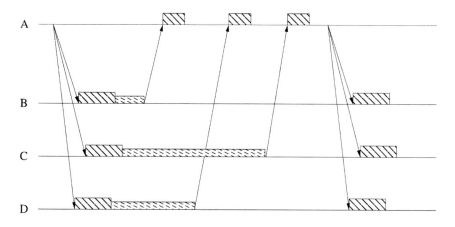

Figure 11.10   ACS scheme [12].

gateway nodes, which translate time from one broadcast neighborhood to another. RBS is applicable to any medium which has broadcast capabilities including wired and wireless networks.

RBS requires close coupling between the synchronization structure and the routing protocol. The route from any node to the sink should contain the translation nodes so that the time translation can be performed on the packet along its way to the sink. With the lack of this coupling, there may not be translation nodes on the route along which the message is relayed. As a result, synchronization services may not be available on some routes. RBS is not suitable for TDMA-based MAC protocols since network-wide synchronization is not provided.

Compared to the traditional sender–receiver synchronization, e.g., TPSN [6], receiver–receiver synchronization necessitates each node exchanging timing information between its neighbors. RBS requires message exchanges with all the neighbors, which translates into $O(n^2)$ message exchanges if there are n nodes in a node's broadcast range. This increases the energy consumption and may lead to frequent collisions when the network density is high. The large number of message exchanges between neighbors also increases the convergence time of RBS because of possible packet errors and collisions in the broadcast wireless channel. Furthermore, the reference sender cannot be synchronized through RBS. This requires additional synchronization rounds for the senders to be synchronized with the network.

11.6   Adaptive Clock Synchronization (ACS)

The disadvantage of excessive message exchanges in RBS can be alleviated by following a hybrid approach between sender–receiver and receiver–receiver synchronization. This is performed by ACS [12]. The synchronization error cannot be deterministically bounded if the communication delay is unbounded or packet losses occur [2], which is usually the case in wireless communication. Accordingly, ACS aims to provide a statistical guarantee on the errors and to reduce the number of transmissions.

ACS uses the receiver–receiver synchronization methodology by following a sender–receiver message exchange policy. The synchronization procedure of this hybrid approach is shown Figure 11.10. Firstly, a sender broadcasts m consecutive synchronization packets to its neighbors within a constant period. Upon receiving each packet, the receivers calculate the relative clock drift by linear regression. This information is sent back to the sender with a random delay to prevent collisions with other neighbors. When the sender receives multiple responses, it calculates its clock skew relative to each of its neighbors. This information is again broadcast to the neighbor nodes.

Consequently, the clock skew relative to each of the neighbors can be determined by using a node's clock skew with respect to the sender node as a reference. The clock skews are initially calculated with respect to the same synchronization packet and the effects of sending delay and access delay are factored out, similar to RBS. Moreover, each node sends a single packet and receives two packets compared to pairwise packet exchanges with each neighbor, which significantly decreases the energy consumption.

The hybrid synchronization structure provides synchronization only in a single broadcast region. For multi-hop networks, ACS follows hierarchical message propagation similar to TPSN. The senders in the network act as level 0 nodes and the synchronization is performed at each level using the same technique. Upon completion of the network-wide synchronization, the nodes can send messages using multiple hops by transforming the time through intermediate nodes, much as in the RBS scheme.

11.6.1   Qualitative Evaluation

ACS exploits the advantages of two different synchronization techniques to provide efficiency and statistical guarantees. This hybrid structure allows for flexible operation in WSNs. The relative clock skew calculation technique reduces the number of messages exchanged between neighbors compared to RBS. As a result, the protocol is scalable for high-density networks since each node sends a single packet and receives two packets for each synchronization attempt irrespective of the density.

The major overhead in ACS, however, is that the senders consume significant amounts of energy because of the coordination functionalities. Consequently, a dynamic operation may be required such that the hierarchical structure and the sender nodes are updated frequently. The synchronization accuracy can only be probabilistically guaranteed. Thus, ACS may not be suitable for applications where 100% synchronization is essential.

11.7   Time Diffusion Synchronization Protocol (TDP)

Both the RBS and TPSN protocols aim to provide multi-hop synchronization by extending a single hop synchronization method and performing this method iteratively. While these solutions may be acceptable for small networks, they are not scalable for very large networks. TDP aims to maintain a common time throughout the network within a certain tolerance [16]. Moreover, this tolerance level can be adjusted depending on the application and the requirements of the network. The equilibrium network time can also be translated into the global time at the sink using a time translation algorithm, which provides a gateway to NTP and the Internet.

TDP assigns three different duties to the nodes in the network to provide multi-hop synchronization. These duties are as follows:

- **Master node:** Master nodes initiate the synchronization messages and a tree-like structure is formed to synchronize a portion of the network. TDP assigns many master nodes in the network to decrease the number of hops to be synchronized and to minimize the convergence time.

- **Diffused leaders:** To propagate the synchronization messages along the tree-like structure, TDP selects diffused leaders, which propagate the synchronization messages further away from the broadcast range of the master nodes.

- **Regular nodes:** Any node that is neither a master nor a diffused leader participates in the synchronization process minimally, i.e., by correcting its clock according to the received synchronization messages.

TDP follows a periodic procedure consisting of *active* and *passive* phases as shown in Figure 11.11. Resembling a TDMA-like operation, the synchronization procedures are performed during the active phase. The protocol then enters the passive phase, where no timing updates are performed. As the duration of the passive phase increases, the network deviates further from the equilibrium time, which necessitates *re*synchronization. On the other hand, a smaller passive period results in more frequent

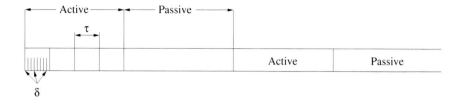

Figure 11.11 Frame structure of TDP [16].

synchronization, increasing the overhead of the protocol. TDP provides an adaptive mechanism to adjust the synchronization schedule according to the timing requirements of the WSN.

The active phase of TDP is further divided into *cycles* of length τ. The master nodes are reelected at each cycle and the synchronization operation is performed. Each cycle of length τ is further divided into *rounds* of length δ, where the master nodes repeatedly broadcast synchronization messages. According to this structure, TDP consists of two major procedures: the *election/reelection procedure* (ERP), where the master nodes and the diffused leaders are selected; and the *time diffusion procedure* (TP), where the synchronization is performed. The ERP is performed at the beginning of each cycle, τ, while the TP is performed at each round, δ, following the ERP.

Election/Reelection Procedure (ERP): The ERP aims to differentiate nodes that are eligible to become master nodes or diffused leaders and performs this selection. The reelection mechanism, which is performed every τ seconds, helps to distribute the load on the master nodes and the diffused leaders. Consequently, the master nodes are elected at the beginning of each τ cycle while the diffused leaders are selected at the beginning of each δ period. This election is performed as a result of two mechanisms: the false ticker isolation algorithm (FIA) and the load distribution algorithm (LDA).

False Ticker Isolation Algorithm (FIA): The FIA ensures that nodes that have high-frequency noise clocks or high access delay fluctuations are prevented from becoming a master or a diffused leader through the peer evaluation procedure (PEP) that will be described next. Since master nodes initiate the synchronization procedure and the clocks are synchronized according to the master node's clock, it is important to select these nodes among the ones that have a consistent clock operation. This ensures that the network reaches an equilibrium time quickly. Using the PEP, each node can self-determine whether it is a *false ticker* and accordingly does not become a diffused leader during the current cycle and a master node in the next cycle.

Load Distribution Algorithm (LDA): Once the false tickers are isolated, the LDA is performed to improve the chances that a node with higher residual energy is selected as the master node. Each node i selects a random number δ_i, which is further corrected by $(1 - \zeta_i)$, where ζ_i is the ratio of the current energy level over the maximum allowed energy level. If the selected number is higher than some threshold, then the node becomes eligible as a master node or a diffused leader.

The TDP automatically self-configures by electing master nodes to synchronize the sensor network. In addition, the election process is sensitive to energy requirements as well as the quality of the clocks. The sensor network may be deployed in unattended areas, and the TDP still synchronizes the unattended network to a common time.

■ **EXAMPLE 11.5**

TDP multi-hop synchronization is illustrated in Figure 11.12, where the elected master nodes are nodes C and G. The synchronization is initiated by the master nodes by sending periodic

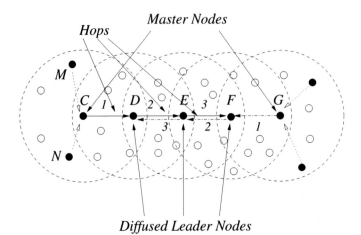

Figure 11.12   TDP concept.

SCAN messages to their neighbors to measure the round-trip times. Once a neighbor i receives the message, it self-determines if it should become the diffused leader node. This is accomplished by calculating the two-sample Allen variance between its local clock and the master node m's clock, which is given as follows:

$$\sigma_{mi}^2(\iota) = \frac{1}{2\iota^2(N-2)} \sum_{g=1}^{N-2} (x_{g+2} - 2x_{g+1} + x_g)^2 \tag{11.4}$$

where ι is the difference in time between two measurements, N is the total number of measurements, and x is the measurement value. The calculated Allen variance is sent by each neighbor back to the master node using REPLY messages. Using the collected Allen variances, the master node m calculates an outlier ratio γ_{mi} of node i as follows:

$$\gamma_{mi} = \left| \frac{\sigma_{mi}^2(\iota) - \sigma_{avg}^2(\iota)}{\sigma_{avg}^2(\iota)} \right| \tag{11.5}$$

where $\sigma_{avg}^2(\iota)$ is the average Allen variance received by the master node. The calculated outlier ratios are then broadcast back to the neighbors so that each node determines whether it is eligible for becoming a diffused leader.

The ones elected to become diffused leader nodes reply to the master nodes and start sending a message to measure the round-trip time to their neighbors. As shown in Figure 11.12, nodes M, N, and D are the diffused leader nodes of node C. Once the replies are received by the master nodes, the round-trip time and the standard deviation of the round-trip time are calculated. Then, the master nodes send a timestamped message containing the standard deviation to the neighbor nodes. The time in the timestamped message is adjusted with the one-way delay. Once the diffused leader nodes receive the timestamped message, they broadcast it after adjusting the time, which is in the message, with their measured one-way delay and inserting their standard deviation of the round-trip time. This diffusion process continues for n times, where n is the number of hops from the master nodes. From Figure 11.12, the time is diffused three hops from the master nodes C and G. Nodes D, E, and F are the diffused leader nodes that diffuse the timestamped messages originating from the master nodes.

The nodes that have received more than one timestamped message originating from different master nodes use the standard deviations carried in the timestamped message as the weighted ratio of their contribution to the new time. In essence, the nodes weight the times diffused by the master nodes to obtain a new time for them. This process aims to provide a smooth time variation between the nodes in the network. A smooth transition is important for some applications such as target tracking and speed estimation.

11.7.1   Qualitative Evaluation

TDP provides network-wide synchronization, where the synchronized time at each node is within a limited range throughout the network. This means that additional protocols such as the TDMA-based MAC protocol can be easily used in conjunction with TDP. The false ticker isolation algorithm eliminates the effects of faulty nodes on the equilibrium time. These nodes are not selected as master nodes or diffused leaders so that the diffused time is not affected. Moreover, the synchronization algorithm is developed considering the multi-hop nature of WSNs. As a result, instead of a step-by-step operation such as in TPSN, each node in the network is involved in synchronization, which improves scalability.

TDP also supports mobile nodes since it does not rely on a static hierarchy. TDP assigns master and diffused leader duties to nodes dynamically based on their clock quality as well as the remaining energy. As a result, the energy consumption is also equally distributed in the network. Finally, TDP does not require an external time server for synchronization. Without time servers, the protocol can reach a network-wide equilibrium time, which can be converted into another frame of reference by interacting with a single node in the network.

The election/reelection procedure, which dynamically elects master and diffused leader nodes, increases the complexity of the protocol because of multiple periods and cycles. If done too frequently, the energy consumption of the protocol also increases. Moreover, TDP adjusts the physical clocks of the sensor nodes according to the synchronization results instead of time translation. This requires higher complexity and increases the overall energy consumption of the network. Furthermore, since it is possible for the clocks to run backward, the software can be severely affected by the synchronization. TDP results in high convergence time if there are no time servers present. Especially in large networks, where multiple master nodes are used, network-wide equilibrium is reached in a long interval.

11.8   Rate-Based Diffusion Protocol (RDP)

TDP aims to diffuse the timing information of the master node using several diffused leaders. As a result, a portion of the network is synchronized to the local clock of the master node. Using a similar approach, RDP aims to synchronize the nodes in the network to the average value of the clocks in the network [9, 10]. Consequently, instead of the timing information, the difference between the clocks of the nodes and their relative importance is diffused in the network.

RDP can be implemented in a synchronous or asynchronous fashion. While the synchronization method is similar, both methods differ by the time in which nodes exchange their timing information. Operation of the diffusion protocol is based on two things: comparison of the local clocks of two nodes and adjusting the clocks accordingly. The diffusion protocol exploits existing synchronization protocols such as RBS for these basic operations and provides a multi-hop synchronization framework.

Denoting the clock value of node i at time t as c_i^t, the vector containing the clock values of each node at time t can be represented as $C = (c_1^t, c_1^t, \ldots, c_n^t)$ for a network of n nodes. The rate-based synchronous diffusion algorithm is initiated by each node exchanging its time information. Consequently, a node i can be informed about the clock readings of its neighbors and constructs a subset of the clock vector C. Since the main goal is to synchronize to the average value of the nodes in the network, each node adjusts its clock proportionally to the time difference between its neighbors j, i.e. $(c_i^t - c_i^j)$. Accordingly, a node i

adjusts its clock by $r_{ij}(c_i - c_j)$ for each of its neighbor j, where r_{ij} is denoted as the *diffusion rate*. By choosing r_{ij} for each neighbor j, node i adjusts its clock as follows:

$$c_i^{t+1} = c_i^t - \sum_{j \neq i} r_{ij}(c_i^t - c_j^t). \tag{11.6}$$

Then, the adjusted clock value and the diffusion rate values are exchanged with the neighbors such that $r_{ij} = r_{ji}$. In other words, after each clock update, a node i decreases its clock value by the amount node j increases its clock value.

As long as $r_{ij} \geq 0$, $\forall i, j$, and $\sum_{\forall j \in \mathcal{N}} r_{ij} \leq 1$, the local clock adjustment policy converges to the global average in the network [10]. In other words, independent of the initial clock vector $C^0 = (c_1^0, c_1^0, \ldots, c_n^0)$, the algorithm converges to $C^s = (c^0, c^0, \ldots, c^0)$, where $c^0 = \sum_{k=1}^{n} c_k^0/n$ is the average time. Consequently, the rate diffusion algorithm establishes a network equilibrium time using local message exchanges.

An important drawback of the rate-based synchronous diffusion algorithm is that the diffusion rate values need to be exchanged between the nodes so that $r_{ij} = r_{ji}$. While this requirement ensures convergence, it increases the overhead of the protocol. A node has either to exchange the diffusion rate value with each of its neighbors one by one, or to broadcast a diffusion rate matrix, which increases the energy consumption.

The asynchronous version of the rate-based diffusion algorithm ensures that each node computes the local average value from the average values computed by its neighbors. Once the node collects the local average values from its neighbors, it calculates its own average value and broadcasts it to its neighbors. Provided that the network is connected, each average operation brings the clock value closer to the global average and the network time converges after several steps.

11.8.1 Qualitative Evaluation

RDP also provides network-wide synchronization through either synchronous or asynchronous message exchanges. This enables TDMA-based MAC protocols to be used in WSNs. Moreover, network-wide equilibrium can be reached without external time servers. As a result, the protocol does not rely on a single point for synchronization and the network equilibrium time can be translated to another frame of reference through a single node in the network.

While both flavors of RDP provide network-wide equilibrium, the overhead of this operation is significantly high. RDP requires several steps to converge to the global average. The frequent clock drifts of the sensor nodes requires frequent synchronization of the network. This also increases the overhead of the protocol. Moreover, the time variations due to access delay have not been considered in the operation of the protocols. As the network density increases, collisions will increase since the protocols require frequent message exchanges. As a result, the energy efficiency of the synchronization operation may increase for highly dense networks.

11.9 Tiny- and Mini-Sync Protocols

One of the most important concerns in the design of synchronization protocols is the complexity of the synchronization algorithm. Tiny-sync and mini-sync protocols have been developed to provide a simple and accurate time synchronization for WSNs [14]. Both protocols are based on a hierarchical structure of sensor nodes, where each node is synchronized with its parent node.

In tiny- and mini-sync protocols, each node estimates the relative clock drift and offset to its parent node in the hierarchical tree structure for synchronization. More specifically, recall that the local clock of a node i can be represented relative to a node j as follows:

$$\tau_i(t) = \alpha_{ij}\tau_j(t) + \theta_{ij} \tag{11.7}$$

where α_{ij} and θ_{ij} are the relative clock drift and offset, respectively. This corresponds to a linear relationship between the clocks of node i and node j. Using several message exchanges, the clock drift and offset can be estimated.

The message exchange between two nodes (node 1 and 2) is explained next. In the first cycle, node 1 sends a packet at time t_{o_1} to node 2, which receives the packet at time t_{b_1} according to its local clock. Upon receiving the packet, node 2 immediately transmits a reply, which is received at time t_{r_1} by node 1. Using this 3-tuple $(t_{o_1}, t_{b_1}, t_{r_1})$, node 1 can determine the following constraints:

$$t_{o_1} < \alpha_{12}t_{b_1} + \theta_{12} \tag{11.8}$$

$$t_{r_1} > \alpha_{12}t_{b_1} + \theta_{12}. \tag{11.9}$$

The 3-tuple $(t_{o_1}, t_{b_1}, t_{r_1})$ defines two data points between which the line defined by (11.7) should pass. By collecting multiple data points through consecutive message exchanges, each node can estimate the relative clock drift and offset. Using the first and third measurements, two lines can be drawn. The first line defines the upper bound for the clock drift $\overline{\alpha_{12}}$ and the lower bound the clock offset θ_{12}. Similarly, the second line defines the lower bound for the clock drift α_{12} and the upper bound for the clock offset $\overline{\theta_{12}}$. Consequently, two measurements are sufficient to estimate these bounds for clock drift and offset.

Exact determination of the clock drift and offset is not possible because of the random delay factors in message exchanges as discussed earlier. However, the upper and lower bounds can be used to estimate these values to a certain degree of accuracy. Accordingly, the clock drift and offset are determined as follows:

$$\alpha_{12} = \widehat{\alpha_{12}} \pm \frac{\Delta\alpha_{12}}{2} \tag{11.10}$$

$$\theta_{12} = \widehat{\theta_{12}} \pm \frac{\Delta\theta_{12}}{2} \tag{11.11}$$

where

$$\widehat{\alpha_{12}} = \frac{\overline{\alpha_{12}} + \alpha_{12}}{2} \tag{11.12}$$

$$\Delta\alpha_{12} = \overline{\alpha_{12}} - \alpha_{12} \tag{11.13}$$

$$\widehat{\theta_{12}} = \frac{\overline{\theta_{12}} + \theta_{12}}{2} \tag{11.14}$$

$$\Delta\theta_{12} = \overline{\theta_{12}} - \theta_{12}. \tag{11.15}$$

As the node gathers more data points, the accuracy of the estimation improves. However, since sensor nodes are constrained in terms of memory, this estimation should be performed with a minimum number of data points.

In the tiny-sync protocol, each node stores only four data points. It can be observed that the upper and lower bounds for the clock drift and offset are determined by drawing lines between data points 1 and 3. Therefore, the data points related to the second measurement can be discarded. Accordingly, in tiny-sync, for each measurement, the best four data points that correspond to a better bound are selected and the estimation is performed accordingly. While this operation limits the number of constraints and the computational complexity, it does not always lead to the optimum result. More specifically, the data points discarded in an earlier round may be more helpful for determining closer bounds in the following rounds. As a result, tiny-sync provides a lightweight operation at the cost of deviating from the optimum clock drift and offset values.

Tiny-sync is extended through the mini-sync protocol which stores more data points to determine the optimum values for the relative clock drift and offset. Instead of selecting four data points for each three measurements and discarding the other two, mini-sync discards the data points only if they are proven not to improve the estimate. The other data points are stored throughout the synchronization. A data point, A_j, is only discarded if it satisfies the following condition:

$$m(A_i, A_j) \leq m(A_j, A_k) \tag{11.16}$$

for $1 \leq i < j < k$, where $m(X, Y)$ is the slope of the line that goes through the points X and Y. For each measurement, mini-sync checks whether this condition is met. If not, the new constraints are stored to improve the estimate of the clock drift and offset values.

11.9.1   Qualitative Evaluation

The mini-sync protocol provides deterministic bounds of synchronization accuracy through message exchanges and simple calculations. The complexity and memory requirements of the protocols are low, making them suitable for WSNs. In particular, the tiny-sync protocol requires only four data points to be stored. Since both protocols rely on multiple measurements, synchronization accuracy is not affected by wireless channel errors, which improves the robustness of the protocols.

Both protocols rely on a hierarchical structure in WSNs. Management of this structure increases the overhead of the protocol. Moreover, synchronization of mobile nodes is not supported. Since the estimation algorithm relies on multiple measurements, the convergence time of the protocols is high. In order to improve the convergence time, more frequent message exchanges need to be performed, which increases the overall energy consumption as well as the chance of collisions in the wireless channel for high-density networks.

11.10   Other Protocols

In addition to the synchronization approaches described in the previous sections, there exist several synchronization protocols developed for WSNs. We briefly describe these protocols next.

11.10.1   Lightweight Tree-Based Synchronization (LTS)

Similar to TPSN described in Section 11.4, LTS relies on a tree structure to perform network-wide synchronization [7]. The protocol is based on a message exchange between two nodes to estimate the clock drift between their local clocks. This pairwise synchronization scheme is extended for multi-hop synchronization.

The multi-hop synchronization mechanism of LTS follows two approaches: centralized and distributed. The centralized design is based on the construction of a spanning tree in the network such that each node is synchronized to the root node. In this mechanism, synchronization accuracy decreases as the depth of the spanning tree increases. Therefore, the tree is constructed such that each branch is of similar length and the network is covered through a tree with minimum depth. To this end, LTS employs two distributed tree construction algorithms. After the tree is constructed, the root of the tree initiates pairwise synchronization with its children nodes and the synchronization is propagated along the tree to the leaf nodes.

The distributed multi-hop synchronization mechanism of LTS does not rely on the construction of a tree. Moreover, contrary to the centralized scheme, where the root node initiates synchronization, the distributed synchronization can be initiated by any node in the network. This provides event-based synchronization capabilities, where each node performs synchronization only when it has a packet to send. In this case, each node is informed about its distance to the reference node for synchronization and adjusts its synchronization rate accordingly. Since synchronization accuracy is inversely proportional to distance, nodes farther apart from the reference node perform synchronization more frequently.

The performance evaluations reveal that the distributed mechanism increases the overhead by up to 100% compared to the centralized mechanism. The increased path length due to the lack of a spanning tree is a reason for this increase. However, the distributed mechanism achieves better synchronization performance when only a portion of the network needs to be synchronized.

11.10.2   TSync

TSync [3] is a hybrid protocol that combines tree-based synchronization with receiver–receiver synchronization such as RBS [4], which is explained in Section 11.5. To this end, two flavors of the protocol are developed: *hierarchical referencing time synchronization* (HRTS) and *individual time request* (ITR). HRTS is a centralized protocol where the synchronization is initiated through the sink node, which is the root node in the hierarchical tree. On the contrary, ITR is a decentralized protocol where the synchronization mechanism is initiated by any node in the network. Consequently, the nodes can be synchronized in the tree using the receiver–receiver synchronization technique. To address the contention and latency for synchronization messages because of additional traffic in the network, TSync provides MAC layer support by reserving a dedicated channel for synchronization messages. Accordingly, these messages are not affected by the data traffic in the network. While this MAC layer enhancement improves the accuracy of synchronization, the complexity required from the sensor nodes is increased, since multi-channel communication is necessary. The centralized version of TSync, HRTS, is shown to perform better than RBS, while the distributed version, ITR, cannot meet the accuracy bounds provided by RBS. However, the number of messages exchanged by the sensor nodes is decreased, which helps improve the energy efficiency of the synchronization procedures.

11.10.3   Asymptotically Optimal Synchronization

One of the few theoretical bounds for synchronization in WSNs is provided by the asymptotically optimal time synchronization framework [8], which focuses on high-density WSNs. The synchronization is initiated by a node in the middle of the network and each node in the network is synchronized to this node through message exchanges in the shape of concentric waves. Assuming that there is no communication delay and contention in the network, i.e., perfect channel conditions, this scheme is asymptotically optimal for WSNs. Furthermore, it can be ensured that the entire network is synchronized to the middle node as a result of this scheme. While theoretically valuable, the practical aspects of synchronization such as communication delays and medium access collisions in the wireless channel limit the applicability of these solutions. Furthermore, the protocol requires several message exchanges among the nodes in the network, which increases the overall energy consumption as well as the complexity of the protocol as network size and density grows.

11.10.4   Synchronization for Mobile Networks

The synchronization protocols discussed so far assume a static topology, where the connectivity of nodes does not change drastically over time. While this assumption is true for most WSN realizations, mobility is another concern that is important, particularly for ad hoc networks. In a mobile network, two nodes can communicate with each other only when they are located close to each other. The time synchronization protocol for ad hoc networks addresses this issue through a delay-tolerant synchronization scheme [13]. In this scheme, the nodes are assumed to be connected long enough such that they can exchange two messages, which constitute the basis of the synchronization algorithm. Moreover, the local clock of each node is assumed to have a bounded clock skew. Accordingly, denoting the local clock of a node i by

$\tau_i(t)$, the clock skew is assumed to be bounded as

$$1 - \rho \leq \frac{d\tau_i(t)}{dt} \leq 1 + \rho. \tag{11.17}$$

Based on this assumption, it can be concluded that a change in the local clock of a node is bounded by $(1 - \rho)\Delta t \leq \Delta \tau_i(t) \leq (1 + \rho)\Delta t$ in terms of *real* time, t. The synchronization is performed by exchanging two messages. The sender node informs the receiver regarding the difference between the received ACK for the first message M_1 and the time the second message, M_2, is sent, i.e., $t_2 - t_1$. Accordingly, the receiver can estimate the delay, d_2, for sending message M_2. This delay can then be used to bound the local clock of the sender according to the receiver's clock. The delay estimation is performed at the receiver as follows:

$$0 \leq d_2 \leq (t_5 - t_4) - (t_2 - t_1)\frac{1 - \rho_r}{1 + \rho_s} \tag{11.18}$$

where ρ_r and ρ_s are the clock skew bounds for the receiver and sender according to (11.17). As the nodes move around the network, these messages are propagated in the network and each node can be synchronized according to the delay estimation. In this way, the local clocks of the nodes are not adjusted and each node is informed about the difference between the local clocks. However, since the delays are estimated according to the upper and lower bounds, the error in estimation increases with the number of hops. Furthermore, if a node cannot communicate with another node for a large amount of time, the clock drifts between each node may exceed the assumed bounds and introduce additional errors.

References

[1] Standard for a precision clock synchronization protocol for networked measurement and control systems. ANSI/IEEE Standard 1588, 2002.

[2] F. Christian. Probabilistic clock synchronization. *Distributed Computing*, 3(3):146–158, September 1989.

[3] H. Dai and R. Han. TSync: a lightweight bidirectional time synchronization service for wireless sensor networks. *ACM SIGMOBILE Mobile Computing and Communications Review*, 8(1):125–139, January 2004.

[4] J. Elson, L. Girod, and D. Estrin. Fine-grained network time synchronization using reference broadcasts. *Proceedings of the Fifth Symposium on Operating Systems Design and Implementation (OSDI'02)*, volume 36(SI), pp. 147–163, Boston, MA, USA, 2002.

[5] Y. R. Faizulkhakov. Time synchronization methods for wireless sensor networks: a survey. *Programming and Computing Software*, 33(4):214–226, July 2007.

[6] S. Ganeriwal, R. Kumar, and M. B. Srivastava. Timing-sync protocol for sensor networks. In *Proceedings of ACM SenSys'03*, pp. 138–149, Los Angeles, USA, November 2003.

[7] J. V. Greunen and J. Rabaey. Lightweight time synchronization for sensor networks. In *Proceedings of the 2nd International ACM Workshop on Wireless Sensor Networks and Applications (WSNA'03)*, pp. 11–19, San Diego, CA, USA, September 2003.

[8] A. Hu and S. D. Servetto. Asymptotically optimal time synchronization in dense sensor networks. In *Proceedings of the 2nd International ACM Workshop on Wireless Sensor Networks and Applications (WSNA'03)*, pp. 1–10, San Diego, CA, USA, September 2003.

[9] Q. Li and D. Rus. Global clock synchronization in sensor networks. In *Proceedings of IEEE INFOCOM 2004*, Hong Kong, China, March 2004.

[10] Q. Li and D. Rus. Global clock synchronization in sensor networks. *IEEE Transactions on Computers*, 55(2):214–226, February 2006.

[11] D. L. Mills. Internet time synchronization: the network time protocol. *IEEE Transactions on Communications*, COM-39(10):1482–1493, October 1991.

[12] S. PalChaudhuri, A. K. Saha, and D. Johnson. Adaptive clock synchronization in sensor networks. In *Proceedings of IPSN'04*, pp. 340–348, Berkeley, CA, USA, April 2004.

[13] K. Romer. Time synchronization in ad hoc networks. In *Proceedings of ACM MobiHoc'01*, pp. 173–182, Long Beach, CA, USA, October 2001.

[14] M. L. Sichitiu and C. Veerarittiphan. Simple, accurate time synchronization for wireless sensor networks. In *Proceedings of IEEE Wireless Communications and Networking, WCNC'03*, volume 2, pp. 1266–1273, New Orleans, LA, USA, March 2003.

[15] F. Sivrikaya and B. Yener. Time synchronization in sensor networks: a survey. *IEEE Network*, 18(4):45–50, July/August 2004.

[16] W Su and I. F. Akyildiz. Time-diffusion synchronization protocol for wireless sensor networks. *IEEE/ACM Transactions on Networking*, 13(2):384–397, April 2005.

[17] B. Sundararaman, U. Buy, and A. Kshemkalyani. Clock synchronization for wireless sensor networks: a survey. *Ad Hoc Networks*, 3(3):281–323, May 2005.

12

Localization

WSNs are closely associated with the physical phenomena in their surroundings. The gathered information needs to be associated with the location of the sensor nodes to provide an accurate view of the observed sensor field. Moreover, WSNs may be used for tracking certain objects for monitoring applications, which also necessitates location information of the sensor nodes being incorporated into the tracking algorithms. Furthermore, as explained in Chapter 7, geographical routing protocols, which provide scalability and localized decisions for multi-hop communication, require the locations of the nodes to be known. These requirements motivate the development of efficient localization protocols for WSNs.

Localization protocols can be broadly classified into two categories as shown in Figure 12.1: *range-based* localization protocols and *range-free* localization protocols [3, 7]. Range-based localization protocols require the existence of beacon nodes that have accurate location information. Using several ranging techniques that will be described in Section 12.2, the remaining nodes in the network estimate their distance to three or more beacon nodes. Based on this information, the location of a node is estimated. Range-free localization protocols, on the other hand, do not require distance estimation. While the existence of a beacon node may still be required, the location of other nodes is estimated through range-free techniques.

In this chapter, we first describe the challenges for localization in WSNs. Several ranging techniques that are used as a building block for localization protocols are then described. Based on these building blocks, we discuss the range-based and range-free localization protocols.

12.1 Challenges in Localization

The localization protocols face several challenges for distributed location estimation such as *physical layer measurement errors*, *computation constraints*, *lack of GPS data*, *low-end sensor nodes*, and *architecture*. Furthermore, WSN applications pose different requirements for localization protocols in terms of scalability, robustness, and accuracy. Coupled with these requirements, the localization protocols need to address the unique challenges in WSNs. The main design challenges and the factors affecting time synchronization are as follows.

12.1.1 Physical Layer Measurements

Most localization protocols rely on *ranging* techniques for localization as we will explain in Section 12.2. These techniques provide local information in terms of distance or orientation related to the neighbors of a node. This local information can then be combined to provide location estimates. Ranging techniques are based on message exchanges between nodes in the network and corresponding signal strength or timing measurements. The received signal strength or the timing information can be converted into distance measurement. Similarly, the direction from which the signal has been received, which can be

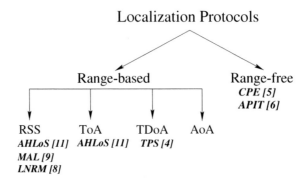

Figure 12.1   Overview of localization protocols.

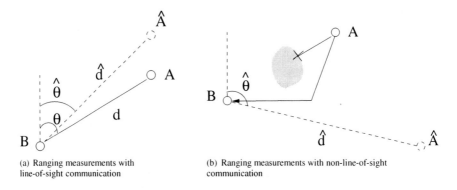

(a) Ranging measurements with
line-of-sight communication

(b) Ranging measurements with non-line-of-sight
communication

Figure 12.2   The effect of multi-path and non-line-of-sight communication on ranging techniques.

measured through sophisticated antenna configurations, provides orientation information for the node with respect to its neighbors.

Regardless of the particular ranging technique, the ranging measurements rely on the same assumption. Let us consider the situation in Figure 12.2(a), where node A transmits a signal to node B which estimates node A's position through ranging techniques. It is assumed that the received signal sent from node A travels a *direct path* to node B. If this assumption holds true, node B can estimate the angle of arrival, $\hat{\theta}$, and the distance, $\hat{d}$ with respect to node A. Accordingly, distance or orientation can be accurately estimated subject to errors due to measurement imperfections. Consequently, node A's position can be estimated as $\hat{A}$ as shown in Figure 12.2(a) within a small error. In practice, however, wireless signals can be subject to multi-path and shadowing, which result in the signal propagating through a longer path and being received from a different angle as shown in Figure 12.2(b). In case the direct path between nodes is obstructed, the remaining multi-paths dominate the ranging measurements. Consequently, distance measurements based on the received signal strength or timing measurements result in highly variable distance error since the signal travels a longer distance. Similarly, the angle of arrival can be measured significantly differently and the node can be projected to locate at a different orientation as shown in Figure 12.2(b).

The significantly different ranging measurements result in high localization error. Moreover, in non-line-of-sight (NLoS) environments, the inaccurate ranging measurements cannot be corrected through averaging techniques. In these cases, the neighbors that lead to inaccurate measurements need to be detected and excluded from localization calculations. Furthermore, more accurate but simple ranging techniques are still necessary for WSNs since NLoS operation is not uncommon.

12.1.2   Computational Constraints

Local information from several neighbor nodes in terms of distance, orientation, or connectivity is usually combined to provide location estimates and minimize erroneous information from a single measurement through multiple measurements. The individual measurements can be regarded as constraints for an optimization problem and the most likely location of the node can be estimated through maximum likelihood (ML) techniques. While these techniques usually result in highly accurate location estimates, the processing required for optimization is significantly large. Many formalizations of the localization problems require high processing power and memory. Consequently, most algorithms cannot be efficiently performed on wireless sensor nodes because of processing or memory constraints. The computational constraints of the individual sensor nodes can be addressed through centralized solutions, where the constraint for a global optimization problem is determined by each sensor node and sent to the sink. Then, the sink can solve this global problem and determine the locations of the nodes. While these centralized techniques are efficient in terms of processing, they incur high communication overhead and create single point-of-failure problems. Consequently, protocols with simple and distributed localization algorithms which will not increase the processing or energy consumption of individual sensor nodes are necessary.

12.1.3   Lack of GPS

GPS is utilized for location services in several embedded systems such as cell phones, navigation systems, or laptops. While GPS provides highly accurate location information, it may not be feasible for most WSN deployments. Firstly, GPS components available for WSNs are very costly, exceeding almost three times the cost of a sensor node [2]. Furthermore, GPS operation has a high energy consumption profile, which may impose additional constraints on the lifetime of a WSN. Furthermore, WSNs are usually static and localization protocols may be required to run only during initialization of the network. Consequently, GPS operation may not be cost effective for many WSN realizations.

While equipping each sensor node with a GPS unit may not be feasible, GPS can still be used with a limited number of nodes to serve as *beacon nodes*. This approach has been adopted by many localization protocols, which aim to localize the WSN through beacon information from a limited number of nodes with accurate location information. Using local information exchanges, the rest of the network can be localized with respect to these beacon nodes.

12.1.4   Low-End Sensor Nodes

Wireless sensor nodes are equipped with low-end components to provide low-cost operation. These imperfect components pose several challenges for localization in WSNs. In addition to the effect of the wireless channel on ranging measurements as explained in Section 12.1.1, the measurement hardware on sensors also introduces errors in distance or orientation estimation. The received signal strength measurements may be noisy due to transceiver imperfections. Similarly, since low-end crystals are used, ranging estimates, which depend on timing measurements, result in erroneous results. Furthermore, because of space limitations on the embedded system board, the required diversity may not be achieved for angle-of-arrival measurements. Consequently, the ranging measurements cannot provide accurate information and this error can propagate throughout the network if not addressed. More importantly, the error incurred in the measurements is non-deterministic in nature and cannot be easily mitigated.

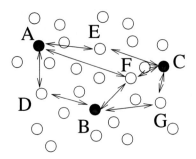

Figure 12.3 Typical architecture of a WSN with beacon and regular nodes for ranging measurements.

Therefore, measurements from multiple neighbors at different times should be combined to reach an acceptable accuracy in localization.

In addition to the measurement of hardware errors, the communication errors caused by the inherently noisy nature of the wireless channel affect localization protocols. Since these protocols depend on the information exchange between nodes, the lost or corrupted packets may influence localization protocols and result in a bias toward information from nodes with better channel quality. Consequently, reliable communication protocols should be an inherent part of localization protocols.

The low-end components in the sensor nodes also cause frequent node failures, which make a node disconnect from the network for some amount of time or permanently. Hence, localization protocols should not rely on a single node for information delivery and localization calculations. Robustness and scalability should be provided, which is the case for most protocols developed for WSNs.

12.2   Ranging Techniques

The range-based localization protocols as will be explained in Section 12.3 necessitate efficient ranging techniques through which the distance or angle to a particular node can be found. The main motivation behind the ranging techniques can be explained through the following example.

■ **EXAMPLE 12.1**

Shown in Figure 12.3 is a network consisting of beacon nodes and regular nodes. The beacon nodes (A, B, and C) have accurate information regarding their location and the remaining regular nodes (D, E, F, and G) estimate their locations by measuring the distance or angle between themselves and the beacon nodes. These techniques use *distance* or *angle* measurements from a set of fixed reference points and apply multilateration or triangulation techniques to estimate the location of a node.

Each ranging technique may require special hardware support, which limits the applicability of these techniques. Moreover, the accuracy of distance or angle estimation varies for each technique. Consequently, the choice of the ranging technique is based on the requirements of the application as well as the hardware capabilities. Next, we describe the four main ranging techniques in WSNs, namely received signal strength, time of arrival, time difference arrival, and angle of arrival.

12.2.1   Received Signal Strength

The most common ranging technique is based on received signal strength measurements. Since each sensor node is equipped with a radio and in most cases is able to report the received signal strength of an incoming packet, this technique has minimal hardware requirements. The main idea is to estimate the distance of a transmitter to a receiver using the following information:

- the power of the received signal;
- knowledge of the transmitted power;
- the path-loss model.

In this scheme, the beacon nodes broadcast periodic beacon messages, which are used to estimate the distance to them. The power of the received signal is communicated by the transceiver circuitry through the RSSI (Received Signal Strength Indicator). The received signal strength from sensor node i at node j at time t is represented by $P_R^{ij}(t)$, which is formulated as

$$P_R^{ij}(t) = P_T^i - 10\eta \log(d_{ij}) + X_{ij}(t) \tag{12.1}$$

where P_T^i is a constant due to the transmitted power and the antenna gains of the sensor nodes, η is the attenuation constant (e.g., 2 or 4), and $X_{ij}(t)$ is the uncertainty factor due to multi-path and shadowing. Since the transmitted power of the beacon, P_T^i, is usually fixed and known by the receivers, the received signal strength can be used to estimate the distance with an error proportional to the uncertainty factor, $X_{ij}(T)$, and the RSSI measurement error.

The accuracy of the RSSI-based ranging techniques is limited. Firstly, the effects of shadowing and multi-path as modeled by the term $X_{ij}(t)$ in (12.1) may be severe and require multiple ranging measurements. Moreover, in cases where there is no line of sight between a beacon and a node, the received signal strength measurements result in a significant error in distance measurements. As shown in Figure 12.2(b), if a node is not in the line of sight of the beacon because of an obstacle, it receives the transmitted signals through reflections from the environment. The signal received by the node traverses a longer path than the direct path and the reflected signal is attenuated more, which results in a longer distance measurement than the actual distance between the beacon and the node. This type of error is more severe than multi-path effects because multiple measurements cannot improve the distance estimate. Since each of the measurements is still based on reflected rays, the error in distance measurements cannot be corrected.

Another major challenge with the RSSI-based distance measurements is the difficulty in estimating the parameters for the channel model in (12.1). While the transmit power may be fixed for a localization application for each beacon node, the parameters associated with antenna gains may differ from node to node. Moreover, the attenuation constant, η, varies with the environment. Hence, pre-deployment calibration may be required to estimate these parameters for localization algorithms.

12.2.2   Time of Arrival

Time-of-Arrival (ToA) techniques rely on accurate measurements of transmit and receive times of signals between two nodes. These measurements are used to estimate the distance based on the propagation time and the speed of the signal. Since timing information is used for distance measurements, synchronization is essential for these techniques. Based on the measurement type, two types of ToA measurements can be performed:

- **Active:** The receiver transmits a ranging packet, which is immediately responded to by a transmitter. The round-trip time is used to estimate the distance between the nodes.
- **Passive:** In this case, the transmitter and receiver measurements are made separately. Accordingly, a transmitter sends a beacon signal, which is used by the receiver to estimate the delay

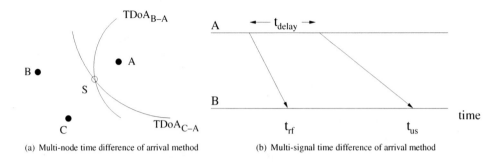

Figure 12.4   Two TDoA methods.

between the two nodes. However, for this technique, the time of signal transmission needs to be known.

ToA techniques can be used to measure the propagation time between two nodes. The nodes then estimate the relative distance to a beacon by applying the measured propagation time to a complex distance formula.

ToA techniques require very accurate hardware to measure the actual received time of the beacon signals. Any small error in time measurement can result in large distance estimate errors because of the high propagation speed of RF signals in air. Therefore, ToA techniques are not practical for traditional WSNs. On the other hand, as we will see in Chapter 17 and Chapter 16, challenging environments such as under water and underground exhibit low propagation speeds because of the properties of the communication media. Therefore, ToA techniques may be more suitable for these environments.

12.2.3   Time Difference of Arrival

While actual propagation time measurements of a signal require sophisticated measurement hardware, the *difference* between the receive times of two separate signals can be used to estimate the distance between nodes. Since less accurate measurements are tolerated, time difference of arrival (TDoA) techniques can be used for WSNs in practice. TDoA techniques can be classified into two:

- **Multi-node TDoA:** These techniques use ToA measurements of signals transmitted from multiple beacon nodes for distance measurement. As shown in Figure 12.4(a), three beacons can be used to accurately locate a node using TDoA measurements. Firstly, the node to be localized, i.e., node S, measures the arrival time of beacons sent from three beacon nodes A, B, and C. If the three nodes are synchronized and transmit the beacon signals at the same time, the difference between a pair of beacon nodes defines a hyperbola on which node S should lie. These lines are shown as $TDoA_{B-A}$ and $TDoA_{C-A}$ in Figure 12.4(a). The intersection of the two hyperbolas is then used to locate the node.

- **Multi-signal TDoA:** These techniques require multiple beacon nodes to perform localization of a node. Moreover, precise synchronization between beacon nodes is essential for accurate time difference measurements. Instead, two different kinds of signals can be used by a node to estimate its distance to another node. By using two signals that have different propagation speeds, the time difference of arrival of these signals can be used to estimate the distance. Multi-signal TDoA is shown in Figure 12.4(b) for a node A, which transmits two signals, and node B, which estimates the distance to node A based on the difference in ToA of the two signals. A common approach for the multi-signal TDoA is to use an RF signal in conjunction with an ultrasound microphone and speaker such as the Crossbow Cricket motes [1]. As shown in Figure 12.4(b), node A first

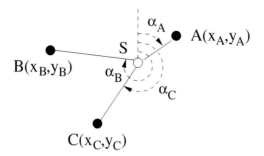

Figure 12.5 AoA measurements for three beacon nodes A, B, and C, and a node S with unknown location.

transmits an RF signal which is received by node B at time t_{rf}. Node A then waits for a time t_{delay}, which is also known by node B, and transmits an ultrasound signal. This signal is received by node B at time t_{us}. Then, the distance between two nodes is calculated as

$$d_{AB} = (s_{rf} - s_{us})(t_{us} - t_{delay} - t_{rf})\qquad(12.2)$$

where s_{rf} is the speed of RF waves and s_{us} is the speed of ultrasound. Since RF waves travel significantly faster than ultrasound waves, the difference between the receive times of these two signals can be used to accurately measure the distance [1]. Using Cricket nodes, localization accuracy up to centimeters is possible [10].

The multi-signal TDoA technique provides high accuracy under line-of-sight conditions. However, similar to the RSSI-based techniques, NLoS communication may lead to high localization errors. Because of the different paths either type of signal may follow, NLoS communication leads to significantly different propagation delays. Consequently, localization protocols based on multi-signal TDoA techniques need to consider the NLoS communication case to improve the accuracy. Moreover, compared to RSSI-based localization, TDoA techniques require additional transmitter–receiver pairs in each node for the second type of signal. Hence, these techniques may not be suitable for some WSN applications.

12.2.4 Angle of Arrival

In addition to signal strength and time, the direction of the received signal can also be exploited for localization. Angle-of-arrival (AoA) techniques rely on directional antennas or special multiple antenna configurations to estimate the AoA of the received signal from beacon nodes. As shown in Figure 12.5, a node S with an unknown location, (x_S, y_S), receives beacon signals from nodes A, B, and C with well-known locations, (x_A, y_A), (x_B, y_B), and (x_C, y_C), respectively. Then, node S estimates the AoA of each beacon, i.e., α_A, α_B, and α_C. AoA measurements are then combined with their locations to estimate the location of node S.

AoA techniques provide high localization accuracy depending on the measurement accuracy. However, higher complexity antenna arrays are required for direction measurement, which increases the cost. Furthermore, antenna arrays require a certain spacing to provide spatial diversity and to accurately measure the AoA. Considering the size of typical sensor nodes, such a separation may not be feasible, which makes AoA techniques impractical for certain WSNs. Finally, AoA techniques also suffer from multi-path and scattering as well as NLoS conditions. Since the direction of arrival is used for localization, the error due to NLoS components of the received signal may be even more severe compared to that of RSSI- or TDoA-based techniques.

Several other localization protocols exist that provide location information in a network of sensor nodes. The localization protocols can be classified into two categories. *Range-based* techniques exploit the ranging techniques we explored in Section 12.2 as a building block. Then, distributed techniques are used to improve the accuracy of these protocols. *Range-free* techniques, on the other hand, do not rely on ranging measurements and provide localization services through interactions between sensor nodes as well as local computations.

12.3   Range-Based Localization Protocols

Range-based localization protocols exploit multiple pairwise range measurements to estimate the locations of nodes. Multiple range measurements between a node and its different neighbors can be used to improve the accuracy of the location estimate. Localization techniques vary, depending on the ranging technique. We first overview the basic localization techniques and then describe the localization protocols based on them. There exist three main localization techniques [11]:

- **Trilateration:** If a node has distance measurements between its neighbors, then trilateration techniques can be used to estimate its location. As shown in Figure 12.6(a), in a 2-D space three neighbor nodes with known locations are sufficient to localize a node through distance measurements. Each distance measurement defines a circle around the neighbor with radius equal to the measurement, on which the node should lie. The intersection of three such circles from three nonlinear neighbors defines the exact location for the node. However, this technique assumes perfect distance measurements, which is not feasible in WSNs because of ranging errors. Hence, more than three nodes are required for localization.

- **Triangulation:** If the AoA technique is used for ranging, then the angle information from two beacon nodes is used to localize the node. As shown in Figure 12.6(b), two beacon nodes and the unknown node define a triangle. Using basic trigonometric relations, the location of the unknown node can be found using the locations of the two beacon nodes and the AoA measurements at the unknown node.

- **Maximum likelihood multilateration:** The trilateration technique fails to provide an accurate estimate of position if the distance measurements are noisy. Instead, maximum likelihood (ML) estimation methods are necessary to incorporate distance measurements from multiple neighbor nodes. In this case, the difference between the distance measurements and the estimated distance is minimized to find the location of the node.

12.3.1   Ad Hoc Localization System

The ad hoc localization system (AHLoS) provides localization service using either RSS or ToA measurements [11]. AHLoS consists of two procedures: firstly, ranging measurements are performed by each node and then location estimation is performed according to the ranging measurements to nodes with known locations. The ML (multilateration) technique is used to estimate the location of unknown nodes in the network.

The basic building block for localization in AHLoS is *atomic multilateration*, which is the simplest form of multilateration as explained next.

■ **EXAMPLE 12.2**

Atomic multilateration is shown in Figure 12.6(c), where an unknown node S receives beacon signals from multiple beacon nodes, A, B, C, and D, with known locations. Let us assume that ToA measurement is performed with ultrasound signals and the time it takes for a signal to reach node S from a beacon i is given by $t_{i,S}$. The location of the node (x_S, y_S) is estimated by calculating the

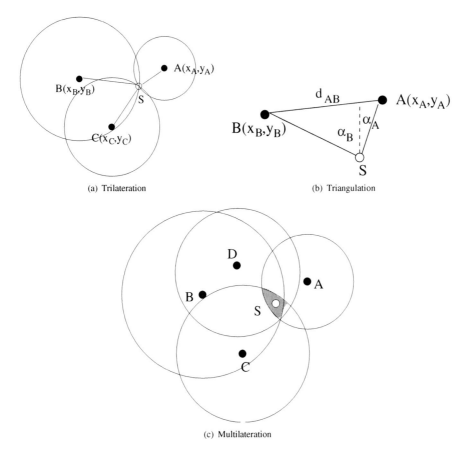

(a) Trilateration (b) Triangulation

(c) Multilateration

Figure 12.6   Range-based localization techniques.

difference between the measured distance and the estimated distance between a beacon node as follows:

$$f_i(x, y, v) = vt_{i,x} - \sqrt{(x_i - x)^2 + (y_i - y)^2} \qquad (12.3)$$

where v is the speed of the ultrasound signal, and (x_i, y_i) is the location of beacon i. Using at least three beacon nodes, the location of the unknown node, S, can then be estimated by taking the MMSE of the system of f_i equations [11].

The atomic multilateration explained in Example 12.2 provides localization of a single node which has at least three neighboring beacon nodes. However, WSNs are usually deployed with a limited number of beacon nodes. Furthermore, because of ranging errors, more than three beacon nodes are required for accurate location estimation. Hence, AHLoS utilizes the location estimates of neighbor nodes to improve the estimation of unknown node locations. Based on atomic multilateration, a multi-hop network of sensors can be localized with the help of a subset of beacon nodes. This procedure is called *iterative multilateration*.

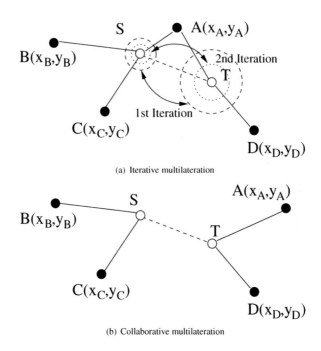

(a) Iterative multilateration

(b) Collaborative multilateration

Figure 12.7   Multilateration techniques of AHLoS.

■ **EXAMPLE 12.3**

Iterative multilateration is illustrated in Figure 12.7(a), where two unknown nodes S and T are surrounded by four beacon nodes, A, B, C, and D. Note that node S is surrounded by three beacon nodes, whereas node T has only two beacon nodes as neighbors. Since node S can communicate with at least three beacon nodes, atomic multilateration can be used to estimate its location. However, atomic multilateration cannot be applied to node T, which requires additional distance information to a node with known location. In this case, iterative multilateration is used, where node S first estimates its location and becomes a *beacon node*. Using the additional information from node S, node T can then perform atomic multilateration and calculate its position. This information can in turn be used by node S to further improve the accuracy of its estimate. As a result, the uncertainty in location estimation as shown by the circles around nodes S and T can be decreased at each iteration.

There may be cases in the network where a node does not have any neighbors with enough beacon nodes as neighbors. This case is illustrated in Figure 12.7(b), where nodes S and T cannot perform atomic multilateration or iterative multilateration since neither of them has at least three neighbor beacon nodes. In this case, *collaborative multilateration* is used where one of the nodes solves a joint set of location estimate functions using multi-hop information from an unknown neighbor node to find the locations of both nodes simultaneously.

Collaborative multilateration leads to unique location solutions only when a participating node has at least three participating neighbors. A *participating neighbor* is defined as a node which is either a beacon node or a node that has at least three participating neighbors.

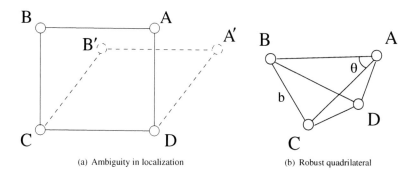

(a) Ambiguity in localization   (b) Robust quadrilateral

Figure 12.8   Ambiguity and robustness in localization.

■ **EXAMPLE 12.4**

Consider the example in Figure 12.7(b). If node S wants to perform collaborative multilateration, it has two beacon neighbors as participating neighbors and node T as an unknown node. Since node T also has two beacon nodes and node S as a participating neighbor, node T is also a participating node. Consequently, node S solves a set of location estimate equations in the form of

$$f(x_u, y_u) = d_{i,u} - \sqrt{(x_i - x_u)^2 + (y_i - y_u)^2} \qquad (12.4)$$

where x_u and y_u are the locations of the unknown nodes S and T, x_i and y_i are the locations of the beacon nodes A, B, C, and D, and $d_{i,u}$ is the estimated distance between a known and an unknown node. As a result, $f(x_u, y_u)$ is the difference between the estimated distance, $d_{i,u}$, and the Euclidean distance found from the location estimate (x_u, y_u).

Qualitative Evaluation

As a result of collaborative multilateration, the number of nodes with location estimates can be increased and these nodes can be used for iterative multilateration. Consequently, using three different multilateration techniques, AHLoS aims to provide distributed localization in WSNs. An important advantage of AHLoS is that it does not rely on a single type of ranging technique. The localization results using both RSS and ToA techniques yield acceptable localization accuracy in the range of tens of centimeters in small-scale networks. This is because of the improvement of localization estimation through iterative multilateration. On the other hand, for large-scale networks, iterative multilateration leads to inaccurate results since the initial estimation errors are propagated through the network.

12.3.2   Localization with Noisy Range Measurements

The location estimation primitive shown in (12.3) aims to minimize the error between measured distances and the distance according to location estimates. However, this distance error minimization metric cannot guarantee the actual *locations* of the nodes to be found. Such an example topology is shown in Figure 12.8(a), where node A can be localized in two different locations, without violating the internode distance constraints. Consequently, minimizing the distance error as in (12.3) will still lead to *ambiguity* in localization.

To prevent ambiguity, the localization protocol with noisy range measurements relies on *robust quadrilaterals* as building blocks of localization [8]. A robust quad is a collection of four nodes, which can be localized unambiguously even in the presence of noisy distance measurements. An example of a robust quadrilateral is shown in Figure 12.8(b) with four nodes, A, B, C, and D. The six distance measurements among these four nodes provide constraints such that the relative locations of these nodes are unique up to a rotation, translation, or reflection. Consequently, the quadrilateral is called *globally rigid*. However, rigidity does not guarantee unique position estimates if the distance measurements are noisy. Hence, the quadrilateral should also be *robust* to provide such a guarantee. This is accomplished by dividing the quad into four triangles (ABC, ABD, ACD, and BCD). A triangle is defined as a *robust triangle* if it meets the following constraint:

$$b \sin^2 \theta > d_{min} \qquad (12.5)$$

where b is the length of the shortest side of the triangle and θ is the smallest angle as shown for triangle ABC in Figure 12.8(b), and d_{min} is the robustness threshold. If all the four triangles of a quad are robust, the quad is defined as a *robust quad* and is used as a building block of the localization protocol.

Two or more robust quads can also be connected to form a larger rigid graph if they share three vertices. All the positions of the nodes in this graph can be localized with a high confidence level even under errors in distance measurements. These subgraphs are defined as *clusters* and the protocol operation is based on these clusters of multiple robust quads. The protocol operation is conducted in three main phases: cluster localization, cluster optimization, and cluster transformation. We describe each phase next, in detail:

- **Cluster localization:** The first phase is initiated by each node to find the largest connected robust subgraph according to the robust quad definition above. Assume a node i initiates the localization process. Firstly, a robust quad is found through neighbor message exchanges. This is accomplished by each neighbor, which broadcasts its distance measurements to the node i. Node i then computes the robust quads and the cluster. Once these clusters are found, localization is performed among the nodes in the cluster [8].
- **Cluster optimization:** The location estimates can also be refined by local computations after all the location information in the cluster has been exchanged by the members of the cluster. This optional phase improves the localization accuracy at the cost of increased processing at each node.
- **Cluster transformation:** Since localization is performed independently, each cluster has a unique coordinate system. In a multi-hop network, these coordinate systems need to be transformed to provide consistent localization. The transformation among clusters is performed if there are at least three non-collinear nodes that are members of both clusters. Using these common nodes, the location at each cluster can be transformed to the coordinate system of other clusters. Consequently, *location transformation* is performed instead of maintaining a common coordinate system within the network. The location of any given node can then be found through these transformations.

Using robust quads as a building block for localization provides probabilistic bounds on the localization error. If the individual distance measurements include an error that is normally distributed with standard deviation σ, then the localization protocol bounds the worst case probability of error, which is found to be

$$P(x > d + d_{err}) = \Phi\left(\frac{d_{err}}{\sigma}\right) \qquad (12.6)$$

where d_{err} is the worst case error. Consequently, if the threshold d_{min} in (12.5) is chosen as a multiple of the standard deviation σ of the distance measurement, then the worst case probability of error is bounded.

12.3.3 Time-Based Positioning Scheme

The time-based positioning scheme (TPS) exploits the TDoA ranging technique described in Section 12.2.3 [4]. The localization scheme is based on a network structure as shown in Figure 12.9(a), where

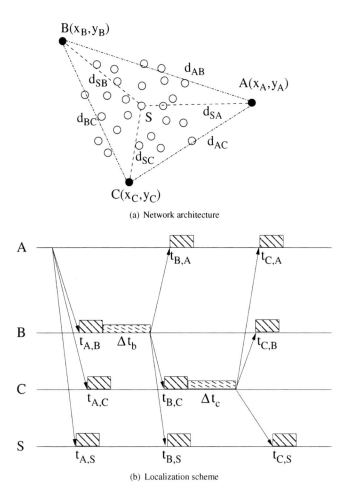

(a) Network architecture

(b) Localization scheme

Figure 12.9   Time-based positioning scheme (TPS) [4].

three beacon nodes are deployed around the sensor network. The beacon nodes are equipped with powerful transceivers and can reach all the nodes in the network.

Localization of the network is performed by each node independently using the beacons broadcast by the beacon nodes. Each beacon node periodically broadcasts a beacon with a period of T seconds. TPS does not require the beacon nodes to be synchronized and the beacons can be transmitted at different times. As shown in Figure 12.9(b), a node S can receive beacons from three beacon nodes A, B, and C with known locations (x_A, y_A), (x_B, y_B), and (x_C, y_C), respectively. Node S tries to estimate its location (x_S, y_S) through TDoA measurements from these beacon nodes.

TPS consists of two steps: *range detection* and *location computation*. In the first step, one of the base stations is used as a master base station and the signal round-trip time between the master base station and the other base stations is used by node S to estimate its location.

■ **EXAMPLE 12.5**

The range detection phase for node S is illustrated in Figure 12.9(b). Firstly, node A broadcasts a beacon, which is received at $t_{A,S}$, $t_{A,B}$, and $t_{A,C}$ at nodes S, B, and C, respectively. Upon receiving the beacon, node B replies back after a delay, Δt_b, to node A. Nodes C and S receive this signal at $t_{B,C}$ and $t_{B,S}$, respectively. Finally, node C replies to nodes A and B after Δt_c, which is received by node S at time $t_{C,S}$.

After node S receives the three beacon messages it calculates the following equations:

$$d_{SB} = d_{SA} + k_1 \tag{12.7}$$

$$d_{SC} = d_{SA} + k_2 \tag{12.8}$$

where d_{SA}, d_{SB}, and d_{SC} are the distances from node S to beacon nodes A, B, and C, respectively, and

$$k_1 = v \cdot (\Delta t_1 - \Delta t_b) - d_{AB}, \quad k_2 = v \cdot (\Delta t_2 - \Delta t_c) - d_{AC}, \tag{12.9}$$

with v the speed of the RF signal, $\Delta t_1 = t_{BS} - t_{AS}$ and $\Delta t_2 = t_{CS} - t_{AS}$.

The second phase of TPS is *location computation*, where the distance from node S to node A, i.e., d_{SA}, and the location of node S, (x_S, y_S), are calculated. To this end, based on the ranging procedure explained above, each node solves a set of equations as follows:

$$(x_S - x_A)^2 + (y_S - y_A)^2 = d_{SA}^2 \tag{12.10}$$

$$(x_S - x_B)^2 + (y_S - y_B)^2 = (d_{SA} + k_1)^2 \tag{12.11}$$

$$(x_S - x_C)^2 + (y_S - y_C)^2 = (d_{SA} + k_2)^2. \tag{12.12}$$

As discussed in Section 12.2.3, various factors such as multi-path and shadowing affect the TDoA measurements. The measurement errors can be minimized by using techniques such as ML multilateration, which is used by AHLoS as explained in Section 12.3.1. However, ML multilateration results in excessive computations because of the nonlinearity of the operations and becomes a significant source of energy consumption. TPS, on the other hand, improves the localization accuracy by using multiple measurements in the first phase of the protocol. More specifically, the parameters k_1 and k_2 are defined in (12.9) for a single beacon interval. Instead of using these values at each interval and estimating the location, multiple consecutive intervals are considered to find an average of these values as follows [4]:

$$k_1 = \frac{v}{I}\left[\sum_{i=1}^{I}(\Delta t_1^i - \Delta t_b^i)\right] - d_{AB} \tag{12.13}$$

$$k_2 = \frac{v}{I}\left[\sum_{i=1}^{I}(\Delta t_2^i - \Delta t_c^i)\right] - d_{AC} \tag{12.14}$$

where I is the number of intervals used for averaging and subscript i denotes the measurement in interval i. Accordingly, the location of the node is estimated at each I interval. Since only simple operations are performed by each node, the computational overhead of the localization protocol decreases while improving the accuracy as we explain next.

Similar to the previous localization protocols, the accuracy of localization in TPS depends on the accuracy of the TDoA measurements. More specifically, for TPS, if the distances between the beacon nodes are large compared to the distances between sensor nodes, the variance of x_S is a function of the variance of k_1, while the variance of y_S is a function of the variance of k_2. Consequently, decreasing the uncertainty in the measurements of k_1 and k_2 by averaging over multiple intervals directly improves the localization accuracy.

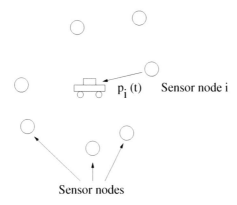

Figure 12.10 Use of mobile robots in sensor networks.

Qualitative Evaluation

TPS is a distributed localization technique which relies on independent TDoA measurements at each sensor node. Compared to AHLoS explained in Section 12.3.1 and the localization protocol with noisy range measurements explained in Section 12.3.2, TPS does not require internode communication for localization. This significantly decreases energy consumption since internode communication increases as the density in the network increases. On the other hand, TPS requires powerful beacon nodes which are capable of broadcasting their messages to the whole network. This requirement may not be valid for some WSN architectures, where such nodes do not exist.

12.3.4 Mobile-Assisted Localization

The range-based localization protocols discussed so far utilize ranging measurements from nodes with *fixed* locations. Consequently, the localization accuracy is inherently limited depending on the network topology and the deployment strategy. Since a uniform network deployment is not feasible in practice, some portions of the network may have a lower density and the neighbors of a node may not be sufficient. Moreover, the locations of the beacon nodes may not be optimal for each node in the network. Since the topology of the sensor network rarely changes because of the limited (or lack of) mobility, the localization accuracy of particular nodes cannot be improved through these techniques.

Mobile-assisted localization (MAL) employs mobile agents to improve the localization accuracy in WSNs [9]. As described in Section 12.3.2, the uniqueness of the multilateration technique depends on forming *globally rigid* subgraphs. The global rigidity of a subgraph is satisfied if (1) the removal of any vertices leaves the graph connected and (2) the removal of any edge leaves the graph locally rigid. The localization with noisy measurements scheme described in Section 12.3.2 performs localization by finding already existing rigid subgraphs in the network through pairwise distance measurements. However, this may not be the case for some portions of the network because of low density. MAL uses a mobile agent to gather additional distance measurements to form globally rigid subgraphs.

Compared to the range-based localization protocols described in earlier sections, in MAL, sensor nodes do not perform distance measurements. Instead, a mobile agent, which travels throughout the network, is used to find the distance between multiple sensor nodes and itself to estimate the distance between these sensor nodes as shown in Figure 12.10. As an example, the mobile agent can find the pairwise distance between four nodes with an additional seven measurements provided that these measurements are not collinear.

Accordingly, MAL is initialized by the mobile agent, which finds four nodes to create a globally rigid subgraph. The mobile agent then selects seven locations and measures its distance to the from nodes from these locations. The pairwise distances between the four nodes are then found by solving a set of equations involving the distance measurements. After the pairwise distance values are estimated, MAL localizes the nodes using multilateration techniques.

The second phase following initialization is to increase the rigid subgraph by adding additional neighbor nodes. From the rigid graph, the mobile agent selects a node and moves around this node. In this way, the mobile agent searches for additional non-localized nodes. If a non-localized node is found, the mobile agent uses the selected localized nodes along with some of its localized neighbors to perform similar localization for the non-localized node. The technique is repeated until all the nodes in the network are localized or the mobile agent cannot improve localization any further.

Qualitative Evaluation

The advantage of MAL is that sensor nodes do not need to perform additional distance measurements or solve complex localization equations. Instead, a mobile agent is used to perform the localization tasks. In this way, for each sensor node, sufficient distance measurements can be made. On the other hand, MAL requires a mobile agent, which can be a robot or a human, to perform the localization tasks. Since a mobile agent may not be available for most applications, MAL has limited applicability. Furthermore, the localization performance depends on the measurements and calculations performed at a single node in the network. Hence, any errors in the measuring algorithm or the localization calculations can affect the whole network. Similarly, such a dependency creates single point-of-failure problems in WSNs. Finally, MAL has been developed for a single mobile agent. Therefore, for large networks, the localization latency may be significantly large because of the time it takes for the mobile node to traverse the whole network.

12.4   Range-Free Localization Protocols

Range-based localization protocols described in Section 12.3 rely on ranging measurements for distributed location estimates. Depending on the protocol, these ranging measurements may need to be performed among each neighbor of a node or with only a smaller number of beacon nodes. While the accuracy of the range-based localization protocols is high, the cost associated with them is also high because of the additional energy consumption for ranging measurements. On the other hand, several applications of WSNs do not require very accurate location information and only coarse localization may be sufficient. The range-free localization protocols provide an efficient way to localize the sensor nodes without specific ranging measurements. Instead, connectivity information between nodes is exploited to determine constrains on the location of the nodes. Similar to the range-based protocols, beacon nodes are also used to provide a reference for localization. Next, we describe two range-free protocols in detail: convex position estimation [5] and the approximate point-in-triangle (APIT) [6] protocol.

12.4.1   Convex Position Estimation

The lack of ranging information poses additional challenges for localization since a limited amount of information is available. However, even without accurate distance measurements, the connectivity information between a node and its neighbors can still be exploited to constrain the location of the node as described in the following example.

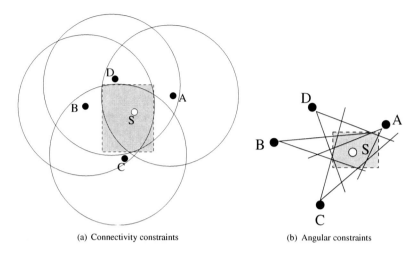

(a) Connectivity constraints   (b) Angular constraints

Figure 12.11   Connectivity and angular constraints can be exploited for convex position estimation.

■ **EXAMPLE 12.6**

Consider the case shown in Figure 12.11(a), where node S is within the communication range of nodes A, B, C, and D. If the location of node B and its transmission range R are known, then node S can deduce that it is located in a circle centered around node B with radius R. Any additional neighbor with location information such as nodes A, C, and D further constrains the possible location of node S. As a result, the connectivity of a node provides a proximity constraint parameter, which can be utilized to solve an optimization problem and estimate the location of the node.

The convex position estimation (CPE) algorithm provides an optimization framework, where the locations of the nodes in a WSN are found as a result of a joint optimization problem [5]. The goal of the algorithm is to estimate the locations of unknown nodes in a WSN, where a subset of the beacon nodes have accurate location information as shown in Figure 12.11(a). Using the location information of the beacon nodes and the connectivity information of all the nodes in the network, the locations of the remaining nodes can be estimated.

The protocol operation is as follows. Each node exchanges messages with its neighbors to populate a connectivity table representing its immediate neighbors. Connectivity information consists of nodes from which a node receives messages. Considering the limited transmission range of wireless sensor nodes, the connectivity information provides natural bounds on the distance between two connected nodes. This information is sent to a central location such as the sink, where a centralized optimization problem, which exploits these constraints, is formulated.

On collecting the connectivity information from the nodes in the network, the central controller formulates a linear program (LP) to find the positions of the nodes in the network, where the connectivity information is used as a constraint. The LP is a feasibility problem, which is to find the unknown locations satisfying the connectivity constraints between unknown and known nodes. The solution to the optimization problem is the vector $\mathbf{x} = [x_1, y_1, \ldots, x_m, y_m, x_{m+1}, y_{m+1}, \ldots, x_n, y_n]$, where the first m elements are the locations of the beacon nodes (known) and the remaining $n - m$ are the estimated locations of the unknown nodes.

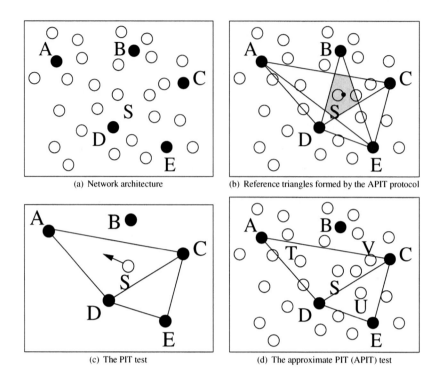

(a) Network architecture

(b) Reference triangles formed by the APIT protocol

(c) The PIT test

(d) The approximate PIT (APIT) test

Figure 12.12   The APIT protocol [6].

As shown in Figure 12.11(a), the maximum communication ranges of the nodes can be used as constraints for the optimization framework. If AoA information is available, angular constraints can also be used for estimation as shown in Figure 12.11(b). By combining the connectivity constraints from multiple neighbors of a node, the central controller creates a *bounding box* in which the node should reside. As shown in Figure 12.11(a) and Figure 12.11(b), the bounding box is a rectangle which includes the possible locations of the node subject to the connectivity or angular constraints, i.e., feasible position sets. Then, the center of the bounding box is selected as the location of the node.

The bounding box is generally generated from the connectivity information from beacon nodes. However, in case of a lack of sufficient beacon nodes, the unknown nodes are also included in the connectivity information to further constrain the location of the node. This is similar to the iterative multilateration technique used in AHLoS. However, the location estimates of the unknown nodes contain uncertainty in the location estimation. Therefore more of these nodes are required for accurate location estimation. More specifically, using the CPE algorithm, close to eight connections to unknown nodes are required to achieve the same level of certainty achieved through a connection to a single beacon node [5]. Exploiting the additional constraints imposed by the neighboring unknown nodes improves the accuracy of the localization algorithm, especially in high-density WSNs.

Although the CPE algorithm provides high accuracy for localization in WSNs, it can be viewed as a guideline for the development of localization solutions for WSNs due to its centralized nature. The CPE algorithm necessitates the connectivity information being collected at a centralized location. Moreover, after the joint optimization problem is solved, the location information needs to be disseminated to each node in the network. This requirement can significantly increase the overhead and the energy consumption of the protocol. Moreover, the localization protocol relies on a single node and it is vulnerable to single points of failure. Furthermore, the connectivity information of a node can change frequently because of the changes in the wireless channel. This necessitates frequent information exchange throughout the network, which makes the solution impractical for large-scale WSNs.

12.4.2   Approximate Point-in-Triangulation (APIT) Protocol

The APIT protocol relies on a network that consists of wireless sensor nodes as well as beacon nodes with known locations [6]. However, localization of the sensor nodes is performed without ranging techniques. Instead, the RSS information from multiple nodes is used to estimate the relative locations of these nodes. Consequently, the APIT protocol does not require any additional hardware such as sounders and microphones, which are required for timing measurements.

The network architecture of the APIT protocol is shown in Figure 12.12(a), where multiple beacon nodes are distributed in the network field. These nodes have known locations and periodically broadcast beacons to the rest of the nodes in the field. Each sensor node can receive beacons from a subset of these beacon nodes and localize itself. The locations of the beacon nodes are used to form triangles by each node.

■ **EXAMPLE 12.7**

Referring to Figure 12.12(a), let us assume that node S receives beacon signals from beacon nodes A, B, C, D, and E. Since node S is aware of the positions of the beacon nodes, it can form triangles using any three of these nodes as shown in Figure 12.12(b). If the node can determine whether it is inside or outside of these triangles, the intersections of these triangles can be used to define a small area in which the node should reside. Consequently, using multiple beacon signals, a node can localize itself up to a certain accuracy.

Based on the architecture in Example 12.7, the APIT protocol consists of four phases: (1) beacon exchange, (2) PIT testing, (3) APIT aggregation, and (4) center of gravity (COG) calculation. Next, we describe each phase of the protocol:

- **Beacon exchange:** Upon receiving beacons from the beacon nodes, a node populates a table with the locations of each beacon node and the value of the received signal strength from that node. This information is then broadcast to the neighbors of the node. Consequently, each node is informed about the connectivity of each of its neighbors to the beacon nodes and the corresponding RSS values. This information is used to estimate whether a node is inside a triangle formed by three beacon nodes in the PIT testing phase.

- **PIT testing:** The PIT test is performed to determine whether a node is inside a triangle formed by three beacon nodes. Ideally, this can be performed by moving the node along any axis as shown in Figure 12.12(c). The node is outside the triangle if there exists a direction such that whenever a node is moved in this direction, its distances to the three vertexes of the triangle increase or decrease simultaneously. Otherwise, the node is inside the triangle. PIT testing is illustrated in the following example based on the topology in Example 12.7.

■ **EXAMPLE 12.8**

As shown in Figure 12.12(c), when node S is moved in the direction shown by the arrow, its distance to nodes C, D, and E increases. Hence, it can be concluded that node S is *outside* the triangle CDE. However, since its distance to node A decreases, but increases for nodes C and D, it is *inside* the triangle ACD. This is referred to as the point-in-triangulation (PIT) test. However, in WSNs, since a node cannot be moved, an approximate PIT (APIT) test is performed instead.

The APIT test is illustrated in Figure 12.12(d), which exploits the high density of WSNs. Instead of moving a node, each node checks the connectivity information of its neighbors. For a given triangle ACD, a node S first determines the neighbor nodes (T, U, V in Figure 12.12(d)) that also receive beacons from these beacon nodes. Then, node S checks whether it has a neighbor that is farther from the three beacon nodes simultaneously. To this end, the RSS values are used. Although the RSS values are not used to calculate the exact distance, the difference between the RSS values of two nodes is used to determine whether a node is farther or closer to the beacon node. If a node finds such a neighbor, then it determines that it is outside the particular triangle.

The APIT test is performed for each triangle and then APIT aggregation is performed to constrain the location of the node.

- **APIT aggregation:** The APIT test determines the triangles in which node S resides. As shown in Figure 12.12(b), the intersection of these triangles defines the region where the node should reside. Following the APIT test, node S calculates the maximum overlapping area to estimate its location through APIT aggregation.

- **COG calculation:** The last phase of the APIT protocol is the COG calculation. In this phase, the node determines its location as the COG of the overlap area found in the previous phase. This location is shown as a dot in Figure 12.12(b).

Qualitative Evaluation

The APIT protocol provides a range-free localization technique through simple geometric calculations using beacon signals from specific nodes. Consequently, the complexity of the protocol is significantly low compared to range-based algorithms. On the other hand, the localization accuracy of the algorithm is also low and is proportional to the number of nonlinear beacon nodes that a node is connected to. However, based on the impact of localization errors on other protocols such as routing and tracking, a localization error less than $0.4R$ is acceptable, where R is the transmission range of the beacon nodes. Consequently, the APIT protocol is applicable to scenarios where high localization accuracy is not required; however, minimizing protocol overhead is of major concern.

References

[1] Cricket indoor location system. http://nms.lcs.mit.edu/projects/cricket.

[2] MTS420 environment sensor board. Available at http://xbow.com.

[3] J. Aspnes, W. Whiteley, and Y. R. Yang. A theory of network localization. *IEEE Transactions on Mobile Computing*, 5(12):1663–1678, December 2006.

[4] X. Cheng, A. Thaeler, G. Xue, and D. Chen. TPS: A time-based positioning scheme for outdoor wireless sensor networks. In *Proceedings of IEEE INFOCOM'04*, volume 4, pp. 2685–2696, Hong Kong, China, March 2004.

[5] L. Doherty, K. Pister, and L. Ghaoui. Convex position estimation in wireless sensor networks. In *Proceedings of IEEE INFOCOM'01*, volume 3, pp. 1655–1663, Anchorage, AL, USA, April 2001.

[6] T. He, C. Huang, B. Blum, J. Stankovic, and T. Abdelzaher. Range-free localization schemes for large scale sensor networks. In *Proceedings of ACM MobiCom 2003*, pp. 81–95, San Diego, CA, USA, September 2003.

[7] K. Langendoen and N. Reijers. Distributed localization in wireless sensor networks: a quantitative comparison. *Computer Networks*, 43(4):499–518, 2003.

[8] D. Moore, J. Leonard, D. Rus, and S. Teller. Robust distributed network localization with noisy range measurements. In *Proceedings of ACM SenSys*, Baltimore, MD, USA, November 2004.

[9] N. B. Priyantha, H. Balakrishnan, E. D. Demaine, and S. Teller. Mobile-assisted localization in wireless sensor networks. In *Proceedings of IEEE INFOCOM'05*, volume 1, pp. 172–183, Miami, FL, USA, March 2005.

[10] N. B. Priyantha, A. K. Miu, H. Balakrishnan, and S. Teller. The Cricket compass for context-aware mobile applications. In *Proceedings of ACM MobiCom'01*, Rome, Italy, July 2001.

[11] A. Savvides, C. Han, and M. Strivastava. Dynamic fine-grained localization in ad-hoc networks of sensors. In *Proceedings of ACM MobiCom 2001*, pp. 166–179, Rome, Italy, July 2001.

13

Topology Management

The topology of a network can be defined according to graph theory as the *locations of the nodes* that are available for communication, i.e., the vertex, and the *wireless links between these nodes* used for communication, i.e., the edges. Topology management solutions generate network topologies that are tailored to guarantee a certain requirement such as connectivity, coverage, or lifetime [1, 9, 12, 13, 15, 17]. The resulting topology directly affects the performance of each individual protocol as well as the overall performance of the network. Therefore, topology management is a crucial part of the WSN communication protocols.

In WSNs, *topology* refers to not only the locations of the nodes but also their activity states, i.e., which nodes are active for communication, as well as the links between them. Hence, several techniques exist for topology management. The four main topology management techniques are shown in Figure 13.1. These techniques can be used individually as well as in conjunction with each other since different properties of the topology are controlled by each one.

- **Deployment:** The first phase of topology management is network deployment as shown in Figure 13.1(a). Deployment techniques determine the locations of the nodes in the network, which plays an important role in the coverage as well as the connectivity of the network. Accordingly, the maximum area that can be covered by the on-board sensors can be maintained. Similarly, deployment affects the links that can potentially be formed given the limitations of the wireless channel.

- **Power control:** The topology is partly defined by the links between the nodes. In a wireless network, these links depend on the capabilities of the transceiver. Power control solutions control the transmit power of the transceivers to maintain the communication range of a node as shown in Figure 13.1(b). Accordingly, the number of neighbors of a node or the hop length in a multi-hop path can be determined for energy efficiency, reliability, or latency goals.

- **Activity control:** In a WSN, the transceiver of the sensor node can be turned off during certain periods to increase energy efficiency. As a result, certain nodes become *disconnected* from the network during their sleep states. Hence, the network topology can also be maintained by managing the activity of the sensor nodes as shown in Figure 13.1(c). Accordingly, redundant nodes can be turned off while still maintaining the connectivity and capacity of the network.

- **Clustering:** In addition to flat topology-based techniques, the network can also be partitioned into clusters as shown in Figure 13.1(d) to improve scalability and energy efficiency. In this case, a group of nodes are placed into clusters, which are controlled through cluster heads (CHs).

Topology management approaches for WSNs are described in the following sections. These approaches are classified into four as shown in Figure 13.2. We provide general rules for deployment in WSNs in Section 13.1. The power control problem and its solutions are described in Section 13.2. In Section 13.3, activity scheduling protocols are discussed. Finally, we explain clustering solutions in Section 13.4.

Wireless Sensor Networks   Ian F. Akyildiz and Mehmet Can Vuran
© 2010 John Wiley & Sons, Ltd

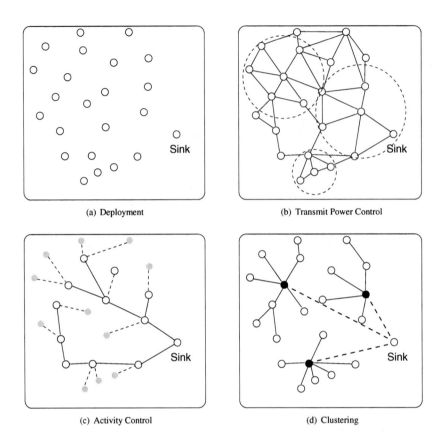

Figure 13.1   Topology management techniques.

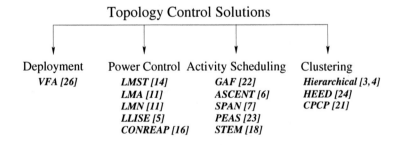

Figure 13.2   Overview of topology control solutions.

13.1   Deployment

The physical location of the nodes in a WSN is crucial in terms of both communication and sensing. Since sensor nodes are equipped with radios of limited range, the communication is performed in a multi-hop fashion [2]. Hence, connectivity is an important factor considered for sensor deployment. Moreover, since a physical phenomenon with spatio-temporal characteristics is observed via the WSN, the deployment of the nodes also affects the accuracy of the samples collected regarding the phenomenon. Hence, coverage is also important for sensor deployment.

A WSN should be deployed according to the coverage and connectivity of the network. Several methods exist for determining network connectivity and coverage given a node reliability model [19, 20]. The node reliability refers to the probability that a node is active in the network. Moreover, given a power budget, an estimate of the minimum required node reliability for meeting a system reliability objective can be found. Using a grid-based topology for WSNs, theoretical bounds and an insight into the deployment of WSNs can be obtained. As an example, as the node reliability decreases, the sufficient condition for connectivity becomes weaker than the necessary condition for coverage. This implies that *connectivity in a WSN does not necessarily imply coverage*. Furthermore, the power required per active node for connectivity and coverage decreases at a rate faster than the rate at which the number of nodes increases. Thus, as the number of nodes increases, the total power required to maintain connectivity and coverage decreases.

The relationship between reduction in sensor duty cycle and redundancy in sensor deployment is also important [8]. Two main approaches, namely random and coordinated sleep algorithms, can be compared in terms of two performance metrics, i.e., extensity and intensity. *Extensity* refers to the probability that any given point is not covered, while *intensity* gives the tail distribution of a given point not covered for longer than a given period of time. As the density of the network is increased, the duty cycle of the network can be decreased for a fixed coverage. However, beyond a certain threshold, increased redundancy in the sensor deployment does not provide the same amount of reduction in the duty cycle. Moreover, coordinated sleep schedules can achieve higher duty cycle reduction at the cost of extra control overhead. Using the intensity analysis of the network, random sleeping schedules can be developed for a satisfactory coverage [8].

The virtual force algorithm (VFA) enhances sensor coverage by moving sensor nodes after an initial random deployment [26]. The algorithm assumes a cluster-based architecture and is executed at the CHs. The VFA uses virtual positive or negative forces between nodes based on their relative locations. Moreover, using this cluster-based architecture, target localization can be performed in a distributed manner. The simulation results reveal that the VFA improves coverage of the sensor network and increases accuracy in the target localization. However, the algorithm requires either mobile sensor nodes or redeployment of nodes accordingly, which is not applicable for all WSNs.

13.2   Power Control

As described in Chapter 4, the wireless channel quality can be improved by increasing the transmit power of a node. As a result of increased transmit power, a node can communicate with more nodes. Similarly, in high-density networks, the transmit power can be decreased to limit the number of neighbors of a node, i.e., the *node degree*. This can reduce interference and increase network lifetime. Consequently, power control algorithms encounter a tradeoff in terms of reducing interference and increasing network lifetime vs. reducing network connectivity. By controlling the transmit power of each node, certain network properties such as connectivity, interference, latency, and lifetime can be maintained. Accordingly, power control mechanisms are developed to achieve the following goals:

- **Connectivity:** The main goal of power control algorithms is to preserve connectivity by choosing minimum transmit power for nodes.
- **Minimum spanning:** Reducing transmit power results in a decrease in the average hop length for multi-hop transmission since longer links that require high transmit power are eliminated from the network. This increases the path length between any node and the sink. The transmit power control algorithm should ensure that the resulting path between any node and the sink in the network should be longer at most by a constant factor than the shortest path achievable through maximum transmit power level [5].

- **Distributed operation:** Since it is infeasible to control the transmit power levels of each node by a single entity in WSNs, the power control algorithms should be distributed. Furthermore, the developed protocols should be scalable so that networks of large numbers of nodes can still be supported.

- **Discrete power levels:** Typical sensor node architectures provide a limited number of transmit power levels. Hence, solutions that choose any power level are not feasible and discrete power levels should be supported.

We discuss the following power control mechanisms next: LMST [14], LMA and LMN [11], interference-aware power control [5], and CONREAP [16].

13.2.1 LMST

The local minimum spanning tree (LMST) algorithm aims to construct a minimum spanning tree (MST) through local decisions for power control [14]. The protocol depends on the locational knowledge of the nodes to make localized decisions. The protocol operation can be divided into three phases: *information collection*, *topology construction*, and *transmission power determination*. Moreover, an additional phase, *topology construction with bidirectional links*, can also be performed.

Similar to most distributed topology control protocols, LMST is initiated with an information collection phase. Each node broadcasts HELLO messages at the maximum transmit power to exchange node-specific information. These messages are periodically broadcast throughout the protocol operation and contain the node ID and the position of the node. As a result, each node is informed about the positions of its neighbors in its *visible neighborhood*. A visible neighborhood of a node i is the set of nodes from which node i can receive information during the information collection phase and is denoted as $NV_i(G)$. Accordingly, the network topology with maximum transmit power is denoted as an undirected graph $G = (V, E)$, where V is the set of nodes and $E = \{(i, j) : d(i, j) \leq d_{max}, i, j \in V\}$, and where d_{max} is the maximum distance that a node can reach using its maximum transmit power.

The visible neighborhood information is used by each node to construct the local MST, $T_i = (V(T_i), E(T_i))$ of G_i, which is the graph that contains the visible neighborhood of node i. For the construction of the topology, the transmission power required for successful communication for each edge is used as the weight of that edge. Neglecting the effects of multi-path fading and shadowing,[1] the required transmission power and, hence, the weight can be directly mapped to the distance between two nodes. In the case of equal distances, node IDs are used to assign unique weights to each edge. Accordingly, a unique MST can be formed by each node.

Based on the MST, the relation between each node and its neighbors is defined by the *neighbor relation*. More specifically, node j is referred to as a neighbor of node i if and only if it is in the MST of node i, i.e., $(i, j) \in E(T_i)$. This relation is denoted as $i \to j$. Accordingly, the neighbor set of i is denoted as $N(i) = \{j \in V(G_i) : i \to j\}$. Using the neighbor relation between each node, the resulting network topology with the LMST protocol is defined as $G_0 = (V_0, E_0)$, where $V_0 = V$ is the set of all nodes in the network and $E_0 = \{(i, j) : i \to j, i, j \in V(G)\}$ is the set of all edges in the MST of each node in the network.

Following the topology construction phase, each node determines the transmit power to all of its neighbors in its local MST. This is performed using the HELLO messages received from each neighbor. As explained above, HELLO messages are sent using the maximum transmit power level, P_{max}. Thus, the received power level, P_r, of each HELLO message represents the path loss, PL, between the nodes that sent and received the HELLO message. More specifically,

$$PL = \frac{P_r}{P_{max}}. \tag{13.1}$$

---

[1] For details see Chapter 4.

Assuming symmetric channels, this path loss can be used to adjust the transmit power level, P_t, so that the neighbor can receive the packets with a receive power level of P_{th}. Accordingly,

$$P_t = P_{th} \cdot PL \tag{13.2}$$

$$= \frac{P_{th} P_r}{P_{\max}} \tag{13.3}$$

The value of P_{th} is determined such that the packet can be received without any errors.

Additionally, the LMST algorithm can be used to ensure that all the links in the constructed topology are bidirectional, i.e., if $i \rightarrow j$ then $j \rightarrow i$. This is performed in two ways. If a unidirectional link $i \rightarrow j$ exists, an additional link $j \rightarrow i$ is added to the constructed graph or is removed from the graph completely. Accordingly, the resulting LMST has bidirectional links, which is important for supporting two-way communication between nodes in the network.

The LMST algorithm provides a connected network by constructing local MSTs and controlling the transmit power at each link. As a result, the overall energy consumption can be decreased compared to the case where maximum transmit power is used. Moreover, the resulting LMST consists of fewer links between the nodes in the network. This reduces the interference in the network, which improves the energy efficiency.

On the other hand, LMST requires accurate location information for path-loss determination. Moreover, a simpler channel model is considered, which neglects the random effects of multi-path fading and shadowing. In case of location errors and practical channel effects, the transmit power determination scheme can result in either higher energy consumption or communication errors. Furthermore, the time-varying nature of the wireless channel has not been considered in LMST, which may lead to fluctuations in communication success and connectivity.

13.2.2   LMA and LMN

The LMST algorithm determines the transmit power level for each link for a node. As a result, for each communication link, a node should change its transmit power level according to the determined value. This results in considerable amounts of delay and energy consumption during transmit power adjustments. The local mean algorithm (LMA) and local mean of neighbors (LMN) algorithm address this overhead by finding the best transmit power for a node to be used throughout the network operation without hampering the connectivity of the network [11]. The objective of these algorithms is to maximize the network lifetime.

The LMA aims to adjust the transmit power such that the number of neighbors of a node is limited. Accordingly, each node periodically broadcasts a life message (LifeMSg) with an initial transmit power level. The neighbors of this node that receive this message reply with a LifeAckMsg. The sender keeps track of the number of LifeAckMsgs and adjusts its transmit power accordingly. The number of LifeAckMsgs is indicated by n_r and LMA aims to maintain this value between two predetermined limits: $n_{\min}$ and $n_{\max}$.

If a node receives responses from a number of neighbors that are less than $n_{\min}$, then it increases its transmit power level to reach more nodes. The new transmit power P_t^{new} is updated as follows:

$$P_t^{new} = \min\{B_{\max} P_t, \ A_{inc}(n_{\min} - n_r) P_t\}. \tag{13.4}$$

Accordingly, the transmit power is compensated for each required neighbor to reach the lower bound, $n_{\min}$. Moreover, the update rule ensures that the transmit power is not increased by more than a factor of $B_{\max}$.

If, on the other hand, the number of received responses is more than $n_{\max}$, then the transmit power level is decreased. Similarly, the update rule is as follows:

$$P_t^{new} = \max\{B_{\min} P_t, \ A_{dec}(1 - (n_r - n_{\max})) P_t\}. \tag{13.5}$$

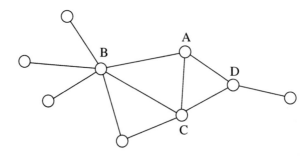

Figure 13.3   Neighborhood information exchange in LMN [11].

Moreover, the transmit power is kept constant if $n_{\min} \leq n_r \leq n_{\max}$.

While LMA maintains the number of neighbors a node has, the second algorithm, i.e., the LMN algorithm, maintains the number of neighbors *of the neighbors* of a node.

■ **EXAMPLE 13.1**

In the LMN algorithm, in addition to the LifeAckMsgs, the responding nodes also include the number of their neighbors. This is shown in Figure 13.3, where node A broadcasts a LifeMsg. Nodes B, C, and D receive this message and reply with LifeAckMsgs indicating the number of neighbors they have, i.e., {6, 4, 3}. Accordingly, node A finds the mean of these values and uses this mean for power control. In our example, $n_r = 4$. The update policy of LMA is used similarly.

In the protocol operation, the limits of $n_{\min} = 4$ and $n_{\max} = 7$ are used. Both protocols generally converge within a limited number of message exchanges. On average, the LMN algorithm results in 0.003% of the nodes not connected, while this number is 0.37% for LMA. This shows that both protocols provide high connectivity whereas the LMN algorithm results in a stronger connectivity. Furthermore, both algorithms perform very close to the global optimum, which can only be achieved through centralized solutions.

13.2.3   Interference-Aware Power Control

The transmit power control schemes presented above (LMST, LMA, and the LMN algorithm) reduce transmit power to improve the network lifetime while maintaining network connectivity. However, the amount of interference caused as a result of the distributed transmit power operation is not considered as a design metric in these solutions. It is generally accepted that reducing the node degree by reduced transmit power results in a reduction in interference. However, this does not result in minimum interference in the network. In this section, we describe the effect of interference on topology control and, more specifically, on transmit power control algorithms.

■ **EXAMPLE 13.2**

In a wireless network, the interference is modeled based on the area that is covered by the communication between two nodes. Consider the case in Figure 13.4, where two nodes i and j are communicating. Based on the distance between these nodes, $d_{i,j}$, the transmit power levels can be adjusted to maintain successful communication with minimum energy expenditure. Considering a circular communication range, the *communication coverage* of an edge $e = (i, j)$ between two

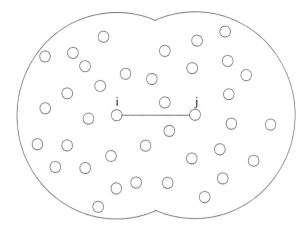

Figure 13.4 Communication coverage of an edge $e = (i, j)$ [5].

nodes i and j is defined as follows [5]:

$$Cov(e) = \{k \in V | k \in D(i, d_{i,j})\} \cup \{k \in V | k \in D(j, d_{i,j})\} \qquad (13.6)$$

where $D(i, r)$ is the disc centered at node i with radius r. Accordingly, the interference of a graph, G, is defined as follows:

$$I(G) = \max_{e \in E} Cov(e). \qquad (13.7)$$

According to the definition of connectivity in (13.7), an interference-aware topology control algorithm can be developed. This is performed by finding a set of edges that minimize the interference constrained by the fact that the resulting topology is connected.

The interference can be minimized to find the interference-optimal spanning forest, which includes the set of trees that maintains connectivity with minimum interference. This is performed by the low-interference forest establisher (LIFE) algorithm [5]. The LIFE algorithm assigns the coverage, $Cov(e)$, of an edge as the weight of each edge and finds the minimum spanning forest (MSF) of the resulting graph.

The interference-optimal topology control algorithm LIFE can be performed only with global knowledge of the network. However, distributed optimization of the interference minimization problem with connectivity is not feasible. The optimization problem can be modified such that, instead of the connectivity requirement, the resulting graph is required to be a *t-spanner*. A t-spanner of a graph includes edges such that the path length between any nodes of the graph is within a constant factor of the shortest path in the original graph. This is distributively solved by the local low-interference spanner establisher (LLISE) algorithm.

■ **EXAMPLE 13.3**

The operation of the LLISE algorithm is illustrated in Figure 13.5. For a given factor t for the t-spanner requirement, each node collects information regarding its $(t/2)$ neighborhood for each of its edges. Let us consider the edge $e = (i, j)$ as shown in Figure 13.4. The $(t/2)$ neighborhood of the edge, e, is defined as the set of edges that can be reached through a path p, where $|p| \leq t/2|e|$, and $|p|$ and $|e|$ are the lengths of the path, p, and the edge, e, respectively. Using the $(t/2)$ neighborhood of the edge, e, node i finds the minimum interference path for e.

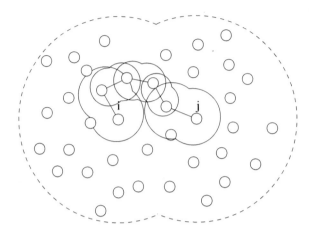

Figure 13.5   Interference-optimal t-spanner of the link $e = (i, j)$ as a result of the LLISE algorithm.

Firstly, node i orders the edges in the $(t/2)$ neighborhood according to their coverage. Then, a subgraph is formed by using these edges in ascending order until a shortest path $p(i, j)$ is found so that $|p| \leq t|e|$. This is performed for each edge of the node. The resulting graph is shown in Figure 13.5, which is interference optimal with the t-spanner property.

The distributed LLISE algorithm performs very close to the centralized LIFE algorithm, especially when t is increased. In other words, interference in the network can be reduced closer to the global optimum by using longer paths compared to the shortest path. Moreover, compared to topology control protocols that consider reducing energy consumption with connectivity constraints, the LLISE algorithm results in lower interference. This may improve the lifetime of the network in the long term. Transmit power control mechanisms generally determine the optimum transmit power between two nodes without considering the interference. Hence, in a typical network operation, competing flows can cause interference, which increases the required power level for successful communication. In this case, the chosen transmit power can lead to retransmissions and an increase in energy consumption as a result. While interference-aware topology control mechanisms do not necessarily find the minimum *power*, the overall *energy consumption* can be decreased by avoiding interference between links.

13.2.4   CONREAP

The transmit power control mechanisms explained so far are based on a rather simplistic channel model. Based on a specific transmit power level, the nodes are considered *connected* if they are within a certain distance, and *disconnected* otherwise. This is based on the unit disc graph model. The wireless channel, however, exhibits a higher uncertainty than the unit disc graph as discussed in Chapter 4 in detail. More specifically, a *transitional region* exists for wireless connectivity, where the nodes in this region are neither fully connected nor fully disconnected. Instead, a percentage of the packets sent can be successfully received. In practical settings, the transitional region can include the same number of nodes (or even more) than the *connected region*, where almost 100% reliability is possible. Hence, considering only the connected region for topology control can significantly limit the energy efficiency of the WSN.

Compared to the previously explained topology control protocols, which consider connectivity as a design parameter, the CONREAP algorithm performs distributed *opportunity-based* topology control, where unreliable links in the transitional region are also considered [16].

The operation of the CONREAP algorithm is based on the concept of *link reachability*. Similar to other topology control mechanisms, the network is modeled as a directed graph, $G(V, E)$, with set of nodes, V, and set of edges, E. Instead of defining edges only if two nodes are connected, a link reachability, λ_e, is assigned to each edge, $e = (i, j)$, where $\lambda_e \in (0.1]$. Moreover, the set of nodes that can receive a packet from the sink is denoted as V_r. Accordingly, the network reachability, $\lambda(G)$, can be defined as a function of the expected number of nodes that can receive a packet from the sink:

$$\lambda(G) = \frac{E(|V_r|)}{|V|} \tag{13.8}$$

which can be found by defining *node reachability* as $\lambda_G(i)$ for a node i. Accordingly, the network reachability is

$$\lambda(G) = \frac{\sum_{i \in V} \lambda_G(i)}{|V|} \tag{13.9}$$

which is a function of the average node reachability over all the nodes in the network.

In addition to the network reachability, the cost associated with the communication is also important. Accordingly, the network energy cost is defined as

$$\epsilon(G) = \epsilon_{Tx}(G) + \epsilon_{Rx}(G) \tag{13.10}$$

where $\epsilon_{Tx}(G)$ and $\epsilon_{Rx}(G)$ are the transmitting and receiving cost, respectively. To determine the efficiency of the algorithm, the reachability-to-energy ratio can be defined as

$$\eta(G) = \frac{\lambda(G)}{\epsilon(G)}. \tag{13.11}$$

■ **EXAMPLE 13.4**

The notion of opportunistic topology control can be illustrated using the definitions above and the sample topology in Figure 13.6(a), where a network of four nodes is shown. The reachability of each link is also shown. The connectivity-based topology control algorithms construct the subgraph G_1 as shown in Figure 13.6(b). Since nodes S and A, and A and B, are connected, only these edges are selected. Since node C does not have 100% connectivity with any of the nodes, it is not included in the subgraph. Considering the node reachability of A, B, and C as 1, 1, and 0, respectively, the network reachability is $\lambda(G_1) = 0.67$. Moreover, considering a single unit for both transmitting and receiving energy, the network energy cost is $\epsilon(G) = 4$. Finally, the reachability-to-energy ratio is $\eta(G_1) = 0.167$.

Instead of considering only connected nodes, the lossy links can also be considered in topology management to construct the topology, G_2, in Figure 13.6(c). In this case, the network reachability, network energy cost, and the reachability-to-energy ratio are $\lambda(G_2) = 0.93$, $\epsilon(G_2) = 4$, and $\eta(G_2) = 0.233$, respectively. Compared to the connectivity-based graph, G_1, the opportunistic graph, G_2, results in the same network energy cost but significantly improves the network connectivity and energy efficiency. This simple example illustrates the advantages of considering lossy links in topology management.

Most WSN applications do not require 100% reliability. As a result, a certain loss rate can be tolerated as long as the network lifetime is high. Accordingly, the network reachability can be lower than 1. The CONREAP algorithm aims to establish a topology, G_R, such that the network energy cost is minimized with the constraint that network reachability is above some threshold, i.e., $\lambda(G_R) \geq \lambda_{Th}$.

The calculation of $\lambda(G_R)$ requires reachability information from *all* the nodes in the network according to (13.8). However, this is not feasible in a distributed protocol. Moreover, node reachability

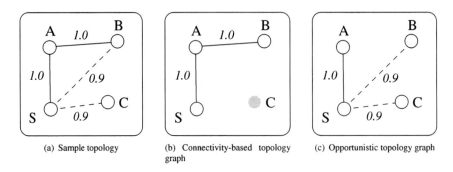

(a) Sample topology   (b) Connectivity-based topology graph   (c) Opportunistic topology graph

Figure 13.6   Opportunistic topology control.

is also not easy to obtain locally. As a result the CONREAP algorithm aims to approximate both node reachability and network reachability through local decisions.

The protocol is initiated through a HELLO message broadcast by each node to its neighbors. Accordingly, the node receives the link reachability information from its neighbors. This information is used as a weight to calculate the shortest path to each neighbor. The information is used iteratively to construct multiple trees until the average reachability is above the threshold, λ_{Th}.

The CONREAP algorithm significantly reduces the energy cost for establishing a topology with a required reachability compared to connectivity-based approaches. Accordingly, the network lifetime can be extended by as much as a factor of 6. The gains are significant when the reachability requirement is not high. However, the CONREAP energy cost approaches connectivity-based approaches as the reachability requirement is close to 1. Since distributed decisions are made to reach the target λ_{Th}, the protocol results in a higher network reachability than required. This suggests that more efficient distributed solutions can still be employed to further improve the energy efficiency of opportunity-based topology control schemes.

13.3   Activity Scheduling

In WSNs, it is clear that, during operation, it would be difficult or even impossible to access the individual sensor nodes [10]. Moreover, sensor topology changes due to node failures and energy depletion. Hence, even when an efficient deployment is in place, the WSN topology should be controlled for longer network lifetime and efficient communication.

Activity scheduling mechanisms control the *active* and *sleep* states of the nodes to maintain certain network properties. Among these, connectivity and coverage are the two main metrics considered in the design of activity scheduling protocols. For connectivity maintenance, distributed construction of a connected dominating set (CDS) or the corresponding unit disc graph is important. As a result, a certain number of nodes are selected to be active in the network so that connectivity is still maintained.

These selected nodes form a *backbone* in the network as shown in Figure 13.1(c). Accordingly, these nodes are referred to as *backbone nodes*, *coordinators*, or simply *active nodes*. The backbone nodes are responsible for relaying traffic throughout the network and maintaining connectivity. The remaining nodes can be completely turned off or activated only to transmit their sensor readings to the closest backbone node. In activity scheduling protocols, the terms *turn off* or *sleep* are generally used to refer to the transceiver component only. Therefore, the sensing and processing activities can still be performed even if a node is not selected as a backbone node. As a result of activity scheduling protocols, the redundancy in the network is exploited to use a limited number of nodes for relaying while extending the network lifetime.

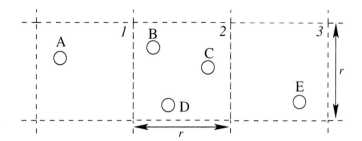

Figure 13.7   Virtual grid structure in the GAF algorithm [22].

The activity scheduling protocols generally reside between the MAC and network layers in the protocol stack. The local interactions between nodes are informed by the MAC layer. Moreover, since the activity scheduling protocols control the active nodes and affect the resulting topology, they should be closely integrated with routing protocols.

In the following, we describe five main activity scheduling protocols: GAF [22], ASCENT [6], SPAN [7], PEAS [23], and STEM [18].

13.3.1   GAF

The geographical adaptive fidelity (GAF) algorithm aims to reduce energy consumption in the network by selecting backbone nodes according to their location in the network [22]. The main goal is to decrease the energy waste because of idle listening by selecting a small number of backbone nodes to maintain connectivity in the network.

The backbone selection mechanism of the GAF algorithm relies on the geographical locations of the nodes. Thus, each node is expected to be aware of its location, which can be provided through an integrated GPS module or with the help of a localization algorithm as discussed in Chapter 12. The GAF algorithm divides the deployment area into *virtual grids* of size r as shown in Figure 13.7. The grid size is selected so that each node in one grid can communicate with any node in the adjacent grids. This requirement results in the following relation

$$r \leq \frac{R}{\sqrt{5}} \tag{13.12}$$

between the grid size r and the communication range of a node R.

■ **EXAMPLE 13.5**

The notion of virtual grid is shown in Figure 13.7, where node A in grid 1 can communicate with any of the nodes B, C, and D in grid 2. Similarly, these nodes are connected to node E in grid 3. Thus, nodes B, C, and D are equivalent in terms of routing if node A has a packet to send to node E. Accordingly, only one of the nodes in grid 2 can be activated while the other two nodes switch to sleep state to save energy.

The GAF algorithm aims to create virtual grids and maintain only a single node in each grid in a distributed way. This preserves the network connectivity while decreasing overall energy consumption and improving the lifetime of the network. The distributed operation is performed through the following three states:

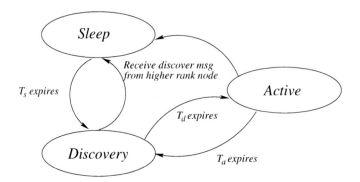

Figure 13.8   Operation states of the GAF algorithm [22].

- **Discovery state:** When a node is initialized, it is in the discovery state. In this state, the node exchanges discovery messages with its neighbors. The discovery message consists of the node ID, grid ID, estimated node active time (*enat*), and node state.

- **Active state:** In this state, the node participates in routing and handles all communication in its grid. The goal of the GAF algorithm is to activate a single node in each grid.

- **Sleep state:** Nodes that are determined as redundant for each grid switch to sleep state, where the transceiver is turned off to save energy.

The state transitions of the GAF algorithm are shown in Figure 13.8. Whenever a node enters the discovery state, it sets a timer T_d. In this state, it receives discovery messages from other nodes and switches to sleep state if a node with a higher *ranking* sends a discovery message. The rank of a node is determined from its state and the expected lifetime, *enat*. A node in the active state has a higher rank than that in the discovery state. Moreover, a node with a higher *enat* has a higher rank and, hence, is used as a backbone node. If the node does not receive any discovery message from a higher ranked node, then it switches to the active state on expiration of timer T_d.

In the active state, each node sets a timer T_a and periodically broadcasts discovery messages with interval T_d. When the timer T_a expires, the node switches back to the discovery state. This lets other nodes in the grid assume the backbone duty. Finally, the nodes that are not selected as backbone nodes switch to sleep state for a duration of T_s. After this timer expires, the node switches back to the discovery state.

The GAF algorithm provides a lifetime extension of 2–4 times compared to the case without activity scheduling. The protocol operation relies on virtual grids, which depend on the locations of the nodes. The main assumption with the virtual grids is that each node in a grid can communicate with the nodes in adjacent grids. Assuming a unit disc graph model, where the communication range is defined as the radius of a circle, the grid size can be determined according to the communication range. However, multi-path fading and shadowing effects impact the communication. As a result, the connectivity between any chosen backbone node in adjacent grids cannot be guaranteed. This reduces the packet delivery ratio by 10% over the lifetime of the network compared to the unit disc graph model.

The performance of the GAF algorithm also relies on accurate location information at each node. Since GPS may not be available in all scenarios, localization protocols are required to provide this information. The inherent error resulting from these protocols may impact the performance of backbone nodes. In cases where the location error is correlated, the GAF algorithm still maintains high accuracy since decisions are made relative to neighbor node positions. Moreover, if the location error is smaller than the virtual grid size, data delivery can still be guaranteed. On the other hand, the GAF algorithm

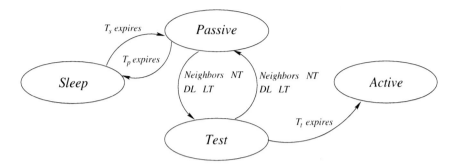

Figure 13.9   Operation states of the ASCENT scheme [6].

is not applicable for cases where location error is higher than the virtual grid size or localization information is not available.

13.3.2   ASCENT

The adaptive self-configuring sensor network topologies (ASCENT) scheme relies on local measurements of nodes for activity scheduling [6]. The communication performance of each node is used to maintain connectivity in the network while decreasing the number of active nodes. The ASCENT scheme has been developed for highly dense networks, where a large number of nodes are deployed. The redundancy provided by the high density is exploited to extend the overall system lifetime by controlling the activities of the nodes. Accordingly, only a small number of *active nodes* participate in forming a backbone for the whole network while the remaining *passive nodes* periodically check the medium to adapt to the network changes.

The ASCENT scheme employs a distributed scheduling mechanism so that each node reacts to the connectivity in its surroundings. Each node is in one of the four states shown in Figure 13.9 and explained below:

- **Test:** Each node is initialized in the test state. In this state, the node exchanges control messages with its neighbors to determine the number of active neighbors in its surroundings.
- **Active:** In this state, the transceiver of the node is on and it acts as a relay for the packets transmitted in the network.
- **Passive:** In this state, the transceiver of the node is also on but it does not participate in communication. Instead, nodes overhear ongoing traffic and gather information about the network condition and data loss rate.
- **Sleep:** If a node is not required for connectivity in the network, it switches to sleep state and turns off its transceiver.

Each node monitors the number of neighbors and the data loss rate in its environment. According to this information, the operation state is determined. The goal of the ASCENT scheme is to maintain the number of active neighbors above a neighbor threshold (NT) and data loss rate above a loss threshold (LT) in any part of the network. The NT controls the connectivity in the network and affects the contention rate. The LT indicates the communication performance around a particular node and may result in an increase of active node if it is low.

As shown in Figure 13.9, the state transitions are performed according to the local measured values. When the node is initialized, it is in the *test* state and exchanges neighbor announcement messages with its neighbors. Moreover, it sets up a timer T_t and switches to the *active* state when the timer expires.

During the message exchange, if the node determines its number of neighbors to be larger than NT, it switches to the *passive* state instead of active state.

In the passive state, a node continues to collect information about its neighborhood but does not transmit any messages. Upon entering this state, the node sets a timer T_p and transmits a *new passive node announcement message*. This message is used by the active nodes in its surroundings to estimate the node density and send it back to the node. Moreover, the data loss rate (DL) is also updated. During this state, if the number of active neighbors is less than NT or DL is higher than LT, the node switches to the test state. Otherwise, with the expiration of timer T_p, the node switches to *sleep* state.

In sleep state, the node turns off its transceiver and sets a timer T_s. In this state, the node is disconnected from the network and cannot participate in communication or gather information regarding the network state. When T_s expires, the node switches back to the passive state. Finally, in the active state, the node continues to communicate until its energy is depleted.

The operation of the ASCENT scheme depends on local measurements of the DL and the number of neighbors. The DL is determined during the passive state. Each node assigns a sequence number to the packets it transmits. The neighbors of this node can assess the DL by monitoring the sequence numbers. Moreover, the DL is monitored according to an exponentially weighted moving average as follows:

$$DL = \rho DL_{measured} + (1 - \rho)DL_{previous} \tag{13.13}$$

where ρ controls the weight of the recently measured DL on the overall DL value.

The number of neighbors is determined according to the communication success between the node and its potential neighbors. This success is based on a *neighbor loss threshold* (NLS). A node is determined as a neighbor if the loss rate from this node does not exceed NLS. The value of NLS is determined from the number of neighbors N already available as follows:

$$NLS = 1 - \frac{1}{N}. \tag{13.14}$$

As a result, NLS increases as the number of neighbors increases. Since a large N implies a larger contention in an area, increasing NLS prevents inaccurate loss rate measurements because of collisions.

In addition to local monitoring, nodes can also request additional neighbors to be active if their loss rate is high. This is explained by the following example shown in Figure 13.10.

■ **EXAMPLE 13.6**

In Figure 13.10(a), two active nodes A and B communicate with each other. The passive neighbors are also shown. If B experiences high DL for packets sent by A, it can request some of the passive nodes to switch to the active state. This is accomplished by broadcasting a *help* message as shown in Figure 13.10(b). The passive nodes that receive this message switch to the test state and broadcast neighbor announcement messages as shown in Figure 13.10(c). When the number of active nodes reaches a certain value, network operation continues as shown in Figure 13.10(d). In this case, nodes A and B can communicate through additional relays C, D, and E, which provide lower DL.

The performance evaluations and testbed experiments show that the ASCENT scheme achieves energy efficiency and high throughput even when the network density is increased. On the other hand, latency is increased since a fixed number of nodes are used for data forwarding in the ASCENT scheme.

13.3.3   SPAN

SPAN addresses activity scheduling through distributed selection of backbone nodes [7]. Activity scheduling schemes such as GAF and ASCENT consider connectivity as the main goal for selecting

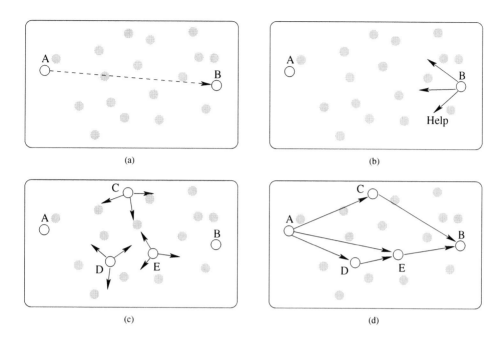

Figure 13.10   ASCENT scheme self-configuration [6].

backbone nodes. However, a connected backbone may not preserve the capacity of the network since the number of potential paths from a sensor node to the sink is decreased. This may lead to congestion at the backbone nodes and decrease the network capacity. Thus, in addition to connectivity, SPAN also considers the capacity of the network for the selection of backbone nodes.

■ EXAMPLE 13.7

The effect of backbone nodes on capacity is shown in Figure 13.11, where white nodes are the backbone nodes, through which the other nodes (A, B, C, D, and E) can communicate with each other. With this selection, the packets between nodes C and D share a portion of the path between nodes A and B. As a result of interference between two paths, the network capacity decreases. However, if node E is selected as a backbone node, the traffic between C and D can be decoupled.

Similar to previous activity scheduling mechanisms, SPAN employs a distributed backbone node selection, where each node determines to become a backbone node according to its local neighborhood information. This information is gathered through periodic exchanges between nodes. Each node broadcasts HELLO messages to its neighbors. The HELLO message consists of the following information: (1) node status (backbone node or not), (2) the list of backbone nodes it is connected to, and (3) its neighbors. As the HELLO messages are exchanged, each node is aware of the backbone nodes in its surroundings and can determine to become a backbone node if required. This information is also used by the routing protocol so that relay nodes are selected only among the backbone nodes.

The exchange of HELLO messages provides a broad view of the neighborhood of a node. Accordingly, the number of backbone nodes in the neighborhood, the number of neighbor nodes, as well as the backbone nodes that these neighbors are connected to, can be communicated to the node. A node determines that it should become a backbone node according to the *coordinator eligibility rule* in SPAN.

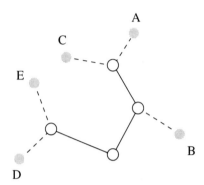

Figure 13.11   The effect of activity scheduling on interference [7].

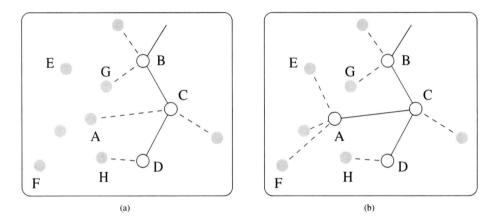

Figure 13.12   Coordinator eligibility criteria in SPAN.

That is, given the current backbone nodes in the neighborhood, if two or more of the neighbors of a node cannot communicate, then that node becomes eligible as a backbone node.

■ **EXAMPLE 13.8**

The eligibility criteria are illustrated in Figure 13.12, where the local neighborhood of a node A is shown with active backbone nodes B, C, and D as well as other neighbors of node A. It can be observed from Figure 13.12(a) that nodes E and F cannot communicate with each other either directly or through one of the backbone nodes. In this case, node A can become a backbone node to improve the capacity of the network.

While a node can determine its eligibility for being a backbone node, there may be multiple such nodes in close proximity because of the high density. As shown in Figure 13.12(a), in addition to node A, nodes G and H also provide connectivity between nodes E and F. To prevent multiple nodes from becoming a coordinator, SPAN employs a backoff mechanism similar to MAC protocols as explained in Chapter 5. If a node determines to become a backbone node, it broadcasts a HELLO message with its updated status. However, this message is delayed according to a random amount of time.

The backoff delay depends on two parameters: the utility and residual energy of the node. The utility of the node is measured according to the number of pairs to which the node provides additional connectivity. If a node i has N_i neighbors, $\binom{N}{2}$ potential links exist. Among these links, if the node i provides additional connectivity to C_i links, then the *utility* of the node is defined as $C_i/\binom{N}{2}$. The second parameter is the residual energy of the node. If a node has E_r units of remaining energy and E_m is the maximum amount of energy, then E_r/E_m is used as a metric in delay calculation. Accordingly, the backoff delay of SPAN is

$$D_i = \left(\left(1 - \frac{E_r}{E_m} \right) + \left(1 - \frac{C_i}{\binom{N_i}{2}} \right) + R \right) N_i T \qquad (13.15)$$

where R is a random number selected within $[0, 1]$ and T is the round-trip delay between two nodes. Through the backoff delay, nodes with higher utility and residual energy have a smaller backoff delay and can broadcast the HELLO message earlier than other candidates.

If a node determines to become a backbone node, it sets a backoff timer to D_i and continues to listen to the channel. If during the backoff duration, it receives a HELLO message from a new backbone node, it reevaluates its coordinator eligibility. This is shown next.

■ **EXAMPLE 13.9**

As shown in Figure 13.12(b), node A becomes a backbone node if it broadcasts a HELLO message earlier than nodes G and H. Then, the other two nodes determine that all of their neighbors are connected and do not try to become backbone nodes. With the help of the backoff delay, SPAN prevents redundant nodes from becoming backbone nodes while preserving the capacity of the network.

Contrary to some of the other schemes, SPAN does not preserve the backbone duty of a node for a long time. Instead, each backbone node periodically checks its neighborhood and withdraws from this duty if every other neighbor can communicate with the help of other backbone nodes. Moreover, when a node becomes a backbone node, it sets a timer to limit its backbone duty. On expiration of the timer, the backbone node checks whether its neighbors can be connected with the help of other *nodes* even if these nodes are not backbone nodes. Accordingly, the node withdraws from backbone duty.

Withdrawal from backbone duty gives other nodes the opportunity to become backbone nodes and distributes the additional energy consumption for backbone duties evenly. The withdrawal of a backbone node starts with an announcement through the HELLO message. Other nodes that hear this announcement contend to become backbone nodes and broadcast their HELLO message if they win the contention. The departing backbone node waits for this HELLO message before switching its radio off. Accordingly, disconnection in the network is prevented.

Simulations with SPAN show that the capacity of the network is preserved compared to an *always-on* network and a slight decrease in latency is observed because of the longer number of hops in each path. On the other hand, SPAN saves energy by a factor of 3.5 or more compared to a network without SPAN. Moreover, network lifetime can be doubled using SPAN.

13.3.4 PEAS

Activity scheduling is generally performed by maintaining neighborhood state information such as number of neighbors, number of backbone nodes, or DL between neighbors as in SPAN and ASCENT. While this information leads to informed decisions regarding backbone node selection, it may lead to

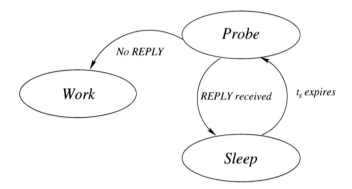

Figure 13.13 Operation states of the PEAS protocol [23].

significant overhead to maintain this information, especially in high-density networks. As the density increases, the number of neighbors of a node increases. This leads to increased memory requirements to store neighborhood information as well as increased communication overhead to exchange this information. Furthermore, the protocol parameters such as time spent in certain states are fixed in SPAN, ASCENT, and GAF. This results in a lack of adaptivity for the activity scheduling protocol.

The Probing Environment and Adaptive Sleeping (PEAS) protocol [23] addresses these factors through a dynamic and distributed scheduling scheme. Each node probes its neighborhood to determine whether any active nodes are available within a given probing range. Based on the activity in the neighborhood, the sleep durations are dynamically updated.

The operation states of the PEAS protocol are shown in Figure 13.13, which consists of the following:

- **Sleep:** The protocol initiates in sleep state. The sleep duration, t_s, is exponentially distributed according to $f(t_s) = \lambda e^{-\lambda t}$, where λ is the probing rate.
- **Probe:** The node enters probe state after waking up. In this state, the node broadcasts a PROBE message to determine any active nodes.
- **Work:** In working state, the node assumes backbone duty and relays messages until it runs out of energy.

The transitions between states are also shown in Figure 13.13. The node switches from sleep state to probe state when the sleep timer t_s expires. In probe state, the node broadcasts a PROBE message. The goal of this message is to determine if any active nodes are in the vicinity of the node. This is determined by the probing range, R_p, which is chosen to be smaller than the transmission range, R_t. If there is any active node within the probing range, R_p, it replies to the PROBE message through a REPLY message. This implies that there is an active node close to the probing node and the probing node can switch to sleep state. Since there may be multiple backbone nodes within R_p distance from the probing node, each backbone node uses a random delay before sending the REPLY message to avoid collisions.

If the probing node does not receive any REPLY message, it switches to working state to act as a backbone node. Once in the working mode, the node continues to relay messages until its energy is depleted.

The sleep duration of the PEAS protocol is randomly distributed with a mean λ. In a WSN, the density may vary at different locations of the sensor field because of the randomness in distribution. Furthermore, nodes may fail during operation of the network resulting in a non-uniform distribution throughout the network. To maintain connectivity, certain nodes will need to sleep for a shorter duration depending on their neighborhood. This is addressed through the *adaptive sleeping* mechanism of the PEAS protocol.

The mean sleep duration, λ, is updated by each node based on activity in its neighborhood to reach a desired mean interval, λ_d.

The mean sleep interval of each node is controlled by the working node associated with that node based on the average probing rate in the neighborhood. Each working node keeps a record of the time it takes to receive a specific number, k, of PROBE messages from any of its probing neighbors. If the working node starts counting the PROBE messages at time $t = t_0$ and receives k PROBE messages at time $t = t_k$, the average probing rate is

$$\hat{\lambda} = \frac{k}{t_k - t_0}. \tag{13.16}$$

The average probing rate, $\hat{\lambda}$, is communicated to each probing node that sends a PROBE message through the REPLY message. Accordingly, each probing node updates its average sleep duration as follows:

$$\lambda^{new} = \lambda \frac{\lambda_d}{\hat{\lambda}} \tag{13.17}$$

which is used to select a new sleeping period.

The PEAS protocol aims to address the unreliable nature of sensors through a probabilistic approach. Each node probes its neighborhood to make local decisions and adapts its protocol parameters based on the activity of its neighbors. Accordingly, dynamic changes of the network such as node failures and non-homogeneous network deployment can be captured through the adaptive protocol operation.

13.3.5 STEM

Topology control schemes such as GAF [22] and ASCENT [6] coordinate the activities of each node to preserve network connectivity. These solutions improve network lifetime by switching redundant nodes in the network to sleep mode. The redundancy is determined from the connectivity of the network. However, these approaches consider only the *reporting state* and aim to preserve the connectivity of the network at all times. The resulting network operation is applicable to monitoring applications, where nodes continuously send their readings to the sink. However, with event-based applications, network traffic is sporadic and connectivity is not required at all times. In other words, the nodes are generally in *monitoring state* until an event of interest occurs. Hence, considering the event-based application requirements, providing connectivity at all times through activity scheduling may decrease the network lifetime.

The Sparse Topology and Energy Management (STEM) protocol addresses this issue by employing a more aggressive activity scheduling mechanism [18]. Instead of providing continuous connectivity, nodes are switched to sleep state most of the time. This decreases the energy consumption but results in a disconnected network. In case an event occurs, the nodes need to be activated to efficiently relay the packets to the sink. For efficient node activation, STEM relies on a dual-radio architecture.

The sensors are assumed to be equipped with two radios: one for scheduling and listening to the channel and another one for actual data communication. Each radio uses a different channel (f_1 and f_2) so that both radios can be used simultaneously. The first channel, f_1, is used for activity scheduling purposes and is referred to as the *wakeup plane*. Whenever the sender and the receiver are activated, data communication is performed by the second radio through the second channel, f_2, which is referred to as the *data plane*. Hence, in the STEM protocol, nodes turn their radio off and periodically listen to the channel to check if any node is trying to communicate. In the case of a communication attempt, the data radio is turned on and the communication takes place.

If there is no traffic in the network, the second radio is turned off. To maintain connectivity with the neighbors of a node, the first radio is periodically turned on for a short listen duration. This is similar to the duty cycle-based MAC protocols explained in Chapter 5. Accordingly, if a node i has a packet destined for a particular neighbor j, node i can *wake up* its neighbor by transmitting a wakeup signal

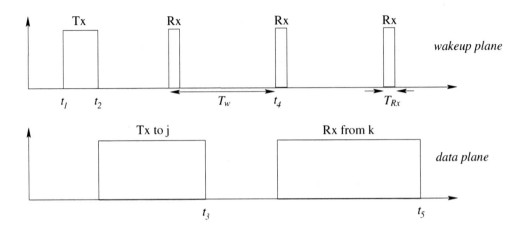

Figure 13.14 Operation timeline for the STEM protocol [18].

during the listen duration of node j. If the intended receiver wakes up, the data plane is used for actual data transfer.

The operation of the STEM protocol for a particular node i is shown in Figure 13.14. The timeline on the top depicts operation on the wakeup plane, whereas the timeline at the bottom shows data communication on the data plane. Each node listens to the wakeup plane for a duration of T_{RX} with a period of T_w. Let us assume node i has a packet to send to one of its neighbors j. At time t_1, node i initiates communication by broadcasting a series of beacons on the wakeup plane until it receives a response from node j at time t_2. Upon receiving a response, node i starts transmitting its data packet(s) on the data plane, which lasts until time t_3. At this time, both nodes (i and j) switch off their second radios and continue the scheduling process on the wakeup plane. Let us also assume that, between t_3 and t_4, a third node k aims to reach node i and starts transmitting beacons. At time t_4, node i listens to the wakeup plane, hears the beacon from node k, sends a reply, and activates its second radio. Accordingly, node k can send its data packet(s), which lasts until time t_5.

Activity scheduling protocols such as SPAN, ASCENT, and GAF maintain active nodes at all times for connectivity. Instead, the STEM protocol activates links on a demand basis. The success of this link activation approach depends on the fact that when a node listens to the wakeup plane, it can detect any other neighbor that is trying to make contact. In other words, the listen duration, T_{RX}, should be long enough to detect incoming beacons. This is illustrated in Figure 13.15. Assuming that a beacon transmit duration is B_{TX} and the interval between each beacons is B_i, the minimum listen interval should be

$$T_{RX} \geq 2B_{TX} + B_i \tag{13.18}$$

to ensure correct detection. Similarly, to reach a particular node, the maximum number of beacons to be transmitted, n_B^{max}, is

$$n_B^{max} \geq \frac{T - T_{RX} + 2B_{TX} + B_i}{B_{TX} + B_i}. \tag{13.19}$$

According to (13.18),

$$n_B^{max} \geq \frac{T}{B_{TX} + B_i}. \tag{13.20}$$

The STEM protocol decreases energy consumption by minimizing it in the monitoring state, which improves the lifetime of a node. However, this improvement comes at the cost of increased latency in switching from the monitoring state to the reporting state. Since the radio is turned off for the majority of

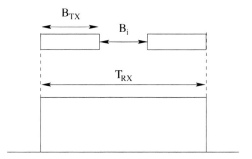

Figure 13.15 The relation between beacon interval, B_i, beacon duration, B_{TX}, and wakeup duration, T_{RX} [18].

the time, route setup latency increases as a penalty for energy conservation. Since node synchronization is not required for the STEM protocol, a node may start to broadcast beacons at any time during the scheduling period T of a node. Accordingly, the average latency in waking up a node is given by

$$\bar{T}_w = \frac{T + B_{TX} + B_i}{2}. \tag{13.21}$$

To find the relative energy gain of the STEM protocol, we need to calculate the energy consumed at both the wakeup plane, E_w, and the data plane, E_d. Accordingly, the energy consumption of the STEM protocol in an interval of t seconds is given by

$$E_{STEM} = E_w + E_d \tag{13.22}$$

where the energy consumption for the wakeup plane is

$$E_w = P_w \cdot t_w + P_{node}(t - t_w). \tag{13.23}$$

Here P_w and t_w are, respectively, the power and time spent for switching from sleep state to active state, whereas P_{node} is given by

$$P_{node} = \frac{P_{sleep}(T - T_{RX}) + P_{RX}T_{RX}}{T} \tag{13.24}$$

where P_{sleep} and P_{RX} are the sleep and receive power, respectively.[2] The second term in (13.22) is the energy consumption at the data plane, which is given by

$$E_d = P_{sleep}(t - t_{data}) + P_{data} \cdot t_{data}. \tag{13.25}$$

Compared to STEM, a protocol without topology management requires a single radio and the energy consumption is given by

$$E_{orig} = P_{idle}(t - t_{data}) + P_{data} \cdot t_{data}. \tag{13.26}$$

Accordingly, the energy consumption gain of the STEM protocol can be found by

$$\Delta E = E_{orig} - E_{STEM} \tag{13.27}$$

from (13.22) and (13.26). In a long enough interval t, assuming the traffic rate is low, we can disregard the time taken for data communication, t_{data}. Moreover, for typical transceivers, wakeup time, t_{wakeup}, and sleep power, P_{sleep}, can also be disregarded. Finally, by letting $P_{idle} = P_{RX}$, the energy consumption

---

[2]More accurately, P_{RX} is a combination of idle, receive, and transmit power but can be approximated by receive power.

gain is approximated as

$$\Delta E \simeq P_{RX} \cdot t \cdot \left(1 - \frac{T_{RX}}{T}\right). \tag{13.28}$$

Accordingly, using (13.21), the tradeoff between the relative gain in energy consumption and the wakeup latency can be shown to be

$$\partial E = \frac{\Delta E}{E_{orig}} = 1 - \frac{T_{RX}}{2\bar{T}_w - B_{TX} - B_i}. \tag{13.29}$$

As a result, as the cost for link activation, $\bar{T}_w$, increases, the gain in energy consumption increases for the STEM protocol. Accordingly, by increasing the listen interval, T, significant energy savings are possible. This, however, increases the end-to-end delay of a packet, since the protocol spends $\bar{T}_w$ on the average at each hop. Accordingly, the advantages of the STEM protocol in terms of energy efficiency are more pronounced in applications where the monitoring state greatly dominates the reporting state. More specifically, energy savings are possible compared to the case without topology management if the time spent in the monitoring state is 50% or more of the total network lifetime.

The STEM protocol can be combined with a topology management protocol such as GAF or SPAN to improve energy efficiency in the reporting state as well. As explained in Section 13.3.1, the GAF algorithm follows a virtual grid structure, where each grid is represented by one node inside this grid. This node is denoted as the *virtual node*. By limiting the size of the grid, each virtual node can successfully communicate with the other virtual nodes in the neighboring grids. Accordingly, connectivity is maintained while decreasing the energy consumption by letting redundant nodes sleep.

STEM and GAF can be combined to exploit the advantages of each protocol. Firstly, the GAF algorithm is run to determine the virtual nodes in the network. Then, each virtual node follows the STEM protocol to further decrease the energy consumption during idle times. By modifying the routing protocol such that virtual nodes are used to create routes, STEM and GAF can successfully be operated on the same nodes. Accordingly, STEM combined with GAF can reduce the network energy consumption by up to 66% compared to the GAF algorithm alone. Similarly, energy savings of up to 7% are possible compared to the STEM protocol alone.

The STEM protocol provides significant savings in energy consumption for the monitoring state. However, its applicability is limited to low-rate event-based applications, where the network is mostly dormant, i.e., in the monitoring state. Furthermore, end-to-end delivery latency is traded off for increased lifetime since nodes need to spend considerable amounts of time at each hop to establish links with their next hop neighbor toward the sink. This increased latency also impacts certain event-based applications, where event information delivery should be performed with minimum delay. As an example, in a forest fire monitoring network, the WSN is in sleep state for a very long time (months or even years) if a fire does not occur. However, in case of a fire, this information should be delivered to the sink immediately to take the required precautions. Furthermore, the energy savings of the STEM protocol are limited in monitoring applications, where the nodes in the network continuously send sampled information to the sink.

13.4 Clustering

In WSNs, high-density deployment is one of the major differences between traditional networks. In the wireless domain, a high density has advantages in terms of connectivity and coverage as well as disadvantages in terms of increased collision and overhead for protocols that require neighborhood information. As a result, scalability is an important problem in WSN protocols as the node numbers increase. The topology control mechanisms discussed so far focus on a flat topology, where each node in the network sends its information to the sink through a multi-hop route. Although these protocols aim to decrease the contention through either power control or node scheduling, scalability is still an issue.

The disadvantages of the flat-architecture protocols can be addressed by forming a hierarchical architecture, where the nodes are grouped in *clusters*. In this section, we discuss a fourth class

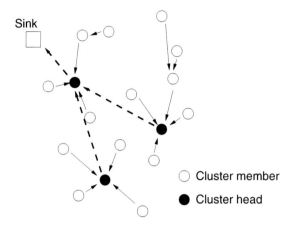

Figure 13.16   Cluster-based topology in WSNs.

of topology control mechanisms: *cluster-based topology control*. Clustering algorithms limit the communication in a local domain and transmit only necessary information to the rest of the network. This structure is depicted in Figure 13.16. A group of nodes form a cluster and the local interactions between cluster members are controlled through a cluster head (CH). Cluster members generally communicate with the cluster head and the collected data are aggregated and fused by the cluster head to conserve energy. The cluster heads can also form another layer of clusters among themselves before reaching the sink.

Overall, clustering protocols have the following advantages in WSNs:

- **Scalability:** Cluster-based protocols limit the number of transmissions between nodes, thereby enabling a higher number of nodes to be deployed in the network.
- **Collision reduction:** Since most of the functionalities of nodes are carried out by the CHs, fewer nodes contend for channel access, improving the efficiency of channel access protocols.
- **Energy efficiency:** In a cluster, the CH is active most of the time, while other nodes wake up only in a specified interval to perform data transmission to the CH. Further, by dynamically changing the CH functionalities among nodes, the energy consumption of the network can be significantly reduced.
- **Local information:** Intracluster information exchange between the CH and the nodes helps summarize the local network state and sensed information of the phenomenon state at the CH [24, 25].
- **Routing backbone:** Cluster-based approaches also enable efficient building of the routing backbone in the network, providing reliable paths from sensor nodes to the sink. Since the information to the sink is initiated only from CHs, route-thru traffic in the network is decreased.

Clustering is an integral part of hierarchical routing protocols as explained in Chapter 7. Hence, several clustering mechanisms have been developed as part of these routing protocols such as LEACH, PEGASIS, TEEN, and APTEEN. In addition to these mechanisms, in the following we explain clustering schemes that can be used with any communication mechanism.

13.4.1   *Hierarchical Clustering*

The energy-efficient hierarchical clustering algorithm is developed to minimize the overall energy consumption of the network by constructing clusters in a distributed manner [3, 4]. The notion of the

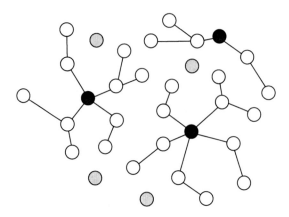

Figure 13.17   Hierarchical clustering.

clustering mechanism is illustrated in Figure 13.17. Each sensor in the network can be selected as a CH according to a probability, p. According to this probability, a node transmits a message indicating its CH duty. In this case, the CH is denoted as the *voluntary cluster head* (black nodes in Figure 13.17). The transmitted message is propagated up to k hops in the network. Each node that receives this message becomes a part of the cluster if it is not a CH. This mechanism is shown in Figure 13.17 for $k = 2$.

As shown in Figure 13.17, there may be several nodes that do not receive a CH message within a given amount of time (gray nodes). These nodes then designate themselves as a CH and advertise their duty. In this case, the CH is called the *forced cluster head*. The resulting topology groups each node in the network into one of the clusters.

The performance of the energy-efficient hierarchical clustering algorithm depends on the selection of the two parameters p and k. The overall energy consumption can be minimized through an appropriate selection of these parameters.

The optimum value of p that minimizes the overall energy consumption can be found by assuming that the network is distributed on a Poisson basis, where the number of nodes N in an area A has mean λA. Given the number of nodes as n, the average number of CHs is np. Firstly, the energy consumption between these CHs and the sink is found.

Since the CHs are evenly distributed in the network, the average distance between a CH and the sink is found as

$$E[D_{ch}|N = n] = \int_A \sqrt{x_i^2 + y_i^2} \left(\frac{1}{4a^2} \right) dA \tag{13.30}$$

$$= 0.765a \tag{13.31}$$

where the sensors are dispersed in a square of size $2a$. Accordingly, the total distance between each CH and the sink is $0.765npa$ and the energy consumption for communication between CHs and the sink is

$$E[E_{ch}|N = n] = \frac{0.765npa}{r} \tag{13.32}$$

where energy consumption for a single hop is assumed in Joules. To find the energy consumption between the cluster members and the CH, we first consider two Poisson processes PP_1 and PP_0 with intensity $\lambda_1 = p\lambda$ and $\lambda_0 = (1 - p)\lambda$, respectively. PP_1 and PP_0 are the distributions of the locations of CHs and cluster members, respectively. Accordingly, the average energy consumption in each cluster is

$$E[E_{cluster}|N = n] = \frac{\lambda_0}{2\lambda_1^{3/2} r}. \tag{13.33}$$

Since there are np clusters on average, the overall energy consumption in the network is given as

$$E[E|N = n] = \frac{np(1 - p)}{r2p^{3/2}\sqrt{\lambda}} + \frac{0.765npa}{r}. \tag{13.34}$$

Finally, removing the condition on N, the average energy consumption is

$$E[E] = \lambda A \left(\frac{np(1 - p)}{r2p^{3/2}\sqrt{\lambda}} + \frac{0.765npa}{r} \right). \tag{13.35}$$

Minimizing this function leads to the optimum p value. Similarly, the optimum value of k is found as follows:

$$k = \frac{1}{r} \sqrt{\frac{-0.917 \ln(\alpha/7)}{p_1\lambda}}. \tag{13.36}$$

Moreover, the clustering algorithm is also extended to form multiple layers of hierarchical clusters. However, the protocol parameters are calculated from only the density of the network. Since a homogeneous distribution is assumed, the energy efficiency of the clustering protocol may not be optimal in the case of a non-uniform distribution of nodes.

13.4.2 HEED

The hybrid energy-efficient distributed (HEED) clustering algorithm combines transmit power control with clustering to form single-hop clusters [24]. The main goal is to minimize the energy consumption for communication by constructing clusters in a distributed fashion. This is performed according to the residual energy of the nodes, where nodes with high residual energy are selected as CHs. Furthermore, the overall communication cost inside a cluster, i.e., *intracluster communication cost*, is also considered in cluster formation.

The HEED clustering mechanism is performed with a period of $T_{CP} + T_{NO}$ seconds, where T_{CP} is the time it takes for the clustering protocol to converge and T_{NO} is the time for normal network activities, i.e., network operation interval. In each period, certain nodes are selected as CHs and the remaining nodes join to each cluster.

The clustering mechanism is initiated by CH selection. This phase is performed by each node according to a probability based on the remaining energy. More specifically, an initial probability, C_{prob}, is determined. Based on this probability, each node determines the probability that it will become a CH:

$$CH_{prob} = \max \left(C_{Prob} \frac{E_{residual}}{E_{max}}, p_{min} \right) \tag{13.37}$$

where $E_{residual}$ and E_{max} are the residual and the maximum energy of a node, respectively. The CH probability, CH_{prob}, depends on the relative lifetime of a node and is lower bounded by a threshold, p_{min}, to ensure convergence within a limited number of iterations.

■ **EXAMPLE 13.10**

The CH selection is performed over several iterations in each phase as shown in Figure 13.18. In each iteration, a node determines to be a CH with probability CH_{prob}. In the first iteration, the CH probability is determined according to (13.37) for each node. Accordingly, nodes determine to be CHs and are denoted as *tentative CHs*. Since CH_{prob} is a function of the remaining energy, the nodes with higher residual energy have a greater chance of becoming tentative CHs. The tentative CHs advertise their duties through a CH message, which includes the node ID, selection status (tentative CH or final CH), and cost. According to the advertised costs, the remaining nodes select the tentative CH with the lowest cost as their CH. As shown in Figure 13.18(a), in the first

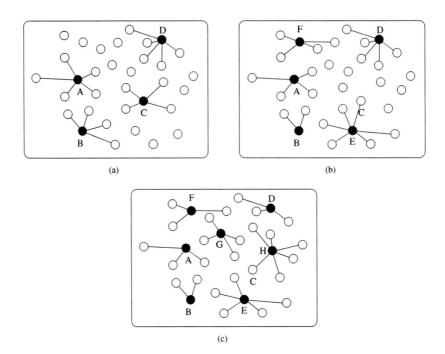

(a)   (b)

(c)

Figure 13.18   CH selection in HEED.

iteration nodes A, B, C, and D are selected as tentative CHs and the nodes in their cluster are shown as connected to these nodes.

In the following iterations, CH_{prob} is doubled for each node and the same procedure is repeated. This allows the remaining nodes to be tentative CHs. Similarly, if a tentative CH receives an advertisement from another tentative CH with a lower cost, it can select that node as its CH and leave CH duty. This is shown in Figure 13.18(b), where nodes E and F are also tentative CHs. While node C is selected as a tentative CH in the first iteration, it is connected to node F in the second iteration since node F has a lower cost. After an iteration, if a node has a CH_{prob} of 1 or larger, it becomes a *final CH* and assumes CH duty for the rest of the clustering period. After a limited number of iterations, HEED results in a topology where each node is either a CH or connected to a CH as shown in Figure 13.18(c).

Cluster membership is established according to the *cost* of each CH as discussed above. HEED utilizes three different cost functions for clustering based on different application requirements. In an application where load distribution is important, the cost function is selected as the *node degree*. Accordingly, a node selects the CH with the fewest neighbors. On the other hand, if dense clusters are required, 1/*node degree* is used as the cost function. Finally, a third cost function, *average minimum reachability power* (AMRP), of a node i is defined as follows:

$$AMRP_i = \frac{\sum_{j=1}^{M} \min(Pwr)_j}{M} \tag{13.38}$$

where M is the number of nodes within the cluster range of node i and $\min(Pwr)_j$ is the minimum power required by a node j to reach node i. The AMRP is used as a measure of the expected intracluster

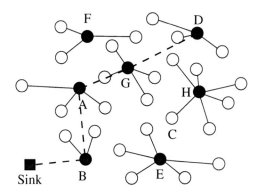

Figure 13.19   Multi-hop routing from CHs to the sink in HEED.

communication energy consumption. As a result, minimizing this cost function to form clusters results in lower energy consumption throughout the network.

If a node receives advertisement messages from multiple CHs, then the cost function is used to select the best CH. Accordingly, the clusters are formed. The iteration procedure ensures that within a limited number of iterations, the clustering algorithm converges.

After the clusters are formed, intracluster communication is maintained by the CHs and each member can forward its data to its CH. Intercluster communication, on the other hand, is performed between CHs using a higher transmit power level. Since clusters are formed with nodes within one hop away from the CH, with an appropriate transmit power level the set of CHs forms a connected network. Consequently, CHs can forward their data to the sink using a multi-hop route through other CHs. As an example, as shown in Figure 13.19, the CH labeled D can reach the sink through a path G–A–B using other CHs.

The HEED algorithm is shown to terminate in a limited number of steps and outperforms generic weight-based clustering protocols.

13.4.3   Coverage-Preserving Clustering

The majority of the clustering protocols for WSNs are designed based on energy efficiency. Cluster formation and CH selection are performed using certain energy cost functions. Accordingly, the goal is to improve network lifetime by distributing the CH duties to nodes with higher residual energy. From network operation considerations, this approach leads to very efficient solutions. However, these solutions generally do not consider the application scenarios for clustering.

In addition to communication, sensor nodes are also responsible for gathering data from the environment through on-board sensors. Similar to communication, each sensor has a limited range in which it can sense events. As a result, each sensor *covers* a certain area that it can sense as shown in Fig 13.20. Since sensor nodes are generally not homogeneously distributed, all sensor nodes do not equally contribute to network coverage. As an example, node A in Figure 13.20 is more significant for network coverage compared to its neighbors B, C, and D since its sensing area includes certain areas not covered by any of the other nodes. On the other hand, node B provides redundant coverage since its sensing area is covered by nodes A, C, and D. In this setting, selecting node A as a CH would be detrimental to network coverage since its energy drains much faster because of both CH and sensing duties. Instead, by selecting node B as the CH, topology management duties can be better distributed. This can be performed by *coverage-aware* clustering mechanisms.

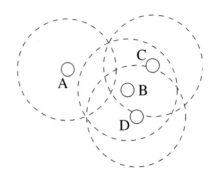

Figure 13.20   Coverage and CH selection in CPCP.

The Coverage-Preserving Clustering protocol (CPCP) uses several coverage-aware clustering metrics to construct a topology that will extend network lifetime without hampering network coverage [21]. The clustering is performed using three different metrics that can be selected according to application requirements. These metrics are: minimum weight coverage cost, weighted sum coverage cost, and coverage redundancy cost.

These clustering metrics are constructed from the following definitions. The sensing area of a node i is denoted as $C(s_i)$. Since a node usually has overlapping sensing areas with other nodes, certain portions of the sensing area of a node can also be sensed by other nodes. Then, the total coverage of a point can be defined as

$$O_{total}(x, y) = \sum_{s_j:(x,y)\in C(s_j)} 1 \tag{13.39}$$

which considers the number of nodes that cover the point (x, y). Similarly, using the remaining energy of each node j, $E(j)$, the total energy $E_{total}(x, y)$ that is available for sensing a specific location (x, y) is given by

$$E_{total}(x, y) = \sum_{s_j:(x,y)\in C(s_j)} E(s_j). \tag{13.40}$$

$E_{total}(x, y)$ depends on two factors: the number of nodes that cover the specific location (x, y) and the remaining energy of these nodes. According to this definition, the *minimum weight coverage cost* is

$$C_{mw}(s_i) = \max \frac{1}{E_{total}(x, y)} (x, y) \in C(s_i). \tag{13.41}$$

As a result, the cost of each node is defined as a function of the available energy to sense the most critically covered point in the sensing area of this node. Similarly, the *weighted sum coverage cost* is

$$C_{ws}(s_i) = \int_{C(S_i)} \frac{dx\, dy}{E_{total}(x, y)} \tag{13.42}$$

which considers the weighted average of the available energy to cover *all* points in the sensing area of a node. Moreover, the *coverage redundancy cost* considers only the number of nodes covering a particular point and is defined as

$$C_{cc}(s_i) = \int_{C(s_i)} \frac{dx\, dy}{O_{total}(x, y)}. \tag{13.43}$$

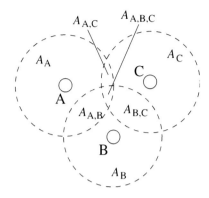

Figure 13.21   Sensing areas of nodes A, B, and C.

■ **EXAMPLE 13.11**

The coverage-aware costs can be found according to the sensing areas and the remaining energy levels of each node in a local neighborhood. As an example, in Figure 13.21, three nodes A, B, and C are shown with respective sensing areas. Assuming the remaining energy of these nodes as $\{E(A), E(B), E(C)\} = \{2, 4, 1\}$ and using the above definitions, the coverage-aware costs for node A are as follows:

$$C_{mw}(A) = \frac{1}{2} \tag{13.44}$$

$$C_{ws}(A) = \frac{|\mathcal{A}_A|}{2} + \frac{|\mathcal{A}_{A,B}|}{6} + \frac{|\mathcal{A}_{A,C}|}{3} + \frac{|\mathcal{A}_{A,B,C}|}{7} \tag{13.45}$$

$$C_{cc}(A) = |\mathcal{A}_A| + \frac{|\mathcal{A}_{A,B}| + |\mathcal{A}_{A,C}|}{2} + \frac{|\mathcal{A}_{A,B,C}|}{3} \tag{13.46}$$

where $|\mathcal{A}|$ is the size of area $\mathcal{A}$.

CH selection is performed based on the coverage-aware costs of each node. Each node that is a member of a cluster sends its information only to its CH. However, CH to sink communication is performed on a multi-hop path from the CH to the sink that involves other sensor nodes. Similar to the coverage-aware cost of a node, the *coverage-aware cost of a path* from node i to the sink can also be defined as follows:

$$C_{final}(s_i) = \sum_{s_j, s_k \in p(s_i, s_{sink})} C_{link}(s_j, s_k) \tag{13.47}$$

where $C_{link}(s_i, s_k)$ is the cost for each link and is given by

$$C_{link}(s_i, s_k) = C(s_i)E_{tx}(s_i, s_k) + C(s_k)E_{rx}(s_i, s_k) \tag{13.48}$$

where $C(s_i)$ is the cost of a node based on one of the cost functions, and $E_{tx}(s_i, s_k)$ and $E_{rx}(s_i, s_k)$ are the energy consumption for transmitting and receiving a packet, respectively. Using (13.47) and (13.48), the minimum cost path from a CH to the sink can be found.

Based on the cost functions, CPCP operation is performed in six phases as shown in Figure 13.22. In Figure 13.22(a), an example network topology shows the coverage areas of each node. In the first phase, nodes exchange residual energy information with their neighbors. At the end of this phase, each node calculates its coverage-aware cost. The second phase is the *CH election* phase. Each node determines an

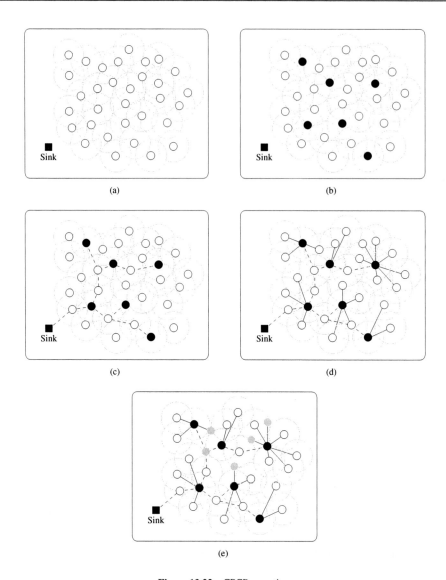

Figure 13.22 CPCP operation.

activation time based on its cost to become a CH. During this time, if a node receives a CH announcement message from any of its neighbors within $R_{cluster}$ distance, it resets its timer and continues listening to any further announcements. If no message is heard during the activation time, the node becomes a CH and broadcasts an announcement informing its neighbors about its CH status. At the end of this phase, nodes with minimum cost in their neighborhood are elected as CHs as shown in Figure 13.22(b). Note that, in most cases, the coverage area of the CH is mostly covered by the neighboring nodes.

In the third phase, end-to-end paths between the CHs and the sink are established as shown in Figure 13.22(c). This is initiated by the sink node through route discovery messages, which are forwarded by each node based on its coverage-aware cost. The coverage-aware cost is used to delay the broadcast of each relay node. As a result, any node in the network receives the route discovery message from the sink by the minimum cost path first. Cluster formation is performed in the fourth phase based on the elected CH. Each non-CH node selects the closest CH and joins the cluster using a *join* message. Accordingly, each node joins the cluster with the CH closest to its location as shown in Figure 13.22(d).

Once the clusters are formed, the sensor activation phase is initiated. In this phase, nodes are activated to collect physical information according to their costs. An activation timer is associated with each node proportional to its coverage-aware cost. If the timer expires, each node broadcasts an ACTIVATION message to indicate its neighbors. If a node receives ACTIVATION messages from some of its neighbors so that its sensing area is covered, the activation timer is reset and the node switches to sleep state. This phase results in a minimum number of active sensors while preserving the coverage as shown in Figure 13.22(e). Finally, in the sixth phase, each active node collects information from its sensor and forwards this to the CH, which can aggregate the data from multiple sensors and send that data to the sink.

The three cost functions provide different advantages for coverage-aware clustering. The minimum weight cost, C_{mw}, and the weighted sum cost, C_{ws}, improve the energy efficiency by providing longer coverage time compared to clustering metrics that consider only energy consumption. Compared to these two metrics, the coverage redundancy cost, C_{cc}, results in a shorter coverage time if 100% is required. However, the advantage of this metric is observed for long-term deployments. If 100% coverage is not essential, C_{cc} maintains coverage at high levels, e.g., > 80% for a longer amount of time compared to the other two metrics. Since redundant nodes are not selected for either clustering or multi-hop communication, over the long term these nodes can be used as the previously selected nodes die.

CPCP maintains high coverage through coverage-aware CH selection and route establishment. Compared to clustering schemes such as HEED, which are based on energy consumption, the balance between coverage and lifetime can be better maintained through coverage-aware schemes. However, CPCP requires precise knowledge of the coverage areas of sensors, which may, in practice, change according to sensor type, node hardware, and even time. Moreover, each node should collect information from its neighbors for CH election. This increases the overhead of the protocol compared to schemes where only local knowledge of the node is used for CH selection.

References

[1] A. A. Abbasi and M. Younis. A survey on clustering algorithms for wireless sensor networks. *Computer Communications*, 30(14–15):2826–2841, October 2007.

[2] I. F. Akyildiz, W. Su, Y. Sankarasubramaniam, and E. Cayirci. Wireless sensor networks: a survey. *Computer Networks*, 38(4):393–422, March 2002.

[3] S. Bandyopadhyay and E. Coyle. An energy-efficient hierarchical clustering algorithm for wireless sensor networks. In *Proceedings of IEEE INFOCOM'03*, volume 3, pp. 1713–1723, San Francisco, CA, USA, April 2003.

[4] S. Bandyopadhyay and E. J. Coyle. Minimizing communication costs in hierarchically clustered networks of wireless sensors. *Computer Networks*, 44(1):1–16, January 2004.

[5] M. Burkhart, P. von Rickenbach, R. Wattenhofer, and A. Zollinger. Does topology control reduce interference? In *Proceedings of ACM MobiHoc'04*, pp. 9–19, Tokyo, Japan, May 2004.

[6] A. Cerpa and D. Estrin. ASCENT: adaptive self-configuring sensor networks topologies. *IEEE Transactions on Mobile Computing*, 3(3):272–285, July 2004.

[7] B. Chen, K. Jamieson, H. Balakrishnan, and R. Morris. SPAN: an energy-efficient coordination algorithm for topology maintenance in ad hoc wireless networks. In *Proceedings of ACM MobiCom'01*, pp. 85–96, Rome, Italy, July 2001.

[8] C. Hsin and M. Liu. Network coverage using low duty-cycled sensors: random and coordinated sleep algorithms. In *Proceedings of ACM IPSN'04*, volume 1, pp. 433–442, Berkeley, CA, USA, April 2004.

[9] S. Jardosh and P. Ranjan. A survey: topology control for wireless sensor networks. In *Proceedings of ICSCN'08*, pp. 422–427, Chennai, India, January 2008.

[10] A. Karnik and A. Kumar. Distributed optimal self-organisation in a class of wireless sensor networks. In *Proceedings of IEEE INFOCOM 2004*, volume 1, pp. 536–547, Hong Kong, China, March 2004.

[11] M. Kubisch, H. Karl, A. Wolisz, L. C. Zhong, and J. Rabaey. Distributed algorithms for transmission power control in wireless sensor networks. In *Proceedings of IEEE WCNC'03*, volume 1, pp. 558–563, New Orleans, USA, March 2003.

[12] P. Kumarawadu, D. J. Dechene, M. Luccini, and A. Sauer. Algorithms for node clustering in wireless sensor networks: a survey. In *Proceedings of ICIAFS 2008*, pp. 295–300, Colombo, Sri Lanka, December 2008.

[13] Mo Li and Baijian Yang. A survey on topology issues in wireless sensor network. In *Proceedings of ICWN'06*, pp. 503–509, Las Vegas, NV, USA, June 2006.

[14] N. Li, J. C. Hou, and L. Sha. Design and analysis of an MST-based topology control algorithm. *IEEE Transactions on Wireless Communications*, 4(3):1195–1206, May 2005.

[15] X. Li, Y. Mao, and Y. Liang. A survey on topology control in wireless sensor networks. In *Proceedings of ICARCV'08*, pp. 251–255, Hanoi, Vietnam, December 2008.

[16] Y. Liu, Q. Zhang, and L. Ni. Opportunity-based topology control in wireless sensor networks. In *Proceedings of the International Conference on Distributed Computing Systems*, pp. 421–428, Santa Fe, NM, USA, June 2008.

[17] P. Santi. Topology control in wireless ad hoc and sensor networks. *ACM Computing Surveys*, 37(2):164–194, June 2005.

[18] C. Schurgers, V. Tsiatsis, and M. B. Srivastava. STEM: topology management for energy efficient sensor networks. In *Proceedings of the IEEE Aerospace Conference*, volume 3, pp. 1099–1108, Big Sky, MT, USA, 2002.

[19] S. Shakkottai, R. Srikant, and N. Shroff. Unreliable sensor grids: coverage, connectivity, and diameter. In *Proceedings of IEEE INFOCOM 2003*, volume 2, pp. 1073–1083, San Francisco, USA, April 2003.

[20] S. Shakkottai, R. Srikant, and N. Shroff. Unreliable sensor grids: coverage, connectivity and diameter. *Ad Hoc Networks*, 3(6):702–716, November 2005.

[21] S. Soro and W. Heinzelman. Cluster head election techniques for coverage preservation in wireless sensor networks. *Ad Hoc Networks*, 7:955–972, 2009.

[22] Y. Xu, J. Heidemann, and D. Estrin. Geography-informed energy conservation for ad hoc routing. In *Proceedings of the 7th Annual ACM/IEEE International Conference on Mobile Computing and Networking (MobiCom'01)*, Rome, Italy, July 2001.

[23] F. Ye, G. Zhong, J. Cheng, S. Lu, and L. Zhang. PEAS: a robust energy conserving protocol for long-lived sensor networks. In *Proceedings of IEEE ICDCS'03*, Los Alamitos, CA, USA, 2003.

[24] O. Younis and S. Fahmy. Distributed clustering in ad-hoc sensor networks: a hybrid, energy-efficient approach. In *Proceedings of IEEE INFOCOM 2004*, Hong Kong, China, March 2004.

[25] O. Younis and S. Fahmy. Distributed clustering in ad-hoc sensor networks: a hybrid, energy-efficient approach. *IEEE Transactions on Mobile Computing*, 3(4):366–379, 2004.

[26] Y. Zou and K. Chakrabarty. Sensor deployment and target localization based on virtual forces. In *Proceedings of IEEE INFOCOM 2003*, pp. 1293–1303, San Francisco, USA, March 2003.

14

Wireless Sensor and Actor Networks

WSNs provide extensive information from the physical world through distributed sensing solutions. Generally, this information is processed at the sink node(s) for the detection of events that occur in the physical world. In this respect, WSNs result in *one-way* information delivery, where information from the physical world is provided to the "cyber world." With the emergence of low-cost actuators and robots that can *affect* the environment, a *two-way* information exchange is possible. As a result, information that is sensed from the environment can be utilized to *act* on the environment. This led to the emergence of distributed wireless sensor and actor networks (WSANs) that are capable of observing the physical world, processing the data, making decisions based on the observations, and performing appropriate actions. A WSAN can be an integral part of systems such as battlefield surveillance and microclimate control in buildings, nuclear, biological, and chemical attack detection, home automation and environmental monitoring [5, 16].

An important example of the application areas of WSANs is fire detection. Distributed sensors can detect the origin and intensity of fire and relay this information to water sprinklers, i.e., actors, to control the fire. Accordingly, the fire can easily be extinguished before it becomes uncontrollable. Through an autonomous architecture of wirelessly connected sensors and actors, this control mechanism can be realized without any human intervention to avoid delays and errors. Similarly, motion and light sensors in a room can detect the presence of people in *smart space* applications. According to the identity and location of the user, the sensors can command the appropriate actors to execute actions based on the pre-specified user preferences such as controlling the electric power in a room or switching lights on or off.

WSANs consists of two classes of components: sensors and actors. The phenomena of sensing and acting are performed by sensor and actor nodes, respectively. These two classes of components have the following distinct characteristics:

- **Sensors:** Sensors are low-cost, low-power devices with limited sensing, computation, and wireless communication capabilities. A sensor node may consist of multiple sensors and observe different physical phenomena through these sensors. Using on-board MCUs, local processing of the data is possible. The processed data are further sent to neighboring sensors, appropriate actors, or the sink through wireless transceivers.

- **Actors:** Actors are resource-rich nodes equipped with higher processing capabilities, higher transmission power, and potentially longer battery life. In addition, actors may be mobile, which improves the effective areas in which they can act. Similar to sensor nodes, an actor node may be embedded with different actuators to perform different tasks. Using their communication and computation capabilities, multiple actors can coordinate to decide on appropriate actions based on information received from multiple sensors.

Wireless Sensor Networks   Ian F. Akyildiz and Mehmet Can Vuran
© 2010 John Wiley & Sons, Ltd

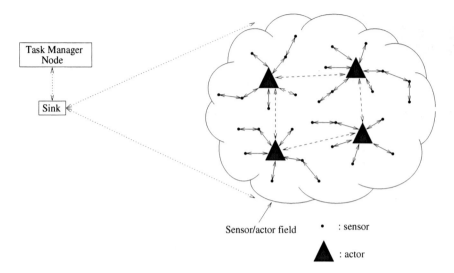

Figure 14.1 The architecture of WSANs including sensors, actors, and a sink.

The typical architecture of WSANs is shown in Figure 14.1, where multiple sensor nodes are deployed in an area to observe a physical phenomenon. The gathered information is relayed to actor nodes that are also deployed in the sensing area. In addition, a sink node can be involved in monitoring, decision making, and task assignment. Based on the different capabilities of sensors and actors, the number of sensor nodes deployed in a target area may be very high. On the other hand, it is usually not necessary for actor nodes to be deployed in large numbers. Since actors generally have higher capabilities, their coverage may be significantly larger than a sensor. Moreover, actors can be mobile robots, which can move around the sensor–actor field based on the information gathered by the sensors. Such a heterogeneous architecture introduces several challenges for the operation of WSANs. Accordingly, WSANs have the following unique characteristics:

- **Heterogeneity:** WSANs consist of heterogeneous components including low-end sensor nodes and more powerful actor nodes. As a result, the communication, computation, and memory capabilities of each component in the network vary significantly. Hence, efficient task allocation schemes that exploit this heterogeneity are required. Moreover, various types of traffic coexist in WSANs, such as sensing information from a large number of sensors to actors, coordination of information between actors, and network maintenance messages from actors to sensors. Accordingly, heterogeneous sets of network protocols are required.

- **Real-time requirement:** WSANs are essentially closed-loop systems, where decisions are made according to the input from the sensors. In many applications where a timely response by actors to the input from sensors is important, real-time guarantees should be provided by the network protocols. As an example, in the fire monitoring application, if a fire is detected, the sensors should relay this information to the appropriate sprinklers to immediately put out the fire. Moreover, the collected and delivered sensor data must still be valid at the time of acting. In the case of the smart space application, if sensors detect a person inside a room, the appropriate actions such as turning on the lights should be performed while the person is still in the room. Therefore, the issue of real-time communication is essential in WSANs.

- **Coordination:** Unlike WSNs where a central entity, i.e., the sink, performs the functions of data collection and coordination, in WSANs, new networking phenomena called *sensor–actor* and *actor–actor coordination* may be required. Since multiple actors that may act on the same phenomenon exist, sensor–actor coordination enables the transmission of event features from sensors to the most appropriate actors. After receiving event information, actors need to coordinate with each other, i.e., actor–actor coordination, to make decisions on the most appropriate way to perform the action.

As we have seen so far, several protocols and algorithms have been proposed for WSNs. However, since the above requirements impose stricter constraints, they may not be well suited for the unique features and application requirements of WSANs. To this end, several solutions that consider the heterogeneity of WSANs have been developed. In this chapter, we provide an overview of the WSAN architecture and the coordination problems. Moreover, key issues in the development of a protocol stack for WSANs are discussed. More specifically, in Section 14.1, we present the architecture of WSANs. We explain the challenges and the solutions to the sensor–actor and actor–actor coordination problems in Section 14.2 and Section 14.3, respectively. In Section 14.4, we investigate the protocol stack of nodes and corresponding challenges for both sensor–actor and actor–actor communications.

14.1 Characteristics of WSANs

WSANs consist of sensor and actor nodes which collect data from the environment and perform appropriate actions based on these collected data, respectively. The architecture of a WSAN is shown in Figure 14.1, where the sensor and actor nodes are distributed in the *sensor/actor field* and the network is monitored through the sink. A user can interface with the sink and hence the network through the task manager, which can communicate with the sink.

14.1.1 Network Architecture

WSANs can be organized into two main architectures. When a sensor detects a phenomenon, either it can send this information to the actor nodes in its surroundings, or the data from all the sensors can be collected at the sink. These two types of architectures are shown in Figure 14.2:

- **Automated architecture:** This type of architecture is shown in Figure 14.2(a), where sensors send their observations to appropriate actors. The actors may coordinate among each other to decide on the appropriate action and perform task assignment. Due to the non-existence of a central controller, e.g., sink or human interaction, this architecture is called automated. In this case, the observations are distributed among actors and they need to coordinate to make decisions.

- **Semi-automated architecture:** In this case, the sink, i.e., central controller, collects data and coordinates the acting process. Sensors detecting a phenomenon route data back to the sink, which may issue action commands to actors.

Depending on the type of application, one of these architectures may be used. The advantage of the semi-automated architecture is that it is similar to the architecture already used in WSN applications. Thus, most of the existing communication solutions can be adopted. However, this architecture results in an increased delay in actions because of the two-way communication between sensors to the sink and the sink to the actors. Moreover, the semi-automated architecture is prone to single points of failure, where the failure of the sink affects all the tasks in the network.

On the other hand, the automated architecture has the following advantages:

1. *Lower latency:* Actors are generally located in the sensor/actor field or nearby. As a result, communication of the observed information from sensors to actors has a much smaller latency compared to sending this information to the sink.

2. *Longer network lifetime:* Information delivery in the semi-automated architecture results in higher load for the sensors located close to the sink. This increases the load on those sensors

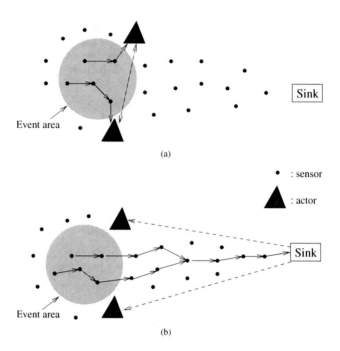

Figure 14.2 (a) Automated and (b) semi-automated architecture.

and introduces additional congestion in the network. If these nodes fail, the network becomes disconnected from the sink. Although data aggregation techniques decrease the probability of these occurrences, sensor nodes around the sink are still more likely to fail than other nodes in the network.

On the contrary, the automated architecture results in localized communication, where the sensor observations are delivered to the actors nearby. Similar to the semi-automated architecture, the nodes within one hop from the actors may still have a higher load of relaying packets. However, this load is generally smaller since multiple actors share the load from an event area. Furthermore, it is more likely that each different event is associated to different actors and, hence, distributes the load. As a result, the automated architecture will have a longer lifetime than the semi-automated architecture.

3. *Lower network resource consumption:* The semi-automated architecture requires network-wide support by the sensors on the path from the event area to the sink. More specifically, sensors that are far from the event area function as relaying nodes. This causes these power-constrained nodes to consume energy and bandwidth in forwarding the information.

On the other hand, event information is transmitted locally through sensor nodes to nearby actors in the automated architecture as shown in Figure 14.2(a). Therefore, the number of sensors involved in the communication process in the automated architecture is smaller than that in the semi-automated architecture. Accordingly, the network load is distributed and is based on the locations of the events. This results in network resource (i.e., energy, bandwidth, etc.) savings in WSANs.

In the remainder of this chapter, we focus on the automated architecture for WSANs because of these advantages and the corresponding additional challenges.

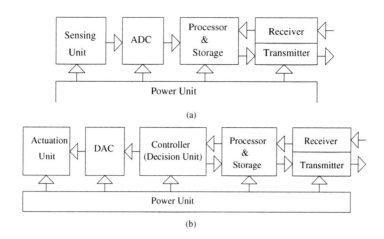

Figure 14.3   (a) The components of (a) sensors and (b) actors.

14.1.2   Physical Architecture

Similar to the physical architecture of the sensor node presented in Chapter 3, sensor and actor nodes also consist of similar components. The main components of sensor and actor nodes are shown in Figure 14.3(a) and Figure 14.3(b), respectively.

As discussed in Chapter 3, sensor nodes are equipped with a power unit, communication subsystems (receiver and transmitter), storage and processing resources, an analog-to-digital converter (ADC) and a sensing unit, as shown in Figure 14.3(a). The sensing unit consists of one or more sensors, through which information regarding a physical phenomenon is collected. The collected analog information is converted to digital data by the ADC, processed by the MCU, and transmitted to nearby sensors or actors.

As shown in Figure 14.3(b), actor nodes are generally composed of similar components as the sensor nodes. However, each component has usually a higher capacity. More specifically, a more powerful MCU, a higher capacity memory, or larger battery power can be integrated. In addition, more powerful transceivers such as WLAN cards can also be included to provide longer communication range between actors and the sink.

In addition to a transceiver and an MCU, an actor node also consists of a decision unit (controller) and an actuation unit. The decision unit (controller) functions as an entity that takes sensor readings collected from nearby sensors as input and generates action commands as output. These action commands are then converted to analog signals by the digital-to-analog converter (DAC) and are transformed into actions via the actuation unit as shown in Figure 14.3(b). The actuation unit may include several actuators based on the application requirements.

Since a distributed operation is employed, a single actor may not have complete information about an event. Instead, sensor readings are generally distributed among multiple actors that reside in the event area. In this case, actor nodes communicate instead of initiating an action alone. The messages that will be exchanged between the actor nodes are constructed by the controller unit and sent through the transceiver. The type of message depends on coordination of the actor nodes.

Sensing and actuation capabilities may not be distributed among different nodes in a WSAN. Instead, integrated sensor/actor nodes that are embedded with both sensors and actuators may replace actor nodes. In this case, the integrated sensor/actor node has a sensing unit and ADC in addition to all the components of an actor node shown in Figure 14.3(b).

(a)   (b)

(c)   (d)

Figure 14.4 Examples of robots: (a) aerial mapping helicopter, (b) robotic mule, (c) sub-kilogram intelligent tele-robots (SKITs), and (d) mini-robot.

A typical example of an integrated sensor/actor node is a robot which consists of an MCU, a transceiver, and on-board sensors and actuators including a mobility unit. Through a network of multiple robots and stationary sensors in the field, the sensing and action capabilities of a single robot can be significantly improved. More specifically, stationary sensors transmit their readings to nearby robots, which process all sensor readings including their own sensor data. In addition, multiple robots can collaborate to improve the reliability of their knowledge of the overall event. Accordingly, the decision unit takes appropriate decisions and the actuation unit performs actions as in an actor node.

Depending on the application, many different actors can be used in a WSAN. Several examples of robots designed by several *robotics research laboratories* and used in distributed WSANs are shown in Figure 14.4. The low-flying helicopter platform shown in Figure 14.4(a) provides ground mapping and air-to-ground cooperation of autonomous robotic vehicles [17]. This platform can be complemented with several additional actuation functionalities such as water sprinkling or disposal of a gas. An example of a *robotic mule*, called an autonomous battlefield robot and designed for the US Army, is shown in Figure 14.4(b). In the USA, there are several autonomous battlefield robot projects sponsored by the Space and Naval Warfare Systems Command [9] and Defense Advanced Research Projects Agency (DARPA) [1]. These developed battlefield robots can detect and mark mines, carry weapons, function as tanks, or perhaps in the future totally replace soldiers on the battlefield. Moreover, the SKITs shown in Figure 14.4(c) are networked tele-robots that have a radio turret which enables communication over UHF frequencies at 4800 kbps[3]. Coordination of these robots is performed through their wireless communication capabilities and a group of SKITs can perform the tasks determined by the application. Finally, possibly the world's smallest autonomous untethered robot (1/4 cubic inch and weighing less than an ounce) has been developed in *Sandia National Laboratories* [2] as shown in Figure 14.4(d).

14.2   Sensor–Actor Coordination

Due to the distributed nature of WSAN communication, effective coordination between the sensors and actors is required. The most important requirement of sensor–actor coordination is to provide low communication delay due to the proximity of sensors and actors. Accordingly, the following three main challenges arise for this coordination:

1. *Requirements of sensor–actor communication:* One of the main requirements of sensor–actor communication is energy efficiency as in WSNs. Moreover, in most WSAN applications, the communication traffic is typically delay sensitive. Therefore, efficient solutions that support real-time traffic for sensor–actor communication are required. An additional requirement for communication in WSANs is the need to ensure the ordering of event data reported to the actors. Consider a multi-hop WSAN, where multiple sensors report different events to a group of actors in overlapping regions. To preserve action integrity, the information related to each event must be delivered to the actors in the order in which it was detected. This is essential to ensure the correctness of the actions. It is referred to as the *ordered delivery* of information collected by the sensors [5].

2. *Actor selection:* As shown in Figure 14.1, in WSANs multiple actors can be in the vicinity of the event area and sensors can use any of these actors to send their information. Therefore, actor selection as a destination for sensor–actor communication is an important challenge. This can be addressed by letting every sensor node decide independently to which actor it will send its readings. However, as a result of these decisions, too many and unnecessary actors can be activated, which hampers the energy efficiency of the network. Instead, sensors detecting an event should coordinate to associate each sensor with the best actor.

3. *Communication technique:* Given the communication requirements and the selected actors, the best communication technique in terms of reliability, delay, and energy should be chosen for the heterogeneous WSAN architecture.

In this section, we focus on the first two challenges of sensor–actor communication in WSANs and defer the last challenge to Section 14.4.

14.2.1   Requirements of Sensor–Actor Communication

Sensor–actor communication provides the necessary tools for the sensors and actors to coordinate to act on a specific event. Hence, energy efficiency, real-time communication, and ordered delivery are important for communication between sensors and actors. Moreover, in a WSAN, if there are multiple sensors reporting an event, the information from different sensors should arrive at the concerned actors approximately at the same time.

Synchronization of event execution can be performed through actor–actor coordination as we describe in Section 14.3. However, it is also conceivable that the sensors can enable this synchronization. In some applications, where the event takes place at different locations, it might also be necessary that the events are passed on to the set of actors not necessarily close to or within the event area when the event was detected, but to the closest set of actors to the event when it is reported to them. In such cases, the sensors must be able to *track* the event and use this information to determine the set of actors to send the information.

Accordingly, the main requirements for sensor–actor communication for WSANs are as follows:

- **Real-time bounds:** Protocols should provide real-time services with given delay bounds according to the application constraints.
- **Energy efficiency:** WSAN protocols should ensure energy-efficient communication among sensors and actors.

- **Event ordering:** The different events should be appropriately ordered as they are reported to the actors.
- **Event synchronization:** Communication protocols should also provide synchronization among different sensors reporting the same event to multiple actors or the same actor. This ensures a coherent response to events in the entire event area.
- **Actor selection support:** The communication protocols should provide the capabilities to track and report the sensed phenomena to a different set of actors not necessarily based on proximity or energy limitations for the case when the events take place at different locations.

Based on these main requirements for sensor–actor communication, the second main challenge is the distributed selection of actors for information delivery. We discuss such actor selection and the corresponding challenges next.

14.2.2   Actor Selection

The actor selection problem is a major challenge in sensor–actor communication due to the dynamic and distributed nature of information delivery in WSANs. The actions and the actors that perform them depend on the events that are sensed by the sensors. As a result, for each event, the appropriate actor group may be different. This requires *event-based* actor selection. Furthermore, an event may occur in a large area such that no single sensor can be informed about all the features. Consequently, collaborative actor selection is required to distributively find the appropriate set of actors.

In general, for the sources/destinations involved in sensor–actor communication, there are four alternatives for actor selection:

- **Minimum actor set:** A minimal set of actors to cover the event region.
- **Minimum sensor set:** The minimum number of sensors to report the sensed event.
- **Minimum actor and sensor set:** Both cases above.
- **Area-based set:** The entire set of actors and sensors in the vicinity of the region.

The first three classifications are referred to as the *redundancy elimination* problem in WSANs [5]. This can be tackled by minimizing the average power consumption of all the sensors and actors that are located in the vicinity of the event, as illustrated in the following example.

■ **EXAMPLE 14.1**

As shown in Figure 14.5, if the minimal set of actors to cover the event area is 9 and if there are 20 actors in that region, then the remaining 11 actors need not act on the environment. In the same example, it might also be desired that only the minimal set of sensors sense and report the environment. As a result, sensors and actors can be chosen such that a minimal set is selected to cover the event.

In addition to the minimum cover set selection, a stricter requirement in some applications might be that the regions covered by different actors are not only a minimal set but also *mutually exclusive*. For example, if there is an application where the sensors report the amount of moisture in the ground and the actors have to irrigate the area uniformly, then the actors should not only cover the entire region but also make sure that the acting regions do not overlap.

The actors can also be selected according to transmission type. More specifically, single hop or multi-hop routes can be created between the sensors and the actor. In WSNs, single hop communication is generally inefficient because of the long distance between the sensors and the sink. However, in WSANs, single hop communication may be preferred since actors are located close to the sensors as discussed in Section 14.1. In fact, the location of the actor determines the effectiveness of the single hop communication.

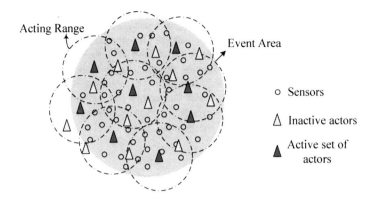

Figure 14.5   Redundancy elimination: minimal set of actors.

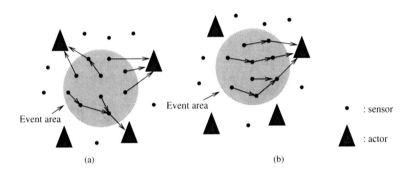

Figure 14.6   (a) Multi-actor (MA) and (b) single actor (SA) selection.

For example, if the event area is small and there is an actor in the middle of it, the nodes located farther away from the actor may still reach the actor without large energy consumption. However, in cases where the event area is large or the actors are at the edge or outside of it, multi-hop communication may be more efficient than single hop communication. Similar to traditional WSNs, the long distance between the actor and the sensor nodes increases the energy consumption if single hop communication is used. Therefore, the type of transmission depends on the deployment and location of actor nodes as well as what sensors are associated with which actors.

Actor selection can be classified into two cases according to the number of actors selected as shown in Figure 14.6:

- **Single actor (SA) selection:** In this case, only one actor receives event features from all the sensors that sense the event, as shown in Figure 14.6(b). One of the main challenges for SA selection is to determine the best SA node to which sensors will send their readings. Selecting an actor node may be based on certain criteria such that (1) the distance between the event area and the actor is minimized so that delays and energy consumption are minimized; (2) the minimum energy path from sensors to the actor is selected; or (3) the actor with the most appropriate action range is selected so that appropriate actions are performed on the event area.

- **Multi-actor (MA) selection:** As shown in Figure 14.6(a), multiple actors can also be selected to receive the information from sensors about the sensed phenomenon. For MA selection, every sensor node can independently decide to which actor it will send its readings. However, a lack of coordination between the sensors can result in too many and unnecessary actors being activated and as a result the total energy consumption of sensor–actor communication can increase. To avoid this situation, sensors should coordinate with each other to form clusters. For each cluster, a single actor should collect the data. These clusters may be formed in such a way that (1) the event transmission time from sensors to actors is minimized; (2) the events from sensors to actors are transmitted through the minimum energy paths; or (3) the action regions of the actors cover the entire event area.

SA selection can be considered as a special case of MA selection. However, for the SA case, there is no guarantee that the action range of the selected actor covers the entire event area. Therefore, instead of considering distance, energy, or timing issues, sensors should try to find the "best" actor for that event, i.e., the actor which has enough action coverage, energy, and capability to perform the required action on the event area. In this situation, the actor which receives the event information will be able to perform the required action itself without coordinating with other actors.

As a result of SA selection, the actor can immediately perform the action if it has a wide action range and sufficient energy. Accordingly, the overall action response latency between sensing and acting can be minimized with SA. However, in cases where one actor is not sufficient for the required action, coordination between additional actors is required. If an actor is not well suited to perform the required action alone because of coverage or energy constraints, it coordinates with other actors through an *announcement message* as will be explained in Section 14.3. Based on feedback from other actors, one or more other actors can be selected to perform the appropriate actions.

Compared to SA selection, MA selection provides more capabilities to actors in terms of determining the event location. The intensity of events may not be uniform inside an event area. As a result, the information received from sensors by each actor may be different. Actors can compare their received samples with each other and determine the areas where the event intensity is high. This can result in more effective actions by moving mobile actors toward the center of the event or selecting a subset of actors that are closer to the event area.

The disadvantage of MA is that actor–actor coordination is mostly based on the *negotiation* (see Section 14.3) among multiple actors unlike the announcement message in SA stated above. Since each actor has partial information regarding the event, actors need to coordinate with each other and exchange event information to take appropriate action decisions. Actor–actor coordination, as will be discussed in Section 14.3, may result in high communication overhead and high latency.

14.2.3   Optimal Solution

The optimal solution to the sensor–actor coordination problem can be analyzed through a centralized view of the network [15]. In this section, a sensor–actor coordination algorithm, which is based on an *event-driven clustering* paradigm, is described. The cluster formation is triggered by an event and the clusters are created *on-the-fly* to optimally react to the event. Accordingly, only the event area is clustered and each cluster consists of those sensor nodes that send their data to the same actor. The resulting architecture is shown in Figure 14.7. Since cluster maintenance can be costly, the event-driven clustering approach eliminates the communication overhead of maintaining clusters when no events occur.

As discussed in Section 14.2.1, sensor–actor communication may have real-time requirements, which should be met through efficient coordination techniques. To characterize these requirements, the following definitions will be used:

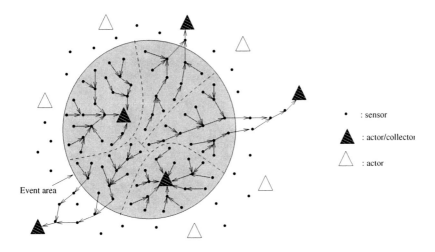

Figure 14.7   Event-driven clustering with multiple actors.

- **Latency bound B:** This is the maximum allowed time between the instant when the physical features of the event are sampled by the sensors and the instant when the actor receives a data packet describing these event features.
- **Expired and reliable packets:** If a data packet is received within the latency bound B, it is said to be *unexpired* and hence *reliable*. On the other hand, if the packet does not meet the latency bound B when it is received by an actor, it is said to be *expired* and hence *unreliable*.
- **Event reliability, r:** This is the ratio of reliable data packets over all the packets generated in a decision interval.
- **Event reliability threshold, r_{th}:** This is the minimum event reliability required by the application.
- **Lack of reliability:** This is the difference $(r_{th} - r)$ between the required event reliability threshold r_{th} and the observed event reliability r at a given time. A negative lack of reliability indicates a reliability above the required threshold and is also referred to as an *excess of reliability*.

According to these definitions, the sensor–actor coordination problem can be stated as follows. What are the best data paths from each sensor residing in the event area to the actors so that (1) the observed reliability r is above the event reliability threshold r_{th} (i.e., $r \geq r_{th}$) and (2) the energy consumption associated with data delivery paths is a minimum? The solution to the sensor–actor coordination problem is referred to as *event-driven clustering with multiple actors*, which can be modeled as an integer linear program (ILP).

Network Model

Let us represent the network of sensors and actors as a graph $\mathcal{G}(\mathcal{S}^{\mathcal{V}}, \mathcal{S}^{\mathcal{E}})$, where $\mathcal{S}^{\mathcal{V}} = \{v_1, v_2, \ldots, v_N\}$ is a finite set of nodes (vertexes) in a finite-dimension terrain, with $N = |\mathcal{S}^{\mathcal{V}}|$, and $\mathcal{S}^{\mathcal{E}}$ is the set of links (edges) among nodes, i.e., $e_{ij} \in \mathcal{S}^{\mathcal{E}}$ iff nodes v_i and v_j (also i and j for simplicity in the following) are within each other's transmission range. Let $\mathcal{S}^{\mathcal{A}}$ represent the set of actors, with $N_A = |\mathcal{S}^{\mathcal{A}}|$. An actor that collects traffic from one or more sources is referred to as a *collector*. Let $\mathcal{S}^{\mathcal{S}}$ be the set of traffic sources, with $N_S = |\mathcal{S}^{\mathcal{S}}|$. This set represents the sensor nodes that detect the event, i.e., the sensors that reside in the event area. Since the set of sources is disjoint from the set of actors, $\mathcal{S}^{\mathcal{A}} \subset \mathcal{S}^{\mathcal{V}}$, $\mathcal{S}^{\mathcal{S}} \subset \mathcal{S}^{\mathcal{V}}$, and $\mathcal{S}^{\mathcal{A}} \cap \mathcal{S}^{\mathcal{S}} = \emptyset$. Accordingly, the source–destination connections are represented as $\mathcal{P} = \{(s, a) : s \in \mathcal{S}, a \in \mathcal{A}\}$. For the solution of the sensor–actor connectivity problem, the traffic matrix can be omitted from the model when all sources generate information at the same data rate. Accordingly, the energy consumption can be modeled as follows based on the models represented in Chapter 3.

The energy consumption per bit at the physical layer is

$$E = E_{elec}^{trans} + \beta d^{\alpha} + E_{elec}^{rec} \tag{14.1}$$

where the first term, E_{elec}^{trans}, is the *distance-independent* term that takes into account the overheads of transmitter electronics (PLLs, VCOs, bias currents, etc.) and digital processing. The second term in (14.1), βd^{α}, is the *distance-dependent* term that accounts for the radiated power necessary to transmit 1 bit over a distance d between source and destination. At the receiver, the third term, E_{elec}^{rec}, models the *distance-independent* term that takes into account the overhead of receiver electronics.

Integer Linear Program

The objective of the optimization problem is to find *data aggregation trees* (da-trees) from all the sensors that reside in the event area (referred to as sources) to the appropriate actors. A da-tree is compiled by aggregating individual *flows*, where a flow is defined as a connection between a sensor and an actor.

The solution to the sensor–actor coordination problem can be divided into two: (1) *actor selection*, i.e., selection of the optimal subset of actors to which sensor readings will be transmitted; and (2) *da-tree construction*, i.e., construction of the minimum energy da-trees toward those selected actors that meet the required event reliability constraint. As shown in Figure 14.7, each resulting da-tree defines a cluster, which is constituted by all source nodes in the tree.

The optimal strategy for event-driven clustering can be found using an *integer linear program* (ILP) formalization [4]. For this formalization, the network topology is assumed to be *1-connected*, i.e., at least one path exists between each sensor and actor. Accordingly, the ILP aims to minimize the total cost, C^{TOT}, which is defined as

$$C^{TOT} = \sum_{k \in \mathcal{S}^{\mathcal{A}}} \sum_{(i,j) \in \mathcal{S}^{\mathcal{E}}} x_{ij}^{k} \cdot c_{ij} + \gamma \cdot Q. \tag{14.2}$$

The first term in (14.2) is the total energy consumption at each path from the sources to the actors, where x_{ij}^{k} equals 1 if the link (i, j) is a part of the da-tree associated with actor k and c_{ij} is the energy consumption for communication between nodes i and j. The second term in (14.2) is the penalty for using source nodes that cannot transmit their information within the latency bound B, where Q is the number of non-compliant nodes and *gamma* is the penalty coefficient. Accordingly, the optimum clusters and da-tree paths are found with minimum energy consumption and number of non-compliant nodes.

The ILP problem is solved according to several constraints [15], which ensure that a sensor node is connected to only one actor node, aggregation is performed in the da-tree, and non-compliant sources are selected according to the latency bound, B. However, this problem is NP-complete and can only be solved for networks of moderate size (up to 100 nodes) and requires global knowledge of the network. While important for the analysis of WSANs, for operational purposes a distributed solution of this problem is required. Next, we describe a distributed suboptimal but scalable algorithm: namely, the Distributed Event-driven Clustering and Routing (DECR) protocol.

14.2.4 *Distributed Event-Driven Clustering and Routing (DECR) Protocol*

Similar to the centralized solution, the DECR protocol aims to construct da-trees between sources and actors to provide the required reliability, r_{th}, with minimum energy expenditure. The routes

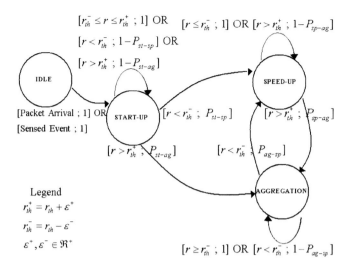

$[r_{th}^- \leq r \leq r_{th}^+ ; 1]$ OR $[r \leq r_{th}^- ; 1]$ OR $[r > r_{th}^+ ; 1-P_{sp-ag}]$

$[r < r_{th}^- ; 1-P_{st-sp}]$ OR

$[r > r_{th}^+ ; 1-P_{st-ag}]$

IDLE

SPEED-UP

[Packet Arrival ; 1] OR   START-UP $[r < r_{th}^- ; P_{st-sp}]$ $[r > r_{th}^+ ; P_{sp-ag}]$

[Sensed Event ; 1]

$[r > r_{th}^+ ; P_{st-ag}]$ $[r < r_{th}^- ; P_{ag-sp}]$

Legend

$r_{th}^+ = r_{th} + \varepsilon^+$

$r_{th}^- = r_{th} - \varepsilon^-$

$\varepsilon^+, \varepsilon^- \in \Re^+$ AGGREGATION

$[r \geq r_{th}^- ; 1]$ OR $[r < r_{th}^- ; 1-P_{ag-sp}]$

Figure 14.8   State transition diagram for a sensor node.

are constructed according to the geographical routing solutions that are explained in Chapter 7. The reliability is maintained through periodic advertisements from each actor related to its observed reliability. Accordingly, each individual sensor updates da-tree formation to meet the reliability requirements with minimum energy consumption. The reliability is controlled by modifying the end-to-end path length from a sensor to an actor through transmit power control.

The DECR protocol consists of four different states, namely *idle, start-up, speed-up,* and *aggregation state* as shown in Figure 14.8. Based on these states, a sensor node adjusts the number of hops and the end-to-end delay to a particular actor according to the observed reliability advertised by that actor. This is achieved by probabilistically modifying the behavior of the sensor nodes at the routing layer, i.e., inducing them to select their next hop so as to increase the delay and reduce the energy consumption when the reliability is high, and, vice versa, reducing the delay at the expense of energy consumption when the reliability is low.

The observed reliability is determined by each actor during a decision interval: each actor computes the event reliability r as the ratio of unexpired packets over all generated packets and periodically broadcasts its value similar to the ESRT and RT^2 protocols described in Chapter 8. The sensor nodes which are associated with a particular collector actor base their state transitions on the reliability observed by the collector, which is broadcast at the end of each decision interval. When the advertised value r is below the so-called *low event reliability threshold* r_{th}^-, where $r_{th}^- = r_{th} - \epsilon^-$, i.e., the lack of reliability $(r_{th} - r)$ is above a certain positive margin ϵ^-, then it is necessary to speed up the data delivery process by reducing the end-to-end delay. Conversely, when the advertised value r is above the so-called *high event reliability threshold* r_{th}^+, where $r_{th}^+ = r_{th} + \epsilon^+$, i.e., the excess of reliability $(r - r_{th})$ is above a certain margin ϵ^+, then there is excess reliability that can be traded off against energy savings. The coefficients ϵ^+ and ϵ^- are used to define a tolerance zone around the required reliability threshold for practical purposes (i.e., reduce oscillations that might lead to instability).

Each sensor node starts in the *idle* state, where it samples the environment for occurring events and monitors the channel for incoming data packets. A sensor enters the *start-up* state when it either senses an event or receives the first data packet from a neighboring sensor. The collective operation of the sensor nodes in the start-up state allows timely establishment of paths to an actor for each source that resides in the event area.

The protocol operation depends on the feedback sent from the actors that indicates the observed reliability. If the reliability is below the low reliability threshold ($r < r_{th}^-$), the sensor node in the start-up state enters the *speed-up* state with probability P_{st-sp}, which is a monotonically increasing function of the *lack of reliability*. If the event reliability r is above the high event reliability threshold r_{th}^+ (i.e., $r > r_{th}^+$), it is possible to save energy. Hence, a node in the start-up state enters the *aggregation* state with probability P_{st-ag}, which is a monotonically increasing function of the *excess of reliability*.

Based on the observed reliability, a sensor can alternate between the speed-up and the aggregation state to respond to the collector. As shown in Figure 14.8, a sensor in the speed-up state enters the aggregation state with probability P_{sp-ag} when $r > r_{th}^+$, while a sensor in the aggregation state enters the speed-up state with probability P_{ag-sp} when $r < r_{th}^-$. A sensor goes back to the idle state if it does not generate or receive packets for *idleTimeout* seconds.

Based on this overall protocol operation, each state is described next:

- **Start-up state:** The main notion of the start-up state is to construct efficient routes from the source nodes to the actors. A sensor node i may be in the start-up state either as a source or as a relay. The next hop is selected based on the *two-hop rule*, where node i selects as the next hop among its neighbors the node j that minimizes the sum of the energy consumption from i to j and the energy consumption from j to the actor c_j closest to j. The cost of energy consumption is computed according to the link energy model in (14.1). Note that the latter link may or may not exist.

  The two-hop rule selects as the next hop the node j associated with the minimum two-hop energy consumption. This procedure is repeated iteratively by each sensor along the path from the source to the actor until the path is established. Once a collector actor receives a packet from one of the source nodes, it transmits its identifier on the reverse da-tree in order to inform the source sensors about its own identity.

- **Speed-up state:** In case of low reliability, the initial paths which have been set up at the end of the start-up state should be improved. This is achieved by applying the greedy routing scheme (GRS) [10] forwarding rule, where each node sends the packet to the node closest to the destination within transmission range. While this rule minimizes the number of hops in the path, energy consumption may increase.

- **Aggregation state:** When the reliability is high, the existing paths can be updated such that the overall energy consumption is decreased. This is performed in the aggregation state by relying on data fusion algorithms to aggregate data packets by any node in the network. Each node routes data to the closest node in its neighborhood that is part of the da-tree.

  As previously discussed, after da-trees are established, each sensor knows which collector actor it is associated with. By overhearing transmissions on the shared medium, each sensor learns which of its neighbors are part of a da-tree (as some neighbor sensors may not even be in the event area) and which da-tree (if any) they are part of, i.e., which collector actor they are associated with. Hence, node v_i in the aggregation state first evaluates the cost of transmitting data to those of its neighbors that are part of a da-tree. This way, it can identify a minimum cost neighbor, i.e., the neighbor v_{min} that requires minimum energy consumption to be reached among those associated with one of the da-trees.

The transitions between different states are driven by feedback from actors. Feedback is sent by each actor periodically, with a period equal to Δf seconds. At each decision instant k, the actor feedback is determined using three different reliability measures, namely the reliability $r[k]$, the *short-term reliability* $r_{sh}[k]$, and the *predicted reliability* $\hat{r}[k+1]$. The actor calculates the reliability $r[k]$ observed during the last decision interval, Δd. Similarly, it calculates the short-term reliability $r_{sh}[k]$ as the reliability observed during the last short decision interval, Δd_{sh}, with $\Delta d > \Delta d_{sh}$. Based on the current reliability $r[k]$ and on the history of past measurements $r[k-1]$, $r[k-2]$, ..., the actor also calculates the predicted reliability $\hat{r}[k+1] = f(r[k], r[k-1], r[k-2], \ldots)$. The feedback packet contains the advertised value of reliability $r_{adv}[k]$, calculated on the basis of these three measures, and is actually sent only if the advertised value of reliability is above the high reliability threshold r_{th}^+ or below the low reliability threshold r_{th}^-. If the sensors receive no feedback, they assume that the reliability lies within r_{th}^+ and r_{th}^-.

14.2.5 Performance

The performance of the distributed protocol compared to the optimal solution is particularly important. To this end, we describe the results of simulations that have been performed for particular network scenarios [15].

As discussed in Section 14.2.3, the optimal solution to the sensor–actor coordination can only be found for smaller networks. Hence, firstly, a network of 70 nodes with a circular deployment area of 20 m radius is considered. For each deployed sensor, the distance from the center of the area and the angle are uniformly distributed random variables. The comparison is performed for different configurations, where the DECR protocol operates in one of the states during the simulation. To this end, start-up and speed-up configurations are designed as well as two aggregation configurations, where the aggregation state is achieved following the start-up state or the speed-up state.

A comparison between the energy consumption profile of the optimal solution and that of the different states of the distributed solution is shown in Figure 14.9 as a function of the event range. As event range increases, the number of sources that send information to the actor(s) increases. More specifically, the overall network cost, i.e., the energy needed to transmit 1 bit from each source to the actors, is shown in Figure 14.9. As expected, the optimal solution results in the lowest energy consumption. Moreover, the energy consumption of the optimal solution is almost independent of the event range. This is related to the tradeoff between the number of sensors and the effect of aggregation. As the number of sources increases with event range, overall energy consumption increases due to the activities of the source nodes. On the other hand, as more nodes are involved in communication, better accuracy can be achieved in aggregation, which improves the energy savings.

Compared to the optimal solution, the DECR protocol is dependent on the event range. An aggregation configuration reached from a start-up configuration leads to an almost-optimal energy consumption, while by reaching the aggregation configuration from a speed-up configuration, the energy consumption can still be decreased consistently, but not as much as in the previous case. Since the initial structure of the da-trees depends on the start-up/speed-up configuration, these states result in higher energy consumption. However, the distributed aggregation operation modifies the structure of the da-trees to reach an energy configuration that is between the speed-up and the aggregation from start-up curves in Figure 14.9.

In a second scenario, a network of 1000 sensors and four actors is considered, where the nodes are randomly deployed in a square area of 100 m × 100 m. Figure 14.10 shows the overall energy consumption for the start-up, speed-up, and aggregation configurations. The speed-up configuration outperforms the start-up configuration in terms of number of hops, and this is achieved with a limited additional energy expenditure. This is also reflected in the distribution of packet delays. Figure 14.11 shows a comparison of packet delays from sensor to actor with 400 nodes, between the speed-up and start-up configurations when the event range is set to 12 m, which corresponds to 20 sources on average. In the start-up configuration, delays are shown to be high in the transient phase at the beginning of the simulation. The average delay (thick line in the figure) converges to a value around 0.3 s. In the speed-up configuration, delays are much smaller and their average is below 0.1 s. Figure 14.12(a) and Figure 14.12(b) show the distribution of delays in the same scenario. In the speed-up configuration (Figure 14.12(b)), the delay is below 0.5 s for almost 100% of the packets, while in the start-up

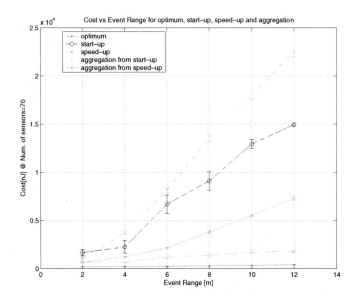

Figure 14.9 Comparison of the optimal solution, speed-up, start-up, and aggregation configurations with 70 nodes.

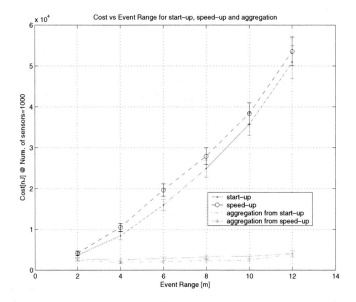

Figure 14.10 Comparison of energy consumption for start-up, speed-up, and aggregation configurations with 1000 sensors.

configuration (Figure 14.12(a)) the variability of delays is much higher and their value can be as high as 2.5 s.

In addition to the performance in various states, the convergence of the DECR protocol is also important. To investigate this, we describe next the simulation results with 100 sensors that are randomly

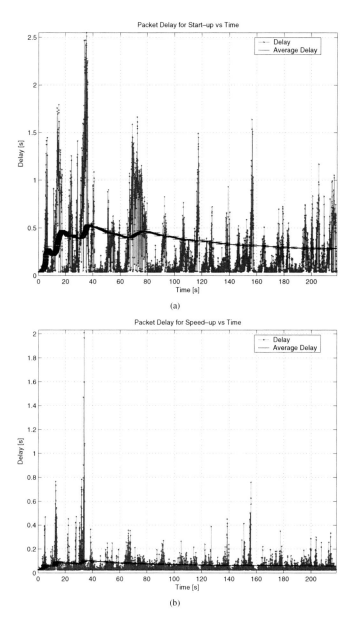

Figure 14.11 Scenario 3: Delays (a–b) for start-up and speed-up configurations, 400 sensors, event range $= 12$ m, sources generating 2 packets/s.

deployed in a 100 m × 100 m terrain [15]. Moreover, the probabilities that govern the transitions among different states are set as follows. The probability P_{st-sp} of moving from the start-up state to the speed-up state is set to 0.5 when the advertised reliability $r < r_{th}^- - (0.1 \cdot r_{th}^-)$ (very low reliability). Otherwise, if the reliability is low but close to the threshold, $P_{st-sp} = 0.1$. The probabilities P_{st-ag} and P_{sp-ag} of moving to the aggregation state from the start-up and speed-up states, respectively, are equally set to 0.05

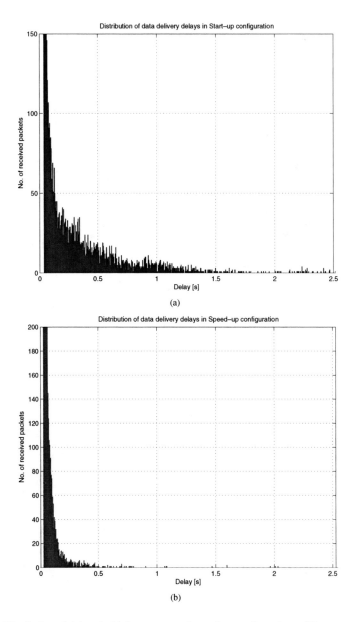

Figure 14.12 Distribution of delays (a–b) for start-up and speed-up configurations, 400 sensors, event range = 12 m, sources generating 2 packets/s.

if the advertised reliability is equal to 1, and 0.02, otherwise. In any case, the probability of switching to the aggregation state needs to be low (less than 0.1), as higher values almost invariably cause instabilities, provoking sudden drops in the observed reliability of the transients. Finally, the transition probability P_{ag-sp} from the aggregation to the speed-up state is set to 0.2 if the current, predicted, and short-term reliabilities are all below the threshold r_{th}^-. Similarly, P_{ag-sp} is set to 0.1 if only the short-term and predicted reliabilities are below the threshold while the current reliability is still above it, while P_{ag-sp} is set to 0.05 if only the short-term reliability is below the threshold while the others are still above it.

The convergence of the DECR protocol is shown in Figure 14.13(a), where the event reliability as observed by the actor is shown. Immediately after the initial phase, when nodes establish data paths to the actor in the start-up state, the reliability drops below the threshold. Hence, the actor advertises low reliability and a higher number of sensors start moving to the speed-up state. As a result, the reliability is increased above the threshold and a small portion of the sensor nodes move to the aggregation state. The network operation converges to a stable state after a few oscillations, where the reliability is within the high and low reliability thresholds. Accordingly, the number of sensors in each of the three active states is shown in Figure 14.13(b), where the transitions between states can be clearly seen.

14.2.6   Challenges for Sensor–Actor Coordination

The following research issues related to actor selection exist for sensor–actor coordination in WSANs:

- In-sequence delivery of different events detected in a region is required for both SA and MA selection to ensure that there are no adverse effects on the target environment. Similarly, synchronization should be ensured in the reporting time of the sensed phenomena between different actors responsible for acting on the event.
- In certain applications where multiple events occur in different locations, it may be necessary for the sensed information to be sent to an actor or to a set of actors determined according to the location of the event.
- In MA, it may sometimes be necessary to address the redundancy in the set of actors to which the sensed information is sent in order to save average energy consumed by the actors in the region. In these cases, it is necessary to send the information only to a subset of actors which cover the entire event region.

14.3   Actor–Actor Coordination

The sensor–actor coordination explained in Section 14.2 results in information related to the events be delivered to the actors. In addition to this communication, in most situations *actor–actor communication* is also required to perform the most appropriate action in WSANs. Actors are resource-rich nodes with high transmission power. Hence, unlike sensor–actor communication, actor–actor communication can be long range. Moreover, a smaller number of (mobile) resource-rich actor nodes are involved in actor–actor communication. Therefore, it is similar to the communication paradigm of ad hoc networks. In this section, we investigate the characteristics and challenges of actor–actor coordination which deals with the actions performed by actors after receiving event information.

Actor–actor communication occurs in the following situations:

- **Task distribution:** In certain cases, the actor that receives the sensor data may not act on the event area due to a small action range or insufficient energy. In this case, the appropriate task should be distributed to nearby actors.
- **Event information exchange:** If multiple actors receive the same event information, only partial information is available at each actor. Hence, these actors should "talk" to each other to gather the overall event information, decide on the appropriate action, and assign the tasks to appropriate actors.
- **Task coordination:** In certain applications, if multiple actors are required to cover the entire event region, it may be necessary to ensure that these regions are non-overlapping or *mutually exclusive* in order to ensure uniform acting behavior over the entire region.
- **Synchronization:** If multiple actors receive information from multiple sensors for the same event, it may be necessary to ensure that these multiple actors act on the environment at the same

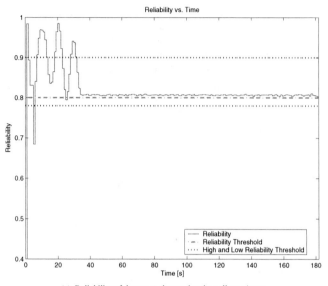

(a) Reliability of the event observed at the collector/actor

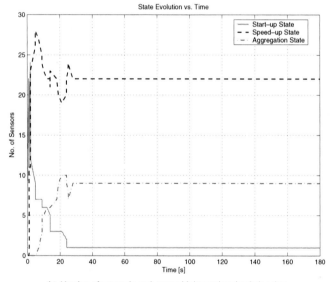

(b) Number of sensors in each state with increasing simulation time

Figure 14.13 Convergence of DECR.

time. This *synchronization* requirement in the execution of the task is required in applications where a partial execution of the task alters the state of the event in the region where it has not been executed.

- **Task execution:** In case multiple events occur simultaneously, task execution can be done via actor–actor communication. Also, it may be desired that the tasks are executed sequentially. This constraint is referred to as *ordered execution* of tasks.

- **Event information relay:** After an actor node receives event information, if the event is spreading to other actors' acting areas, the actor node can relay the sensor data or action command to those actors. In this way, there will be no need for sensors in those areas to send information to the nearby actors as it will be forwarded by the initial set of actors.

All of the above situations indicate the necessity of actor–actor coordination and condense to the following question: "*What actor(s) should execute which action(s)?*"

The answer to this question can be given by exploiting the *coordination* between actor nodes. Actors should, whenever possible, coordinate strongly with each other in order to maximize their overall task performance [11]. Then, the above question can be restated as follows: "*How should MA task assignment be done?*"

The task assignment problems in WSANs can be classified using the following two axes:

- **Single actor task (SAT) vs. multi-actor task (MAT):** SAT refers to the cases where each task requires exactly one actor. On the other hand, MAT refers to the cases where a task requires multiple actors. Accordingly, MAT assignment problems involve tasks that require the combined effort of multiple actors.
- **Centralized decision (CD) vs. distributed decision (DD):** In WSANs there is a need to take a decision on the action to be performed according to the event. The decision can be performed in a centralized way (called CD) or in a distributed way (DD). DD allows neighboring actors to coordinate locally, which provides timely actions and network size-independent coordination. On the other hand, CD provides action decisions to be taken in an organized way since the decision is taken only at one actor node which may be equipped with more powerful communication facilities.

In the following, we describe the task assignment problem as well as its optimal and distributed solutions. Further challenges related to actor–actor coordination are also discussed.

14.3.1   Task Assignment

Event information delivery and task execution are two distinct operations in WSANs that involve sensors and actors. While information delivery is performed by sensors *to* the actors, task execution is performed *by* the actors. However, the set of actors that receives event information and the set of actors needed to perform the task associated with the event may be different. This requires actor–actor communication to find the best set of actors and the task assignments for each actor. Accordingly, task assignment can be accomplished in different ways depending on (1) the set of actors that receives event information, i.e., SA vs. MA, (2) the set of actors that is required to perform the task, i.e., SAT or MAT, and (3) the decision methodology, i.e., DD or CD.

MAT

For MATs, the coordination depends on the set of actors that receives the event information and the decision mechanism. If multiple actors receive event information from sensors and DD is performed, the actors *negotiate* with each other and coordinate locally to select the "best" actors for the task. The "best" actor generally refers to the one which (1) is close to the event area, (2) has high capability and residual energy, or (3) has small action completion time.

On the other hand, in the CD case, the actors directly transmit the specifications of the event such as location, intensity, etc., to the predetermined actor node which functions as the decision center. This decision center, which already has information about the actors in the network, selects the "best" actors for that task and triggers them to initiate the action.

The selected actors (through both DD and CD) may not be the same set of actors that receives sensor data via sensor–actor coordination. As discussed earlier, this is based on the fact that the actors that receive the event information may not be the "best" actors for that task. For example, they may not be close enough to the event area, or they may not be capable of performing the required task.

In case only one actor receives event information at the end of the sensor–actor coordination phase, there is still a need for coordination among actors in order to determine the actors that will act on the event area. However, in this case, since all sensor data are collected at one actor, this actor can function as the central decision unit. It then broadcasts an *announcement message* to other actors, which contains details about the event and the task. Based on the feedback from other actors, it selects the "best" actors and assigns the action task to them.

Following task assignment, each selected actor performs the associated actions within its action range. Note that, depending on the characteristics of actors or the task assignment rules, if an actor is chosen, it either acts on its action range or may act selectively on part of its action area. However, in order to react to a certain event, the union of the action ranges of the selected actors should cover the entire *event area*.

Another challenge for distributed task assignment is the relation between the area in which the action should be performed and the action range of the actors. Depending on the actuator capabilities, certain portions of the event area may require more than one actor to perform an action. This depends on the event intensity and the capabilities of actors in that area. On the other hand, if the action range of the actors is much larger than the task specifications, the actions may be performed outside the event area. Depending on the application, this may lead to unfavorable results as well as unnecessary consumption of actor resources. Similarly, if the action ranges of several actors overlap with each other and all actors act at the same time, the resulting action may surpass the requirements of the task. Thus, while assigning tasks to actors, considerations of the action coverage must also be taken into account.

Overall, regardless of the number of actors receiving sensor data, the objective of MAT is to select the "best" actors while meeting the action coverage requirements of the corresponding application and the event. The type of actor selection (i.e., MA or SA) in the sensor–actor coordination phase only affects the coordination mechanism selecting the "best" actors.

SAT

Similar to MAT, the main objective of the SAT assignment problem is to select the "best" actor. This actor is selected so that the resulting action range covers exactly or most of the entire event area for the action task. In fact, if the type of actor selection is MA, the coordination for SAT can be considered as a special case of MAT, where actors that receive the sensor data coordinate either in DD or CD and select the "best" actor.

On the other hand, if a SA is informed about the event features, the major challenge is selection of the best actor. More specifically, the SA can make a decision in an isolated fashion and thus initiates the action by itself. Instead, the actor may also communicate with other actors or the decision center to find the best actor. Intuitively, if sensor–actor communication takes a long time and the application is delay intolerant (the delay bound of received sensor data is low), then as long as the actor can provide the minimum requirements of the task (e.g., it should be able to act on the whole event area and to have enough energy), initiating an action immediately is reasonable in order to perform the action on time. This way, perhaps the action is not performed by the "best" actor; however, it is guaranteed that the action is completed in a timely manner. On the other hand, if the delay bound of the data is not very low or the actor does not provide the minimum requirements of the task, it should not immediately start to perform the action by itself; instead, in CD it should communicate with the decision center and allow it to choose the appropriate actor, or in DD it should broadcast an announcement message, as explained above, to inform the other actors about the task and then should select the "best" one according to their responses.

14.3.2 Optimal Solution

The optimal solution to the actor–actor coordination problem is based on the objective of optimally allocating tasks to the different actors to collaboratively achieve a global goal. More specifically, the actor–actor coordination problem can be defined as a task assignment problem and formulated as a *mixed integer nonlinear program (MINLP)* [7]. The objective of the actor–actor coordination is to select the best actor(s) to perform the appropriate action in an event area. To this end, the following definitions will be used:

- **Action area:** An event area is identified according to the information sent by the sensor nodes. This information defines the *action area*, i.e., the area where the actors should act.

- **Action range:** The action range defines the circular area where an actor is able to act. In particular, each collector receives data from a subset of the sources. However, the collector may not be able to act on the entire action area, i.e., this area may not be totally within the collector's action range.

- **Action completion time:** This is the time it takes for an actor to complete an action. In addition to the action range, action completion time is also important for satisfying real-time bounds related to a particular event–action pair.

- **Action completion bound:** This is the maximum allowed latency between the time when the event is sensed and the time the action is completed.

In addition, the collector may not be the "best" actor for the task in terms of action completion time and/or energy consumption. For these reasons, actor–actor coordination is required before initiating the action. The particular challenges in this problem are illustrated in Example 14.2.

■ **EXAMPLE 14.2**

A topology with several actors and their action ranges are shown in Figure 14.14. It can be observed that multiple actors can act on a certain area, which is referred to as an *overlapping area*. The overlapping areas are numbered from 1 to 8 in Figure 14.14. In an overlapping area, the actor–actor coordination problem consists of selecting a subset of the actors and their action powers to optimally divide the action workload, so as to maximize the *residual energy* of the actors while respecting the *action completion bound*. Moreover, certain areas are covered by only a single actor, which are referred to as *non-overlapping areas*. The non-overlapping areas are shown as unshaded regions in Figure 14.14. For such an area, the coordination problem simplifies to selecting the power level for the actor that minimizes energy consumption while respecting the action completion bound.

Based on these definitions, we describe the action area and the actor models next and represent the formulation of the actor–actor coordination problem as a mixed integer nonlinear program (MINLP) in Section 14.2.

Action Area and Actor Model

Let $\mathcal{S}^A$ be the set of actors, with $N_A = |\mathcal{S}^A|$, and let $\mathcal{S}^C$ be the set of collectors ($\mathcal{S}^C \subseteq \mathcal{S}^A$). An actor is identified by $a \in \mathcal{S}^A$ with coordinates $C_a^A = (x_a^A, y_a^A)$ and action range R_a, which defines the circular area A_a^A where it can act. Accordingly, A_c^C is the cluster area that is under the responsibility of *collector* $c \in \mathcal{S}^C$, and the whole event area is thus $\bigcup_{c=1}^{|\mathcal{S}^C|} A_c^C$. $A_{c,nov}^{C,h}$ and $A_{c,ov}^{C,m}$ are the hth *non-overlapping* and the mth *overlapping* areas, respectively, inside the portion of area under the responsibility of collector c. H_c represents the number of non-overlapping areas, while M_c represents the number of overlapping areas. $\mathcal{S}_{c,ov}^{A,m}$ is the set of actors that can act on the mth overlapping area $A_{c,ov}^{C,m}$ that is under the responsibility of collector c.

Each actor a is characterized by the following parameters: R_a [m] is the action range of a and P_a^{Max} [W] is the maximum power that actor a can use to perform the action. Actors can select their power from L different levels

$$P_{a,p} = \frac{P_a^{\text{Max}}}{L} \cdot p, \quad p = 1, 2, \ldots, L \tag{14.3}$$

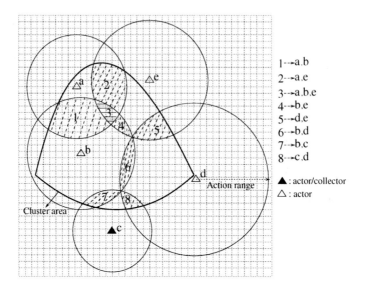

Figure 14.14   Cluster area for collector c.

where $P_{a,p}$ is the pth power level for actor a. Moreover, a higher power corresponds to a lower action completion time. Moreover, the *efficiency* of actor a is denoted as η_a and E_a^{Av} [J] is the available energy of actor a, discounted with the energy needed to act on non-overlapping areas where only actor a can act.

Mixed Integer Nonlinear Program (MINLP)

The actor–actor coordination problem can be formulated as a MINLP. The objective is to find, for each portion of the event area, the subset of actors that maximizes the average residual energy of all actors involved in the action, under the constraint of the action completion bound. For this formalization, the energy required to perform the action is assumed to be orders of magnitude higher than the energy required for communication. Moreover, it is assumed that actors are able to *selectively* act on part of their action area. Accordingly, the MINLP aims to maximize the average residual energy per actor, E_{Avg}^{Res}, which is defined as

$$E_{Avg}^{Res} = \frac{\sum_{a=1}^{N_A} h_a E_a^{Res}}{\sum_{a=1}^{N_A} h_a}. \tag{14.4}$$

The MINLP is solved according to several constraints [15], which define the energy required for actor a to complete the action on the overlapping areas where it is involved and ensure that each actor uses only one power level from its power levels. Moreover, each overlapping area is ensured to be acted on by at least one actor. If multiple actors act on the same area, the time to complete the action is reduced. Accordingly, the overall action completion time for multiple actors acting on the area is limited to be smaller than the action completion bound, for each overlapping area. Solution of the MINLP provides important insight into the actor–actor coordination problem. However, a distributed solution is required for efficient operation in a WSAN. Next, the Localized Auction Protocol (LAP) is described.

14.3.3   Localized Auction Protocol

The distributed solution to the actor–actor coordination problem is based on the behavior of agents in a *market economy*, where the allocation of resources occurs as a result of interactions between buyers and sellers [14, 11]. The Localized Auction Protocol (LAP) is based on a *real-time auction protocol*, where the objective of the auction is to select the best set of actors to perform the action on each overlapping area. Following the analogy with the market economy, overlapping areas are *items* that are traded by the actors. Accordingly, an actor assumes one of the following roles:

- **Seller:** A seller is the actor that is responsible for a portion of event area, i.e., the actor that receives event features for that area. It corresponds to a collector.
- **Auctioneer:** An auctioneer is the actor that is in charge of conducting the auction on a particular overlapping area. It is selected for each overlapping area by the collector/seller responsible for that area.
- **Buyer:** A buyer is one of the actors that can act on a particular overlapping area.

A localized auction takes place in each overlapping area. The *bid* of each actor participating in the auction consists of a power level and the corresponding *action completion time*, i.e., the time needed by that actor to complete the action on the whole area as well as the available energy of the actor. The objective is to maximize the total *revenue* of the team, where the team consists of the actors participating in the auction and the revenue depends on the *residual energy*, i.e., E_{Avg}^{Res}.

Multiple localized auctions take place in parallel under the responsibility of different auctioneers. When seller c (the collector) receives the event features from the sensors, it decides whether an action needs to be performed on the area it is responsible for and computes all the non-overlapping and overlapping areas. The coordination problem arises for the overlapping areas where more than one actor can act, while for the non-overlapping areas the seller directly assigns the action task to the corresponding actor.

Seller c selects M_c *auctioneers*, one for each overlapping area, from the actors that can act on each of these areas. Let $s^{(m)} \in \mathcal{S}^A$ be the auctioneer selected by seller c to conduct the auction for the mth overlapping area. This auctioneer is selected to be the closest actor to the center of the overlapping area. This way, since the auctioneer is close to each actor in the overlapping area, the energy spent on communication and the auction time are reduced. After selecting the auctioneer $s^{(m)}$, the seller c provides it with the area $\mathcal{A}_{c,ov}^{C,m}$ where the auction should take place, the *action completion bound* δ, and the *auction time bound* τ_c, which is the maximum allowed time for the auction. The auctioneer determines the winners of the auction based on the bids it receives from the buyers.

At the beginning of the auction, the auctioneer sends a `JOIN_AUCTION` message to all the buyers competing for the area. After a buyer a hears this announcement, it submits its available energy, E_a^{Av}, and L two-dimensional bids $\underline{b_a} = \{b_a^1, b_a^2, \ldots, b_a^L\}$, where $b_a^{(p)} = [P_{a,p}^{(m)}; T_{a,p}^{(m)}]$, $p = 1, 2, \ldots, L$. If the bids are sorted such that $T_{a,1}^{(m)} \le T_{a,2}^{(m)} \le \cdots \le T_{a,L}^{(m)}$, the auctioneer determines the winners by calculating the optimal solution for the residual energy maximization problem $\mathbf{P}_{Max}^{Res}$ defined in Section 14.2. However, in this case the problem is limited to the overlapping area that the auctioneer is responsible for. This way, since the bids are submitted to the auctioneer only once, signaling overhead is reduced [13].

14.3.4   Performance Evaluation

In this section, the performance of the optimal and distributed solutions to the actor–actor coordination problem is described.

The average residual energy for three different solution approaches are shown in Figure 14.15(b). More specifically, the *optimal, 1-actor,* and *localized auction* solutions are compared. In the optimal

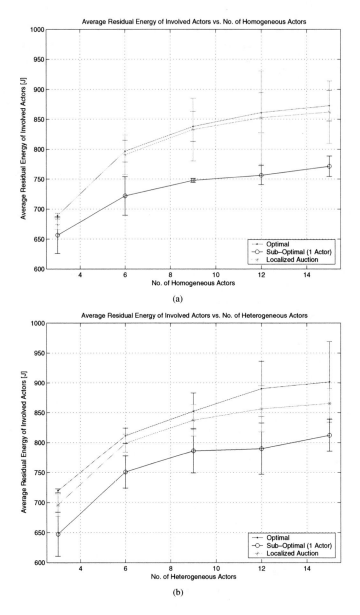

Figure 14.15 Average residual energy of the involved actors in the homogeneous (a) and heterogeneous (b) case.

solution, the best set of actors is chosen so that the average residual energy of the involved actors is maximized, while guaranteeing that the action is completed before the action completion time. In the 1-actor heuristic, the action is performed by one actor only for each overlapping area. In the localized auction, each overlapping area is taken care of by an auctioneer that divides it among the actors based on their bids.

For the experiments, a heterogeneous network structure is considered such that half of the actors are low-efficiency actors ($\gamma_a = 0.6$) and half are high-efficiency actors ($\gamma_a = 0.9$). As shown in Figure 14.15(b), LAP leads to near-optimal residual energy, as each auctioneer calculates the optimal solution separately for its overlapping area. However, this greatly simplifies the problem and can be achieved with local communications among actors. Moreover, in the heterogeneous scenario, the proposed localized solution effectively exploits the high-efficiency actors, thus reducing the dissipated energy to complete the action.

14.3.5   Challenges for Actor–Actor Coordination

Actors may coordinate either in a centralized or a distributed way to solve the task assignment problem in WSANs. Accordingly, actor–actor coordination encounters the following challenges:

- A communication model is needed between actors, which is valid for both SAT and MAT cases. Although, as mentioned in Section 14.1, actors can perform long-range communication and thus generally can communicate directly with their neighbor actors, if the distance between neighbor actors is larger than the transmission range of actors, they cannot directly communicate with each other. In those situations, actors use sensor nodes as intermediaries, which means that actor–actor coordination is performed via sensor nodes.

- In DD, for both SAT and MAT cases, in-sequence execution of different events detected in a region may be required to ensure that there are no adverse effects on the target environment. We refer to this requirement as the *ordered* execution of tasks for a series of events.

- In both DD and CD, for MAT, some applications may require *synchronization* of actors to act on the event at the same time. In this case, the actors have to coordinate either in a distributed or centralized fashion to determine the time of execution of the task.

- For both CD and DD, and both SAT and MAT cases, when the events are in different locations it may be necessary for the task to be executed by a set of actors that are not necessarily close to the location of an event when it was first sensed. In these cases, based on the location of the event, the actors receiving the event information forward it to a different set of actors corresponding to the estimated new position of the event.

- In both CD and DD, for MAT, it may be necessary to address the redundancy in the set of actors that performs a task in order to save on the average energy consumed by the actors in the region. In these cases, it is necessary for only a subset of actors covering the entire event region to be selected to carry out the task.

- In CD, the challenge is to select the actor which will function as a decision unit. Moreover, there is a need for an effective mechanism to provide the decision unit actor with knowledge of the current characteristics (location, capability, etc.) of other actors in the network so that it can trigger the most appropriate actors for the task.

- One of the most important requirements of actor–actor coordination is to minimize the task completion time. Thus, coordination and communication protocols should support real-time properties of WSANs.

14.4   WSAN Protocol Stack

The protocol stack for sensor and actor nodes consists of three planes: the communication plane, coordination plane, and management plane, as shown in Figure 14.16.

- **Communication plane:** This plane consists of five layers, namely the application, transport, routing, MAC, and physical layers. The information exchange among the nodes of the network is enabled through this plane.

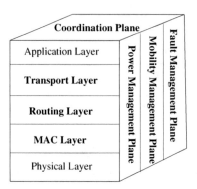

Figure 14.16   WSAN protocol stack.

- **Coordination plane:** The data received by a node at the communication plane are submitted to the coordination plane which decides how the node acts on the received data. Moreover, the coordination plane provides nodes to be modeled as a social entity, i.e., in terms of the coordination and negotiation techniques it possesses.
- **Management plane:** This plane is responsible for monitoring and controlling a sensor/actor node so that it operates properly. It also provides information needed by the coordination layer to make decisions.

In the following, we discuss the requirements and characteristics of each plane for both sensor–actor and actor–actor coordination problems.

14.4.1   Management Plane

The functions performed by the management layer can be categorized into the following three areas:

- **Power management plane:** This plane manages how a node uses its power. For example, when the power level of a node is low, this plane informs the coordination plane so that the node will not participate in sensing, relaying, or acting activities.
- **Mobility management plane:** This plane detects and registers the movements of nodes so that network connectivity is always maintained.
- **Fault management plane:** This plane refers to the detection and resolution of node problems. For example, when the sensitivity of a sensing unit or the accuracy of the actuation unit degrades, the fault management plane informs the coordination plane about this situation.

14.4.2   Coordination Plane

The coordination plane determines how a node behaves according to the data received from the communication plane and management plane. After sensing an event, sensors communicate their readings with each other. At each sensor node, the exchanged data are submitted to the coordination plane for decisions. Accordingly, sensors are able to coordinate among themselves on a higher level sensing task. Moreover, sensor–sensor coordination may also be required to determine nodes which will not transmit data (due to low power or applied MAC protocol), to perform multi-hop routing and data aggregation, and, most importantly, to select actor(s) to which sensor data will be transmitted.

The existence of the coordination plane is much more critical for actors than for sensors, since actors may need to collaborate with each other in order to perform appropriate actions. When an event occurs, the common goal of all actors is to provide the required action on that event. Thus, social abilities, i.e., sophisticated coordination and negotiation abilities, are necessary in WSANs to ensure coherent behavior in the community of actors. These required social abilities of an actor are defined in the coordination plane. Specifically, which layer in actor–actor coordination is responsible for making decisions about which actors act on what part of the event area and whether to have these actors act concurrently or, if sequentially, then in what order [12].

14.4.3 Communication Plane

The communication plane receives commands from the coordination plane regarding the decision on how the node will behave. Accordingly, the link relation between nodes is established by using communication protocols. Specifically, the communication plane deals with the construction of physical channels, access of the node into the medium (MAC), the selection of routing paths through which the node transmits its data, and the transport of packets from one node to another.

In addition to the *conventional reliability*, the new transport protocols must also support real-time requirements in WSANs. For instance, when the transport protocol for sensor–actor communication detects low reliability, the transport protocol for actor–actor communication regulates the traffic between actors so that the actor receiving low reliable event information can inform the other nearby actors about this situation as soon as possible. Since sensor–actor and actor–actor communications occur consecutively in WSANs, a unified transport protocol is needed which works well for both cases.

In WSANs, when sensors detect an event, there is no specific actor to which a message will be sent. This uncertainty occurs due to the existence of multiple actors and causes problems in terms of routing solutions. Selecting the appropriate actor node is one of the challenges for a source sensor node. The source data should then be routed toward the selected actor in an energy-efficient way. While the source data are transmitted through relaying sensors toward an actor node, the data may be aggregated or forwarded in order to achieve high efficiency. In addition to determining path selection and data delivery, the routing protocol should support real-time communication by considering different deadlines due to different validity intervals. Moreover, the routing protocol should also consider the issue of prioritization and should provide data with low delay bounds to reach the actor on time.

For actor–actor communication, routing protocols developed for ad hoc networks such as DSR, AODV, OLSR [6] can be used as long as they are improved so that real-time requirements are met and the communication overhead occurring at sensor nodes due to actor-actor communication is low.

In order to effectively transmit the event information from a large number of sensors to actors there is a need for the MAC protocol. Moreover, in some applications (i.e., distributed robotics) actors may be mobile. As they move, they may leave the transmission regions of some sensors and enter other sensors' regions or they may become disconnected from the network. Therefore, another function of the MAC protocol in WSANs is to maintain network connectivity between sensors and mobile actors. Furthermore, as discussed earlier, timely detection, processing, and delivery of information are indispensable requirements in a sensor/actor network application.

Classical contention-based protocols are not appropriate for real-time sensor–actor communication since contention-based channel access requires handshaking, which increases the latency of the data. Instead, reservation-based protocols are mostly used for WSAN traffic. Such protocols are suitable for WSANs because they can potentially reduce the delay and provide real-time guarantees as well as save power by eliminating collisions.

For actor–actor communication, the existing MAC protocols developed for ad hoc networks cannot be directly used. They should be improved so that they support real-time traffic, since in WSANs, depending on the application, interaction with the world may impose a real-time constraint on computation and communication.

In addition to the layered protocol stack, cross-layer solutions are also adopted for WSANs. One of the main factors which causes low *event reliability* is network congestion. In the case of high congestion, the MAC layer reacts locally by exponential backoff, while the transport layer reacts by lowering the transmission rates of sensors. However, normally these two layers act independently of each other, which causes inefficiencies due to the duplication of functions. By using a cross-layering approach, each protocol shares its data with other protocols, which avoids those inefficiencies. In addition to the interactions among the transport, MAC, routing, and physical layers, in WSANs, there should also be interdependency between the application layer and those lower layers. The application layer must adapt to time-varying QoS parameters offered by the lower layers. While the network provides the best possible QoS to the application, this QoS will vary with time as channel conditions and network topology change. Thus, applications must also adapt to the QoS offered. The basic ideas of cross-layer communication are also valid for actor–actor communication.

The effectiveness of the sensor networking can experience a profound leap if the actors are also an integral part of the deployed network. When the sensor field is complemented with actors, there will be one more option called acting as well as sensing and deciding on a human controller. On the other hand, the realization of WSANs needs to satisfy the requirements introduced by the coexistence of sensors and actors.

Based on the characteristics and challenges of WSANs described in this chapter, several communication protocols have been developed for WSANs. We refer interested readers to [5, 8].

References

[1] DARPA tactical mobile robotics. Available at http://www.darpa.mil/sto/smallunitops/tmr.html.

[2] Sandia National Laboratories. Available at http://www.sandia.gov/media/NewsRel/NR2001/minirobot.htm.

[3] Sub-kilogram intelligent tele-robots (SKITs). Available at http://robotics.usc.edu/?l=Projects:FormerProjects:SKIT:index.

[4] R. K. Ahuja, T. L. Magnanti, and J. B. Orlin. *Network Flows: Theory, Algorithms, and Applications.* Prentice Hall, 1993.

[5] I. F. Akyildiz and I. H. Kasimoglu. Wireless sensor and actor networks: research challenges. *Ad Hoc Networks,* 2(4):351–367, October 2004.

[6] M. Conti, S. Giordano, G. Maselli, and G. Turi. Cross-layering in mobile ad hoc network design. *IEEE Computer,* 37(2):48–51, February 2004.

[7] J. Czyzyk, M. P. Mesnier, and J. Moré. The NEOS server. *IEEE Journal on Computational Science and Engineering,* 5(3):68–75, July–September 1998.

[8] F. Dressler. *Self-Organization in Sensor and Actor Networks.* John Wiley & Sons, Ltd, 2007.

[9] H. R. Everett and D. W. Gage. A third generation security robot. In *SPIE Mobile Robot and Automated Vehicle Control Systems,* volume 2903, pp. 118–126, Boston, MA, USA, November 1996.

[10] G. G. Finn. Routing and addressing problems in large metropolitan-scale internetworks. Technical report ISU/RR-87-180, ISI, March 1987.

[11] B. P. Gerkey and M. J. Mataric. Sold! Auction methods for multirobot coordination. *IEEE Transactions on Robotics and Automation,* 18(5):758–768, October 2002.

[12] V. Lesser. Reflections on the nature of multi-agent coordination and its implications for an agent architecture. *Autonomous Agents and Multi-Agent Systems,* 1:89–111, January 1998.

[13] P. Maille and B. Tuffin. Multibid auctions for bandwidth allocation in communication networks. In *Proceedings of IEEE INFOCOM'04,* volume 1, p. 65, Hong Kong, China, March 2004.

[14] R. P. McAfee and J. McMillan. Auctions and bidding. *Journal of Economic Literature,* 25(2):699–738, June 1987.

[15] T. Melodia, D. Pompili, V. C. Gungor, and I. F. Akyildiz. A distributed coordination framework for wireless sensor and actor networks. In *Proceedings of ACM MobiHoc'05,* p. 110, Urbana–Champaign, IL, USA, May 2005.

[16] E. M. Petriu, N. D. Georganas, D. C. Petriu, D. Makrakis, and V. Z. Groza. Sensor-based information appliances. *Instrumentation & Measurement Magazine, IEEE,* 3(4):31–35, December 2000.

[17] S. Thrun, M. Diel, and D. Hähnel. Scan alignment and 3-D surface modeling with a helicopter platform. In *Proceedings of the International Conference on Field and Service Robotic (FSR'03),* pp. 287–297, Lake Yamanaka, Japan, 2003.

15

Wireless Multimedia Sensor Networks

Recent progress in CMOS technology has enabled the development of single chip camera modules that could easily be embedded into inexpensive transceivers. Moreover, microphones have for long been used as an integral part of wireless sensor nodes. The interconnection of multimedia sources with inexpensive communication devices has fostered research in the networking of multimedia sensors. Consequently, wireless multimedia sensor networks (WMSNs) became the focus of research in a wide variety of areas including digital signal processing, communication, networking, control, and statistics in recent years [13, 29, 44]. The union of the experience gained from each of these areas independently helps to develop WMSNs that allow the retrieval of video and audio streams and still images, as well as delivering this information in real time. In addition to the ability to retrieve multimedia data, WMSNs are also be able to store, process in real time, correlate, and fuse multimedia data originating from heterogeneous traffic sources.

This chapter focuses on these important aspects and provides a broad overview of the protocols, algorithms, applications, and research challenges for WMSNs. The realization of these networks will enable a wide variety of application areas that were not possible before with scalar sensor networks. The most important application areas can be summarized as follows:

- **Multimedia surveillance:** Video, image, and audio collection capabilities of sensors in WMSNs enable broader and enhanced surveillance applications. Using feeds from multiple heterogeneous sensors, extensive coverage of a particular area can be provided. Moreover, it is possible to convey much more detailed information to the user compared to traditional WSNs.

- **Traffic avoidance, enforcement, and control:** Traffic is a major problem in metropolitan areas. Easily deployable wireless multimedia devices can help to solve many problems related to traffic jams by rerouting individuals according to information from a network of camera sensors or indicating empty lots in parking areas in real time. Moreover, multimedia sensors may monitor the flow of vehicular traffic on highways and retrieve aggregate information such as average speeds and the number of cars. This helps for remote detection of violations and identification of violators through video feeds.

- **Automated assistance for the elderly and family monitors:** WMSNs provide novel techniques for monitoring and studying the behavior of elderly people without disturbing their daily routines. As a result, the causes of illnesses that affect them, such as dementia [49], can be identified. Furthermore, remote assistance to elderly people through video or audio sensors is also possible.

- **Environmental monitoring:** Visual information content provided through either still images or video streams improves the capabilities of environmental monitoring applications. In addition to scalar data such as temperature or humidity, visual information helps also for closer monitoring of the habitat without significant interactions or interference.

Wireless Sensor Networks   Ian F. Akyildiz and Mehmet Can Vuran
© 2010 John Wiley & Sons, Ltd

- **Industrial process control:** Control and monitoring of the industrial environment are crucial for the efficiency of an industrial plant. Multimedia content may be used for time-critical industrial process control. A manufacturing process for products such as semiconductor chips, automobiles, food, or pharmaceuticals can be monitored and analyzed through WMSNs. Integration of machine vision systems with WMSNs can simplify and add flexibility to these systems while decreasing the costs for maintenance and operation.

Distributed vision systems that will emerge from the research in WMSNs have potential advantages in addition to the applications described above. In particular, traditional monitoring and surveillance systems can be improved through three main advancements: namely, enlarging the view, enhancing the view, and providing multiple viewpoints of the same phenomenon [21]. When a single camera is considered for a surveillance application, the coverage of the application is only limited by the field of view (FoV) of a fixed camera or the field of regard (FoR) of a pan–tilt–zoom camera. On the other hand, deploying multiple, low-cost visual sensors both improves the coverage of the application and provides a more robust operation. Similarly, multiple cameras provide necessary redundancy to enhance the observations that would not be possible through a single camera. This also prevents single points of failure or problems related to obstruction of a camera's line of sight. Furthermore, multiple visual sources provide the flexibility to adaptively extract data depending on the requirements of the application. Accordingly, a multi-resolution description of the scene and multiple levels of abstraction can be provided.

While developments and solutions in WSNs provide a great deal of information for designing the multimedia counterparts of these networks, many aspects of the sensor network paradigm need to be rethought. Especially, the need for various quality of service (QoS) demands of multimedia applications is the major challenge. Furthermore, experience from several research areas is needed to converge to efficient and flexible solutions for WMSNs. More specifically, energy-constrained wireless communication theory and image and video processing techniques need to be combined for advanced multimedia communication.

In this chapter, the state of the art in algorithms, protocols, and hardware for the development of WMSNs is discussed. Furthermore, several open research issues are highlighted. Section 15.1 points out the characteristics of WMSNs as well as their differences from traditional WSNs. In Section 15.2, various architectures that could be adopted for WMSNs are described. The existing platforms as well as the deployed research testbeds are introduced in Section 15.3. Sections 15.4, 15.5, 15.7, 15.8, and 15.9 focus on existing solutions at the physical, link, network, transport, and application layers of the protocol stack, respectively. Section 15.10 provides cross-layer design aspects in WMSNs. Finally, in Section 15.11 additional complementary research areas are discussed.

15.1 Design Challenges

The design of WMSNs needs expertise from a wide variety of research areas including communication, signal processing, control theory, and embedded systems. The convergence of these areas will enable more powerful networks that are able to interact with the environment and provide even more meaningful data than the traditional data-only WSNs. There are several design challenges to realizing WMSNs as we explain below.

15.1.1 Multimedia Source Coding

The generation, processing, and transmission of multimedia traffic such as audio, video, and still images result in major challenges. Transmission of multimedia content has for long been associated with multimedia source coding techniques because of the large amounts of traffic generated by multimedia sources such as cameras. For example, a single monochrome frame in the NTSC-based *Quarter Common*

Intermediate Format (QCIF, 176×120), requires around 21 kB, and at 30 frames per second (fps) a video stream requires over 5 Mbps bandwidth. As a result, the raw data have to be compressed through sophisticated coding techniques that exploit the redundancy in the video or still images.

Multimedia source coding aims to reduce the information to be sent through the wired or wireless media by exploiting the redundancy in the video or still image. In a still image, this could be done by extracting the interrelations between each pixel in the picture. In video, two types of source coding techniques are used. Firstly, at each frame, the correlation between pixels is extracted much like the still images. This is called *intraframe compression*. Furthermore, each consecutive frame in a video stream is inherently correlated because of the large portion of background image that stays consistent as well as the movement of certain objects between frames. These correlation aspects are exploited through *interframe compression* to further reduce the amount of data to be transmitted. Interframe compression is also known as *predictive encoding* or *motion estimation*.

The compression techniques certainly decrease the information to be transmitted to convey a video stream. This compression, however, comes at the cost of degradation of the video quality that is usually referred to as *distortion*. Effective compression techniques developed so far result in good rate-distortion levels to efficient multimedia transmission.

Despite their significant compression capabilities and good rate-distortion performance, traditional source coding techniques are not directly applicable to resource-constrained WMSNs. This is mainly related to the fact that predictive encoding requires complex encoders and powerful processing algorithms, which significantly increase the energy consumption. Hence, these techniques are not directly suitable for low-cost multimedia sensors. To this end, the traditional balance of complex encoders and simple decoders can be reversed through distributed versions of the traditional encoding techniques or within the framework of so-called *distributed source coding* [28], which exploits the source statistics at the decoder, and by shifting the complexity at this end allows the use of simple encoders. Clearly, such algorithms are very promising for WMSNs and especially for networks of video sensors, where it may not be feasible to use existing video encoders at the source node due to processing and energy constraints. These techniques will be covered in more detail in Section 15.9.

15.1.2   High Bandwidth Demand

Although multimedia coding techniques significantly decrease the transmitted information, the compressed information still surpasses the current capabilities of wireless sensor nodes. More specifically, video streams require transmission bandwidth that is orders of magnitude higher than what is supported by currently available sensors. This high bandwidth demand requires new transmission techniques that provide larger bandwidth at acceptable energy consumption levels. Furthermore, novel hardware architectures for multimedia-capable transceivers need to be developed. To this end, the ultra-wide-band technique can be exploited as discussed in Section 15.4.

15.1.3   Application-Specific QoS Requirements

The networking and communication techniques proposed so far for WSNs generally follow a best effort service approach. In other words, no tight guarantees in terms of energy consumption, delay, jitter, or throughput are provided. On the other hand, multimedia applications specifically require these types of guarantees for efficient delivery of the sensed phenomenon. Moreover, the type of application (still image transmission or video streaming) affects the QoS requirements in the network. Furthermore, as will be discussed in Section 15.9, each multimedia stream contains information that requires different levels of QoS guarantees. Consequently, the design of WMSNs requires the development of algorithms that support application-specific QoS requirements. These requirements may be in terms of bounds on energy consumption, delay, reliability, distortion, or network lifetime.

15.1.4 Multimedia In-network Processing

In-network processing has been exploited for energy-efficient data delivery in WSNs. The same is also true for WMSNs but the characteristics of the multimedia content require novel approaches for this area. An important difference is that, in WSNs, in-network processing has usually been used for aggregation operations. Considering the correlated information sensed by nearby sensors, aggregating this information is an easy solution for data traffic. Moreover, with scalar data, it is straightforward to perform linear operations such as additions or taking averages. This is mainly because a single packet contains meaningful data to process in scalar data-based WSNs. Aggregation is not straightforward in WMSNs since the multimedia information is carried through multiple packets in a stream. As a result, to infer meaningful information from a stream of traffic, the whole stream needs to be collected and the image or video needs to be decoded. This requires significant storage and processing capabilities at intermediate nodes. Only after decoding can an intermediate node perform aggregation tasks on multiple traffic streams. This branch of in-network processing is therefore not practical for WMSNs. On the other hand, resource-hungry encoding operations can be distributed among a group of nodes to leverage the energy consumption for a multimedia flow. This idea has been used in distributed JEPG encoders developed for WMSNs as will be discussed in Section 15.9.

15.1.5 Energy Consumption

Energy consumption is of major concern in traditional WSNs. This fact is even more pronounced in WMSNs due to two fundamental differences. Multimedia applications generate high volumes of traffic, which require longer transmission times for the battery-constrained sensor devices. While transmission is usually mitigated through in-network processing solutions in traditional WSNs, the extensive processing requirements of multimedia data may make these techniques unsuitable for WMSNs. Solutions for WMSNs need to guarantee the QoS requirements of applications while minimizing the energy consumption.

15.1.6 Coverage

Scalar sensors are usually indifferent to the direction of data acquisitions. This results in relatively circular sensing ranges as discussed in Chapter 13. Multimedia sensors, however, are characterized by their directivity. Especially, video and image sensors have a limited field of view that results in conic coverage areas substantially different from those of scalar sensors. Moreover, the sensing range in this field of view is significantly longer than scalar sensors. These fundamental differences in coverage motivate novel approaches in topology control and design in WMSNs.

15.1.7 Resource Constraints

Similar to traditional WSNs, each component in WMSNs is constrained in terms of battery, memory, processing capability, and data rate [14]. The increased traffic volume as well as the significantly higher processing requirements of encoders signify the importance of energy-efficient operation in WMSNs. Consequently, the existing resources need to be efficiently consumed for multimedia delivery.

15.1.8 Variable Channel Capacity

Multimedia communication has been investigated for wired and wireless networks during the last few years. Providing QoS guarantees in terms of delay, jitter, and throughput has been the main focus of research in this area. While many lessons can be learned from previous research, the characteristics of WMSNs create novel challenges.

15.1.9   Cross-layer Coupling of Functionalities

Cross-layer design has been driving the research in WSNs as discussed in Chapter 10. It is important to exploit and consider the cross-layer effects of each layer functionality on every other layer and develop communication protocols accordingly. While the cross-layer design principles also apply to WMSNs, an additional dimension exists because of the direct influence of the application layer on the multimedia traffic. Joint source and channel coding algorithms have been developed for wireless multimedia communication in cellular and ad hoc networks. Coupled with the close interaction of other protocol layers in WMSNs, a complete cross-layer design is necessary for energy-efficient communication.

15.2   Network Architecture

WMSNs are envisioned to be composed of heterogeneous sensor devices that have various capabilities in terms of sensing, processing, and communication. Consequently, various types of traffic can be generated by these heterogeneous sensor devices, such as still images, audio, video, or scalar data. To handle this variety, a heterogeneous architecture must be designed.

Architecture design issues in traditional WSNs have been focused on *scalable network architectures*. The scalability is usually achieved through flat, homogeneous architectures in which every sensor has the same physical capabilities and can only interact with neighboring sensors.

Contrary to the homogeneous nature of traditional WSNs, the inherent heterogeneity due to various technologies that support different sensor types as well as the traffic generated by these sensors require a different approach. As an example, the large amount of traffic generated by multimedia sensors may not be handled by flat topologies consisting of low-end transceivers. Likewise, the requirements for processing power for data processing and the power for communication may be different for every different sensor. A sensor node embedded with only a microphone may generate relatively low data that can be handled by existing low-end sensor nodes. On the other hand, the excessive power required for video processing and the bandwidth required for even encoded video frames may require different processor and transceiver technologies. Consequently, to process and transfer data simultaneously may not be possible at each device in the network.

These intrinsic differences of WMSNs led to the realization of heterogeneous network architectures. These architectures are mainly classified into: *single tier* and *multi-tier*. Single tier architectures are based on a flat topology, where the network can be composed of either homogeneous or heterogeneous components. Multi-tier architectures, on the other hand, exploit the higher processing and communication capabilities of high-end nodes and yield a hierarchical network operation. An overview of these possible architectures for WMSNs is shown in Figure 15.1, where three sensor networks with different characteristics are shown. We describe each of these architecture types next. Furthermore, we explain in detail coverage issues that are fundamentally different in WMSNs.

15.2.1   Single Tier Architectures

The first cloud on the left of Figure 15.1(a) shows a single tier network of video sensors. In this architecture, a subset of the deployed sensors may have higher processing capabilities. These nodes are referred to as *processing hubs*, which can be used for local processing of the sensed media information. Moreover, multiple processing hubs can be used to perform a single processing job distributively, which constitutes a distributed processing architecture.

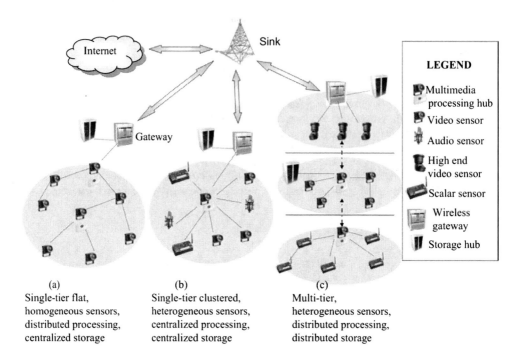

(a)   (b)   (c)
Single-tier flat,   Single-tier clustered,   Multi-tier,
homogeneous sensors,   heterogeneous sensors,   heterogeneous sensors,
distributed processing,   centralized processing,   distributed processing,
centralized storage   centralized storage   distributed storage

Figure 15.1   Reference architecture of a WMSN.

In the single tier architecture, the multimedia content recorded and/or processed by the sensor nodes is relayed to a *wireless gateway* through a multi-hop path. The wireless gateway is interconnected to a *storage hub*, which is in charge of storing multimedia content locally for subsequent retrieval. As a result, recorded and processed data can be stored at a central location to relieve the storage restrictions on the sensor nodes and to perform more sophisticated processing jobs offline.

Instead of a centralized storage architecture, a distributed storage scheme, where all the nodes inside the network store a part of the collected information, can also be implemented. Such a scheme may result in energy savings since, by storing it locally, the multimedia content does not need to be wirelessly relayed to remote locations. The wireless gateway is also connected to a central *sink*, which implements the software front-end for network querying and tasking.

The single tier architecture can also be deployed for networks consisting of heterogeneous components. Such a reference architecture is shown in Figure 15.1(b). In this case, nodes can be organized into clusters, where a central cluster head controls other video, audio, and scalar sensors in the cluster. Moreover, the cluster head can be used to perform more resource-hungry procedures such as intensive multimedia processing, or aggregation. Consequently, in this architecture, the cluster heads can also be used as *processing hubs*. The information gathered by the cluster head is then relayed to a wireless gateway or a storage hub for further processing and storage.

15.2.2   Multi-tier Architecture

The last cloud shown in Figure 15.1(c) represents a multi-tiered network, with heterogeneous sensors. This hierarchical structure provides the flexibility to use network resources adaptively. The multi-tier architecture employs low-end scalar sensors at the lower hierarchical levels to perform simpler

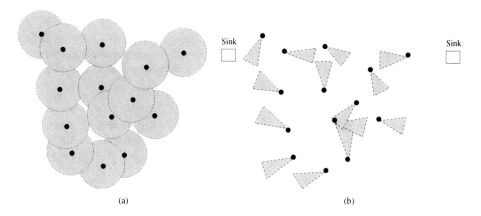

Figure 15.2 Illustration of the coverage of (a) scalar sensors and (b) video sensors.

tasks. The information gathered by these sensors can, then, be used to initiate more complex sensing tasks such as real-time video monitoring to adapt to and communicate any events that occur in the surroundings. These tasks are accomplished by high-end sensor nodes equipped with video cameras at higher hierarchical levels. Furthermore, image and video processing and storage can be initiated only when there is sufficient interest in the sensed phenomenon reported by the low-end devices. Such a hierarchical architecture extends the lifetime of the network while improving the adaptivity and efficiency of the deployed heterogeneous sensors. As shown in Figure 15.1(c), the multi-tier architecture may include clusters in each tier. As a result, autonomous operation for each cluster can be performed, which decreases the energy consumption due to communication between these sensors. Moreover, necessary information to be communicated to the higher tiers may be aggregated at the cluster heads minimizing the traffic in the network.

15.2.3 Coverage

Defining an architecture for WMSNs also depends on the coverage of the network. In order to provide sufficient coverage for a specific application, the architecture should be defined by considering the coverage issues for each of the components in the network. An important challenge in this area stems from the different coverage properties of multimedia sensors such as cameras and microphones compared to the scalar sensor types such as temperature or humidity sensors.

When traditional WSNs are considered, the sensing capability and hence the *sensing range* of a sensor are characterized as omnidirectional in nature. In other words, the sensor is generally indifferent to the direction of sensing. As a result, the coverage area of a scalar sensor can generally be modeled as a circle centered around the sensor as shown in Figure 15.2(a). Then, the coverage of a sensor can easily be modeled using its sensing range. As explained in Chapter 13, the recent work on coverage in WSNs mainly investigates the tradeoff between sensing range and communication range, which defines the range that a node can reach through wireless communication. Consequently, the coverage problem in WSNs is generally tied to connectivity in the network [17].

Coverage for multimedia sensors, on the other hand, exhibits fundamentally different characteristics in terms of directivity, range, line of sight, and dynamic view as explained next.

Directivity

Contrary to the omnidirectional nature of the scalar sensors, audio and especially video sensors collect information based on its direction. For example, in video sensors, the coverage area concept is exchanged with the *field of view* (FoV), which defines the area that a video sensor can "see."

■ **EXAMPLE 15.1**

The difference in coverage for scalar sensors and that for visual sensors is illustrated in Figure 15.2. As shown in Figure 15.2(a), the coverage of a scalar sensor is generally circular. On the other hand, in Figure 15.2(b), the FoVs of the video sensors in a WMSN are shown. It can therefore be seen that the coverage of a WMSN depends on the direction of the sensors in the network as well as the characteristics of the sensors that define the shape of the FoV. As shown in Figure 15.2(b), different types of cameras result in different FoV shapes and sizes.

This fundamental difference in coverage incurs additional information requirements for the coverage optimization problem in this area. While, in WSNs, the location and the sensing range of a sensor are sufficient to characterize its coverage, in WMSNs, many parameters such as FoV shape and size and direction of the sensor should also be considered. Furthermore, each multimedia sensor/camera perceives the environment or the observed object from a different and unique viewpoint, given the different orientations and positions of the cameras relative to the observed event or region.

Range

Sensing range in WSNs is usually fixed due to the omnidirectional nature of the scalar sensors. Moreover, the coverage area of scalar sensors is closely related to the location of the sensor. However, due to the directional changes, the sensing range changes based on the direction of a sensor. Moreover, multimedia sensors such as cameras generally have larger sensing ranges compared to scalar sensors. Therefore, the location of a sensor can be unrelated to its FoV. As an example, in a distributed surveillance application, where multiple video sensors are deployed in a room, multiple sensors that are located farther away can be monitoring the same area. Due to the long range of the multimedia sensors, conventional topology control protocols as well as coverage optimization solutions are not applicable to WMSNs. Another important factor in the coverage range of multimedia sensors is that the range of a multimedia sensor depends on the application type. An object that is very close to a camera can be viewed with high resolution, while a farther object cannot be clearly monitored due to the limited resolution of the camera lens. For an application where the monitored object is required to be visualized with high precision, the range of this sensor is limited. On the other hand, if only the presence of an object is important for a different application, the coverage area and, hence, the range of a sensor would be much larger.

Line of Sight

The coverage of WMSNs also depends on the obstacles in the surroundings. A camera sensor can only capture an image of an object only if there is no obstacle in between. Therefore, the coverage of a sensor and, hence, of the network is closely related to its line of sight. This fact results in different challenges for coverage problems in WMSNs.

Dynamic Coverage

The pan and tilt capabilities of today's cameras, as well as their zoom capabilities, make the coverage of WMSNs dynamic. This adds flexibility to the operation of WMSNs but also a challenge for their design. The ability to change the FoV of sensors in the network results in a more adaptive operation depending on the conditions in the surroundings. During the lifetime of the network, a group of video sensors can be directed to a location of interest to gather more information using the pan and tilt properties of video sensors. Similarly, the FoV of a video sensor can be focused onto a smaller area using its zoom capability. These local changes in the orientation and the focus of the video sensors dynamically change the coverage of the network. While controlling the coverage of the network provides several advantages, this dynamic change should be communicated efficiently in the network to provide complete coverage of the interested area. However, in WMSNs, the FoVs of video sensors are usually unrelated to their location. As a result, two sensors with overlapping FoVs can be located at distant positions. Consequently, the change in view of one of these sensor should be effectively communicated to the other sensor through a multi-hop path.

15.3   Multimedia Sensor Hardware

The fundamental difference in the design of WMSNs is the capabilities of the multimedia sensor devices. Embedding audio or video capture devices in resource-constrained wireless sensor devices provides a significantly diverse set of functionalities for WMSNs. However, the hardware architecture as well as the components of the wireless devices need to accommodate the higher bandwidth and processing power requirements of multimedia sensing. In this section, we overview the existing platforms in this field. In particular, audio sensors and low- and medium-resolution video sensors are described.

The existing hardware for WMSNs can be classified into three categories. The first category is the *audio sensors*. These sensors include microphones that have already been included in the earlier versions of wireless sensor boards. The second and third categories are *low-resolution sensors* and *medium-resolution video sensors*, respectively. The low-resolution video sensors are built specifically for WMSNs and are part of an embedded system which may consist of low-end transceivers as well as specialized processor and memory units that are required for low-power image processing and storage. The high-resolution video sensors usually include a webcam connected to a high-end wireless device for communication. In the following, we will describe the existing hardware in these categories. Furthermore, currently existing research testbeds composed of these components are described.

15.3.1   Audio Sensors

Audio sensors already exist in conventional sensing boards for WSNs. Although audio sensors are common components in these cases, not many applications have so far been developed for audio signal detection and transmission in WSNs. Microphones used in these boards are usually low-end devices with a low sensitivity and can detect audio signals with frequencies usually lower than 5 kHz. Two examples of audio sensors are the Panasonic WM-62A used in Crossbow sensor boards (MTS300/310, MTS510) and the wireless, battery-less wearable microphones.

Mic-on-board

The audio sensor used in one of the most common sensor boards, Crossbow's MTS300/310 and MTS510, is the Panasonic WM-62A microphone (Figure 15.3). It draws a small current of $500 \, \mu$A. The microphone circuitry is composed of a preamplifier and a second-stage amplifier with a digital-pot control [1]. The output of the preamplifier can be used for recording audio or measurements. The output of the second-stage amplifier is used for tone detection and to perform more complex audio processing tasks. The microphone works in an omnidirectional manner such that sounds coming from any direction can be recorded. Sound waves with a frequency smaller than 5 kHz can be successfully received by these audio sensors.

The most common usage of the on-board microphones is acoustic ranging. The main idea is to use the difference between the propagation speed of a radio wave and an acoustic signal to estimate the

Figure 15.3   MTS310 sensor board with on-board microphone seen at the top left corner of the board.

distance between two nodes. In these applications, a node transmits a packet and activates its sounder simultaneously. Another node that receives the packet initiates a timer and records the time difference between reception of the packet and the tone sent by the other node. This time difference can be used to roughly estimate the distance between these nodes.

Another promising application for audio sensors is speech recognition and audio signal transmission. However, the limitation that signals only up to 5 kHz range can be detected is a major problem for these applications. This limitation is basically due to the processing speed of the Mica2 or MicaZ nodes that these boards are connected to. Considering that the human voice has frequency components higher than 5 kHz, providing high-quality voice recording may not be feasible with mic-on-boards. However, many potential applications such as sound detection are still possible with the low-end microphones.

Wireless, Battery-less Wearable Microphones

A wireless and battery-less microphone for health monitoring has been implemented at MIT [18]. The microphone is in the shape of a mole and can be attached to any part of the human body to capture respiratory, lung, or surrounding sounds. Multiple sensors can be attached to different parts of the body to collect internal and external sounds. Communication is performed between a wearable reader and the sensors through magnetic induction. The transceiver in the reader generates an electromagnetic field and the sensor modulates this field by changing its capacitance according to the observed acoustic wave. Consequently, only the reader is battery powered and the sensors operate battery-less. This provides flexibility for operation on the human body since no wires or heavy batteries are necessary for the sensors.

15.3.2   Low-Resolution Video Sensors

Traditionally, charge-coupled device (CCD) technology has been used for the development of video cameras. CCD chips consist of millions of photosensitive diodes called *photosites* that, when exposed to light, can record the intensity of light for a short period of time. This information is then processed by an image processor to construct an image. The intrinsic properties and the processing requirements of the CCD technology, however, prevent the construction of single chip cameras. Instead, CCD cameras

Table 15.1   Specifications of video sensors.

Name	Resolution	Frame rate (fps)	Processor speed (MHz)	Embedded transceiver	Data rate (kbps)
Cyclops	352 × 288	N/A	7.3	No	N/A
CMUcam	80 × 143	16.7	75	No	N/A
CMUcam2	176 × 255	26	75	No	N/A
CMUcam3	352 × 288	50	60	No	N/A
MeshEye (low-res.)	30 × 30	N/A	50	Yes	250
MeshEye (high-res.)	352 × 288	15			
Panoptes	320 × 240	20	206	Yes	1024
Acroname Garcia	640 × 480	30	400	Yes	250, 1024

have usually been implemented in three- to eight-chip packages. This requirement increases the size of the camera as well as the processing and energy consumption requirements.

Recently, a new technology called CMOS (Complementary Metal Oxide Semiconductor), which is generally used for the manufacture of computer processors, has been used for video cameras. Using CMOS technology, image sensors can be implemented with a lens, an image sensor, and an image processing circuit on the same chip. This significantly reduces the scale and cost of image sensors. Consequently, CMOS image sensors have been used in many recent electrical devices such as cell phones, PDAs, and digital cameras. Also, low-resolution, small-size video sensors have been embedded into systems with wireless transceivers to manufacture stand-alone multimedia sensors. Since CMOS image quality closely follows CCD quality for low- and medium-resolution sensors, the reduction in size does not affect the quality. Finally, CMOS sensors consume much less energy than their CCD counterparts, which makes them suitable candidates for WMSNs (see Table 15.1). Next, we describe the recent embedded multimedia sensor nodes built specifically for WMSNs.

Cyclops

Cyclops is a sister-board developed for Mica2 and MicaZ nodes [48]. Much like the sensor boards, Cyclops is connected to a Mica2 or MicaZ node for communication purposes. It includes an image sensor, a microcontroller unit (MCU), a complex programmable logic device (CPLD), an external SRAM, and an external flash memory. This board separates the complexity of the vision algorithms from the network node. The separate MCU has two advantages for Cyclops. Firstly, image processing tasks can be performed separately on the sister-board without loading the main MCU on the main board that is also responsible for communication tasks. Secondly, tasks such as capturing and interpreting an image to produce an inference as well as simple image processing tasks such as background subtraction are usually time consuming. Due to the serial task scheduling principle of TinyOS, which is generally used in Mica-family nodes, performing such tasks on the main board leads to a high delay for other tasks and then to starvation problems. Cyclops prevents this problem by parallel processing using the MCU and the CPLD on the sister-board (Figure 15.4).

The microcontroller on the Cyclops board is used to control the imager and run local computation on the captured images to produce an inference. The CLDP has a higher processing speed, which is necessary to perform more complex image processing tasks as well as provide synchronization and memory control for image capture. The external SRAM and the flash memory provide the necessary memory for storage and further processing of captured images. The Cyclops firmware includes TinyOS libraries that are also compatible with the Mica-family motes. The libraries provide simple manipulation

Figure 15.4   Cyclops video sensor mote.

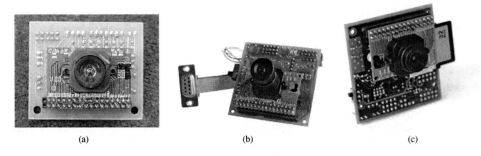

(a)   (b)   (c)

Figure 15.5   CMUcam family of embedded camera systems: (a) CMUcam, (b) CMUcam2, and (c) CMUcam3.

capabilities as matrix operations as well as more advanced processes such as background subtraction and coordinate conversion for image compression and analysis.

CMUcam

Another platform for image sensors is the CMUcam family of embedded camera systems. So far CMUcam, CMUcam2, and CMUcam3 have been developed as shown in Figure 15.5(a), 15.5(b), and 15.5(c), respectively. CMUcam consists of a CMOS camera, a microcontroller, and a level shifter for the RS232 interface [50]. An important feature of CMUcam is that a second microcontroller can be connected to perform image processing tasks in parallel. CMUcam is a camera module that needs to be connected to a transceiver unit for communication.

CMUcam2 is an improved version of the first generation that has a more powerful processor in terms of memory (RAM, 252 vs. 136 bytes; ROM, 4096 vs. 2048 words) and a camera chip with a higher resolution. Moreover, a frame buffer is added to the sensor to enable more complex image processing tasks on-board.

The latest version, CMUcam3, is an embedded camera endowed with a CIF resolution (352×288) RGB color sensor that can load images into memory at 26 fps. CMUcam3 has software for JPEG

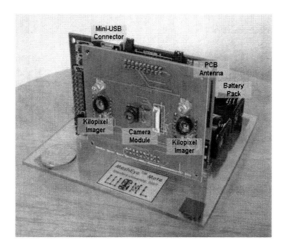

Figure 15.6   MeshEye video sensor mote.

compression and a basic image manipulation library, and can be interfaced with an IEEE 802.15.4-compliant TelosB mote [4].

MeshEye

MeshEye sensor motes (Figure 15.6) have been developed for applications that may require multiple cameras on a single sensor mote [22]. The mote can accommodate a maximum of two CIF (352 × 288) image sensors and four low-resolution (30 × 30) optical sensors. A transceiver is also embedded into the board that is compatible with the IEEE 802.15.4 standard with a data rate of 250 kbps similar to Mica-family and Telos motes. The expansion interface on the sensor board allows up to six cameras to be used concurrently. A external FRAM and flash memory can also be interfaced with the mote for storage and computation.

15.3.3   Medium-Resolution Video Sensors

In addition to the embedded platforms that include low-power, low-resolution cameras that are specifically designed for WMSNs, high-power solutions that usually include webcams also exist. These solutions are usually based on platforms such as the ones used in PDAs or the Crossbow Stargate boards that possess higher processing, storage, and communication capabilities than typical sensor motes. The enhanced processing capabilities of these devices enable an off-the-shelf webcam to be integrated to form a stand-alone visual sensor. These devices can be used in applications where higher resolution and image processing are necessary. Here, we explain two sample architectures that use sophisticated system boards and webcams for the implementation of a medium-resolution video sensor.

Panoptes

Panoptes is one of the first stand-alone visual sensor platforms (Figure 15.7) to be implemented using mostly off-the-shelf components [26]. The Panoptes video sensor uses the Intel StrongARM embedded platform that is used in Compaq IPAQ PDAs. StrongARM has an embedded processor that runs at 206 MHz with 64 MB of memory, which significantly surpasses the capabilities of the low-resolution

Figure 15.7   Panoptes video sensor platform.

platform explained above. The Linux operating system is used in these embedded platforms. These capabilities make the Panoptes video sensors suitable for use as high-end devices in a heterogeneous sensor network. Panoptes can communicate with other nodes through IEEE 802.11 wireless cards, which have significantly higher communication bandwidth compared to the traditional sensor motes. Consequently, these platforms can also be used as gateways to collect and transfer high volumes of information gathered from several low-end sensors. The Panoptes video sensor is interfaced with a USB-based video camera that can capture video of 320×240 resolution at 18–20 fps. The Panoptes video sensor is also equipped with a software architecture that includes many image processing tools including video capture, JPEG and differential JPEG compression, filtering for compression, buffering, and streaming.

GARCIA Mobile Video Sensor

The GARCIA mobile video sensor incorporates mobility into the high-resolution video sensors. This high-end video sensor consists of a pan–tilt camera installed on an GARCIA robotic platform [2, 5] as shown in Figure 15.8. The GARCIA robot is equipped with two separate 40 MHz processors to handle motion control, sensor inputs, serial interface, and infrared (IR) communication. The robot is equipped with eight IR sensors that are located on its front, sides, and bottom. This provides environmental awareness of the robot for automated motion control, obstacle detection, and maneuvering. A microprocessor can be connected to the main board to provide sensing tasks as well as communication. The GARCIA robot can also be interfaced with a pan–tilt camera bloom to connect a webcam on top of it, which constitutes a mobile video sensor as shown in Figure 15.8. This platform can be used for adaptive sampling in a heterogeneous sensor network.

15.3.4   Examples of Deployed Multimedia Sensor Networks

There have been several recent experimental studies, as follows, mostly limited to video sensor networks.

SensEye

The SensEye testbed is an example of how heterogeneous components can be used in a WMSN to provide a surveillance application [32]. This application aims to accomplish three tasks: object detection, recognition, and tracking. The SensEye network architecture follows a multi-tier architecture as discussed in Section 15.2. Heterogeneous components with various sensing and processing capabilities are organized into a hierarchical structure. This structure provides continuous and adaptive surveillance

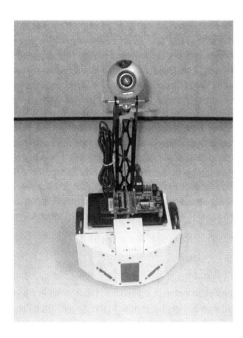

Figure 15.8 Acroname GARCIA, a mobile robot with a mounted pan–tilt camera and endowed with IEEE 802.11 as well as ZigBee interfaces.

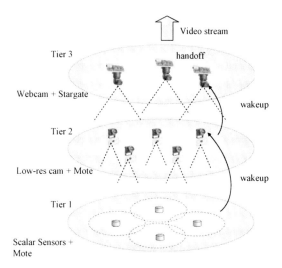

Figure 15.9 The multi-tier architecture of SensEye [32].

operation while minimizing energy consumption. The multi-tier architecture of SensEye is shown in Figure 15.9, which is composed of three tiers.

The lowest tier of the SensEye testbed consists of low-end motes that provide continuous sensing of the surveillance area through vibration sensors. Any activity recorded by the lowest tier sensors is communicated to the second tier for object recognition through low-resolution video sensors. In this tier, the low-resolution video sensors are used for object detection and recognition. The objects which are detected through the vibration sensors on the lowest tier can then be visualized by adaptively waking up the video sensors on this tier. The third tier consists of medium-resolution webcams interfaced with Crossbow Stargate boards. These boards are capable of communicating with both the lower tier components through the low-data-rate communication interface and also a control center through the IEEE 802.11 interface. The higher storage and processing capabilities of these nodes on the third tier are also exploited to perform gateway functionalities as well as handling resource-hungry image processing tasks. Furthermore, higher resolution images of an interested area can be recorded on this tier, when requested by the low-resolution video sensors on the second tier. In addition to this architecture, a fourth tier consisting of high-resolution, pan–tilt–zoom cameras and PCs can be used for more complex object tracking tasks. The hierarchical tasks improve energy efficiency by reducing the energy consumption by an order of magnitude when compared to a flat architecture. This is achieved by using the higher end devices and their resources only when necessary while still providing complete coverage for the surveillance application.

Meerkats Testbed

The Meerkats testbed has been constructed to investigate the energy consumption profile of WMSNs of medium-resolution video sensor nodes [39]. Similar to the SensEye testbed, Stargate boards interfaced with Logitech webcams are used in the testbed. Communication between nodes in the network is performed through IEEE 802.11 wireless network cards connected to each Stargate board.

The Meerkats testbed is used to measure the energy consumption for different types of operations that are typical of a WMSN. More specifically, five main categories are investigated: idle, processing, storage, communication, and visual sensing. These categories build up a benchmark for energy consumption in WMSNs. Both steady state and transient energy consumption behavior are obtained by direct measurements of current with a digital multimeter. Shown in Figure 15.10 is the energy consumption for the five major operations as well as sleep mode, where all the circuits on the sensor are switched to sleep. Furthermore, the breakdown of the energy consumption in the processor, sensor, and radio can also be observed. An important observation is that image processing applications as well as reading from or writing to flash memory are more energy consuming than communication. This contradicts the conventional balance of energy consumption for processing and communication in WSNs, which favors processing. Moreover, visual sensing consumes almost as much energy as communication, which is also an important characteristic of WMSNs. In WMSNs, delays and additional amounts of energy consumed due to transitions (e.g., to go to sleep mode) are not negligible and must be accounted for in the network and protocol design.

IrisNet

IrisNet (Internet-scale Resource-Intensive Sensor Network Services) [45] is an example software platform for deploying heterogeneous services on WMSNs. IrisNet allows a global, wide area sensor network to be harnessed by performing Internet-like queries on this infrastructure. Video sensors and scalar sensors are spread throughout the environment, and collect potentially useful data. IrisNet allows users to perform Internet-like queries to video sensors. The user views the sensor network as a single unit that can be queried through a high-level language. Each query operates on data collected from the global sensor network, and allows simple Google-like queries as well as more complex queries involving arithmetic and database operators.

The architecture of IrisNet is two tiered: heterogeneous sensors implement a common shared interface and are called *sensing agents* (SAs), while the data produced by sensors is stored in a distributed database that is implemented on *organizing agents* (OAs). Different *sensing services* run simultaneously

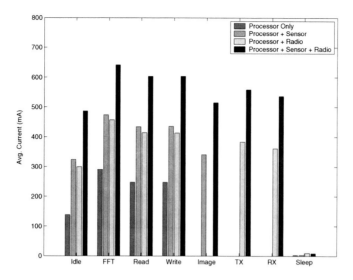

Figure 15.10 The average current consumed for various tasks in a multimedia testbed and the distribution among node components [39].

on the architecture. Hence, the same hardware infrastructure can provide different sensing services. For example, a set of video sensors can provide a parking space finder service, as well as a surveillance service. Sensor data are represented in the Extensible Markup Language (XML), which allows easy organization of hierarchical data. A group of OAs are responsible for a sensing service, collect data produced by that service, and organize the information in a distributed database to answer the class of relevant queries. IrisNet also allows sensors to be programmed with filtering code that processes sensor readings in a service-specific way. A single SA can execute several such software filters (called *senselets*) that process the raw sensor data based on the requirements of the service that needs to access the data. After senselet processing, the distilled information is sent to a nearby OA.

15.4 Physical Layer

The physical layer for currently existing multimedia sensors consists of IEEE 802.11 WLAN cards, Bluetooth, or IEEE 802.15.4 ZigBee-compatible transceivers. Due to the high volume of traffic generated by multimedia applications, some of these physical layer technologies may not be the preferred technique for WMSNs. Especially, providing multimedia communication through low-data-rate ZigBee-compatible transceivers is a major challenge. On the other hand, while WLAN cards provide excessive bandwidth that is sufficient for the transmission of still images and video, the energy consumed by this technology is significantly high. As a result, high-density deployments with a high number of video sensors would not be feasible for the realization of WMSNs. Hence, it is imperative that new physical layer technologies are adapted to the low-power, high-data-rate, multimedia communication needs of WMSNs.

In this section, we discuss the ultra-wide-band technology that has the potential to support higher data rates at low power. Ultra wide band (UWB) is a wireless radio technology designed to transmit data at very high bandwidths ($\sim$ 480 Mbps) within short ranges ($\sim$ 10 m) using little power [43]. UWB has been used in radars for decades and recently utilized for wireless personal area networks (WPANs) to exchange data between consumer electronics, home theater systems, PCs, PC peripherals, and mobile

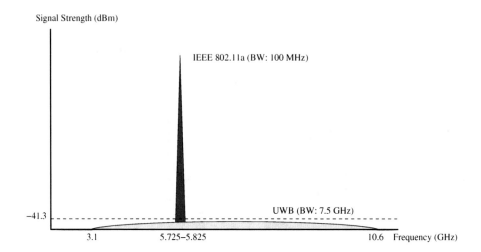

Figure 15.11   Illustration of the spectrum of UWB and the coexistence with IEEE 802.11a at 5 GHz [37].

devices. The low-power operation and the high bandwidth provided by this technology also make it a potential candidate for multimedia sensor communication.

The realization of UWB technology was made possible through the recent FCC Report and Order issued in February 2002 that allowed a spectrum of 7.5 GHz for unlicensed use of UWB devices [11]. This spectrum was chosen as the 3.1–10.6 GHz band and is the largest allocation for unlicensed usage so far. The UWB signal is defined by the FCC to have a bandwidth of at least 500 MHz with effective isotropic radiated power (EIRP) below −41.3 dBm. The large bandwidth lends its name to the UWB signal. Moreover, the limited EIRP of the UWB signal minimizes interference with other devices in this huge spectrum.

The interference limitation governed by the FCC regulations enables UWB devices to operate in a large spectrum band that is occupied by existing radio services without causing interference. Most existing techniques depend on dividing the spectrum into bands and avoid transmitting at the same bands simultaneously. Instead, the UWB technology relies on transmitting signals at low power over a large spectrum such that the effect of transmission is regarded as noise by other devices. This is illustrated in Figure 15.11, where the coexistence of UWB communication and IEEE 802.11a that operates in the 5 GHz band is shown. Because of the very low power at a given spectrum band, UWB communication minimizes interference to other users. This ensures that existing wireless networks and services are protected.

In addition to preventing interference with other networks, the realization of UWB communication also requires multi-access techniques that govern multiple UWB transceivers transmitting in close proximity. There exist two main variants of UWB that are classified according to the multi-access scheme deployed: time-hopping impulse radio UWB (TH-IR-UWB) and multicarrier UWB (MC-UWB). We will describe these two schemes next.

15.4.1   Time-Hopping Impulse Radio UWB (TH-IR-UWB)

In this scheme, the transmission scheme is based on a time-slotted structure much like a TDMA scheme. The time is divided into frames, each of which consists several chips. The information is transmitted during these very short-duration chips. In order to prevent collisions with multiple transmitters, each transmitter selects a different chip at each frame in a pseudo-random manner.

An important requirement of TH-IR-UWB is the extremely precise synchronization for transmitting very short pulses. This requires transceivers capable of very fast switching. A major advantage of TH-IR-UWB systems is that simple transceivers can be very inexpensive to construct.

15.4.2   Multicarrier UWB (MC-UWB)

MC-UWB is based on orthogonal frequency division multiplexing (OFDM) such that information is transmitted through multiple carriers simultaneously [23]. Consequently, the vulnerability to interference can be reduced by choosing carriers such that narrow-band interference is avoided. Moreover, MC-UWB provides more flexibility and scalability, but requires an extra layer of control at the physical layer. Implementing a MC-UWB front-end can be challenging, compared to TH-IR-UWB, due to the continuous variations in power over a wide bandwidth. This is particularly the case for the power amplifier.

15.4.3   Distance Measurements through UWB

The high bandwidth produced through UWB communication provides many capabilities to transceivers. One such feature is very precise distance measurements through UWB signals. Positioning capabilities are needed to associate physical meaning to the information gathered by sensors. Moreover, knowledge of the position of each network device allows for scalable routing solutions. As we discussed in Chapter 12, angle-of-arrival localization techniques and signal strength-based techniques usually suffer from non-line-of-sight and multi-path effects. In this respect, UWB techniques allow ranging accuracy on the order of centimeters [27], which is crucial in WMSNs.

15.5   MAC Layer

The main challenges in the MAC layer and solutions for WSNs were discussed in Chapter 5. The peculiarities of multimedia traffic, however, impose additional challenges for MAC protocol development. Specifically, the MAC protocols need to enable reliable data transfer with minimum energy expenditure while supporting application-specific QoS requirements. While protocols in WSNs are usually tailored for a single type of traffic, multimedia traffic, namely audio, video, and still images, requires service differentiation. Even a flow associated to a single image transfer contains a separate class of information due to the different types of information generated by the state-of-the-art compression techniques. As an example, the JPEG 2000 standard, which was developed for image compression [6], provides the flexibility to encode different parts of an image at different resolutions, i.e., region of interest (ROI). Consequently, even from a single image, different types of traffic can be generated that require various prioritization support throughout the network. Consequently, the MAC layer should support prioritization mechanisms and QoS.

In the following, we investigate the channel access policies for WMSNs. As explained in Chapter 5, the main causes of energy loss in sensor networks are related to *packet collisions* and subsequent retransmissions, *overhearing packets* destined for other nodes, and *idle listening*, a state in which the transceiver circuits remain active even in the absence of data transfer. Thus, regulating access to the channel assumes primary importance and several solutions have been proposed in the literature. Next, we overview three major channel access policies and how they can be applied to WMSNs. Furthermore, we will describe two major solutions that have been developed for WMSNs:

- **Contention-based protocols:** Contention-based protocols developed for WSNs usually assume a single radio architecture. To improve energy efficiency, the devices are put to sleep whenever they are not needed. However, the applicability of existing contention-based protocols to multimedia traffic is limited, primarily because of the strict QoS requirements of this type of traffic.

Many contention-based protocols developed for WSNs aim to reduce energy consumption at the cost of latency and throughput degradation. While energy consumption is still a major constraint in WMSNs, protocols need to provide strict bounds on latency and throughput for efficient delivery of multimedia traffic. This requires new duty cycle calculations based on permissible end-to-end delay needs. Furthermore, current compression algorithms for still image and video provide variable output rates, which mainly depend on the contents of the image or video. Consequently, fixed duty cycle-based operations are not applicable for this type of dynamic traffic.

- **Contention-free protocols:** Contention-free protocols usually exploit reservation schemes such as a slotted time frame to coordinate the traffic in a localized region through clusters. The bursty nature of the multimedia traffic and the high-bandwidth requirements impose additional challenges for protocol design for contention-free protocols.

  The strict latency requirements of real-time streaming video require scheduling policies such that packets with shorter delay tolerance are given priority in the network. Consequently, the MAC protocol should also provide prioritization to nodes according to their packets. Based on the allowed packet loss of the multimedia stream, the dependencies between packet dropping rate, arrival rate, and delay tolerance [51] can be used to decide the TDMA frame structure and schedules of individual nodes. Especially, variable TDMA (V-TDMA) schemes are preferable when heterogeneous traffic is present in the network. Since real-time streaming media are delay bounded, the link layer latency introduced in a given flow in order to satisfy data-rate requirements of another flow needs to be analyzed well when V-TDMA schemes are used.

- **Multi-channel protocols:** Existing hardware platforms developed for WSNs usually provide data rates on the order of hundreds of kbps. The fact that this bandwidth is also shared by multiple competing devices in the wireless channel also degrades the available bandwidth for a specific flow. Furthermore, multi-hop communication has adverse affects on the bandwidth that a traffic experiences. Considering these effects, the available bandwidth provided by conventional WSNs is not sufficient for multimedia traffic. While developing new low-cost communication technologies is a clear-cut solution, an imminent solution can also be found through multi-channel protocols. Current wireless sensor devices such as the Crossbow Mica-family [3] and Rockwell's WINS [10] provide the capability to operate in multiple frequency bands and change these bands in real time. By using multiple channels in a spatially overlapped manner, existing bandwidth can be efficiently utilized for supporting multimedia applications. An important challenge in this aspect is to develop distributed channel assignment protocols to adapt to both the dynamic nature of the multimedia traffic and topology considerations. Furthermore, the increased latency due to switching to a different channel [33] needs to be carefully accounted for, as its cumulative nature at each hop affects real-time media.

Another common technique is to use a low-cost transceiver in addition to the communication radio as we saw in Chapter 5 to develop wakeup schemes. These schemes can be used to develop wakeup mechanisms that will adapt to the dynamic and bursty nature of the multimedia traffic. For example, enforcing wakeups based on the packet backlog is a better design approach compared to those with static timer-based schemes [40].

The MAC layer solutions developed for WMSNs mainly exploit contention-free multi-channel access principles to provide the QoS guarantees required by WMSNs. In the following, we explain two variations of this idea, i.e., the FRASH [16] and RICH [35] protocols. Furthermore, exploitation of multiple input, multiple output (MIMO) technology at the MAC layer is also discussed.

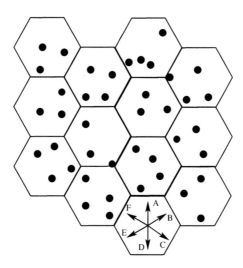

Figure 15.12 Architecture assumed for the FRASH MAC protocol.

15.5.1 Frame Sharing (FRASH) MAC Protocol

One of the major challenges in contention-free protocols is the design of the frame structure and the schedules of each node in the time frames. Moreover, since MAC protocols for WMSNs should also provide QoS requirements for different flows, scheduling becomes a major challenge in this area. The FRASH MAC protocol [16] aims to address this challenge by adapting the earliest deadline first (EDF) scheduling scheme distributively.

EDF scheduling has been used in real-time multimedia protocols in the Internet [34]. The main idea behind the EDF approach is to label each packet in the network with a deadline to reach a destination. This deadline can be specified according to the requirements of the application. As the packets are inserted into the network, each node schedules the received packets according to their deadlines. As a result, as the name implies, the packets with earlier deadlines have higher priority over nodes with later deadlines even if they have been received later by the node.

The FRASH MAC protocol exploits the EDF scheduling algorithm in a TDMA-based medium access mechanism. The network structure is designed such that each node in the network is organized into cells motivated by the architecture of cellular networks as shown in Figure 15.12. Accordingly, each node in a cell can communicate with any node inside the cell. Moreover, the traffic generated in the cell for network-wide usage is transmitted to other cells through specialized nodes called *router nodes*. Inter-cell interference in this architecture is prevented by assigning different operating frequencies to each cell in the network following the spatial reuse property generally used in cellular networks. As an example, seven different channels are used in the entire network such that no neighbor cell uses the same frequency. The nodes inside the cell communicate through multicast messages at the same frequency. Inter-cell communication is performed through router nodes that are able to communicate in various frequency channels to the neighbor router nodes.

Based on the cellular architecture, the FRASH MAC protocol is designed for periodic multimedia traffic. Each node in the network schedules its transmission according to the EDF scheme. An important assumption in the FRASH protocol is that each node generates sensory information in a periodic manner. Consequently, each node in a cell can determine the transmission schedule of other nodes. Based on this information, each node *replicates a global EDF schedule* for packet transmission. The periodic nature of the traffic assumed in this algorithm removes the need for explicit contention control mechanisms.

A direction	Intra-cell Frame	B direction	Intra-cell Frame	C direction	Intra-cell Frame	D direction	Intra-cell Frame	E direction	Intra-cell Frame	F direction

Figure 15.13   The frame structure of the FRASH protocol.

The frame structure of the FRASH protocol is shown in Figure 15.13. The time is divided into frames such that inter-cell and intra-cell communication are divided. Moreover, the inter-cell communication frames are allocated according to the communication direction in the hexagonal cell structure. In each intra-cell frame, the time is further divided into slots. In this frame, nodes are scheduled according to the EDF scheme. In case a node does not use its allocated slot, other nodes can use this slot for aperiodic messages. Furthermore, the information generated in the cell is then transferred to other cells by the router nodes during the intra-cell frame.

The FRASH protocol provides explicit delay bounds for the generated traffic in the network. The EDF mechanism is exploited for QoS provision. The drawback of the protocol is that the periodic traffic is assumed to distributively implement the EDF mechanism. Therefore, event-based applications, where the network generates significant amounts of data only in the case of object detection, cannot be supported. Another drawback of the FRASH protocol is that strict synchronization between nodes is required to implement the TDMA structure.

15.5.2   Real-Time Independent Channels (RICH) MAC Protocol

The RICH MAC protocol [35] aims to provide QoS support to WMSNs by exploiting CDMA communication. The drawback of the FRASH protocol, i.e., that synchronized operation is required for the TDMA-based scheme, is overcome by exploiting the CDMA principles. Consequently, each node in the network is assigned a specific pseudo-random CDMA code such that interference between neighbor nodes is minimized.

The RICH protocol is also based on the cellular architecture assumed by the FRASH protocol. Each node is assumed to be equipped with seven different radios such that a node can receive from six nodes and transmit at a different channel simultaneously. Consequently, simultaneous transmissions between nodes are possible. Accordingly, each node follows a distributed EDF scheduling algorithm. In this algorithm, the sampling rates and the transmission frequency of each node are optimized according to the available bandwidth of its neighbors.

The RICH protocol exploits the information between each neighbor to perform a localized optimization at each node for EDF scheduling. Accordingly, fast convergence to the centralized optimal result can be achieved. Based on these principles, the RICH protocol provides strict deadlines for different traffic classes in a WMSN. However, the main drawback of this protocol is that very complex transceiver architectures are required to support simultaneous CDMA communication at each node. While this provides great flexibility to the MAC protocol, many WMSN applications that require simple and low-cost components may not be amenable to this approach.

15.5.3   MIMO Technology

The high data rate required by multimedia applications can be addressed by spatial multiplexing in MIMO systems that use a single channel but employ interference cancellation techniques. Recently, *virtual* MIMO schemes have been proposed for sensor networks [30], where nodes in close proximity form a cluster. Each sensor functions as a single antenna element, sharing information and thus simulating the operation of a multiple antenna array. A distributed compression scheme for correlated sensor data that specially addresses multimedia requirements is integrated with the MIMO framework in [31]. However, a key consideration in MIMO-based systems is the number of sensor transmissions and the required signal energy per transmission. As the complexity is shifted from hardware to sensor coordination, further research is needed at the MAC layer to ensure that the required MIMO parameters, like channel state and desired diversity/processing gain, are known to both the sender and receiver at an acceptable energy cost.

15.5.4   Open Research Issues

The MAC layer provides many capabilities that are crucial for efficient communication in WMSNs. In addition to the already existing solutions, there are still open research issues to be addressed in this area.

- In TDMA-based schemes, bounds on unequal slot/frame lengths for differentiated services should be decided by the allowed per-hop delay (and consequently end-to-end delay). Schedules, once created, should also be able to account for a dynamically changing topology due to nodes dying off or new ones being added.

- Multi-channel protocols utilize bandwidth better, and thus may perform favorably in those cases when applications demand a high data rate [33]. An open question is whether switching delay can be successfully hidden with only one interface per node. If this is possible, it may greatly simplify the design of communication protocols while also performing as a multi-channel, multi-interface solution. Also, the sleep–awake schedule of the radios should be made dynamic in order to accommodate the varying frame rate for video sensors.

- Multi-channel protocols are not completely collision free because of their susceptibility to control packet collisions [19, 42, 52]. All available channels cannot be assumed to be perfectly non-overlapping. This may require dynamic channel assignment, taking into account the effect of adjacent channel interference, in order to maintain the network QoS.

- Coordinating link and transport layer error recovery schemes is a challenge that remains to be addressed. In order to ensure that buffer overflow conditions do not occur, mechanisms that detect increased MAC level contention and regulate data generation rate could be implemented.

- In-network storage schemes [15], in which data are stored within the sensor node itself and accessed on demand, may further reduce available buffer capacity of the sensor nodes. Thus allocation of optimum buffer sizes truly merits a cross-layer approach, spanning application layer architecture to existing channel conditions seen at the physical layer.

15.6   Error Control

The low-power wireless channel results in high error rates that need to be minimized for efficient multimedia transfer. This increased effect of the wireless channel has been considered in many networking protocols developed for WSNs by incorporating error control schemes such as forward error correction (FEC) to multi-hop communication. While these effects should still be considered, streaming of real-time multimedia data over a sensor network brings new challenges. Since the QoS requirements of audio or video streams as well as still images need to be guaranteed, highly unreliable communication links need to be prevented. As we saw in Chapter 6, the selection of a suitable error control scheme mainly depends on the requirements of the application as well as the characteristics of the network.

Error control for WMSNs exhibits fundamentally different characteristics due to the content of the data to be transmitted. Contrary to WSNs, where each packet of a flow is equally important for the application, the multimedia traffic includes *unequal importance* between packets. More specifically, not all packets have the same importance for correct reconstruction of the multimedia content. Moreover, a packet that contains some errors can still be used in reconstructing the multimedia content and the original content can be reconstructed with tolerable distortion. The effects of unequal importance can be overcome through *unequal protection*, where variable strength FEC code is applied to different parts of the video stream depending on their relative importance.

Delivering error-resilient multimedia content and minimizing energy consumption are contradicting objectives. Additionally, the multimedia traffic type also influences the error control scheme to be employed. Although multimedia transfer over wireless channels has been investigated comprehensively, the energy constraints of the sensor nodes make most of these solutions impractical. Therefore, energy-efficient error control schemes are yet to be developed in this area. As we will see next, energy consumption can be jointly reduced by optimizing source coding, channel coding, and transmission power control parameters. The image coding and transmission strategies are adaptively adjusted to match current channel conditions by exploiting the characteristics of multimedia data, such as unequal importance.

15.6.1   Joint Source Channel Coding and Power Control

The Joint Source Channel Coding and Power Control (JSCCPC) protocol, which considers a single hop WMSN, is an initial step in providing error control in multi-hop networks [56]. More specifically, energy-efficient image transfer is considered through an adaptive error control scheme, where the QoS requirements are also met. This is accomplished through determining source and channel coding parameters as well as transmit power for the transmission.

In WMSNs, image data need to be transmitted in a compressed form due to the significantly low bandwidth. One of the efficient compressing schemes is JPEG 2000, which uses wavelet-based image compression such that the image is compressed into several resolution levels each of which consists of subbands. This layering provides dynamic packetization such that packets are generated according to the bit-rate constraint and the image distortion requirement. In other words, by using a combination of different layers, higher compression can be performed at the cost of increased distortion. The packets associated to each layer are then further encoded using FEC codes for channel error protection.

JSCCPC leverages channel information for the design of an optimal number of layers used for transmission as well as the FEC code power, and the transmission power. The design is realized such that the energy consumption is minimized and the distortion constraints are met. More specifically, JSCCPC determines a class of configurations called *layers* such that, at each layer, a combination of encoding quality for the image compression, the FEC code power for protection, and transmission power level are determined. The packets are protected through rate-compatible punctured convolutional (RCPC) codes that enable multiple levels of protection through the same coding structure.

A transmission strategy for a layer in JSCCPC is denoted by $s_l = \langle s_l^s, s_l^c, s_l^r, s_l^p \rangle$, where

- $s_l^s \in S_l^s$ is the level of source error resilient coding,
- $s_l^c \in S_l^c$ is the RCPC code rate,
- $s_l^r \in S_l^r$ is the header protection strength,
- $s_l^p \in S_l^p$ is the transmit power level,

and $\bar{s} = \langle s_1, s_2, \ldots s_L \rangle$ is the transmission strategy of the whole image. The goal of JSCCPC is to find the best strategy $\bar{s}$ such that

$$\arg \min_s E(\bar{s}) \text{ s.t. } D(\bar{s}) \le D_{\max} \tag{15.1}$$

where $E(\bar{s})$ is the overall energy consumption with the strategy $\bar{s}$, and $D(\bar{s})$ is the resulting distortion. Next, we describe how these are determined.

Considering an image that contains L layers, the overall expected gain in applying L levels of encoding to the image is given as

$$G(\bar{s}) = \sum_{l=1}^{L} G_l(s_l). \tag{15.2}$$

Therefore, the overall distortion is the difference between the expected gain, $G(\bar{s})$, and the distortion, D_{total}, when no layers are received:

$$D(\bar{s}) = D_{total} - G(\bar{s}).$$ (15.3)

Similarly, the energy consumption is the sum of the amount of energy consumed for each layer l

$$E(\bar{s}) = \sum_{l=1}^{L} E_l(s_l).$$ (15.4)

The operation of JSCCPC depends on the estimation of the energy consumption and the distortion for each layer l. Consequently, the most efficient energy consumption is selected by an iterative approach. The feasible transmission strategies are first ordered according to their energy consumption. Consequently, for layer l, switching from strategy s_l to s'_l results in an increase in energy consumption and a decrease in distortion given by

$$\Delta E_l(s_l, s'_l) = E_l(s'_l) - E_l(s_l),$$ (15.5)

$$\Delta G_l(s_l, s'_l) = G_l(s'_l) - G_l(s_l),$$ (15.6)

respectively.

Defining the normalized gain as

$$g_l(s_l, s'_l) = \frac{\Delta G_l(s_l, s'_l)}{\Delta E_l(s_l, s'_l)},$$ (15.7)

then the feasible strategies are selected according to the following conditions:

$$\min_{k<i} g_l(s^k, s^i) > 0,$$ (15.8)

$$\min_{k<i} g_l(s^k, s^i) > \max_{j>i} g_l(s^i, s^j),$$ (15.9)

which ensure that the selected strategy is within the acceptable energy and gain values for a particular layer. As a result, a strategy for each layer is determined by selecting the strategy that results in minimum energy consumption while meeting both the feasibility and distortion constraints.

The dynamic nature of JSCCP provides energy-efficient and reliable image transmission over single hop WMSNs. Many degrees of freedom are provided in terms of source and channel coding as well as transmission power control. The lessons learned through this algorithm can help design error control mechanisms that are also efficient in multi-hop scenarios.

15.6.2   Open Research Issues

There exist several open research issues for the development of MAC protocols for WMSNs, as outlined below:

- Error control schemes developed for WSNs usually disregard the content of data while focusing on the peculiarities of the low-power communication. In WMSNs, there is a need to develop models and algorithms to integrate source and channel coding schemes in existing error control schemes.
- Since multimedia data are usually error tolerant, new packet dropping schemes for multimedia delivery have to be delivered for buffer management. New schemes should be capable of selectively dropping packets that will not impact the perceived quality at the end-user.

- Energy-constrained sensor networks naturally call for selective repeat ARQ techniques. However, this can introduce excessive latency. There is a need to study tradeoffs between the degree of reliability required (i.e., acceptable packet error rate) and the sustainable delay at the application layer.

- Packet length clearly has a bearing on reliable link level communication and may be adjusted according to application delay sensitivity requirements. More specifically, the problems associated with burst-error channel and the adaptation of video quality have to be considering in determining an optimal packet length. Depending on the information content of the data and the channel conditions, variable length FEC codes can be used to reduce the effects of transmission errors at the decoder.

15.7 Network Layer

Research on the network layer, especially on routing protocols, has always been the driving force behind the development of WSNs. The same is true for WMSNs, considering many of the challenges of delivering multimedia traffic over a multi-hop network. Especially, delivering QoS guarantees for different types of traffic is a challenging task considering the energy and processing constraints of sensor nodes and the unreliability of communication through the wireless channel. In this section, we discuss the existing research directions for the network layer functionalities, while stressing their applicability to delay-sensitive and high-bandwidth needs.

The concerns of routing in WSNs differ significantly from the specialized service requirements of multimedia streaming applications. The desired data rate of an application may not be provided by any path in the network, which requires the routing algorithm to revert to multi-hop routing techniques. Similarly, the latency requirements of a streaming application require accurate resource estimation and reservation schemes at the routing layer with support to provide this estimated bound on latency. In this section, we discuss various routing protocols for WMSNs. Before going into the details of each protocol, it is important to list the features which must be possessed by a routing protocol for WMSNs. More specifically, the network layer for WMSNs should consider the following three major points in route establishment:

- **Network conditions:** A common issue for any routing protocol is that the network layer should consider the network state in many respects. Network conditions may include interference seen at intermediate hops, the number of backlogged flows along a path, the residual energy of the nodes, as well as channel conditions. A routing decision based on these metrics can avoid paths that may not support high-bandwidth applications or introduce retransmission because of bad channel conditions.

- **Traffic classes:** As we saw in Section 15.3, a WMSN may consist of heterogeneous types of nodes that are able to collect audio, video, and still images. The nature and requirements of the types of data are different for efficient "reconstruction" at the sink. Furthermore, the same types of sensors can generate various types of data depending on the queries associated to the user. Consequently, the content and nature of the sensed data also vary. This requires different priorities assigned to video data, audio traffic, and images. In parallel, different types of route establishment techniques may be applied to these types of traffic to guarantee heterogeneous QoS requirements.

- **Support for streaming:** The real-time nature of some applications such as audio and video transfer imposes additional challenges on the network layer that have not been considered before. Although solutions from multimedia delivery in wireless cellular or ad hoc networks can be applied, the unique properties of wireless sensor nodes prohibit many of these applications. Consequently, strict QoS guarantees need to be provided instead of the best effort approaches used so far in WSNs.

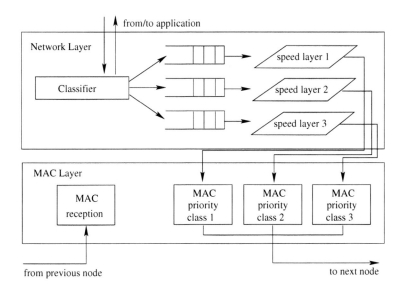

Figure 15.14   The MMSPEED protocol architecture [25].

In the following, we describe the MMSPEED protocol that exploits many of these features to provide multimedia delivery in WMSNs.

15.7.1   Multi-path and Multi-speed Routing (MMSPEED) Protocol

Providing reliability and end-to-end delay guarantees for different types of flows is a major challenge in the design of routing protocols for WMSNs. An important step in this area is the MMSPEED protocol, which provides probabilistic bounds on both reliability and delay for different traffic types [24, 25].

MMSPEED builds on the design principles of the SPEED protocol, which was described in Chapter 7. The SPEED protocol provides an end-to-end delay guarantee for a flow based on the *speed* concept. The speed associates the distance advanced by a packet in a hop with the delay in the transmission of the packet. Consequently, based on the location of a node, an end-to-end guarantee can be provided. However, SPEED can only support a single type of traffic and trades off reliability for delay guarantees. On the other hand, multimedia traffic requires priority support for various traffic flows as well as reliability guarantees.

MMSPEED provides multiple levels of prioritization for different flows through virtually isolated *speed layers*. At each speed layer, the SPEED protocol is performed with different speed guarantees. On top of this mechanism, three prioritization mechanisms are implemented to provide different QoS guarantees for different flows. The protocol architecture spans the network and MAC layers as shown in Figure 15.14 and the priority mechanisms are implemented at these layers.

The first prioritization mechanism is implemented at each node queue. Each node has separate queues for each priority level. The packets at the highest priority queue are transmitted before the lower priority queues. This eliminates the effect of lower priority packets on the end-to-end delay of higher priority traffic.

The second prioritization mechanism is performed at the MAC layer as shown in Figure 15.14. This mechanism provides a higher probability of channel access to nodes with higher priority packets. For each priority level, different interframe spacings (IFS) and backoff window lengths are determined.

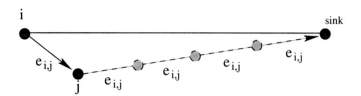

Figure 15.15 At each hop, the local error rate and the hop distance are assumed to be the same for the following hops to estimate the reachability probability [25].

Consequently, a node with a higher priority packet performs backoff for a shorter duration, which increases its chance to access the channel compared to nodes with lower priority packets. In addition, the medium access delay for higher priority packets is also decreased through this prioritization mechanism.

The first two prioritization mechanisms of MMSPEED improve the delay performance for different types of traffic. However, since the SPEED mechanism explained in Chapter 7 is used for each speed layer, reliability is not provided. This problem is addressed by assigning multiple routes for high-priority traffic. This is accomplished first by determining a target reliability level for the traffic type. This level can be determined based on the QoS requirements of each traffic type. For example, a real-time video streaming application may compensate for loss frames and hence a lower target reliability level can be chosen, e.g., 0.2. In this case, single path routing would be sufficient to achieve the low target reliability level without high energy consumption. On the other hand, for compressed image transfer, a higher reliability level, e.g., 0.8, is necessary since the redundancy of the image to be transferred is already filtered by the compression. In this case, MMSPEED exploits multi-path routing to improve the reliability of the packet. More specifically, a node forwards the packet to more than one node according to the loss rates of its neighbors.

Once a target level is set, the originator of the traffic, e.g., node i, can determine the number of multiple paths according to the reliability levels of each of its neighbors, e.g., node j. To this end, each node j maintains the recent average of packet loss ratios and informs its neighbors. As a result, node i has the average packet loss ratio between itself and node j, $e_{i,j}$. Accordingly, an end-to-end *reachability probability* from node i to the sink s via node j can be calculated as follows:

$$RP^s_{i,j} = (1 - e_{i,j})(1 - e_{i,j})^{\lceil d_{j,s}/d_{i,j} \rceil} \tag{15.10}$$

where $e_{i,j}$ is the average packet loss percentage between nodes i and j, $d_{j,s}$ and $d_{i,j}$ are the distances between node j and the sink, and between nodes i and j, respectively. This estimate for the end-to-end reachability probability has the following assumptions as illustrated in Figure 15.15: (1) the error rate at each hop from node i to the sink s is the same as the error rate at the first hop, i.e., between nodes i and j, $e_{i,j}$; (2) the hop distance at each hop is the same as the first hop distance, $d_{i,j}$, hence the number of hops between node j and the sink is $d_{j,s}/d_{i,j}$.

Using the reachability probability estimate in (15.10), the number of multiple paths that will ensure the target reliability level can be found iteratively by adding additional paths. To this end, the *total reaching probability* (TRP) of a packet is given by

$$TRP = 1 - (1 - TRP')(1 - RP^s_{i,j}) \tag{15.11}$$

where TRP' is the total reaching probability before adding the path with node j. By calculating the TRP, the target reliability level can be found by iteratively adding a path through one of the neighbors in the vicinity.

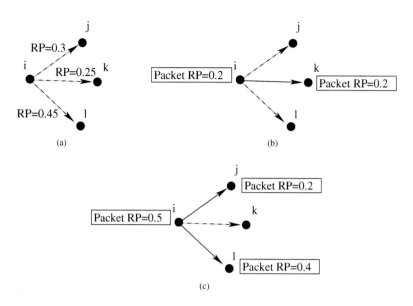

Figure 15.16   MMSPEED protocol operation: (a) an example topology with three potential neighbors, (b) single path route selection for low-reliability traffic, (c) multi-path route selection for high-reliability traffic.

■ **EXAMPLE 15.2**

An example of MMSPEED protocol operation is shown in Figure 15.16(a), where node i has three neighbors j, k, and l with estimated reachability probabilities of $0.3, 0.25$, and $.0.45$, respectively. If node i has a packet with a target reliability level of 0.2, one of the neighbor nodes can be used to forward the packet as shown in Figure 15.16(b), where node k is selected with a required reliability level of 0.2. However, if a packet with a target reliability level of 0.5 needs to be forwarded, none of the immediate neighbors provide such a reliability alone and multiple paths should be selected. Accordingly, node i selects nodes j and l since the total reaching probability is $1 - (1 - 0.3)(1 - 0.45) > 0.5$ as shown in Figure 15.16(c).

The reachability probability RP given in (15.10) assumes a fixed packet loss percentage and hop length for each hop in the end-to-end path. However, dynamic changes in both hop length and packet loss percentage can occur. MMSPEED reacts to these changes by assigning new target reliability levels for each chosen branch so that the original target reliability level can still be met. According to the scheme in Example 15.2, the new target probabilities are calculated as 0.2 for node j and 0.4 for node l. The changes in reachability probability and the hop distance for the following hops for these two nodes are then compensated dynamically. With this dynamic compensation, the end-to-end reaching probability is likely met, although the packet forwarding decision is made locally.

■ **EXAMPLE 15.3**

The dynamic compensation property of MMSPEED is illustrated in Figure 15.17. In this case, the source node i_1 generates a packet that needs to be transmitted with a reachability target of $P^{req} = 0.8$. Now, suppose that this node has two neighbors i_2 and i_3 with reachability probabilities of $RP^s_{i_1,i_2} = 0.7$ and $RP^s_{i_1,i_3} = 0.6$, respectively. Accordingly, choosing these nodes satisfies the

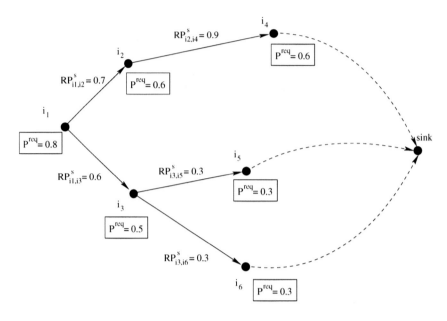

Figure 15.17   An example for the dynamic compensation in MMSPEED [25].

reachability target since $1 - (1 - 0.7)(1 - 0.6) = 0.88 > 0.8$. When node i_1 transmits the packet to nodes i_2 and i_3 the updated reachability targets are assigned as 0.6 and 0.5 respectively.

At the second hops, nodes i_2 and i_3 make routing decisions according to their local estimates of reachability probability for their neighbors. Accordingly, node i_2 finds node i_4 with $RP_{i_2,i_4} = 0.9 > 0.6$, which satisfies the reachability target. However, node i_3 cannot ensure the assigned reachability target 0.5 through any of its neighbors i_5 and i_6 since they both have a RP of 0.3. Then, node i_3 creates another branch and forwards the packet to both of its neighbors. Accordingly, each following node dynamically compensates the previous wrong decision as the packet travels to the final destination, branching more paths throughout the progress of the packet as shown in Figure 15.17.

To decrease the delay and energy consumption associated with creating multiple paths from a single node, MMSPEED uses multicast at the MAC layer to transmit the packet to several next hops simultaneously.

Overall, MMSPEED provides delay and reliability support for different traffic types through the virtual layering mechanism. A cross-layer approach including buffer management, routing, and MAC is developed. Although strict delay and jitter bounds, which are crucial for multimedia streaming, cannot be supported, MMSPEED is an important step in providing QoS in WMSNs.

15.7.2   Open Research Issues

There exist many open research issues for the development of efficient routing solutions for WMSNs, as outlined below:

- The identification of the optimal routing metrics is an important subject for research. Most routing protocols that consider more than one metric, such as energy, delay, etc., form a cost function that is then minimized. The choice of the weights for these metrics needs to be undertaken, and is often subject to dynamic network conditions. Thus, further work is needed to shift this decision-making process and network tuning from the user end into the network.

- As the connectivity between different domains improves, end-to-end QOS guarantees are complicated by the inherent differences in the nature of the wired and wireless media. When sensed data from the field are sent via the Internet, a single routing metric is unsuitable for the entire path between source and end-user. Decoupling reliability and routing parameters at such network boundaries and a seamless integration of schemes better suited to wired or wireless domains, respectively, need to be explored.

15.8 Transport Layer

The high-data-rate traffic generated by multimedia applications makes the transport layer solutions an important part of communication in WMSNs. More specifically, providing end-to-end reliability and congestion control mechanisms in an energy-constrained network is a challenging task. Particularly in WMSNs, several additional considerations are in order to accommodate both the unique characteristics of the WSN paradigm and multimedia transport requirements.

Conventional transport protocols that have been developed for WSNs as explained in Chapter 8 focus on event-based reliability. In other words, information regarding a geographical area is important rather than the individual items of data from each of the sensors in this area. Consequently, a collective reliability notion is generally adopted in WSNs [12]. Although this notion may still be true for WMSNs, the contents of the transmitted packets are also important. Consequently, a transport protocol should consider the interdependencies and the priority levels of transmitted packets to provide a high-quality reconstruction of the images or videos at the sink.

In traditional WSNs, when congestion occurs, packets are randomly dropped. However, in WMSNs, transmitted packets usually depend on other packets for reliable reconstruction at the sink node. As we will explain in Section 15.9, image and video coding algorithms usually result in multiple priority levels. As an example, the MPEG-4 algorithm generates multiple P-frames that depend on an I-frame. Consequently, congestion in a WMSN should be avoided considering these interdependencies and priority levels inherent in the multimedia traffic. Based on these unique properties of multimedia traffic, the transport layer should consider the following main topics in WMSNs:

- **Congestion control:** The high load due to multimedia transfer makes congestion control very important in WMSNs. While typical transmission rates for sensor nodes are between 40 and 250 kbps, indicative data rates of constant bit-rate voice traffic may be 64 kbps while video traffic, on the other hand, may be bursty and on the order of 500 kbps [57]. The effect of congestion is even more pronounced considering also that multiple simultaneous traffic may be flowing through the network. Consequently, if congestion occurs in the network, it leads to both performance degradation and energy depletion in each node. Traditional congestion control mechanisms react to congestion *after* it occurs in the network. For WMSNs, where congestion has adverse effects, control mechanisms are necessary which prevent congestion *before* it occurs.

- **Reliability:** Reliability has always been the primary concern of transport protocols in both traditional wireless networks and WSNs. The content of multimedia streams, however, requires the notion of reliability to be rethought. This is because the reliability notion in a multimedia stream directly depends on the type of encoding performed. As an example, MPEG-4 encodes certain frames based on previously sent frames called I-frames. This makes I-frames more important than the intermediate frames. On the other hand, sometimes losing even a whole frame may not degrade the quality of a video depending on the importance of the particular frame. Thus, it is important that reliability is enforced on a per-packet basis to best utilize the existing networking resources. Clearly, this requires cross-layer interactions with the multimedia encoders.

- **Use of multiple paths:** In WMSNs, the channel conditions and contention in the channel may not permit a high data rate for the entire duration of multimedia flows. By allowing multiple paths, the effective data rate at each path is reduced and the application can be supported. The availability of multiple paths between source and sink can be exploited by opening multiple connections for multimedia traffic [46]. In this case, the order of packet delivery is strongly influenced by the characteristics of the route chosen. However, in real-time video/audio feeds or streaming media, information that cannot be used in the proper sequence becomes redundant. Hence, efficient techniques for packet reordering are necessary for the transport layer.

Based on these considerations, novel transport protocols need to be developed for WMSNs. We describe next representative solutions for video and image transfer.

15.8.1   Multi-hop Buffering and Adaptation

In multi-hop networks, one way to provide reliability is to use efficient buffer management schemes. The multi-hop buffering and adaptation scheme is an example that aims to provide reliability in WMSNs characterized by video traffic. More specifically, non-real-time video transfer is considered. Consequently, high-quality reconstruction of the frames generated by the sensor nodes is crucial instead of the end-to-end delay requirements. Furthermore, 100% reliability is not required in these types of traffic since scalable encoding can be performed, exploiting the redundancy in the transmitted information.

In the event of congestion, the multi-hop buffering and adaptation scheme acts in two steps. Firstly, the congestion is minimized by dropping the lower priority packets in the node's buffer. As a result, high-priority packets are not affected by the congestion in the network and a certain level of quality can be achieved at the sink during reconstruction. Secondly, a backpressure message is sent toward the data generator. Consequently, the nodes along the path are informed to stop transmitting lower priority packets. The rationale behind the backpressure message is explained through the following example.

■ **EXAMPLE 15.4**

In a distributed priority queuing mechanism, each node in the network responds to congestion by dropping the lowest priority packets in its queue. In a distributed network, each node may have packets of different priority level. Consider the case in Figure 15.18, where a multi-hop path composed of several nodes is shown. The direction of flow and the packets with priority levels stored in each node's buffer are also shown in Figure 15.18. In this case, a lower number represents a higher priority level. In case congestion occurs at node D, this node drops the lowest priority level packet in its queue, i.e., a packet with priority level 2. However, if the other nodes on this path are not informed of this action, they will continue to send their stored packets. This results in priority level 2 packets at node B still being delivered to node D. To prevent this, the multi-hop buffering and adaptation scheme deploys a coordination mechanism such that when a node drops a packet of a particular priority level, e.g., 2, this is communicated to the upstream nodes in the route. As a result, node B drops level 2 packets and node A drops the level 3 packet. This backpressure message also helps to inform the source node such that the encoding rate can be decreased to send lower quality data. Using the specific characteristics of the transmitted data helps to develop efficient transport layer protocols.

15.8.2   Error Robust Image Transport

While video data have important characteristics to be exploited in transport protocols, image transfer is very different due to its intrinsic characteristics. An important difference is the relatively low amount of

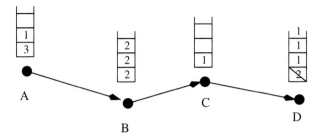

Figure 15.18 Priority mismatch in a multi-hop network.

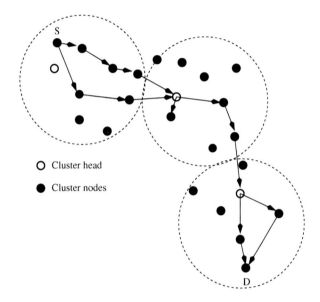

Figure 15.19 Forwarding scheme in error robust image transport.

traffic of image transmission compared to video, which consists of multiple frames. On the other hand, the correlation between frames in a video transmission can be exploited to decrease the amount of traffic transmitted. In image transmission, however, only the pixel correlation in the image can be exploited to reduce the information content. This relatively smaller redundancy requires each image produced by the sensor nodes to be reliably transmitted to the sink.

The reliability of a multi-hop communication protocol can be improved through in-network processing and multi-path transmission. The error robust image transport scheme achieves this through a cross-layer solution in a cluster-based network architecture [55] as shown in Figure 15.19. The network is partitioned into clusters, each of which is controlled by a cluster head. The coordination and route selection inside a cluster is performed by cluster heads.

Traditional multi-path routing protocols generate uncorrelated routes such that the reliability of the transmission is improved. Consequently, no common nodes usually exist in these multiple paths. While efficient, such a strategy for the transmission of images in WMSNs leads to inefficiency in terms of end-to-end reliability. The reason behind this is that image data are usually in high volume and require FEC codes to be used for energy-efficient transfer in a hop. As a result, any error in transmission may propagate through the multi-hop network until the sink reconstructs the image from the multiple routes.

In the error robust image transport scheme, on the other hand, the multiple paths are constructed specifically to intersect at cluster heads as shown in Figure 15.19, leading to a cross-layer reliability mechanism that involves the link, routing, and transport layers. Each image source encodes the captured image and creates multiple copies to be sent to different nodes in the cluster. These nodes are selected by the cluster head randomly. At each link FEC is used to further improve the error resiliency. The relay nodes then transport the copies to the next cluster head for in-network processing. At each cluster head, multiple copies of the image are collected and processed to regenerate the original image. Any errors during the transmission are corrected through multiple copies. Once the image is regenerated, it is encoded again and multiple copies are sent. As a result, redundancy is created through both FEC codes and multiple paths for reliable communication. This redundancy also helps to eliminate the losses due to node failures, which is also common in WSNs.

An important drawback of the error robust image transport scheme is that significant resources are consumed in reliably transporting an image. Firstly, multiple copies of the same image traverse the network, increasing the consumed energy for an image. Secondly, cluster heads need to perform image decoding and encoding for in-network processing. For this, cluster heads need to wait for each copy to arrive in order to perform encoding, which adds to the end-to-end delay. These costs are definitely important for the operation of the network. On the other hand, in networks with heterogeneous components, where cluster heads are resource-rich devices and image delivery has no delay bounds, the error robust image transport scheme provides high reliability.

15.8.3   Open Research Issues

In summary, the transport layer mechanisms that can simultaneously address the unique challenges posed by the WMSN paradigm and multimedia communication requirements must be incorporated. While several approaches were discussed, some open issues remain and are outlined below:

- In WMSN applications, the data gathered from the field may contain multimedia information such as target images, acoustic signals, and even video captures of a moving target, all of which enjoy a permissible level of loss tolerance. The presence or absence of an intruder, however, may require a single data field but needs to be communicated without any loss of fidelity. Thus, when a single network contains multimedia as well as scalar data, the transport protocol must decide whether to focus on one or more functionalities so that the application needs are met without unwarranted energy expense.

- Despite the existence of reliable transport solutions for WSNs as discussed in Chapter 8, none of these protocols provide real-time communication support for applications with strict delay bounds. Therefore, new transport solutions which can also meet certain application deadlines must be developed.

- The success in energy-efficient and reliable delivery of multimedia information extracted from the phenomenon directly depends on selecting the appropriate coding rate, number of sensor nodes, and data rate for a given event [53]. However, to this end, the event reliability should be accurately measured in order to efficiently adapt the multimedia coding and transmission rates. For this purpose, new reliability metrics coupled with application layer coding techniques should be investigated.

15.9 Application Layer

The multimedia content of the traffic in WMSNs requires additional capabilities at the application layer. Especially, the heterogeneity in traffic as well as the dynamic resource availability in a multi-hop network make efficient control a challenging task. Consequently, the application layer for WMSNs needs to consider the following two services: (1) providing *traffic management and admission control* and (2) performing *source coding* according to the application requirements and hardware constraints, by leveraging advanced multimedia encoding techniques. In this section, we provide an overview of these challenges.

15.9.1 Traffic Management and Admission Control

Providing QoS guarantees in a network, in general, requires sophisticated traffic management and admission control procedures. This requirement is even more important in networks of low-power, low-data-rate sensor nodes, where network resources are scarce and dynamic. Hence, differentiation of traffic classes and resource management are important in WMSNs. Because of the heterogeneity in multimedia traffic, WMSNs will need to provide support and differentiated service for several different traffic classes. These traffic classes are characterized according to two major metrics, namely reliability and delay. Moreover, since multimedia information coexists with scalar data such as information from other sensors or control information in WMSN, they should also be differentiated. Consequently, the following traffic classes can be defined in WMSNs:

- **Real-time, loss-tolerant, multimedia streams:** Video and audio streams usually constitute this class. In addition, more heterogeneous streams consisting of audio/video and other scalar data (e.g., temperature readings), as well as metadata associated with the stream, can also be considered. A major requirement is that the information needs to reach the sink within strict delay bounds. One reason is that in real-time monitoring applications, timely information retrieval is crucial. Moreover, since data may be flowing through multiple paths, a general delay bound on each path is required for efficient decoding of the multimedia stream. On the other hand, since video streams generally allow a certain level of distortion, this traffic class is loss tolerant. Traffic in this class usually has a high-bandwidth demand.

- **Delay-tolerant, loss-tolerant, multimedia streams:** For applications such as video storage for offline processing, real-time delivery is not important. Hence, strict delay bounds are not necessary as long as the multimedia stream is delivered to the sink with acceptable loss. An important requirement of this class is to provide sufficient bandwidth throughout the network so that the limited buffers of the sensors are not exceeded.

- **Real-time, loss-tolerant data:** Real-time applications also require scalar data to complement the multimedia information generated by the sensors. In the case of densely deployed networks, where the collected data include a high degree of redundancy due to spatial correlation, the information gathered by the nodes needs to be timely delivered to the sink. In this case, packet losses may be tolerated due to the correlation between nodes. Hence, sensor data have to be received timely but the application is moderately loss tolerant.

- **Real-time, loss-intolerant data:** Control data or scalar sensor monitoring data need to be reliably delivered to the sensors in the network or the sink in a timely fashion. Therefore, the application layer has to provide necessary resources for this type of traffic.

- **Delay-tolerant, loss-intolerant data:** This class includes data from critical monitoring processes that require some form of offline postprocessing.

- **Delay-tolerant, loss-tolerant data:** This class includes environmental data from scalar sensor networks, or non-time-critical snapshot multimedia content, with a low- or moderate-bandwidth demand.

Admission control has been investigated in traditional WSNs as we discussed in Chapter 9. While these approaches address application-level QoS considerations, they fail to consider multiple QoS requirements (e.g., delay, reliability, and energy consumption) simultaneously. Especially, in WMSNs, packets that belong to even the same stream may require different QoS requirements. While real-time and orderly delivery may be important for all packets, the reliability requirements may differ for different types of packets. Therefore, cross-layer admission control techniques that consider both the information content of the traffic and the peculiarities of communication in WMSNs need to be developed.

15.9.2   Multimedia Encoding Techniques

Multimedia encoding techniques focus on two main goals for efficient transmission of multimedia data. Firstly, the correlation between pixels of an image/frame or between frames of a video stream can be exploited to significantly reduce the information content to be transmitted without major quality degradation. This is usually referred to as *source coding*. Secondly, the compressed data should be efficiently represented to allow reliable transmission over lossy channels. This is usually referred to as *channel coding* or *error-resilient coding*. Furthermore, multimedia encoding techniques for WMSNs should also consider minimizing the processing and transmission power necessary to effectively deliver the multimedia content. Consequently, the main design objectives of a multimedia encoder for WMSNs are as follows:

- **High compression efficiency:** It is necessary to achieve a high ratio of compression to effectively limit bandwidth and energy consumption.
- **Low complexity:** The resource constraints of multimedia sensors in terms of processing and energy consumption require encoder techniques to be of *low complexity* to reduce cost and form factors, and *low power* to prolong the lifetime of the sensor nodes.
- **Error resiliency:** Low-power wireless communication increases the errors in the channel. As a result, the encoders should be designed to account for these effects and provide robust and error-resilient coding.

Multimedia encoding techniques have been investigated extensively in the context of the Internet and wireless networks. The results of this research have led to powerful standards for both still image and video encoding. In the following two sections, applications of the state-of-the-art solutions to WMSNs are presented.

15.9.3   Still Image Encoding

Still image capture and transmission is one of the most promising applications for WMSNs. While video streaming can demand high resources and complex QoS-aware protocols, still image transmission stays as a transition between scalar data transmission in WSNs and multimedia transmission in WMSNs. The most important encoding schemes for still images are JPEG and the recent JPEG 2000 standard. Since it is both more recent and more superior to JPEG, we will explain the applicability of the JPEG 2000 standard to WMSNs.

JPEG 2000 in WMSNs

The generation of a bit stream from an image requires many computations that increase the energy consumption during compression of an image. Hence, it is important to assess the cost of using JPEG 2000 while enjoying the features provided by it.

JPEG 2000 contains several features that are directly applicable to WMSNs. Some of these features that make JPEG 2000 a strong candidate for WMSNs are as follows [20, 38]:

- **Low-bit-rate performance:** JPEG 2000 is superior to other still image standards at low bit rates, which refers to the number of bits used to encode a pixel on the image, i.e., bits per pixel (bpp). JPEG 2000 provides high-quality images even at bit rates lower than 0.25 bpp, which is important for WMSNs equipped with low-data-rate transceivers.

- **Progressive transmission:** This refers to the fact that, using JPEG 2000, an image can still be constructed before all the encoded packets are received. Depending on how the packets are ordered in the network, as more packets are received the decoder can increase the resolution of the image, the size of the image, or the focus of the image. This provides great flexibility for communication in WMSNs since the amount of multimedia traffic to be sent can be tailored according to the application needs.

- **Parsing:** Packets in a JPEG 2000 stream are identical to each other in format. Hence, the change in the progression technique is possible only by changing the order of the packets. This provides excessive in-networking capabilities while the image is being transmitted without decoding.

- **Error resiliency:** Compared to current standards, JPEG 2000 provides higher resiliency. This means that even if some bits or packets of an image stream are lost, the image can still be decoded at a high enough quality.

JPEG 2000 provides many advantages for image compression in low-data-rate applications. However, processing cost is still high, primarily because of wavelet coding. Hence, image compression using the JPEG 2000 scheme may be infeasible for resource-constrained WMSNs. An alternative for performing compression in the network is to divide the workload among multiple sensor nodes so that distributed compression is performed [54]. The energy- efficient distributed JPEG 2000 image compression scheme provides two methods for distributing the compression tasks among multiple sensors: parallel wavelet decomposition and tiling.

Parallel wavelet decomposition: The first method is based on the parallel wavelet transform, where an image is partitioned into blocks of multiple rows and a 1-D wavelet transformation is performed on each of the blocks independently. In a distributed WMSN, this technique is adopted in a cluster through a two-round compression architecture. The compression is controlled by the cluster head. As shown in Figure 15.20, the image source first distributes each block of the image to the nodes in the cluster. After the first 1D discrete wavelet transform (DWT), the outputs are sent to the cluster head, which combines the results and divides them into columns for the second-round DWT. Then, each node sends its results to the next cluster head. This process is repeated at each cluster until the required number of layers is processed. While the processing burden on each node is significantly decreased, this scheme leads to a higher number of transmissions because of the dissemination and collection of the image data. On the other hand, since the parallel wavelet transform does not introduce any artifacts, the output of the distributed compression is not affected.

Tiling: The second scheme divides the original image into tiles and multiple nodes compress a part of the image independently. Instead of distributing the wavelet decomposition process, the image can be partitioned into tiles to make each node operate on a small image. In this method, the image is first partitioned into tiles and each tile is sent to several nodes in a cluster. Each tile is compressed separately and sent to the next cluster head. This process decreases the energy consumption at each node while providing an acceptable level of communication for the distributed operation. A major drawback is that if the number of tiles is large, the quality loss and blocking artifacts increase.

Both schemes are based on a clustered network architecture, where each cluster is responsible for performing a layer of the JPEG compression. Each cluster head manages the distributed compression operation and directs the nodes to forward their computations to the next cluster. As a result, the raw image generated by a single source is distributively encoded while the image is transmitted to the sink.

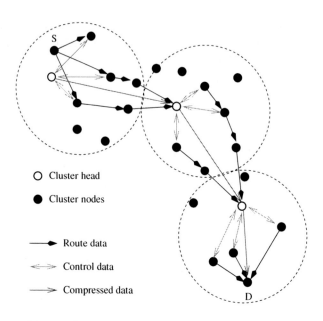

Figure 15.20   The parallel wavelet decomposition architecture.

15.9.4   Distributed Source Coding

Video encoding requires higher processing capabilities to extract the redundancy in frames of a video feed as well as between each consecutive frame. In this section, we present the applications of distributed source coding in video encoding in WMSNs. The traditional high-complexity encoder/low-complexity decoder architecture generally used for multimedia applications is not applicable for WMSNs. This is due to two main reasons: firstly, each node in a WMSN that will be responsible for encoding has scarce energy resources, frequently preventing high-complexity and energy-consuming operations. Secondly, the distributed nature of the WMSNs allows novel approaches that can also consider the relationship between multiple multimedia sources. To accomplish this major goal, the encoder/decoder complexity has to be swapped for more energy-efficient encoding without affecting the compression efficiency.

Distributed source coding (DSC) techniques have been explained in Chapter 9. Among these, Wyner–Ziv coding is a promising solution for distributed compression in WMSNs. There are two main aspects, where the results of DSC can be used to transmit still image or video in resource-constrained WMSNs:

- **DSC with spatial correlation:** The existence of multiple sources for multimedia results in spatially correlated information content residing in WMSNs. As we discussed in Chapter 5, spatial correlation in WSNs is a common concept, which is usually exploited in the design of communication protocols. In WSNs, spatial correlation is directly related to the location of the nodes. In WMSNs, due to the directivity of the video cameras, spatial correlation is related to the FoV of cameras. As a result, nodes with overlapping FoVs observe correlated information that can be exploited to encode the multimedia content. This case is particularly applicable to still image encoding.

- **DSC with temporal correlation:** Video frames are inherently temporally correlated. This correlation at each encoder can be exploited to develop low-complexity encoders based on the results of Wyner–Ziv coding. This results in the encoding of reference frames and intermediate frames very similar to the MPEG standards with a significantly lower complexity at the encoders.

Despite the theoretical advantages, practical solutions for DSC have not been developed until recently. Clearly, such techniques are promising for WMSNs and especially for networks of video sensors. The encoder can be simple and low power, while the decoder at the sink will be complex and loaded with most of the processing and energy burden. The objective of a Wyner–Ziv video coder is to achieve lossy compression of still images or video streams and a performance comparable to that of interframe encoding (e.g., MPEG), with a complexity at the encoder comparable to that of intraframe coders (e.g., JPEG). Next, we introduce some of the applications of DSC in WMSNs.

Pixel Domain Wyner–Ziv Coding

As explained above, the temporal correlation between frames in a video can be exploited to apply the results of Wyner–Ziv coding. One way to implement this idea is to use a combination of intraframe compression at the encoder side and exploit the temporal correlation through interframe decoding at the decoder side [8, 9].

The encoder side of this scheme consists of two different encoding techniques. For regularly spaced frames, interframe compression is performed to reduce the information content using the spatial correlation between the pixels of the frame. These frames are called *key frames* and sent to the decoder as *side information*. Other frames between each key frame are called *Wyner–Ziv frames* and are coded using Wyner–Ziv encoding.

At the decoder side, each key frame is extrapolated and used as side information for the decoding of Wyner–Ziv frames. The decoder also uses previously decoded Wyner–Ziv frames as side information. To apply the Wyner–Ziv coding techniques, a correlation function is chosen between the side information and the information to be decoded. If reliable decoding is impossible, the decoder requests additional parity bits from the encoder buffer.

The pixel domain Wyner–Ziv encoder exploits the DSC techniques. The interframe coded key frames are used to generate side information for the Wyner–Ziv frames. A major difference between the pixel domain Wyner–Ziv encoding and DSC is the following. In DSC, encoding of the correlated information X is performed without the knowledge of side information, Y. In video encoding, this information is present at the encoder since all the frames are generated by the same encoder. However, since exploiting this correlation for encoding requires motion prediction techniques, which are computationally expensive, it is left to the decoder to exploit it.

One of the main disadvantages of the pixel domain Wyner–Ziv coding is that the decoder relies on feedback from the encoder for proper decoding. This incurs additional delay and energy consumption that result from the multi-hop communication requirement of each feedback. This prevents offline processing of video since encoding cannot provide the necessary feedback in case of errors.

Transform Domain Wyner–Ziv Coding

Spectral information can be exploited instead of pixels to reduce the information content in image or video encoding. This is successfully accomplished through the discrete cosine transform (DCT) or wavelet coding in conventional compression schemes. A similar approach is also possible in DSC through *transform domain Wyner–Ziv coding* [7, 47].

In the transform domain Wyner–Ziv coding scheme, the frames are decomposed into spectral coefficients by the DCT. These coefficients are then independently quantized, grouped into coefficient bands, and compressed. As in the pixel domain encoder described in the previous section, the decoder generates a side information frame based on the previously reconstructed frames. Based on the side information, a bank of decoders reconstructs the quantized coefficient bands independently.

Compared to the pixel domain case, the transform domain Wyner–Ziv coding has a better performance in terms of video quality. Furthermore, rate-distortion performance is between conventional intraframe transform coding and conventional motion-compensated transform coding.

Power-Efficient Robust High-Compression Syndrome-Based Multimedia Coding (PRISM)

The solutions for DSC that we have described so far are based on the classical Wyner–Ziv coding techniques. A major drawback of these schemes is the feedback mechanism required at the decoder. In order to correctly decode the received bits, the decoder may request additional parity bits from the encoder. In multi-hop networks, this leads to significant delays and energy consumption based on the location of the encoder in the network.

The need for feedback from the encoder for decoding is removed through the PRISM scheme [47]. The main goal is to shift the motion compensation task from the encoder to the decoder. As a result, the complexity is shifted to the decoder instead of the encoder. Similar to the Wyner–Ziv coding schemes, side information is generated to be exploited at the decoder for motion compensation. At the encoder, the input X, which is a spatial region in a frame, and the side information Y, which is the highest correlated spatial region of the reference frame, are assumed to be correlated in the following form:

$$Y = X + N \tag{15.12}$$

where N is the *correlated noise*. The decoder tries to estimate the correlated noise to correctly decode the information, X, sent by the encoder. As a result, estimating the correlated noise, N, serves as motion compensation at the decoder.

15.9.5 Open Research Issues

Application layer protocols tailored specifically for WMSNs are still missing for the efficient realization of WMSNs. Hence, the following research topics need to be established:

- While theoretical results on Slepian–Wolf and Wyner–Ziv coding have existed for 30 years, there is still a lack of practical solutions. The net benefits and the practicality of these techniques still need to be demonstrated.

- It is necessary to fully explore the tradeoffs between the achieved fidelity in the description of the phenomenon observed and the resulting energy consumption. As an example, the video distortion perceived by the final user depends on source coding (frame rate, quantization) and on channel coding strength. For instance, in a surveillance application, the objective of maximizing the event detection probability is in contrast to the objective of minimizing the power consumption.

- There is a need for simple yet expressive high-level primitives for applications to leverage collaborative, advanced, in-network multimedia processing techniques.

15.10 Cross-layer Design

As described in Chapter 10, in WSNs, there is a strict interdependence among functions handled at different layers of the communication stack. This interdependence is even more pronounced in WMSNs because the performance of multimedia encoding that is performed at the application layer directly depends on the channel coding that is performed at the physical layer. Furthermore, in multi-hop networks, the encoding performance can be affected by the operation of the layers in between. Consequently, the application layer has to perform source coding based on information from the lower layers to maximize the multimedia performance. At one end, there have been many improvements in cross-layer communication protocols as we discussed in Chapter 10. However, the influence of multimedia coding has not been considered as part of this work. At the other end, cross-layer approaches developed for most of the multimedia compression or encoding schemes consider only a point-to-point channel, including only the effects of the wireless channel.

WMSNs require the sensor network paradigm to be rethought in view of the need for mechanisms to deliver multimedia content with a predefined level of QoS. We describe next a new cross-layer

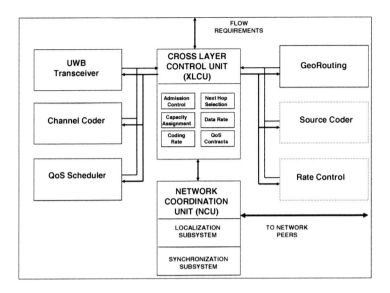

Figure 15.21   Architecture of the cross-layer controller.

communication architecture based on the TH-IR-UWB technology, whose objective is to reliably and flexibly *deliver QoS to heterogeneous applications in WMSNs,* by leveraging and controlling interactions among different layers of the protocol stack according to the requirements of the application.

15.10.1   Cross-layer Control Unit

The cross-layer control unit (XLCU) [40, 41] is based on a cross-layer communication architecture, where the main goal is to reliably and flexibly deliver QoS to heterogeneous applications in WMSNs, by controlling the functionalities of different layers. The communication protocol is based on the TH-IR-UWB transmission technique. Accordingly, the following approaches have been unified in the XLCU architecture:

- **Network layer QoS support:** The XLCU resides at the network layer to provide packet-level service differentiation. The XLCU architecture is shown in Figure 15.21, where the networking functionalities of the physical, MAC, and network layers are controlled. The operational decisions are based on the application requirements as well as the information from each individual component in the architecture.
- **Geographical forwarding:** The accurate positioning capabilities of the UWB technology are leveraged to provide geographical forwarding mechanisms according to the QoS requirements.
- **Hop-by-hop QoS contracts:** To provide end-to-end QoS support, the XLCU employs hop-by-hop QoS contracts such that the next hops are selected according to locally determined criteria. These criteria are determined according to a localized optimization framework.
- **Receiver-centric scheduling:** To mitigate interference from neighbor nodes, receiver-centric scheduling is performed such that each receiver acts according to local measurements and consequently control loss recovery and rate adaptation.
- **Dynamic channel coding:** Since power control is not beneficial in TH-IR-UWB, adaptation to interference at the receiver is achieved through dynamic channel coding. The encoder at node i receives a block of L uncoded bits, and selects the encoding rate $R_{E,i}$ among the set $\underline{R_E} =$

$|R_E^1, R_E^2, \ldots, R_E^P|$. Accordingly, when code R_E^p is selected, the encoder produces a block of coded bits of length L/R_E^p.

To provide service differentiation between different flows in the network, the XLCU defines various classes of traffic according to the delay requirements (real time or delay tolerant), reliability requirements (loss intolerant or loss tolerant), and type of flow (multimedia stream or data flow). Each class is identified through its requirements, which are described as a set of tuples $\Psi^A = \{\psi^a(\delta^a, \beta^a, \zeta^a), a \in 1, \ldots, N_\psi^A\}$, where ψ^a, $a \in 1, \ldots, N_\psi^A$, are N_ψ^A different subflows of the flow generated by the application A. For each subflow ψ^a, δ^a is the maximum allowed end-to-end delay for packets associated with the subflow, β^a is the required bandwidth, and ζ^a is the end-to-end packet error rate (PER) that can be allowed by the application. The communication decisions are made according to these requirements.

The application's requirements are used to determine hop-by-hop QoS contracts for packet transmission. The global end-to-end requirements are then guaranteed by the individual guarantees of each device along the path to the destination. Considering a flow $\psi^a(\delta^a, \beta^a, \zeta^a)$ generated at node i, a multi-hop path from i to the destination N needs to be established, with maximum end-to-end delay δ^a, minimum guaranteed bandwidth β^a, and maximum end-to-end packet error rate ζ^a.

The required bandwidth β^a needs to be provided at each hop. Moreover, the delay and packet error rate requirements are determined according to the geographical advance of the packet. More specifically,

$$\delta_{ij}^a = \left(\frac{\langle d_{ij} \rangle_{iN}}{d_{iN}} \right) \cdot \delta^a \qquad (15.13)$$

and

$$\zeta_{ij}^a \leq 1 - (1 - \zeta^a)^{\lceil \hat{N}_{ij}^{Hop} \rceil^{-1}} \qquad (15.14)$$

where $\langle d_{ij} \rangle_{iN}$ is the advance of the packet if it has been transmitted from node i to node j and d_{iN} represents the distance between i and the destination. According to the single hop distance, in (15.14), the end-to-end path is assumed to consist of $\hat{N}_{ij}^{Hop}$ hops and the packet error rate requirement is determined accordingly.

Based on the local requirements, the next hop is selected according to an *admission control protocol*. A node i with a packet to send first broadcasts a short CONTRACT_REQUEST packet, which describes the requirements of the service at i, i.e., $\mathcal{F}_i$. If a neighbor j of i (1) has positive advance toward the sink N with respect to i and (2) is able to provide the requested service with the required QoS, it replies with an ADM_GRANTED packet. This initial handshake informs the sender node about the potential neighbors for the service. Accordingly, the sender node performs a local optimization process to select the optimal relay node. This is communicated to the selected node through a CONTRACT_REQUEST packet and the handshake is concluded through a CONTRACT_ESTABLISHED message. The hop-by-hop admission control mechanism helps forward the packet until the destination.

The admission control protocol relies on a local optimization problem, which is solved at each hop. Accordingly, the problem can be cast as follows:

p^{dist}: Distributed Admission Control, Routing, and Channel Coding Problem

Given: $i, N, S_i, \mathcal{P}_i^N, E^{pulse}, \mathcal{F}_i$

Find: $j^* \in S_i \cap \mathcal{P}_i^N, R_{E,i}^a \ \forall a \in \mathcal{F}_i$

Minimize: $E_{(i,j)}^{bit} = \dfrac{1}{\beta^{tot}} \displaystyle\sum_{a \in \mathcal{F}(i)} \dfrac{E^{pulse} \cdot \hat{N}_{ij}^{TX,a} \cdot \hat{N}_{ij}^{Hop} \cdot \beta^a}{R_{E,j}^a}$ (15.15)

Subject to:

packet error rate:

$$R_{E,i}^a \leq \min\left(\frac{E_i^{(r)}}{\gamma_{C,j}^a(\zeta_{ij}^a)|\eta_j + (\sigma^2/T_{f,j})\sum_{k\in\mathcal{F}(i),k\neq i} E_k^{(r)}|}, 1\right), \quad \forall a \in \mathcal{F}_i; \qquad (15.16)$$

rate admission control:

$$\sum_{a\in\mathcal{F}_j} \frac{\beta^a}{R_{E,\mathcal{N}_j}^a(\gamma_{C,\mathcal{N}_j}^a(\zeta_{j\mathcal{N}_j}^a))}$$

$$+ \sum_{a\in\mathcal{F}_j} \frac{\beta^a}{R_{E,\mathcal{U}_j^a}^a(\gamma_{C,j}^a(\zeta_{\mathcal{U}_j^a j}^a))} + \frac{R_j^{sched,up}}{R_{E,j}^{sched}} + \frac{R_j^{sched,down}}{R_{E,j}^{sched}} \leq R_{0,j}; \qquad (15.17)$$

delay admission control:

$$\sum_{a\in\mathcal{F}_i} \frac{LR_{0,j}}{R_{E,\mathcal{N}_j}^a} + T^{sched,up} + T^{sched,down}$$

$$+ \sum_{a\in\mathcal{F}_i} L\left(1 + \frac{b_j^a}{\phi_j^a}\right) \cdot \frac{1}{R_{0,j} R_{E,j}^a} + \frac{L}{R_{0,j} R_{E,j}^a} \leq \delta_{ij}^a, \quad \forall a \in \mathcal{F}_i \qquad (15.18)$$

where the following parameters are used:

- $E^{pulse} = 2 \cdot E_{elec}^{pulse} + P^{TX} \cdot T_{f,i}$ [J/pulse] is the energy to transmit one pulse from node i to node j, where E_{elec}^{pulse} is the energy per pulse needed by transmitter electronics and digital processing; P^{TX} [W] and $T_{f,i}$ [s] are the average transmitted power and the frame length, respectively.
- $\hat{N}_{ij}^{TX,a}$ is the average number of transmissions of a packet from flow a required for the packet to be correctly decoded at receiver j.
- $\hat{N}_{ij}^{Hop} = \max(d_{iN}/\langle d_{ij}\rangle_{iN}, 1)$ is the estimated number of hops from node i to the destination N when j is selected as the next hop.
- S_i is the *neighbor set* of node i, while $\mathcal{P}_i^N$ is the *positive advance set* of i, i.e., $j \in \mathcal{P}_i^N$ iff $d_{jN} < d_{iN}$.
- $\mathcal{F}_i$ is the set of incoming or generated flows at node i.
- The bandwidth requirement β^a of application a can be expressed as $\beta^a = R_{0,i}^a \cdot R_{E,i}^a$, where $R_{0,i}^a$ [pulses/s] represents the raw pulse rate for application a required to achieve the rate β^a, when a coding rate $R_{E,i}^a$ is used.
- $\beta^{tot} = \sum_{a\in\mathcal{F}_i} \beta^a$ represents the total bandwidth requirement, in bps, for flows incoming or generated at i.

Once the next hop is selected, *rate assignment* is performed such that the signal-to-interference plus noise ratio (SINR) of the communication is bounded by

$$SINR_i^a \geq \gamma_{C,i}^a(\zeta_i^a), \qquad (15.19)$$

for an allowed PER ζ_i^a at receiver i, where $\gamma_{C,i}^a(\zeta_i^a)$ is the SINR threshold that guarantees the PER ζ_i^a. The rate is selected such that $R_i^a = R_{E,i}^a R_{0,i}$, and is bounded by

$$R_{E,i}^a \leq \min\left(\frac{E_i^{(r)}}{\gamma_{C,i}^a(\zeta_i^a)|\eta_i + (\sigma^2/T_{f,i})\sum_{j\in\mathcal{F}(i),j\neq i} E_j^{(r)}|}, 1\right). \qquad (15.20)$$

Hence, the optimal coding rate for flow a is selected as

$$R^a_{E,i} = \max_{1 \le p \le P} R^p_E \text{ s.t. (15.20) holds.} \tag{15.21}$$

After the next hop is selected with the optimal rate, finally, *receiver-centric scheduling* is performed to determine the transmission times for the contracted flow. This is performed according to a *schedule*, which is a vector of *appointments*, i.e., tuples $(a, u, t^a_k, R^a_{E,u})$, where a is an application flow, u is a node, $u \in \mathcal{F}_i$, t^a_k is the starting time for transmission of the kth packet from flow a at u, and $R^a_{E,u}$ is the required coding rate. The scheduling mechanism of the XLCU is based on a procedure inspired by the wireless fair scheduling (WFS) paradigm, which is designed to guarantee delay-bounded and throughput-guaranteed access in single hop, single rate wireless packet networks. More specifically, the wireless fair service approach [36] is extended for a UWB multi-rate, multi-hop environment.

The XLCU architecture is based on an innovative design that aims at providing differentiation in the domains of throughput, delay, reliability, and energy consumption, based on a modular cross-layer controller that performs admission control, routing, scheduling, bandwidth assignment, and coding to satisfy application requirements. This solution provides delays that are very low and with low jitter, and the throughput is fairly constant in time, while energy consumption is structurally much lower than CSMA/CA systems due to the lower transmitted power and no idle listening.

15.11   Further Research Issues

Most of the challenges in realizing a practical implementation of WMSNs can be classified into layer-specific considerations as we have discussed so far in this chapter. Nevertheless, there are still additional areas that need to be addressed for a complete realization of WMSNs. In this section, we overview the still open research issues in many contexts for the realization of WMSNs.

15.11.1   *Collaborative In-network Processing*

In-network processing has long been the main theme for networking in traditional WSNs. Considering the heavy processing requirements and increased bandwidth needs of multimedia applications, it is clear that WMSNs will also require efficient in-network processing algorithms. Consequently, it is necessary to flexibly perform these functionalities *in-network* with minimum energy consumption and limited execution time. Similar to image or video compression, the objective of in-network processing is to further reduce the amount of injected information into the network and communicate only necessary information.

The information communicated in the network is also not always of the same context in distributed networks. The content requested from the same set of sensor nodes can vary significantly depending on the application type as well as the query. Consequently, both networking protocols and compression techniques should be developed to be highly adaptive to efficiently respond to this dynamic nature of the queries. Consequently, *application-specific querying and processing* is crucial in WMSNs. Moreover, it is necessary to develop efficient querying languages, and also to develop distributed compression, filtering, and in-network processing architectures, to allow real-time retrieval of useful information.

Aggregation and fusion of both inter-media and intra-media are necessary considering the funda-mental differences between WMSNs and WSNs. Especially, the computational burden exposed through complex multimedia operations prevents well-established distributed processing schemes that have been developed for WSNs being used in this context. In WSNs, it is well known that *processing is always cheaper than communication*. This is, however, not completely true in WMSNs. In particular, the tradeoffs between compression at end-nodes and communicating raw data have yet to be clearly analyzed. This leads to fundamental architecture decisions such as whether this processing can be done on sensor nodes (i.e., a flat architecture of multifunctional sensors that can perform any task), or if the need for specialized devices, e.g. *computation hubs*, arises.

Next, we will introduce two main directions for in-network processing in WMSNs. It is clear that these directions will likely drive research on architectures and algorithms for the distributed processing of multimedia data.

Data Alignment and Image Registration

The distributed nature of WMSNs results in multiple observations of probably the same phenomenon from different sensor nodes, at different times. Similar to WSNs, the central location, however, is interested in observing the *phenomenon* that is of interest to the network instead of the individual *observations* of the sensor nodes. Consequently, information gathered from different sources should be combined wisely to provide better information than what is possible from individual observations.

In image or video processing, the concept of merging information from multiple sources is investigated under the umbrella of *data alignment* approaches. More specifically, image registration, which is widely used in areas such as remote sensing, medical imaging, and computer vision, is a promising field for WMSNs. The basic idea of image registration is to align different images of the same scene taken at different times, from different viewpoints, and/or by different sensors.

Image registration consist of four steps as we will describe next:

- **Feature detection:** Image registration depends solely on determining common objects in different images to determine the viewpoints or locations that are associated with each image. To extract this information, distinctive objects are detected. This is accomplished by defining closed-boundary regions, edges, contours, line intersections, corners, etc.

- **Feature matching:** Features detected in each image should be compared among different images to determine the relationship between images. Feature matching techniques try to match the same objects on different images to provide this relationship.

- **Transform model estimation:** Once correlated images are determined, accurate models to transform each image into a common reference need to be used. This is accomplished by defining *mapping functions* and estimating their parameters. Using these functions, each image can be transformed to produce a common frame of information using multiple sources.

- **Image resampling and transformation:** In this last step, the sensed image is transformed by means of the mapping functions.

Data alignment and image registration operations generally consume large amounts of processing power. Therefore, exploiting these techniques in-network is a major challenge for WMSNs. Moreover, efficient information transmission methods need to be devised such that these techniques are exploited at resource-rich processing hubs or at the sink.

WMSNs as Distributed Computer Vision Systems

Efficient transmission of still image and video frames has been the primary focus of research in WMSNs. While many important insights have been learned so far from these approaches, the distributed nature of WMSNs is still open to different approaches in visual systems. Especially, transmitting images and video from a distributed network of nodes can still be seen as *raw* information transfer. The in-network processing approaches developed so far mainly focus on distributing the high processing load of compression techniques through a parallel processing architecture composed of multiple sensors. However, raw data transfer is still of major concern. Then, it is left solely to the central location to collect a high volume of information from multiple sensors and process it centrally.

As with WSNs, intelligent information rather than raw data is important in WMSNs. This corresponds to local processing of visual information and the transmission of information of interest. Spreading the processing load on every sensor in the network provides the transition from a distributed camera network to a distributed computer vision system. This research topic is very important in WMSNs.

Visual observations across the network can be performed in WMSNs through distributed computations on observations from multiple nodes. Consequently, simple but efficient tools are necessary to express data as an observation that is of interest to the user. To this end, it is necessary to coordinate computations across the nodes in the network efficiently and return only the results of interest to the end-user.

15.11.2 Synchronization

Synchronization is an important aspect of networking in WSNs and is required by many algorithms for efficient operation as we discussed in Chapter 11. It is clear that the need for synchronization also exists in WMSNs for the same reasons. Moreover, in the case of WMSNs, synchronization becomes even more pronounced because of the information content of distributed nodes and the dependence of this multimedia information on time. This increases the need for accurate synchronization solutions in WMSNs as well.

In-network processing tools for WMSNs require the whole image/video be decoded at an aggregate point to further decrease the content of information. In addition, the whole information content from multiple flows needs to be available at the same time. Since storing this huge amount of information is very costly, multiple flows to be aggregated need to be generated at closely intervals and received within certain delay bounds. This prevents buffer overflows and makes aggregation of multiple streams possible.

Multimedia data returned by sensors can be heterogeneous, and possibly correlated. As an example, video and audio samples could be collected over a common geographical region, or still pictures may complement scalar data containing field measurements. While the temporal correlation between these multiple streams can be estimated from the content of the data at a central location, this requires extensive computation. This might not be suitable for applications that require fast and accurate responses for events occurring in a surveillance system. Therefore, inter-media synchronization schemes that can be implemented inside the network would improve the efficiency of such applications. Providing a common frame for markers that can be associated with each multimedia flow gives the end-user the flexibility to process these data. The problem of multimedia synchronization is cross-layer in nature and influences primarily the physical and application layers. At the physical layer, data from different media are multiplexed over shared wireless connections, or are stored in common physical storage. The application layer is concerned with the inter-media synchronization necessary for presentation or playout, i.e., with the interactions between the multimedia application and the various media.

References

[1] MTS/MDA Sensor Board Users Manual.

[2] Acroname GARCIA robotic platform. http://www.acroname.com/garcia/garcia.html.

[3] Crossbow MICA2 mote specifications. http://www.xbow.com.

[4] Crossbow TelosB mote specifications. http://www.xbow.com.

[5] SENSONET sensor network testbed. Available at http://www.ece.gatech.edu/research/labs/bwn/.

[6] JPEG 2000 requirements and profiles. ISO/IEC JTC1/SC29/WG1 N1271, March 1999.

[7] A. Aaron, S. Rane, E. Setton, and B. Girod. Transform-domain Wyner–Ziv codec for video. In *Proceedings of the Society of Photo-Optical Instrumentation Engineers – Visual Communications and Image Processing*, pp. 520–528, San Jose, CA, USA, January 2004.

[8] A. Aaron, S. Rane, R. Zhang, and B. Girod. Wyner–Ziv coding for video: applications to compression and error resilience. In *Proceedings of the IEEE Data Compression Conference (DCC'03)*, pp. 93–102, Snowbird, UT, USA, March 2003.

[9] A. Aaron, E. Setton, and B. Girod. Towards practical Wyner–Ziv coding of video. In *Proceedings of the IEEE International Conference on Image Processing (ICIP'03)*, volume 3, pp. 869–872, Barcelona, Spain, September 2003.

[10] R. Agre, L. P. Clare, G. J. Pottie, and N. P. Romanov. Development platform for self-organizing wireless sensor networks. In *Proceedings of the Society of Photo-Optical Instrumentation Engineers*, volume 3713, pp. 257–268, Orlando, FL, USA, April 1999.

[11] G. R. Aiello and G. D. Rogerson. Ultra-wideband wireless systems. *IEEE Microwave Magazine*, 4:36–47, June 2003.

[12] Ö. B. Akan and I. F. Akyildiz. Event-to-sink reliable transport in wireless sensor networks. *IEEE/ACM Transactions on Networking*, 13(5):1003–1016, October 2005.

[13] I. F. Akyildiz, T. Melodia, and K. Chowdury. A survey on wireless multimedia sensor networks. *Computer Networks*, 51(4):921–960, March 2007.

[14] I. F. Akyildiz, W. Su, Y. Sankarasubramaniam, and E. Cayirci. Wireless sensor networks: a survey. *Computer Networks*, 38(4):393–422, March 2002.

[15] R. Biswas, K. R. Chowdhury, and D. P. Agrawal. Data-centric attribute allocation and retrieval (DCAAR) scheme for wireless sensor networks. In *Proceedings of the IEEE International Conference on Mobile Ad-hoc and Sensor Systems (MASS'05)*, pp. 422–429, Washington DC, USA, November 2005.

[16] M. Caccamo, L. Y. Zhang, L. Sha, and G. Buttazzo. An implicit prioritized access protocol for wireless sensor networks. In *Proceedings of the IEEE Real-Time Systems Symposium (RTSS'02)*, pp. 39–48, Austin, TX, USA, December 2002.

[17] Y. Charfi, N. Wakamiya, and M. Murata. Visual sensor networks: opportunities and challenges. In *Information and Communication Technologies International Symposium (ICTIS'07)*, Sydney, Australia, April 2007.

[18] K. J. Cho and H. H. Asada. Wireless, battery-less stethoscope for wearable health monitoring. In *Proceedings of the IEEE Northeast Bioengineering Conference*, pp. 187–188, Philadelphia, PA, USA, April 2002.

[19] K. R. Chowdhury, N. Nandiraju, D. Cavalcanti, and D. P. Agrawal. CMAC – a multi-channel energy efficient MAC for wireless sensor networks. In *Proceedings of the IEEE Wireless Communications and Networking Conference (WCNC'06)*, volume 2, pp. 1172–1177, Las Vegas, NV, USA, April 2006.

[20] C. Christopoulos, A. Skodras, and T. Ebrahimi. The JPEG 2000 still image coding system: an overview. *IEEE Transactions on Consumer Electronics*, 46(4):1103–1127, November 2000.

[21] R. Cucchiara. Multimedia surveillance systems. In *Proceedings of the ACM International Workshop on Video Surveillance and Sensor Networks*, pp. 3–10, Singapore, November 2005.

[22] I. Downes, L. B. Rad, and H. Aghajan. Development of a mote for wireless image sensor networks. In *Proceedings of COGnitive systems with Interactive Sensors (COGIS'06)*, Paris, France, March 2006.

[23] Anuj Batra et al. Multi-band OFDM physical layer proposal for IEEE 802.15 task group 3a. IEEE P802.15 Working Group for Wireless Personal Area Networks (WPANs), March 2004.

[24] E. Felemban, C.-G. Lee, E. Ekici, R. Boder, and S. Vural. Probabilistic QoS guarantee in reliability and timeliness domains in wireless sensor networks. In *Proceedings of IEEE INFOCOM 2005*, volume 4, pp. 2646–2657, Miami, FL, USA, March 2005.

[25] E. Felemban, C.-G. Lee, and Eylem Ekici. MMSPEED: multipath multi-speed protocol for QoS guarantee of reliability and timeliness in wireless sensor networks. *IEEE Transactions on Mobile Computing*, 5(6):738–754, June 2006.

[26] W.-C. Feng, B. Code, E. Kaiser, M. Shea, W.-C. Feng, and L. Bavoil. Panoptes: Scalable low-power video sensor networking technologies. In *Proceedings of the 11th ACM International Conference on Multimedia*, pp. 562–571, Berkeley, CA, USA, November 2003.

[27] S. Gezici, Z. Tian, G. B. Giannakis, H. Kobayashi, A. F. Molisch, H. V. Poor, and Z. Sahinoglu. Localization via ultra-wideband radios. *IEEE Signal Processing Magazine*, 22(4):70–84, July 2005.

[28] B. Girod, A. Aaron, S. Rane, and D. Rebollo-Monedero. Distributed video coding. *Proceedings of the IEEE*, 93(1):71–83, January 2005.

[29] E. Gürses and Ö. B. Akan. Multimedia communication in wireless sensor networks. *Annals of Telecommunications*, 60(7-8):872–900, July–August 2005.

[30] S. K. Jayaweera. An energy-efficient virtual MIMO communications architecture based on V-BLAST processing for distributed wireless sensor networks. In *Proceedings of the IEEE International Conference on Sensor and Ad-hoc Communications and Networks (SECON'04)*, pp. 299–308, Santa Clara, CA, USA, October 2004.

[31] S. K. Jayaweera and M. L. Chebolu. Virtual MIMO and distributed signal processing for sensor networks – an integrated approach. In *Proceedings of the IEEE International Conference on Communications (ICC'05)*, volume 2, pp. 1214–1218, Seoul, South Korea, May 2005.

[32] P. Kulkarni, D. Ganesan, P. Shenoy, and Q. Lu. SensEye: a multi-tier camera sensor network. In *Proceedings of ACM Multimedia*, Singapore, November 2005.

[33] P. Kyasanur and N. H. Vaidya. Capacity of multi-channel wireless networks: impact of number of channels and interfaces. In *Proceedings of ACM MobiCom'05*, Cologne, Germany, August 2005.

[34] C. L. Liu and J. W. Layland. Scheduling algorithms for multiprogramming in hard real time environment. *Journal of the ACM*, 20(1):40–61, 1973.

[35] X. Liu, Q. Wang, W. He, M. Caccamo, and L. Sha. Optimal real-time sampling rate assignment for wireless sensor networks. *ACM Transactions on Sensor Networks*, 2(2):263–295, 2006.

[36] S. Lu, T. Nandagopal, and V. Bharghavan. Design and analysis of an algorithm for fair service in error-prone wireless channels. *ACM Wireless Networks*, 6(4):323–343, January 2000.

[37] K. Mandke, H. Nam, L. Yerramneni, and C. Zuniga. The evolution of UWB and IEEE 802.15.3a for very high data rate WPAN. *High Frequency Electronics, Technology Report*, 2(5):22–32, September 2003.

[38] M. W. Marcellin, A. Bilgin, M. J. Gormish, and M. P. Boliek. An overview of JPEG-2000. In *Proceedings of the Conference on Data Compression*, p. 523, Snowbird, UT, USA, 2000.

[39] C. B. Margi, V. Petkov, K. Obraczka, and R. Manduchi. Characterizing energy consumption in a visual sensor network testbed. In *Proceedings of the IEEE/Create-Net International Conference on Testbeds and Research Infrastructures for the Development of Networks and Communities (TridentCom'06)*, Barcelona, Spain, March 2006.

[40] T. Melodia and I. F. Akyildiz. Cross-layer quality of service support for UWB wireless multimedia sensor networks. In *Proceedings of the IEEE INFOCOM Mini-Conference'08*, Phoenix, AZ, USA, April 2008.

[41] T. Melodia and I. F. Akyildiz. Cross-layer QoS-aware communication for ultra wide band wireless multimedia sensor networks. *IEEE Journal on Selected Areas in Communications*, 2010.

[42] M. J Miller and N. H. Vaidya. A MAC protocol to reduce sensor network energy consumption using a wakeup radio. *IEEE Transactions on Mobile Computing*, 4(3):228–242, May/June 2005.

[43] F. R. Mireles and R. A. Scholtz. Performance of equicorrelated ultra-wideband pulse-position-modulated signals in the indoor wireless impulse radio channel. *Proceedings of IEEE Communications, Computers and Signal Processing*, 2:640–644, August 1997.

[44] S. Misra, M. Reisslein, and G. Xue. A survey of multimedia streaming in wireless sensor networks. *IEEE Communications Surveys & Tutorials*, 10(4):18–39, 2008.

[45] S. Nath, Y. Ke, P. B. Gibbons, B. Karp, and S. Seshan. A distributed filtering architecture for multimedia sensors. In *Proceedings of BaseNets'04*, San Jose, CA, USA, October 2004.

[46] T. Nguyen and S. Cheung. Multimedia streaming with multiple TCP connections. In *Proceedings of the IEEE International Performance Computing and Communications Conference (IPCCC'05)*, Phoenix, AZ, USA, April 2005.

[47] R. Puri and K. Ramchandran. Prism: a new robust video coding architecture based on distributed compression principles. In *Proceedings of the Allerton Conference on Communication, Control, and Computing*, Allerton, IL, USA, October 2002.

[48] M. Rahimi, R. Baer, O. Iroezi, J. Garcia, J. Warrior, D. Estrin, and M. Srivastava. Cyclops: in situ image sensing and interpretation in wireless sensor networks. In *Proceedings of the ACM Conference on Embedded Networked Sensor Systems (SenSys)*, San Diego, CA, USA, November 2005.

[49] A. A. Reeves. Remote monitoring of patients suffering from early symptoms of dementia. In *International Workshop on Wearable and Implantable Body Sensor Networks*, London, UK, April 2005.

[50] A. Rowe, C. Rosenberg, and I. Nourbakhsh. A low cost embedded color vision system. In *Proceedings of the IEEE/RSJ International Conference on Intelligent Robots and Systems (IROS)*, Lausanne, Switzerland, October 2002.

[51] C. Santivanez and I. Stavrakakis. Study of various TDMA schemes for wireless networks in the presence of deadlines and overhead. *IEEE Journal on Selected Areas in Communications*, 17(7):1284–1304, July 1999.

[52] C. Schurgers, V. Tsiatsis, and M. B. Srivastava. STEM: topology management for energy efficient sensor networks. In *Proceedings of the IEEE Aerospace Conference*, volume 3, pp. 1099–1108, Big Sky, MT, USA, 2002.

[53] M. C. Vuran, Ö. B. Akan, and I. F. Akyildiz. Spatio-temporal correlation: theory and applications for wireless sensor networks. *Computer Networks*, 45(3):245–261, June 2004.

[54] H. Wu and A. Abouzeid. Energy efficient distributed image compression in resource constrained multihop wireless networks. *Computer Communications*, 28(14):1658–1668, September 2005.

[55] H. Wu and A. A. Abouzeid. Error resilient image transport in wireless sensor networks. *Computer Networks*, 50(15):2873–2887, October 2006.

[56] W. Yu, Z. Sahinoglu, and A. Vetro. Energy-efficient JPEG 2000 image transmission over wireless sensor networks. In *Proceedings of the IEEE Global Communications Conference (GLOBECOM)*, pp. 2738–2743, Dallas, TX, USA, January 2004.

[57] X. Zhu and B. Girod. Distributed rate allocation for multi-stream video transmission over ad hoc networks. In *Proceedings of the IEEE International Conference on Image Processing (ICIP)*, pp. 157–160, Genoa, Italy, September 2005.

16

Wireless Underwater Sensor Networks

Wireless underwater sensor networks (UWSNs) are envisioned to enable applications for a wide variety of purposes such as oceanographic data collection, pollution monitoring, offshore exploration, disaster prevention, assisted navigation, and tactical surveillance [7]. Multiple unmanned or autonomous underwater vehicles (UUVs, AUVs), equipped with underwater sensors, will also find application in exploration of natural undersea resources and scientific data gathering in collaborative monitoring missions. To make these applications viable, there is a need to enable communications among underwater devices. Similar to the requirements of WSNs, underwater sensor nodes and vehicles must possess self-configuration capabilities, i.e., they must be able to coordinate their operations by exchanging configuration, location, and movement information, and to relay monitored data to an onshore station.

The traditional approach for ocean-bottom or ocean-column monitoring is to deploy underwater sensors that record data during the monitoring mission, and then recover the instruments [29]. This approach has the following disadvantages:

- **No real-time monitoring:** The recorded data cannot be accessed until the instruments are recovered, which may be several days, weeks, or months after beginning the monitoring mission. In surveillance or environmental monitoring applications such as seismic monitoring, however, real-time data retrieval is crucial.

- **No online system reconfiguration:** Interaction between onshore control systems and the monitoring instruments is not possible. This impedes any adaptive tuning of the instruments; nor is it possible to reconfigure the system after particular events occur.

- **No failure detection:** If *failures* or *misconfigurations* occur, it may not be possible to detect them before the instruments are recovered. This can easily lead to the complete failure of a monitoring mission.

- **Limited storage capacity:** The amount of data that can be recorded during the monitoring mission by every sensor is limited by the capacity of the onboard storage devices (memories, hard disks).

These disadvantages of traditional underwater monitoring techniques limit the possible applications for this environment. On the other hand, the distributed sensor network paradigm may provide capabilities that significantly surpass the existing underwater applications. Therefore, there is a need to deploy underwater networks that will enable real-time monitoring of selected ocean areas, remote configuration, and interaction with onshore human operators. All this can be obtained by connecting underwater instruments by means of wireless links and forming underwater sensor networks as we will explain in this chapter. The above features enable a broad range of underwater applications:

Wireless Sensor Networks   Ian F. Akyildiz and Mehmet Can Vuran

- **Ocean sampling networks:** Networks of sensors and autonomous underwater vehicles (AUVs), such as the Odyssey-class AUVs [1], can perform synoptic, cooperative adaptive sampling of the 3-D coastal ocean environment [3]. Recent underwater experiments demonstrate the advantages of bringing together sophisticated new robotic vehicles with advanced ocean models to improve the ability to observe and predict the characteristics of the oceanic environment [5].

- **Environmental monitoring:** Underwater sensor networks can perform pollution monitoring (chemical, biological, and nuclear). For example, it may be possible to detail the chemical slurry of antibiotics, estrogen-type hormones, and insecticides to monitor streams, rivers, lakes, and ocean bays (*water quality in-situ analysis*) [43]. Monitoring ocean currents and winds, improved weather forecast, detecting climate change, understanding and predicting the effect of human activities on marine ecosystems, biological monitoring such as tracking fishes or microorganisms, are other possible applications. More specifically, underwater sensor networks can be used to detect extreme temperature gradients (thermoclines), which are considered to be a breeding ground for certain marine microorganisms [44].

- **Undersea explorations:** Underwater sensor networks can help detect underwater oilfields or reservoirs, determine routes for laying undersea cables, and assist in the exploration for valuable minerals.

- **Disaster prevention:** Sensor networks that measure seismic activity from remote locations can provide *tsunami* warnings to coastal areas [34], or study the effects of submarine earthquakes (*seaquakes*).

- **Assisted navigation:** Sensors can be used to identify hazards on the seabed, to locate dangerous rocks or shoals in shallow waters, mooring positions, and submerged wrecks, and to perform bathymetry profiling.

- **Distributed tactical surveillance:** AUVs and fixed underwater sensors can collaboratively monitor areas for *surveillance, reconnaissance, targeting,* and *intrusion detection* systems. As an example, a 3-D underwater sensor network can realize a tactical surveillance system that is able to detect and classify submarines, small delivery vehicles (SDVs), and divers based on the sensed data from mechanical, radiation, magnetic, and acoustic microsensors. With respect to traditional radar/sonar systems, underwater sensor networks can reach a higher accuracy, higher coverage, and robustness as well as enable detection and classification of low-signature targets by also combining measures from different types of sensors.

- **Mine reconnaissance:** The simultaneous operation of multiple AUVs with acoustic and optical sensors can be used to perform rapid environmental assessment and detect mine-like objects.

The potential capabilities of underwater sensor networks and the wide variety of new applications that will be enabled by them motivate the development of communication techniques for the underwater environment. While the vast amount of solutions for terrestrial WSNs that we have discussed in the previous chapters provides a valuable insight into networking in this environment, there exist many challenges unique to underwater communication. Especially, the significantly different characteristics of communication in water require many networking paradigms to be revisited. Consequently, the major challenges in the design of underwater sensor networks are as follows:

- The available bandwidth is severely limited.
- Propagation delay under water is five orders of magnitude higher than that in RF terrestrial channels and extremely variable.
- The underwater channel is severely impaired, especially because of multi-path and fading problems.
- High bit error rates and temporary losses of connectivity (shadow zones) can be experienced due to the extreme characteristics of the underwater channel.
- Battery power is limited and usually batteries cannot be recharged; also, solar energy cannot be exploited.

- Underwater sensors are prone to failure because of fouling and corrosion.

In this chapter, we discuss several fundamental aspects of underwater communications and sensor networks. The remainder of this chapter is organized as follows. In Sections 16.1 and 16.3 we introduce the communication architecture and design challenges, respectively, of underwater sensor networks. In Section 16.4, we investigate the underwater acoustic communication channel and summarize the associated physical layer challenges for underwater networking. In Sections 16.5, 16.6, 16.7, 16.8, and 16.9, we discuss physical, data link, network, transport, and application layer issues in underwater sensor networks, respectively. In Section 16.10 we highlight cross-layer approaches for communication in UWSNs.

16.1 Design Challenges

Firstly, we present the main differences between terrestrial and underwater sensor networks and detail the key design issues and deployment challenges for underwater sensors.

16.1.1 Terrestrial Sensor Networks vs. Underwater Sensor Networks

While many aspects of networking under water have similarities with the terrestrial sensor networks, there exist many differences that require communication protocols to be tailored for underwater sensor networks.

The fundamental difference between terrestrial and underwater sensor networks is the communication medium. As explained in more detail in Section 16.4, the underwater environment results in different propagation, attenuation, and fading characteristics than in air. In fact, despite the fact that underwater networking is a rather unexplored area, underwater communications have been experimented on since World War II, when, in 1945, an underwater telephone was developed in the USA to communicate with submarines [30]. The results of research in this area reveal that acoustic communication is the typical physical layer technology for underwater networks.

The propagation characteristics of water dictate that electromagnetic waves propagate at long distances through conductive sea water only at extra low frequencies (30–300 Hz). This requirement requires large antennae and high transmission power, which are not suitable for the deployment of sensor networks. For example, the Berkeley Mica2 mote, which is one of the most popular experimental platforms in the terrestrial sensor networking community, has a transmission range of 120 cm under water at 433 MHz, according to the experiments performed at the Robotic Embedded Systems Laboratory (RESL) at the University of Southern California. Therefore, electromagnetic wave communication is not the best candidate for underwater communication.

In contrast to electromagnetic waves, optical waves do not suffer from such high attenuation. However, optical wave communication under water suffers from significant scattering. Moreover, the transmission of optical signals requires high precision in pointing the narrow laser beams. Therefore, optical waves are only used for short-range communication in the underwater environment. Thus, links in underwater networks are based on *acoustic wireless communications* [37].

In addition to the communication techniques, the following differences between terrestrial and underwater sensor networks are also important:

- **Size and cost:** While terrestrial sensor nodes are expected to become smaller in size and increasingly inexpensive, underwater sensors are expensive devices. This is especially due to the more complex underwater transceivers and to the hardware protection needed in the extreme underwater environment.
- **Deployment:** While terrestrial sensor networks are densely deployed, the deployment is deemed to be more sparse under water, due to the cost involved and to the challenges associated with the deployment itself.

- **Power:** The power needed for acoustic underwater communications is higher than that in terrestrial radio communications due to the greater distances and to more complex signal processing at the receivers to compensate for the impairments of the channel. Therefore, more energy is consumed and, hence, higher battery capacity is required.

- **Memory:** While terrestrial sensor nodes have very limited storage capacity, underwater sensors may need to be able to do some data caching as the underwater channel may be intermittent.

16.1.2   Real-Time Networking vs. Delay-Tolerant Networking

As in terrestrial sensor networks, depending on the application there may be very different requirements for data delivery. For example, surveillance applications may need very fast reactions to events and thus networking protocols that provide guaranteed delay-bounded delivery are required. Hence, it is necessary to develop protocols that deal with the characteristics of the underwater environment in order to quickly restore connectivity when lost and that react to unpaired or congested links by taking appropriate action (e.g., dynamical rerouting) in order to meet the given delay bound. Conversely, other applications may produce large bundles of data to be delivered to the onshore sink without particular delay constraints. In this respect, the Delay-Tolerant Networking Research Group (DTNRG) [2, 14] developed mechanisms to resolve the intermittent connectivity, long or variable delays, asymmetric data rates, and high error rates by using a *store and forward* mechanism based on middleware between the application layer and the lower layers. Similar methodologies may be particularly useful for applications, such as those that record seismic activity, that have very low duty cycle and produce, when activated, large bundles of data that need to be relayed to a monitoring station where they can be analyzed to predict future activity. On the other hand, sensor networks intended for disaster prevention such as those that provide earthquake or tsunami warnings require immediate delivery of information and hence real-time protocols. Therefore, the design of networking solutions for underwater acoustic sensor networks should always be aware of the difference between real-time and delay-tolerant applications and jointly tune existing solutions to the application needs and to the characteristics of the underwater environment.

16.2   Underwater Sensor Network Components

The design challenges and the unique characteristics of underwater sensor networks require different components for the realization of these networks. In this section, we describe these components in two parts, i.e., the underwater sensors that are used to collect information about the underwater environment, and the autonomous underwater vehicles that are used to provide many capabilities to underwater sensor networks such as mass data retrieval, connectivity establishment, and network management.

16.2.1   Underwater Sensors

Based on the various unique challenges posed by the underwater environment, several underwater sensors have been developed. The typical internal architecture of an underwater sensor is shown in Figure 16.1. It consists of a main controller/CPU which is interfaced with an oceanographic instrument or sensor through sensor interface circuitry. The controller receives data from the sensor and can store the data in the on-board memory, process them, and send them to other network devices by controlling the acoustic modem. The electronics are usually mounted on a frame which is protected by a PVC housing. Sometimes all sensor components are protected by bottom-mounted instrument frames that are designed to permit azimuthally omnidirectional acoustic communications, and protect sensors and modems from the potential impact of trawling gear, especially in areas subject to fishing activities. The protecting frame may be designed so as to deflect trawling gear on impact, by housing all components beneath a low-profile pyramidal frame [12].

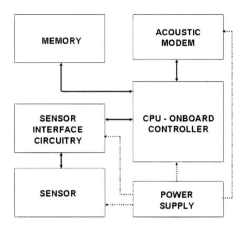

Figure 16.1   Internal organization of an underwater sensor node.

In addition to the main controller and the transceiver circuitry, underwater sensor devices are equipped with a vast variety of sensors. These devices include sensors to measure the quality of water and to study its characteristics such as temperature, density, salinity (interferometric and refractometric sensors), acidity, chemicals, conductivity, pH (magnetoelastic sensors), oxygen (Clark-type electrode), hydrogen, dissolved methane gas (METS), and turbidity. Disposable sensors, which detect ricin – the highly poisonous protein found in castor beans and thought to be a potential terrorism agent – also exist. DNA microarrays can be used to monitor both abundance and activity-level variations among natural microbial populations. Other existing underwater sensors include hydrothermal sulfide, silicate, voltammetric sensors for spectrophotometry, gold-amalgam electrode sensors for sediment measurements of metal ions (*ion-selective analysis*), amperometric microsensors for H_2S measurements for studies of anoxygenic photosynthesis, sulfide oxidation, and sulfate reduction of sediments. In addition, force/torque sensors for underwater applications requiring simultaneous measurements of several forces and moments have also been developed, as well as quantum sensors to measure light radiation and sensors for the measurement of harmful algal blooms.

Examples of currently available underwater sensors and modems are shown in Figure 16.2. The Aquacomm underwater modem, shown in Figure 16.2(a), provides reliable communication on the order of 10^{-6} bit error rate at 100–480 bps data rates. The modem can be operated down to a depth of 120 m with a communication range of 3 km. The LinkQuest underwater acoustic modems, shown in Figure 16.2(b), are short-range sensor nodes with a range of 300 m and can work at a depth of 200 m. The modems can support a communication rate of 7 kbps.

16.2.2   AUVs

In addition to static sensor nodes, several types of AUVs exist as experimental platforms for underwater experiments. Examples of existing AUVs are shown in Figure 16.3. Some of the AUVs resemble small-scale submarines (such as the Odyssey-class AUVs [1] developed at MIT shown in Figure 16.3(a)). Other type of AUVs are simpler devices that do not encompass such sophisticated capabilities. For example, *drifters* and *gliders* are oceanographic instruments often used in underwater exploration. Drifter underwater vehicles, an example of which is shown in Figure 16.3(b), drift with local current and have the ability to move vertically through the water column, and are used for taking measurements at preset depths [18]. Underwater gliders [13] are battery-powered AUVs that use hydraulic pumps

(a)   (b)

Figure 16.2 Examples of underwater sensor nodes: (a) Aquacomm underwater modem and (b) LinkQuest underwater sensor nodes.

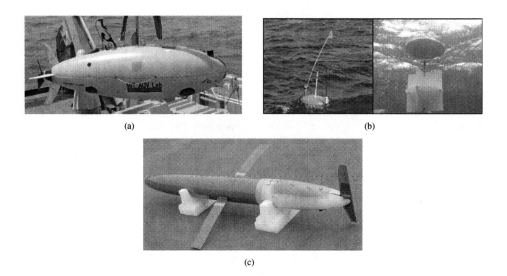

(a)   (b)

(c)

Figure 16.3 Examples of AUVs: (a) Odyssey AUV, (b) view of a drifter from above (left) and from below (right), and (c) Spray glider.

to vary their volume by a few hundred cubic centimeters in order to generate the buoyancy changes that power their forward gliding. When they emerge on the surface, GPS is used to locate the vehicle. This information can be relayed to the onshore station while operators can interact by sending control information to the gliders. Depth capabilities range from 200 to 1500 m while operating lifetimes range from a few weeks to several months. These long durations are possible because gliders move very slowly, typically 25 cm/s (0.5 knots). Spray glider, shown in Figure 16.3(c), is an example of an underwater glider.

While these existing products provide extensive communication capabilities, the still existing challenges related to the deployment of low-cost, low-scale underwater sensors can be listed as follows:

- **Size:** It is necessary to develop less expensive, robust "nano-sensors", e.g., sensors based on nanotechnology, which involves the development of materials and systems at the atomic, molecular, or macromolecular levels in the dimension range of approximately 1–500 nm.

- **Maintenance:** It is necessary to devise periodic cleaning mechanisms against corrosion and fouling, which may affect the lifetime of underwater devices. For example, some sensors for pCO_2, pH, and nitrate measurement, and fluorometers and spectral radiometers, may be limited by bio-fouling, especially on a long time scale.

- **Self-configuration:** There is a need for robust, stable sensors at a high range of temperatures since sensor drift of underwater devices may be a concern. To this end, protocols for in-situ calibrations of sensors to improve the accuracy and precision of sampled data must be developed.

- **New sensor types:** There is a need for new integrated sensors for *synoptic* sampling of physical, chemical, and biological parameters to improve the understanding of processes in marine systems.

16.3   Communication Architecture

We now introduce the reference architectures for 2-D and 3-D underwater networks, as well as 3-D networks of AUVs, which can enhance the capabilities of underwater sensor networks [7].

The network topology is in general a crucial factor in determining the *energy consumption*, the *capacity*, and the *reliability* of a network. Hence, the network topology should be carefully engineered and post-deployment *topology optimization* should be performed, when possible.

Underwater monitoring missions can be extremely expensive due to the high cost of underwater devices. Hence, it is important that the deployed network be highly reliable, so as to avoid unsuccessful missions due to the failure of single or multiple devices. For example, it is crucial to avoid designing the network topology with single points of failure that could compromise the overall functioning of the network.

The network capacity is also influenced by the network topology. Since the capacity of the underwater channel is severely limited, as will be discussed in Section 16.4, it is very important to organize the network topology in such a way that no *communication bottleneck* is introduced.

The communication architectures introduced here are used as a basis for discussing the challenges associated with underwater sensor networks (UWSNs). In the remainder of this section, we discuss the following architectures:

- **Static 2-D UWSNs for ocean bottom monitoring:** These are constituted by sensor nodes that are anchored to the bottom of the ocean, as will be discussed in Section 16.3.1. Typical applications may be environmental monitoring, or monitoring underwater plates in tectonics [16].

- **Static 3-D UWSNs for ocean-column monitoring:** These include networks of sensors whose depth can be controlled by means of various techniques and will be discussed in Section 16.3.2, and may be used for surveillance applications or monitoring of ocean phenomena (ocean bio/geo/chemical processes, water streams, pollution).

- **The 3-D networks of AUVs:** These networks include fixed portions composed of anchored sensors and mobile portions constituted by autonomous vehicles, as will be detailed in Section 16.3.3.

Next, we present these three different architectures for UWSNs in detail. However, these are not the only ways to deploy these networks. Depending on the application, users may have different deployment options for these networks under water. Also none of these architectures have yet been standardized.

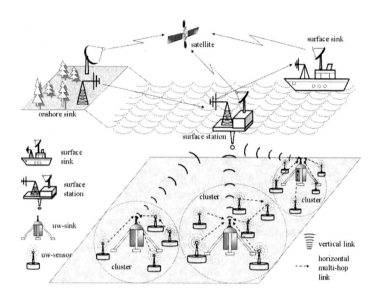

Figure 16.4   Architecture for 2-D UWSNs.

16.3.1   The 2-D UWSNs

A reference architecture for 2-D underwater networks is shown in Figure 16.4. A group of sensor nodes are anchored to the bottom of the ocean with deep-ocean anchors. Underwater sensor nodes are interconnected to one or more *underwater sinks* (uw-sinks) by means of wireless links. The uw-sinks, as shown in Figure 16.4, are network devices in charge of relaying data from the ocean-bottom network to a surface station. To achieve this objective, uw-sinks are equipped with two acoustic transceivers, namely a *vertical* and a *horizontal* transceiver. The horizontal transceiver is used by the uw-sink to communicate with the sensor nodes in order: (1) to send commands and configuration data to the sensors (uw-sink to sensors) and (2) to collect monitored data (sensors to uw-sink). The vertical link is used by the uw-sinks to relay data to a surface station. In deep-water applications, vertical transceivers must be long-range transceivers as the ocean can be as deep as 10 km. The surface station is equipped with an acoustic transceiver that is able to handle multiple parallel communications with the deployed uw-sinks. It is also endowed with a long-range RF and/or satellite transmitter to communicate with the *onshore sink* (os-sink) and/or to a *surface sink* (s-sink).

Sensors can be connected to uw-sinks via direct links or through multi-hop paths. In the former case, each sensor directly sends the gathered data to the selected uw-sink. However, in UWSNs, the power necessary to transmit may decay with powers greater than 2 of the distance [36], and the uw-sink may be far from the sensor node. Consequently, although direct link connection is the simplest way to network sensors, it may not be the most energy-efficient solution. Furthermore, direct links are very likely to reduce the network throughput because of increased acoustic interference due to high transmission power. In the case of multi-hop paths, as in terrestrial sensor networks [8], the data produced by a source sensor are relayed by intermediate sensors until the data reach the uw-sink. This results in energy savings and increased network capacity, but increases the complexity of the routing functionality as well. In fact, every network device usually takes part in a collaborative process whose objective is to diffuse topology information such that efficient and loop-free routing decisions can be made at each intermediate node. This process involves signaling and computation. Since energy and capacity are precious resources in underwater environments, as discussed above, in UWSNs the objective is to deliver event features by

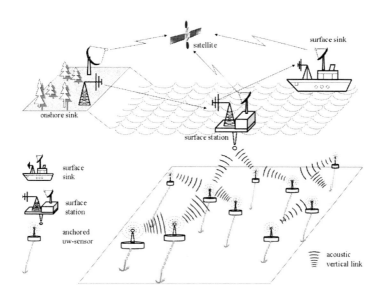

Figure 16.5   Architecture for 3-D UWSNs.

exploiting multi-hop paths and minimizing the signaling overhead necessary for constructing underwater paths at the same time.

16.3.2   The 3-D UWSNs

The 3-D underwater networks are used to detect and observe phenomena that cannot be adequately observed by means of ocean-bottom sensor nodes, i.e., to perform cooperative sampling of the 3-D ocean environment. In 3-D underwater networks, sensor nodes float at different depths in order to observe a given phenomenon. One possible solution would be to attach each uw-sensor node to a surface buoy, by means of wires whose length can be regulated so as to adjust the depth of each sensor node [11]. However, although this solution allows easy and quick deployment of the sensor network, multiple floating buoys may obstruct ships navigating on the surface, or they can be easily detected and deactivated by potential enemies in military settings. Furthermore, floating buoys are vulnerable to weather and tampering or pilfering.

For these reasons, a different approach may be to anchor sensor devices to the bottom of the ocean. In this architecture, depicted in Figure 16.5, each sensor is anchored to the ocean bottom and equipped with a floating buoy that can be inflated by a pump. The buoy pushes the sensor toward the ocean surface. The depth of the sensor can then be regulated by adjusting the length of the wire that connects the sensor to the anchor, by means of an electronically controlled engine that resides on the sensor. A challenge to be addressed in such an architecture is the effect of ocean currents on the described mechanism to regulate the depth of the sensor.

Many challenges arise with such an architecture, which need to be solved in order to enable 3-D monitoring, including:

- **Sensing coverage:** Sensors should collaboratively regulate their depth in order to achieve 3-D coverage of the ocean column, according to their sensing ranges. Hence, it must be possible to obtain sampling of the desired phenomenon at all depths.

- **Communication coverage:** Since in 3-D underwater networks there may be no notion of uw-sink, sensors should be able to relay information to the surface station via multi-hop paths. Thus, network devices should coordinate their depths in such a way that the network topology is always connected, i.e., at least one path from every sensor to the surface station always exists.

Sensing and communication coverage in a 3-D environment are rigorously investigated in [31]. The diameter and the minimum and maximum degree of the reachability graph that describes the network are derived as a function of the communication range, while different degrees of coverage for the 3-D environment are characterized as a function of the sensing range. These techniques could be exploited to investigate coverage issues in UWSNs.

16.3.3 Sensor Networks with AUVs

AUVs can function without tethers, cables, or remote control, and therefore have a multitude of applications in oceanography, environmental monitoring, and underwater resource study. Previous experimental work has shown the feasibility of relatively inexpensive AUV submarines equipped with multiple underwater sensors that can reach any depth in the ocean [1]. Hence, they can be used to enhance the capabilities of UWSNs in many ways. The integration and enhancement of fixed sensor networks with AUVs is an almost unexplored research area which requires new network coordination algorithms such as:

- **Adaptive sampling:** This includes control strategies to command the mobile vehicles to move to places where their data will be most useful. This approach is also known as *adaptive sampling* and has been used in monitoring missions such as in [5]. For example, the density of sensor nodes can be adaptively increased in a given area when a higher sampling rate is needed for a given monitored phenomenon.

- **Self-configuration:** This includes control procedures to automatically detect connectivity holes due to node failures or channel impairment and request the intervention of an AUV. Furthermore, AUVs can be used either for installation and maintenance of the sensor network infrastructure or to deploy new sensors. They can also be used as temporary relay nodes to restore connectivity.

One of the design objectives of AUVs is to make them rely on local intelligence and less dependent on communications from onshore [19]. In general, control strategies are needed for autonomous coordination, obstacle avoidance, and steering strategies. Solar energy systems allow the lifetime of AUVs to be increased, i.e., it is not necessary to recover and recharge the vehicle on a daily basis. Hence, solar-powered AUVs can acquire continuous information for periods of time on the order of months [20].

The distributed coordination of AUVs can be controlled through the *virtual bodies and artificial potentials* (VBAP) methodology [24]. VBAP controls the movement of multiple objects in a virtual body shape through control commands referred to as *artificial potentials*. The virtual body is created by linked, moving reference points called *virtual leaders*. Hence, an AUV is selected as a virtual leader and the movement of each AUV in the group is controlled according to the virtual leader.

VBAP defines three artificial potentials. The potential between each vehicle i and j is denoted as $V_I(x_{ij})$, which is a function of locations of the nodes with respect to the center of mass of the virtual body. Moreover, the potential between a particular vehicle i and a virtual leader k is denoted as $V_h(h_{ik})$, which depends on the relative distance of the vehicle to the virtual leader. Finally, $V_r(\theta_{ik})$ controls the angle between the vehicle and the virtual leader. The movement of a vehicle is defined as minus the gradient of the sum of these potentials. As a result, each vehicle is located not too close to another but close enough to the virtual leader. Moreover, the speed of each vehicle is controlled based on the formation error. Therefore, a group of AUVs slow down if the formation error grows. The group is constrained to maintain a uniform distribution as needed, but is free to spin and possibly wiggle with the

Table 16.1 Available bandwidth for different ranges in UW-A channels.

	Range (km)	Bandwidth (kHz)
Very long	1000	<1
Long	10–100	2–5
Medium	1–10	≈10
Short	0.1–1	20–50
Very short	<0.1	>100

current. The experiments in Monterey Bay (California) in 2003 using a fleet of autonomous underwater gliders proved the validity of the VBAP methodology [15, 5].

16.4 Basics of Underwater Acoustic Propagation

Underwater acoustic communications are mainly influenced by *path loss, noise, multiple paths, Doppler spread*, and *high and variable propagation delay*. All these factors determine the *temporal and spatial variability* of the acoustic channel, and make the available bandwidth of the *underwater acoustic channel* (UW-A) limited and dramatically dependent on range, time, and frequency. Long-range systems that operate over several tens of kilometers may have a bandwidth of only a few kilohertz, while a short-range system operating over several tens of meters may have more than 100 kHz of bandwidth. In both cases these factors lead to a low bit rate [10], on the order of tens of kilobits per second, for existing devices.

Underwater acoustic communication links can be classified according to their range as *very long, long, medium, short*, and *very short* links [37]. Table 16.1 shows typical bandwidths of the underwater channel for different ranges. Acoustic links are also roughly classified as *vertical* and *horizontal*, according to the direction of the sound ray with respect to the ocean bottom. As will be shown later, their propagation characteristics differ considerably, especially with respect to time dispersion, multi-path spreads, and delay variance. In the following, *shallow water* refers to water less than 100 *m* deep, while *deep water* is used for the deeper oceans.

Hereafter we analyze the factors that influence acoustic communications in order to state the challenges posed by the underwater channels for underwater sensor networking. These include:

Path loss

- **Attenuation:** This is mainly provoked by absorption due to the conversion of acoustic energy into heat. The attenuation increases with distance and frequency. Figure 16.6 shows the acoustic attenuation with varying frequency and distance for a short-range shallow-water UW-A channel, according to the Urick propagation model, which will be discussed in Section 16.4.1. The attenuation is also caused by scattering and reverberation (on the rough ocean surface and bottom), refraction, and dispersion (due to the displacement of the reflection point caused by wind on the surface). Water depth plays a key role in determining the attenuation.

- **Geometric spreading:** This refers to the spreading of sound energy as a result of the expansion of the wavefronts. It increases with propagation distance and is independent of frequency. There are two common kinds of geometric spreading: *spherical* (omnidirectional point source), which characterizes deep-water communications; and *cylindrical* (horizontal radiation only), which characterizes shallow-water communications.

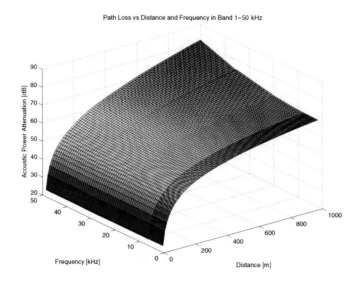

Figure 16.6   Path loss of short-range shallow UW-A channel vs. distance and frequency in the band 1–50 kHz.

Noise

- **Human-made noise:** This is mainly due to machinery (pumps, reduction gears, power plants) and shipping activity (hull fouling, animal life on hull, cavitation), especially in areas of heavy vessel traffic.

- **Ambient noise:** This is related to hydrodynamics (movement of water including tides, currents, storms, wind, and rain) and to seismic and biological phenomena. In [17], boat noise and snapping shrimps have been found to be the primary sources of noise in shallow water by means of measurement experiments on the ocean bottom.

Multiple Paths

- Multi-path propagation may be responsible for severe degradation of the acoustic communication signal, since it generates intersymbol interference (ISI).

- The multi-path geometry depends on the link configuration. Vertical channels are characterized by little time dispersion, whereas horizontal channels may have extremely long multi-path spreads.

- The extent of the spreading is a strong function of depth and the distance between the transmitter and receiver.

High Delay and Delay Variance

- The propagation speed in the UW-A channel is five orders of magnitude lower than in the radio channel. This large propagation delay (0.67 s/km) can reduce the throughput of the system considerably.

- The very high delay variance is even more harmful for efficient protocol design, as it prevents accurate estimation of the round-trip time (RTT), which is the key parameter for many common communication protocols.

Doppler Spread

- The Doppler frequency spread can be significant in UW-A channels [37], causing a degradation in the performance of digital communications: transmissions at a high data rate cause many adjacent symbols to interfere at the receiver, requiring sophisticated signal processing to deal with the generated ISI.
- The Doppler spread generates a simple frequency translation, which is relatively easy for a receiver to compensate for, and a continuous spreading of frequencies, which constitutes a non-shifted signal, which is more difficult to compensate for.
- If a channel has a Doppler spread with bandwidth B and a signal has symbol duration T, then there are approximately BT uncorrelated samples of its complex envelope. When BT is much less than unity, the channel is said to be *underspread* and the effects of the Doppler fading can be ignored, while, if BT is greater than unity, the channel is said to be *overspread* [22].

Most of the above factors are caused by the chemical–physical properties of the water medium such as temperature, salinity, and density, and by their spatio-temporal variations. These variations, together with the waveguide nature of the channel, cause the acoustic channel to be *highly temporally and spatially variable*. In particular, the horizontal channel is by far more rapidly varying than the vertical channel, in both deep and shallow water. The main factor is the propagation loss for underwater acoustic communications, which is generally described by the Urick propagation model. We will describe this model next, then the characteristics of shallow- and deep-water channels will be described in detail.

16.4.1 Urick Propagation Model

The acoustic channel is characterized by the Urick path-loss formula which is as follows:

$$TL(d, f) = \chi \cdot \log(d) + \alpha(f) \cdot d + A \tag{16.1}$$

where the path loss, $TL(d, f)$, is shown in dB as a function of internode distance d and operating frequency f. The term χ is the geometric spreading, which can be spherical for deep water and cylindrical for shallow water. The last term A is the transmission anomaly.

The underwater propagation has three components as shown in (16.1). The first component is the *distance-dependent* attenuation, which is common in many transmission media. The term χ accounts for the *geometric spreading* under water and varies depending on the depth of the water. For shallow-water environments, cylindrical spreading is observed, where the signal spreads horizontally due to low depth. As a result, a higher percentage of the signal is preserved as the distance increases. Therefore, usually $\chi = 10$ is used. On the other hand, in deep water, the signal spreads spherically leading to higher attenuation as the distance increases. For deep water, $\chi = 20$ is used.

The second component in (16.1) is the *frequency-dependent* attenuation, where $\alpha(f)$ [dB/m] is the *medium absorption*, which is a function of the operating frequency f. The determination of $\alpha(f)$ is usually done in three ways: theoretical calculation, Fisher&Simmon's model, and Thorp's model, which are based on experiments. The medium absorption is shown in Figure 16.7 as a function of f.

Finally, the last component, A [dB], accounts for the transmission anomaly, which results from degradation of the acoustic intensity caused by multiple path propagation, refraction, diffraction, and scattering of sound. This value is usually between 5 and 10 dB and is higher for shallow-water horizontal links, where multi-path effects are higher.

Moreover, the propagation speed of acoustic signals under water is given as follows:

$$q(z, S, t) = 1449.05 + 45.7 \cdot t - 5.21 \cdot t^2 + 0.23 \cdot t^3$$
$$+ (1.333 - 0.126 \cdot t + 0.009 \cdot t^2) \cdot (S - 35) + 16.3 \cdot z + 0.18 \cdot z^2 \tag{16.2}$$

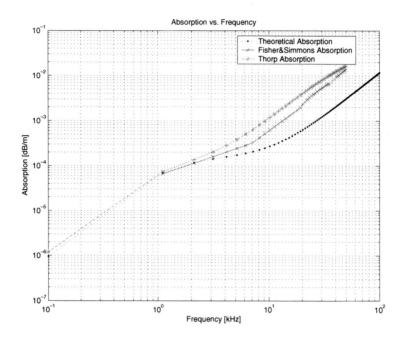

Figure 16.7   The medium absorption coefficient, α, as a function of operating frequency, f.

where $t = T/10$ (T is the temperature in °C), S is the salinity in ppt, and z is the depth in km. The above expression provides a useful tool to determine the propagation speed, and thus the propagation delay, in different operating conditions, and yields values in [1460, 1520] m/s, centered around 1500 m/s.

16.4.2   Deep-Water Channel Model

The deep-water acoustic channel is not severely affected by multiple paths. Instead, a higher attenuation is present due to the spherical spreading of the acoustic signal. The path of the acoustic signals under water is also unique. The signals bend toward the region of *slower* sound speed, i.e., *sound is lazy*. As shown in (16.2), the propagation speed of acoustic signals depends on the depth and temperature. As a result, the propagation path depends mainly on the location of the transmitter and also on the thermal structure of the particular medium. Figure 16.8 shows the four main types of thermal structures and the resulting signal paths. These four types of thermal structures are *isothermal gradient, negative gradient, positive gradient*, and *negative gradient over positive*.

In isothermal environments, water temperature is uniform at the surface and decreases below a certain depth. This causes the signal split as shown in Figure 16.8(b). For negative gradient thermal structure, the water temperature decreases with increasing depth. This causes the acoustic signals to bend down as shown in Figure 16.8(c). The opposite is true for positive gradient as shown in Figure 16.8(d), where acoustic signals bend toward surface. For the last thermal structure, the temperature gradient also changes with depth. In this case, a channel forms close to the surface with long ranges and the signal travels through this channel as shown in Figure 16.8(e). The unique propagation characteristics of acoustic signals under water cause communication regions as well as shadow zones where acoustic signals cannot reach. The deployment and operation of UWSNs significantly depend on these propagation properties.

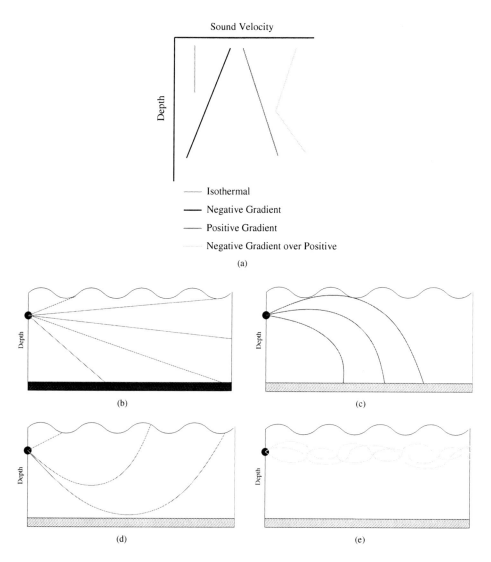

Figure 16.8 Sound propagation paths underwater: (a) depth vs. sound speed profiles; and propagation paths for (b) isothermal gradient, (c) negative gradient, (d) positive gradient, and (e) negative gradient over positive.

In addition to ray bending, the deep-water wireless channel also exhibits randomness in the received signal. This randomness in the channel is usually modeled using the Rayleigh fading channel model, where the envelope of the signal is modeled as a Rayleigh distributed random variable, α. Consequently, the received energy per bit per noise power spectral density is given by $\gamma = \alpha^2 E_b/N_0$, which has a distribution as follows:

$$f_\Gamma(\gamma) = \frac{1}{\gamma_0} \exp\left(\frac{\gamma}{\gamma_0}\right) \tag{16.3}$$

where $\gamma_0 = E[\alpha^2]E_b/N_0$ and E_b/N_0 is the received energy per bit per noise power spectral density, which can be directly found from the signal-to-noise ratio (SNR) of the channel. The SNR is given in

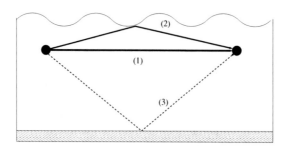

Figure 16.9 Possible paths for communication in shallow-water channels.

dB in underwater acoustic channels as

$$\psi_{dB}(d, f) = SL_{dB\ re\ uPa} - TL_{dB}(d, f) - NL_{dB\ re\ uPa} \tag{16.4}$$

where $SL_{dB\ re\ uPa}$ and $NL_{dB\ re\ uPa}$ are the signal level at the transmitter and the noise level given in dB with reference to μPa and $TL_{dB}(d, f)$ is the path loss given in (16.1). Then, $E_b/N_0 = \psi\ B_N/R$, where $\psi = 10^{\psi_{dB}(d,f)/10}$, B_N is the noise bandwidth, and R is the data rate. The signal level, SL, in (16.4) can be related to the intensity, I_t, and hence the transmit power, P_t, of the transceiver as follows:

$$I_t = \frac{P_t}{2\pi\ 1m\ H} \tag{16.5}$$

$$SL = 10 \log\left(\frac{I_t}{0.67 \times 10^{-18}}\right) \tag{16.6}$$

where P_t is the transmit power in W, and H is the depth in m.

16.4.3 Shallow-Water Channel Model

The shallow-water channel is greatly affected by multiple paths due to, firstly, multiple reflected rays from either the sea bottom or surface, and, secondly, ray bending as a result of sound speed variations with respect to depth. Depending on the depth of the environment, communication in shallow water is performed through three paths as shown in Figure 16.9: (1) direct path, (2) reflection from surface, and (3) reflection from bottom. For configurations where the sensor network is deployed close to the surface in a location of high depth, the reflection from the sea bottom is usually neglected due to the high attenuation in this path. Moreover, reflections from other objects under water also affect the communication performance.

The reflections from the surface or sea bottom constitute the so-called macro-multipath, which can deterministically be modeled. However, the surface and sea bottom are not perfectly level. The waves on the surface and the uneven texture of the sea bottom introduce imperfections into the received signals on a particular path. These time-dependent variations are usually called the micro-multipath and modeled statistically. In shallow-water communication, the wireless channel is modeled as a multi-ray Rayleigh fading channel, where each one of the three main paths is modeled as a Rayleigh fading channel. The composition of these rays at the receiver then governs the success of the communication.

16.5 Physical Layer

Until the beginning of the last decade, due to the challenging characteristics of the underwater channel, underwater modem development was based on *non-coherent* frequency shift keying (FSK) modulation.

Table 16.2 Evolution of underwater modulation techniques.

Type	Year	Rate (kbps)	Band (kHz)	Range (km)^a
FSK	1984	1.2	5	3_s
PSK	1989	500	125	0.06_d
FSK	1991	1.25	10	2_d
PSK	1993	0.3–0.5	0.3–1	200_d–90_s
PSK	1994	0.02	20	0.9_s
FSK	1997	0.6–2.4	5	10_d–5_s
DPSK	1997	20	10	1_d
PSK	1998	1.67–6.7	2–10	4_d–2_s
16-QAM	2001	40	10	0.3_s

^{a} The subscripts d and s stand for *deep* and *shallow* water, respectively.

Non-coherent FSK relies on energy detection and, hence, does not require phase tracking, which is a very difficult task mainly because of the Doppler spread in the UW-A channel, described in Section 16.4. FSK modulation schemes developed for under water try to suppress the multi-path effects by inserting time guards between successive pulses to ensure that the reverberation, caused by the rough ocean surface and bottom, vanishes before each subsequent pulse is received. Dynamic frequency guards can also be used between frequency tones to adapt the communication to the Doppler spread of the channel. Although non-coherent modulation schemes are characterized by a high *power efficiency*, their low *bandwidth efficiency* makes them unsuitable for high-data-rate multi-user networks. Hence, *coherent modulation* techniques have been developed for long-range, high-throughput systems. In the last few years, *fully* coherent modulation techniques, such as phase shift keying (PSK) and quadrature amplitude modulation (QAM), have become practical due to the availability of powerful digital processors. Channel equalization techniques are exploited to leverage the effect of ISI, instead of trying to avoid or suppress it. Decision-feedback equalizers (DFEs) track the complex, relatively slowly varying channel response and thus provide high throughput when the channel is slowly varying. Conversely, when the channel varies faster, it is necessary to combine the DFEs with a phase-locked loop (PLL) [38], which estimates and compensates for the phase offset in a rapid, stable manner. The use of DFEs and PLLs is driven by the complexity and time variability of ocean channel impulse responses. Table 16.2 shows the evolution from non-coherent modems to recent coherent modems.

Differential phase shift keying (DPSK) serves as an intermediate solution between non-coherent and fully coherent systems in terms of bandwidth efficiency. DPSK encodes information relative to the previous symbol rather than to an arbitrary fixed reference in the signal phase and may be referred to as *partially coherent modulation*. While this strategy substantially alleviates carrier phase-tracking requirements, the penalty is an increased error probability over PSK at an equivalent data rate.

With respect to Table 16.2, it is worth noticing that early phase-coherent systems achieved higher bandwidth efficiencies (bit rate/occupied bandwidth) than their non-coherent counterparts, but they have not yet outperformed non-coherent modulation schemes. In fact, coherent systems had lower performance than incoherent systems for long-haul transmissions on horizontal channels until ISI compensation via DFEs for optimal channel estimation was implemented [39]. However, these filtering algorithms are complex and not suitable for real-time communications, as they do not meet real-time constraints. Hence, suboptimal filters have to be considered, but the imperfect knowledge of the channel impulse response that they provide leads to channel estimation errors, and ultimately to decreased performance.

Another promising solution for underwater communications is the orthogonal frequency division multiplexing (OFDM) spread-spectrum technique, which is particularly efficient when noise is spread over a large portion of the available bandwidth. OFDM is frequently referred to as multicarrier modulation because it transmits signals over multiple *subcarriers* simultaneously. In particular, subcarriers that experience higher SNR are allotted a higher number of bits, whereas fewer bits are allotted to subcarriers experiencing attenuation. This concept is referred to as *bit loading* and requires channel estimation. Since the symbol duration for each individual carrier increases, OFDM systems perform robustly in severe multi-path environments, and achieve a high spectral efficiency.

Many of the techniques discussed above require underwater channel estimation, which can be achieved by means of probe packets [21]. An accurate estimate of the channel can be obtained with a high probing rate and/or with a large probe packet size, which, however, results in a high overhead and in the consequent drain of channel capacity and energy. Moreover, the latency in channel estimation due to high propagation delay introduces inconsistencies in the adaptive modulation schemes.

16.6   MAC Layer

In this section, we discuss techniques for multiple access control in UWSNs. As explained in Chapter 5, MAC constitutes one of the major challenges in sensor networks. However, in addition to the many challenges discussed in Chapter 5, MAC in UWSNs poses additional challenges due to the peculiarities of the underwater channel as explained in Section 16.4. In particular, MAC protocols for UWSNs are tailored to consider the very limited available bandwidth and high and variable propagation delay of the acoustic channel. Furthermore, the existence of communication regions and shadow zones as explained in Section 16.4 requires sophisticated methods for MAC to alleviate these effects.

Among many MAC techniques available in terrestrial wireless networks, only two major techniques, i.e., CSMA and CDMA, are applicable to communication under water. Especially, frequency division multiple access (FDMA) is not suitable for UWSNs due to the narrow bandwidth in UW-A channels. Since FDMA systems further divide the channel into smaller bands, the FDMA communication greatly suffers from fading and multi-path effects.

Similarly, time division multiple access (TDMA), alone, results in a limited bandwidth efficiency. This is related to the high RTT due to very long propagation delay and the delay variance due to multiple paths under water as discussed in Section 16.4. As a result, long time guards must be designed between each time slot in order to minimize packet collisions from adjacent time slots. Moreover, the variable delay makes it very challenging to realize precise synchronization with a common timing reference, which is required for TDMA. On the other hand, in hybrid MAC schemes, TDMA is mainly used in conjunction with CDMA, where multiple clusters are designed. In these schemes, TDMA is used for intercluster communication, where the propagation delay and delay variance are at manageable levels. These schemes will be described in Section 16.6.3.

In the following, we will describe the existing MAC schemes used in UWSNs. These schemes are classified into three main groups: CSMA-based (Section 16.6.1), CDMA-based (Section 16.6.2), and hybrid (Section 16.6.3) MAC protocols.

16.6.1   CSMA-Based MAC Protocols

The CSMA scheme is the most used technique in terrestrial sensor networks as we discussed in Chapter 5. CSMA techniques aim to prevent collisions with the ongoing transmissions by introducing a carrier sense mechanism and a backoff procedure at the transmitter side. These procedures, however, should be adjusted according to the propagation delay in the network. In the case of UWSNs, where very long and variable propagation delays exist, these procedures are dramatically affected, leading to inefficient operation.

The use of contention-based techniques that rely on handshaking mechanisms may be impractical under water for the following reasons:

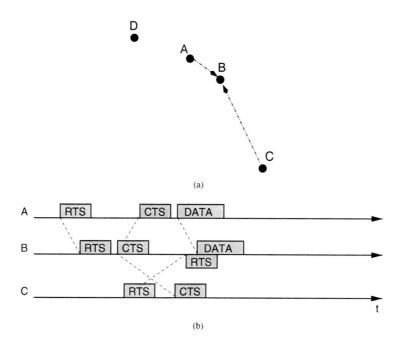

(a)

(b)

Figure 16.10 A potential problem for CSMA-based MAC schemes in the underwater environment: (a) topology and (b) transmission scenario.

- Due to the high propagation delay of UW-A channels, when carrier sense is used, it is more likely that the channel will be sensed idle while a transmission is ongoing, since the signal may not have reached the receiver.
- The high variability of delay in handshaking packets makes it impractical to predict the start and finish times of the transmissions of other stations. Thus, collisions are highly likely.
- Large delays in the propagation of RTS/CTS control packets lead to low throughput.

As we discussed in Chapter 5, CSMA techniques result in fairly efficient operation in terms of collision avoidance. However, this efficiency is achieved only in networks where the propagation delay can be very small compared to the transmission time of a control packet, and, hence, can be neglected. Therefore, the main reason for the inefficiency of CSMA-based MAC schemes in environments with long propagation delays is the high dependency between the transmission time and the distance between two nodes. Consider four underwater sensor nodes, A, B, C, and D, as shown in Figure 16.10(a), where A, B, and D are very close to each other, while C is farther apart and can only hear the transmissions of node B. As shown in Figure 16.10(b), when node A wants to transmit a packet to node B, it first sends an RTS packet. This RTS packet reaches node B quickly since node A and node B are closely located and node B sends back a CTS packet indicating that it can receive the packet. The only way node C is informed about this transmission is if it receives the CTS packet from node B. As shown in Figure 16.10(b), however, the CTS packet reaches node C very late and during the DATA transmission from node A to node B. If node C has a packet to send, it may sense the channel as idle before receipt of the CTS packet and send an RTS to node B, which would collide with the DATA transmission from node A.

A solution for this problem has been provided in terrestrial networks through the Floor Acquisition Multiple Access (FAMA) protocol, which requires certain conditions for RTS and CTS packet sizes.

Accordingly, the RTS packet length should be larger than the maximum propagation delay, while the CTS packet length should be larger than the RTS length plus the maximum round-trip propagation delay plus the transition time of the transceiver from transmit to receive mode. In underwater networks, however, these conditions require very long RTS and CTS packet sizes since the propagation delay is on the order of seconds. Therefore, the bandwidth efficiency can be significantly decreased.

Another potential problem with the CSMA schemes is the virtual sensing operation, which coordinates the backoff times according to the duration of the ongoing transmission through RTS or CTS packets as explained in Chapter 5. Since the propagation delay is variable based on the distance between two nodes, the start and end times of a DATA packet transmission are hard to predict from the receipt of a RTS or CTS packet. In order to alleviate this variability, long guard times are necessary, which, on the other hand, significantly decrease the efficiency of the MAC protocols.

Finally, the high propagation delay in underwater acoustic communication decreases the efficiency as the number of control packets needed for handshaking increases. As an example, sending RTS and CTS packets before the transmission requires nodes that are in the vicinity of the transmission to wait for a longer time. This additional time is on the order of a round-trip propagation time in addition to the transmission of DATA packet. As a result, both the time and the energy consumed for transmission of a packet almost double considering that the transmission time of a data packet is shorter than the propagation delay. Furthermore, since a longer time is spent for a single packet, the collision rate in a region increases, leading to inefficient operation and increased medium access delays.

Many novel access schemes have been designed for terrestrial sensor networks, whose objective, similar to UWSNs, is to prevent collisions in the access channel, thus maximizing the network efficiency. These similarities would suggest tailoring those efficient schemes for the underwater environment. However, the impact of the propagation delay on the CSMA mechanism, as described above, requires other novel functionalities to be included in the MAC protocol design. Furthermore, the main focus in MAC in terrestrial WSNs is on energy–latency tradeoffs. Most proposed schemes developed for terrestrial sensor networks aim at decreasing the energy consumption by using sleep schedules with virtual clustering as explained in Chapter 5. However, these techniques may not be suitable for an environment where dense sensor deployment cannot be assumed, as discussed in Section 16.3. Next, we describe three main solutions for CSMA-based MAC protocols in UWSNs.

Slotted Floor Acquisition Medium Access (FAMA)

Slotted FAMA [23] is based on the channel access discipline of FAMA as described above and combines both carrier sensing (CS) and RTS/CTS exchange between the source and receiver prior to data transmission. As noted above, the main drawback of FAMA is that it requires very long RTS and CTS packets to guarantee collision avoidance in the wireless channel with long propagation delays. Slotted FAMA, which has been developed specifically for UWSNs, overcomes this problem by slotting the time and preventing asynchronous transmission of control and data packets. As a result, each node is aware of the time when a node can start sending a RTS, CTS, or DATA packet and controls its carrier sensing duration according to the slot time.

The operation of Slotted FAMA is shown in Figure 16.11, where the time is divided into slots. Each node can only transmit packets at the beginning of each slot. The slot duration is determined to account for the maximum propagation delay in the network. Consequently, the slot length is $\tau + \gamma$, where τ is the maximum propagation delay and γ is the transmission time of a RTS or CTS packet. Accordingly, when a RTS or CTS packet is transmitted at the beginning of a slot, it is received by all the nodes that can hear this transmission by the end of the slot. This slotting mechanism addresses the hidden terminal problem that occurs due to variable delays explained at the beginning of this section. In this scheme, the duration of the DATA packet is announced in terms of number of slots so that each neighbor can defer its transmission accordingly.

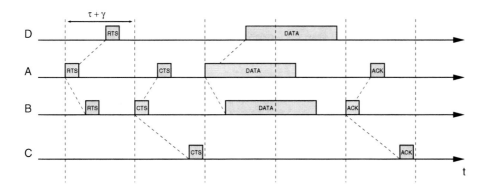

Figure 16.11   The operation of Slotted FAMA.

The communication between two nodes is carried out through a four-way handshake similar to the IEEE 802.11 protocol explained in Chapter 5. The only difference is that each packet is sent at the beginning of a slot as shown in Figure 16.11. The successful receipt of a DATA packet is indicated by transmitting an ACK packet, while a NACK packet is sent in case the DATA packet is not received correctly. In this case, DATA is retransmitted.

The operation of Slotted FAMA can be explained by referring to Figure 16.10(a). When a node sends a RTS packet, the nodes that receive this packet and are not the intended receiver (node D) defer their transmissions for two slots to wait for the DATA packet. If a node hears a CTS packet that is not destined for itself (node C), it waits for the duration of the DATA packet and an additional slot to receive the ACK packet. If a NACK packet is received, then the nodes further defer their transmissions to allow the retransmission of a DATA packet. In Slotted FAMA, the case where a node may receive a corrupted packet or senses interference in the channel is also considered. Consequently, the node assumes the worst case so that a CTS packet, which is not destined for itself, is received and defers transmission to allow a DATA and ACK transmission.

The requirement that very long RTS and CTS packets are sent is overcome by the slotting technique in Slotted FAMA. The slotting technique prevents collisions due to variable propagation delays under water. However, the protocol still suffers from inefficiency since the nodes have to wait for the duration of the slot for each packet transmission, which introduces significant energy consumption due to idle listening. Slotted FAMA aims to improve the efficiency by introducing *trains of packets* during the communication. More specifically, during a transmission, if a node has more than one packet to send to a receiver, it sends multiple packets during a single handshake. In this case, the sender sets a flag in the DATA packet that is being sent to indicate the existence of another packet to be sent. The receiver acknowledges this packet through the ACK packet it sends for each DATA packet. Consequently, the neighbor nodes that receive either of these packets defer their transmission further to allow for a consecutive packet. With the *trains of packets*, the inefficiency caused by sending RTS/CTS packets for each DATA packet is improved since, now, a single RTS/CTS exchange accounts for multiple DATA packets.

Although time slotting eliminates the asynchronous nature of the protocol and the need for excessively long control packets, thus providing savings in energy, guard times should be inserted in the slot duration to account for any system clock drift. In addition, due to the high propagation delay of underwater acoustic channels, the handshaking mechanism may lead to low system throughput, which is particularly important in UWSNs, where the available bandwidth is severely limited.

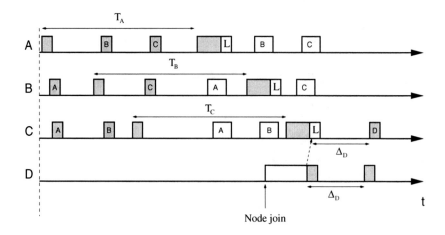

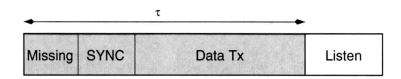

Figure 16.12   The operation of the UWAN-MAC mechanism.

Figure 16.13   The transmission packet structure and the listen duration for the UWAN-MAC mechanism.

Energy-Efficient MAC for Underwater Acoustic Wireless Networks (UWAN-MAC)

UWAN-MAC [32] also deploys CSMA-based MAC and has been primarily developed for high-density UWSNs. Rather than bandwidth optimization, UWAN-MAC is focused on energy efficiency by introducing sleep schedules similar to its terrestrial counterparts.

Each node has a sleep schedule such that each node wakes up periodically in the network to transmit its data. At the beginning of each cycle, a node broadcasts a SYNC packet indicating its period of the sleep schedule. As a result, the neighbor nodes that receive this packet wake up at the next scheduled time to listen to the node. Consequently, every node wakes up for each of its neighbors to receive data in addition to its scheduled wakeup time to transmit data. Note that since relative time information is exchanged by the SYNC packets, UWAN-MAC does not require the propagation delay to be known by each node. As long as the propagation time stays constant, the information about the sleep duration of a node helps keep each node synchronized. The operation of the UWAN-MAC synchronization mechanism is shown in Figure 16.12. When node A broadcasts a SYNC packet, it indicates its sleep period as T_A. Accordingly, when node A's neighbors receive this SYNC packet, they schedule to wake up T_A seconds after receipt of the SYNC packet. Similarly, node A also receives SYNC packets from its neighbors and schedules wakeup times for them.

The data transmission packet structure of each node is shown in Figure 16.13, which consists of *Missing, SYNC, Data Tx,* and *Listen* periods. The *SYNC* period is used to broadcast SYNC packets as explained before, while *Data Tx* is used to transmit the DATA packets. Since each of the neighbors of node A is listening to the transmission period of the node, it can transmit its DATA without any collisions. The *Missing* and *Listen* periods are used to handle node failures/removals and node joins.

At each sleep period, a node collects the list of its neighbors that it has received SYNC messages from. In case there is a change in this list (a SYNC message from a particular node is not received), the node creates a missing node list and broadcasts this information during the *Missing* period shown in Figure 16.13. This list serves as notification to the nodes in the missing list that a communication error may have occurred earlier. If a node does not hear from its neighbors in the missing list for a couple of consecutive cycles, it deletes this node from its neighbor list. On the other hand, the node that is in the missing list replies back to the sender of the SYNC message as if it is a newcomer node. The procedure for newcomer nodes is explained next.

The *Listen* period in the transmission period shown in Figure 16.13 is used to include newcomers to the network. This situation is illustrated in Figure 16.12, where node D joins the network while node C is transmitting a SYNC packet. When node D joins the network, it listens to the channel for the SYNC packets from its neighbors. When it receives a SYNC packet from node C, it replies to this packet with a HELLO packet to indicate its existence. The *Listen* period at the end of each transmission period ensures that node C receives this HELLO packet. Then, node C includes the newcomer node D in its list of neighbors. In the HELLO packet, node F also indicates the time left for its next wakeup time, i.e., Δ_D. Node C can then wake up for the scheduled wakeup time of node D and receive its SYNC packet as shown in Figure 16.12. Node D indicates its schedule to other nodes similarly.

The operation of UWAN-MAC so far assumes that the propagation delay between two nodes does not change. This enables the relative wakeup announcements by the SYNC packets to synchronize nodes. However, as explained in Section 16.4, the underwater acoustic channel suffers from high variable propagation delays and channel fluctuations due to many reasons such as drifts of nodes, scattering objects in the water, and multi-path effects. As a result, the propagation delay fluctuates randomly. In order to account for this fluctuation, UWAN-MAC introduces guard times before and after each listen duration for each of its neighbors. This ensures that a packet is correctly and fully received even if it arrives earlier or later than expected. A potential problem with this extension is that a node's transmission schedule and its listen period may overlap because of the guard times. In this case, the node changes its transmission schedule and broadcasts this via a SYNC packet.

UWAN-MAC achieves significant energy consumption through the sleep schedules. Since it is developed for delay-tolerant applications, the sleep schedules may induce very high medium access delays for communication. Moreover, the throughput of the protocol is decreased due to the overhead in maintaining schedules and the sleep schedule operation.

16.6.2   CDMA-Based MAC Protocols

CDMA is the most promising physical layer and multiple access technique for UWSNs. It is quite robust to frequency-selective fading caused by underwater multiple paths. This advantage depends on the fact that CDMA distinguishes simultaneous signals transmitted by multiple devices by means of pseudo-noise codes that are used for spreading the user signal over the entire available band. This allows the time diversity in the UW-A channel to be exploited by leveraging *Rake filters* [35] at the receiver. These filters are designed to match the pulse spreading, the pulse shape, and the channel impulse response, so as to compensate for the multi-path effects. As a result, CDMA allows a reduction in the number of packet retransmissions, which results in decreased battery consumption and increased network throughput.

There are two main approaches for the CDMA techniques, i.e., direct sequence spread spectrum (DSSS) and frequency hopping spread spectrum (FHSS). In this section, we describe the CDMA-based MAC protocols based on these techniques.

Underwater Medium Access Control (UW-MAC)

UW-MAC [28] is a transmitter-based CDMA scheme that incorporates a novel closed-loop distributed algorithm to set the optimal transmit power and code length to minimize the near–far effect.

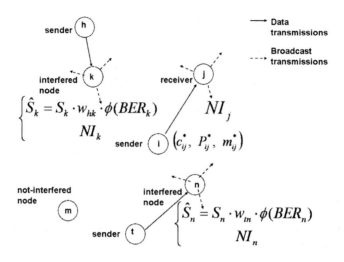

Figure 16.14 Data and broadcast message transmissions.

It compensates for the multi-path effect by exploiting the time diversity in the underwater channel, thus achieving high channel reuse and a low number of packet retransmissions, which result in decreased battery consumption and increased network throughput. UW-MAC leverages a multi-user detector on resource-rich devices such as surface stations, uw-gateways and AUVs, and a single user detector on low-end sensors. UW-MAC aims at achieving a threefold objective, i.e., to guarantee (1) high network throughput, (2) low access delay, and (3) low energy consumption. UW-MAC relies on the distributed power and code self-assignment optimization, where the transmitter nodes select the transmission power and the spreading codes according to the neighbor information broadcast by other nodes periodically. Consequently, in order to allow node i to set its transmit power, P_{ij}, and spreading factor, w_{ij}, associated with the data transmission toward node j, it needs to leverage information on the multiple access interference (MAI) and normalized receiving spread signal of neighboring nodes. This information is broadcast periodically by active nodes, as depicted in Figure 16.14. In particular, to limit such broadcasts, a generic node n transmits only significant values of NI_n and $\hat{S}_n$, i.e., out of predefined tolerance ranges. Consequently, each node performs power and code self-assignment optimization. This optimization problem is investigated for deep-water and shallow-water channels as follows.

Deep Water

For the deep-water acoustic channel, the transmission loss TL_{ij} between nodes i and j is modeled according to (16.1), where multi-path affects are not considered.

The power constraints that node i needs to respect to limit the near–far effect when it transmits to j, and avoid impairing ongoing communications, are

$$
\begin{cases}
\dfrac{N^0 + I_j}{P_{ij}/TL_{ij}} \le w_{ij} \cdot \Phi(\hat{BER}_j) \\[4mm]
\dfrac{N^0 + I_k + P_{ij}/TL_{ik}}{S_k} \le w_{l_k k} \cdot \Phi(\hat{BER}_k), \quad \forall k \in \mathcal{K}_i,
\end{cases}
\tag{16.7}
$$

where, N^0 |W| is the average noise power, I_j and I_k |W| are the MAI at nodes j and $k \in \mathcal{K}_i$, $\mathcal{K}_i$ being the set of nodes whose ongoing communications may be affected by node i's transmit power, and

w_{ij} and $w_{t_k k}$ are the bandwidth spreading factors characterizing the ongoing transmissions from i to j and from t_k to k, respectively, t_k being the node which k is receiving from. In addition, P_{ij} [W] is the power transmitted by i in the communication toward j when an ideal channel (without multi-path, i.e., $A = 0$ dB) is assumed, i.e., when no power margin is considered to face the fading dips, TL_{ij} and TL_{ik} are the transmission losses from i to j and from i to $k \in \mathcal{K}_i$, respectively, S_k [W] is the user signal power that receiver k is decoding, and $\Phi()$ is the MAI threshold which depends on the target bit error rate ($\hat{BER}$) at the receiver node.

The first constraint in (16.7) states that the power P_{ij} transmitted by node i needs to be sufficiently high to allow receiver j to successfully decode it, given its current noise and MAI power level ($NI_j = N^0 + I_j$); the second constraint in (16.7) states that the power P_{ij} transmitted by node i should not be so high as to impair the ongoing communications toward nodes $k \in \mathcal{K}_i$, given their normalized received user spread signals ($\hat{S}_k$), and noise and MAI level (NI_k). Denoting the noise and MAI power of a generic node n as $NI_n = N^0 + I_n$, and the normalized received spread signal, i.e., the signal power after despreading, as $\hat{S}_n = S_n \cdot w_{t_n n} \cdot \Phi(\hat{BER}_n)$, the constraints in (16.7) can be represented as

$$\frac{NI_j \cdot TL_{ij}}{w_{ij} \cdot \Phi(\hat{BER}_j)} \leq P_{ij} \leq \min_{k \in \mathcal{K}_i} |(\hat{S}_k - NI_k) \cdot TL_{ik}|. \tag{16.8}$$

In order to save energy, node i will select a transmit power P_{ij} and a code length c_{ij} in such a way that constraints (16.8) are met, and the energy per bit $E_{ij}^b(P_{ij}, c_{ij}) = (P_{tx} + P_{ij}) \cdot c_{ij}/r$ [J/bit] be minimized, where P_{tx} [W] is a *distance-independent* component accounting for the power needed by the transmitting circuitry, and r [cps] the *constant* underwater chip rate, which depends on the available acoustic spectrum B [Hz] and on the modulation spectrum efficiency η_B, i.e., $r = \eta_B \cdot B$. Since E_{ij}^b decreases as the transmit power and code length decrease, and the relation between the spreading factor w_{ij} and the code length c_{ij} depends on the family of codes, i.e., $w_{ij} = \mathcal{W}^C(c_{ij})$, the optimal solution is $c_{ij}^* = c_{min}$ and $P_{ij}^* = NI_j \cdot TL_{ij}/[\alpha \cdot c_{min} \cdot \Phi(\hat{BER}_i)]$, where the spreading factor is assumed to be proportional to the code length, i.e., $w_{ij} = \alpha \cdot c_{ij}$. Note that this solution achieves the threefold objective of minimizing the energy per bit E_{ij}^b that i needs in order to successfully communicate with j in the minimum possible time, i.e., minimize the energy consumption while transmitting at the highest possible data rate r/c_{min}.

Shallow Water

For shallow-water communication, the multi-path effects should also be considered. In UW-MAC, the channel fading is modeled by a Rayleigh random variable, and the transmission loss between i and j is $TL_{ij} \cdot \rho$, where $TL_{ij} = d_{ij} + 10^{[\alpha(\bar{f}) \cdot d_{ij} + A]/10}$, with $A \in [5, 15]$ dB, and ρ has a unit mean Rayleigh cumulative distribution $D_\rho(\rho) = 1 - \exp(-\pi\rho^2/4)$. A *transmission margin* for link (i, j) is also defined as m_{ij}, where $P_{ij}^* \cdot m_{ij}$ [W] is the actual transmit power, while P_{ij}^* [W] represents the optimal transmission power in an ideal channel, as computed previously. The packet error rate PER_{ij} experienced on link (i, j) when sender i transmits power $P_{ij}^* \cdot m_{ij}$ can be defined as the probability that the received power at node j be smaller than that required in an ideal channel where no multi-path is experienced, i.e.,

$$PER_{ij} = \Pr\left\{ \frac{P_{ij}^* \cdot m_{ij}}{TL_{ij} \cdot \rho} < \frac{P_{ij}^*}{TL_{ij}} \right\} = \Pr\{\rho \geq m_{ij}\}$$

$$= 1 - D_\rho(m_{ij}) = \exp\left(-\frac{\pi m_{ij}^2}{4}\right). \tag{16.9}$$

Hence, the average number of transmissions of a packet such that receiver j correctly decodes it when sender i chooses transmission margin m_{ij} is $N_{ij}^T(m_{ij}) = [1 - PER_{ij}]^{-1} = D_\rho(m_{ij})^{-1}$. For shallow water, the power and code optimization self-assignment problem in a Rayleigh channel is cast as follows:

P: Power and Code Self-assignment Problem

$$\text{Given:} \quad P^{\max}, r, TL_{ij}, NI_j, B\hat{E}R_j, \hat{S}_k, NI_k, \forall k \in \mathcal{K}_i$$

$$\text{Find:} \quad c_{ij}^* \in [c_{\min}, c_{\max}], \; m_{ij}^*, \; P_{ij}^* \cdot m_{ij}^* \in (0, P^{\max}]$$

$$\text{Minimize:} \quad E_{ij}^b(c_{ij}, P_{ij}, m_{ij}) = \frac{(P_{tx} + P_{ij} \cdot m_{ij}) \cdot c_{ij}}{r} \cdot N_{ij}^T(m_{ij})$$

Subject to:

$$N_{ij}^T(m_{ij}) = D_\rho(m_{ij})^{-1} = \left[1 - \exp\left(-\frac{\pi m_{ij}^2}{4}\right)\right]^{-1}; \tag{16.10}$$

$$P_{ij}^{\min}(c_{ij}) = \frac{NI_j \cdot TL_{ij}}{\alpha \cdot c_{ij} \cdot \Phi(B\hat{E}R_j)} = \frac{\Gamma_{ij}}{c_{ij}}; \tag{16.11}$$

$$P_{ij}^{\max} = \min_{k \in \mathcal{K}_i} \left[(\hat{S}_k - NI_k) \cdot TL_{ik}\right]; \tag{16.12}$$

$$P_{ij}^{\min}(c_{ij}) \le P_{ij} \le \min\left[P_{ij}^{\max}, P^{\max}\right]; \tag{16.13}$$

$$P_{ij}^{\min}(c_{ij}) \le P_{ij} \cdot m_{ij} \le \min\left[P_{ij}^{\max}, P^{\max}\right]. \tag{16.14}$$

In constraints (16.13) and (16.14), the transmit power *lower bound*, $P_{ij}^{\min}$, is a *function* that depends on the chosen code length c_{ij}, which is a solution variable of **P**, whereas the transmit power *upper bound*, $\min[P_{ij}^{\max}, P^{\max}]$, is a *constant* only depending on the node maximum transmit power ($P^{\max}$) and on the broadcast MAI (NI_k) and normalized received spread signal ($\hat{S}_k$).

A low-complexity solution for **P** can be found by noting that the minimum energy per bit E_{ij}^b monotonically decreases as P_{ij} and the code length c_{ij} decreases. **P** *may* admit a feasible solution if in (16.13) $P_{ij}^{\min}(c_{ij}) \le \min[P_{ij}^{\max}, P^{\max}]$ holds, i.e., if $c_{ij} \ge \Gamma_{ij}/\min[P_{ij}^{\max}, P^{\max}]$. Consequently, to minimize the objective function, the optimal code length c_{ij}^* is

$$c_{ij}^* = \max\left[\min\left[\frac{\gamma \cdot \Gamma_{ij}}{\min[P_{ij}^{\max}, P^{\max}]}, c_{\max}\right], c_{\min}\right] \tag{16.15}$$

where γ is a margin on the code length aimed at absorbing information inaccuracy. By substituting (16.15) into (16.11), given (16.13), the optimal transmit power *before* applying the margin on the channel, P_{ij}^*, can be found as

$$P_{ij}^* = \min\left[\frac{\Gamma_{ij}}{c_{ij}^*}, P^{\max}\right]. \tag{16.16}$$

Finally, by substituting (16.15) and (16.16) into the objective function, the energy per bit can be obtained as function of the margin only,

$$E_{ij}^b(m_{ij}) = \frac{P_{tx} \cdot c_{ij}^* + \Gamma_{ij} \cdot m_{ij}}{r \cdot [1 - \exp(-\pi m_{ij}^2/4)]}, \tag{16.17}$$

which can then be easily minimized to obtain the optimal margin m_{ij}^* as a numeric solution of the following equation:

$$\frac{dE_{ij}^b}{dm_{ij}} = 0 \Rightarrow -\frac{\pi}{2} m_{ij}^{*2} + \frac{\pi P_{tx} c_{ij}^*}{2\Gamma_{ij}} m_{ij}^* + 1 = \exp\left(-\frac{\pi m_{ij}^{*2}}{4}\right). \tag{16.18}$$

Note that $\mathbf{P}$ is feasible iff the optimal solution $(c_{ij}^*, P_{ij}^*, m_{ij}^*)$ meets constraint (16.14), i.e., iff $P_{ij}^* \cdot m_{ij}^* \leq$ min $[P_{ij}^{max}, P^{max}]$. Otherwise, an energy-efficient suboptimal solution is $c_i^+ = c_{max}$ and $P_{ij}^+ \cdot m_{ij}^+ =$ min $[P_{ij}^{max}, P^{max}]$.

Differently from the deep-water case, the energy per bit in a Rayleigh channel "skyrockets" when an adequate power margin is not used, due to the high number of packet retransmissions, as can be seen in (16.10). Moreover, a tradeoff between the optimal transmit power and code length occurs, which suggests that it is not always possible to *jointly* achieve the highest data rate and the lowest energy consumption, as it is possible in a channel that is not affected by multi-path effects. As previously anticipated, when the MAI at the receiver side is higher than a certain threshold $(NI_j \geq 1.24$ mW$)$ it is no longer possible to select the highest data rate, i.e., the shortest code, to achieve the minimum energy per bit. Conversely, with low MAI at the receiver, this twofold objective can still be achieved.

It is shown that UW-MAC manages to simultaneously meet the three objectives in deep-water communications, which are not severely affected by multi-path effects, while in shallow-water communications, which are heavily affected by multi-path effects, UW-MAC dynamically finds the optimal tradeoff among high throughput, and low access delay and energy consumption, according to the application requirements. The main features of UW-MAC are: (1) it provides a *unique and flexible solution* for different architectures such as *static* 2-D deep water and 3-D shallow water, and architectures with *mobile* AUVs; (2) it is fully *distributed*, since code and transmit power are distributively selected by each sender without relying on a centralized entity; (3) it is intrinsically *secure*, since it uses chaotic codes; (4) it efficiently *supports multicast transmissions*, since spreading codes are decided at the transmitter side; and (5) it is *robust* against inaccurate node position and interference information caused by mobility, traffic unpredictability, and packet loss due to channel impairment.

16.6.3 Hybrid MAC Protocols

Centralized and distributed solutions for MAC provide different advantages in terms of deployment and network operation. Considering the advantages of each technique, a hybrid solution is the multi-cluster MAC protocol, which we will explain next.

Multi-cluster MAC

The multi-cluster MAC protocol [33] has been developed for networks of mobile AUVs such that the network is organized in multiple clusters, each composed of adjacent vehicles. Instead of a flat topology, where each AUV is connected to each of its neighbors, clusters are formed to decrease the packet transmission delay as well as scalability of the MAC protocol. These clusters are formed by closely located AUVs. This formation decreases the effects of high propagation delays in the acoustic wireless channel. Consequently, inside each cluster, TDMA is used. Each time slot in the TDMA frame is followed by long guard bands to overcome the effect of propagation delay. Since vehicles in the same cluster are close to one another, the negative effect of very high underwater propagation delay is limited. Moreover, the relatively short propagation delay between closely located nodes requires shorter time guards, which minimize the effects of using TDMA techniques in long propagation delay networks.

The nodes in the network have three types of duties in this protocol. Very similar to other cluster-based protocols such as S-MAC or LEACH, which were discussed in Chapter 5, a node can be a *cluster head*, *cluster connector* (border node), or *ordinary* node. Because of the mobility in the network, each node can assume one of these duties at any time.

Cluster heads in the multi-cluster MAC protocol can communicate directly with any node in the cluster. Hence, in each cluster, nodes are at most two hops away from each other. This property decreases the propagation delay in the cluster. Cluster management is performed by the cluster heads in each cluster. The cluster connectors are part of more than one cluster and perform intercluster communication. Finally, the ordinary nodes are connected to only one cluster head in the network.

In each cluster, TDMA slots are used to coordinate the traffic. The TDMA frames are composed of multiple time slots. The first and last slot is used by the cluster head, while the other slots are assigned to the members of the cluster. The cluster head uses the first slot for its communication and the last slot is used for cluster management information. Since the network is composed of mobile AUVs, the cluster head uses the last slot in each frame to inform the members of the cluster about cluster formation. During each frame, each node in the cluster estimates its location and informs the cluster head, which then determines the nodes that should be in the cluster based on their locations. This information is broadcast at the end of the TDMA frame. Consequently, the dynamic topology changes are captured in cluster maintenance.

Since there may be multiple clusters in close proximity, interference among different clusters needs to be minimized. This is achieved by assigning different spreading codes to different clusters. Consequently, a hybrid TDMA–CDMA communication scheme is applied. The code selection is again performed by the cluster head from the information from the cluster connectors. Nodes that have information about the neighbor clusters inform the cluster head about their code selection. Accordingly, the cluster head selects the spreading code among the unused codes and informs the cluster members.

Although it is designed for mobile AUVs, the multi-cluster MAC protocol exploits the advantages of many medium access techniques in a hybrid fashion. While centralized control is used for the clusters, this control is limited inside the cluster, improving the scalability of the protocol. Similarly, the cluster formation enables the use of TDMA techniques, which are otherwise inefficient for UWSNs. Similarly, potential interference between clusters is avoided through CDMA techniques.

16.7 Network Layer

The network layer is in charge of determining the path between a source (the sensor that samples a physical phenomenon) and a destination node (usually the surface station). In general, while many impairments of the underwater acoustic channel are adequately addressed at the physical and data link layers, some other characteristics, such as the extremely long propagation delays, are better addressed at the network layer.

As we saw in Chapter 7, many routing protocols have been developed for terrestrial wireless sensor networks. However, the intrinsic properties of the underwater applications and the unique characteristics of the acoustic channel as explained in Section 16.4 prevent these solutions from being directly applicable to UWSNs. Especially, proactive protocols, which require paths from one node to another node to be established, require high control overhead. Consequently, the effective throughput of the communication, which is already affected by the limited bandwidth of the acoustic channel, is diminished. Similarly, reactive protocols, which initiate a route discovery process only when required, also incur high latency in path establishment since end-to-end paths need to be constructed. Reactive protocols are deemed to be unsuitable for UWSNs because they also cause a high latency in the establishment of paths, which may even be amplified under water by the slow propagation of acoustic signals. Furthermore, links are likely to be asymmetrical, due to the characteristics of the bottom and variability in the sound speed channel. Hence, protocols that rely on symmetrical links, such as most of the reactive protocols, are unsuited for the underwater environment. Another class of routing protocols is the geographical routing protocols, where the location information of sensor nodes is exploited to forward packets through the networks. These protocols are favored in terrestrial sensor networks due to their scalability and low control overhead. However, the localization required by these protocols is an important problem in UWSNs, where precise location management may not be possible. Firstly, GPS receivers do not work properly under water. Moreover, the precise synchronization requirements of localization protocols may not be guaranteed in high and variable latency acoustic channels.

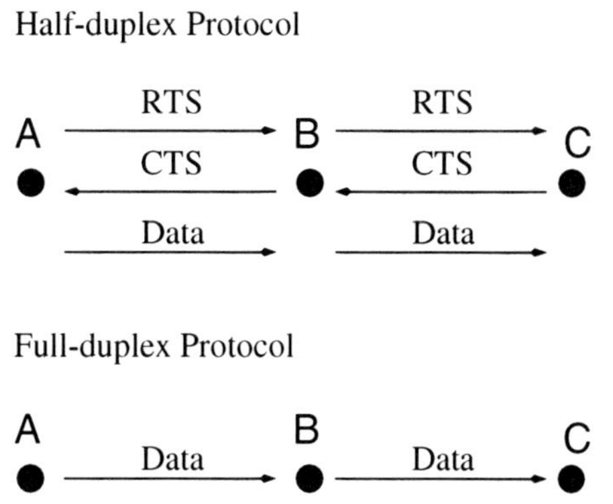

Half-duplex Protocol

Full-duplex Protocol

Figure 16.15 Delay in half-duplex and full-duplex communication under water.

Various routing solutions exist for these challenges in UWSNs. Mainly, the solutions can be classified, according to the locations where routing decisions are made, as *centralized*, *distributed*, and *hybrid*. More specifically, *centralized* solutions exploit the fact that in many deployments for UWSNs, a centralized manager exists as a gateway at the surface to collect information from underwater nodes. The high processing power and large storage capacity of this entity are exploited in many routing protocols, where the routing decisions are made centrally at this manager and disseminated to the network for operation. These solutions also rely on the fact that the network topology is fairly static except for the effects of drift under water. *Distributed* solutions, on the other hand, aim to decrease the latency and high energy consumption in making centralized decisions for routing by enabling each node to construct routes. Finally, *hybrid* solutions combine the advantages of these two types of approaches to construct routes under water. Next, we examine each of these solutions in the context of this classification.

16.7.1 Centralized Solutions

Full-Duplex Communication-Based Routing

Wireless communication in WSNs is usually characterized by a half-duplex channel, where a transceiver can only transmit or receive at a given time. This restriction is usually due to the single channel operation in terrestrial WSNs and the low-cost requirements of sensor nodes. In UWSNs, however, propagation latency is significantly higher and half-duplex communication introduces additional delay since nodes need to wait for the packets to be received before responding. Channel utilization is severely affected if a three- or four-way handshaking mechanism is used. On the other hand, full-duplex communication, where a node can both transmit and receive at the same time, improves channel utilization.

A multi-hop situation is shown in Figure 16.15, where node A needs to transmit a packet to node C. In the case of half-duplex communication, the communication delay is more than three times the total propagation delay between nodes A and C, i.e., $d \geq 3(d_1 + d_2)$. On the other hand, if full-duplex communication is deployed, the communication delay is almost equal to the total propagation delay between nodes A and C, i.e., $d \approx (d_1 + d_2)$.

Using the full-duplex communication principles, a centralized routing protocol is feasible even in the underwater environment that is characterized by high propagation delay. The full-duplex communication-based routing scheme [41] relies on a master node that dynamically manages the topology of the network and constructs routes between any pair of nodes. This communication architecture is motivated by the fact that in underwater applications, generally there exists a gateway

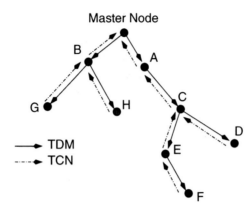

Figure 16.16 Topology discovery message (TDM) and topology completion notice (TCN) propagation for full-duplex communication-based routing.

on the water surface that has higher processing power than the sensor nodes under water. The routes are established through the topology discovery message (TDM) that is generated by the master node and broadcast to the network. TDMs propagate through the network from top to bottom as shown in Figure 16.16. When a node receives a TDM, it first checks the allocated channels indicated in the TDM. Then, it selects a free channel. This channel is also entered into the TDM and broadcast further down the network. As a result, each node selects a channel to transmit during the topology discovery. If more than one node randomly selects the same channel, the parent resolves the conflict by sending a conflict resolution message to all but one of the nodes.

The TDM terminates at the leaf nodes of the network as shown in Figure 16.16. A node determines that it is a leaf node if it does not receive a response to the TDM in a specific time. When this occurs, the leaf node generates a topology completion notice (TCN) and transmits this notice up the network. The nodes that receive this message insert the information about their neighbors into this message and broadcast it toward the central manager. The TCNs received by the central manager from the leaf nodes are then used to manage traffic sessions across the network. This information is exploited by the manager to establish efficient data delivery paths in a centralized fashion, which allows congestion to be avoided and provides some form of QoS guarantee.

UWA Routing

A similar approach for centralized routing is based on the observation that, similar to the terrestrial case, multi-hop routing under water saves energy compared to single hop communication, especially when the communication distance are on the orders of kilometers. Underwater acoustic (UWA) routing [36] exploits this fact by constructing routes through a central manager. The protocol works in two phases: initialization and route establishment. The difference to full-duplex communication-based routing is that topology discovery is initiated by each node in the network instead of the central manager.

During initialization, each node in the network transmits random polling messages to its neighbors. The replies to these messages are used to create neighbor tables at each node such that a list of all the neighbors and the channel quality between them are stored. The neighbors' tables are then sent to the central manager, where the routes are established. The manager determines primary and secondary routes from itself to each node in the network. The secondary routes are used as an alternative route in case the primary routes are not reachable due to dynamic changes in the network and the channel conditions. In this respect, the robustness of the network is improved. The central manager informs each node in the network about the established routes, which ends the initialization phase.

During network operation, if a node determines that its connection with a neighbor is lost, it informs the central manager. Accordingly, the routes are recalculated and communicated to the nodes that are affected by this change.

16.7.2   Distributed Solutions

AUV Assisted Routing

AUV assisted routing [40] relies on mobile AUVs to collect information from underwater sensors. This way of routing is similar to data muling, which has been used for terrestrial WSNs. AUV assisted routing also exploits two communication techniques available for UWSN, i.e., optical and acoustic communication. As we discussed in Section 16.1.1, optical communication provides higher bandwidth than acoustic communication. However, optical communication suffers from a limited range and the line-of-sight requirement. Hence, the mobility of AUVs can be exploited to provide a short-range line-of-sight communication between sensor nodes, and thus to improve the throughput of the network.

The AUV assisted routing protocol is based on a hybrid architecture such that optical communication is used for communication between an AUV and a sensor node, while acoustic communication is used for signaling and communication of small amounts of data. The AUVs move around the network to collect data from nodes. The nodes inform the AUVs through their acoustic interface and when the AUVs come close enough, optical communication is used to upload large amounts of sensor data. These data can then be communicated to a central location by the AUV.

The AUV assisted routing protocol is composed of five stages: (1) locating the first node, (2) locating the next nodes in the sequence, (3) controlling hover mode, (4) data transfer, and (5) clock synchronization. The first node is located through GPS-aided surface navigation and once the AUV is close enough, optical communication is used, where the AUV performs a spiral search to locate the transceiver of the node. The next nodes are located in two different ways. One solution is to provide a map of the network to the AUV. As the second solution, acoustic communication can be exploited to provide accurate localization between nodes in the network.

Once an AUV is close to a node, it uses two ways to locate the transceiver and station such that a line of sight is established with the node. In case ambient light exists, the AUV can locate the nodes by visual servoing to a calibrated color model. On the other hand, localization can be performed through active beaconing using optical communication without the need for ambient light. Once the AUV is located in sight of the node's transceiver, data transfer is performed using optical communication. Furthermore, during this transfer, the clocks of the nodes are synchronized to the AUV to provide consistent data retrieval.

AUV assisted routing minimizes the communication requirement at the nodes by implementing the routing intelligence of the AUVs. However, the drawback of this scheme is the requirement for a complex transceiver which is capable of both acoustic and optical communication. Furthermore, complex maneuvering algorithms and hardware capabilities are necessary at the AUVs to provide accurate localization and prevent perturbations created by currents.

Vector-Based Forwarding (VBF)

Vector-based forwarding (VBF) [42] focuses on mobile 3-D underwater network topologies and limits the nodes in a route through a location-based approach. Packets are forwarded along multiple paths from a source to a destination to increase the robustness in dynamic underwater channel conditions. VBF relies on state information in the packets and, hence, does not require each node in the network to store state information about its neighbors. VBF constructs multiple paths from a sender to the receiver such that the

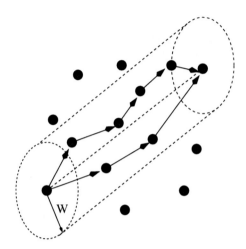

Figure 16.17   The routing pipe and the routes generated by vector-based forwarding (VBF).

packets are forwarded inside a *routing pipe* as shown in Figure 16.17. The principal axis of the routing pipe is the vector from the sender to the destination, hence the name vector-based forwarding. The radius of the pipe (W in Figure 16.17) is indicated by the packet sent by the sender. Moreover, each packet carries the positions of the sender, the destination, and the forwarder. Upon receiving a packet, a node computes its position relative to the forwarder by measuring its distance to the forwarder and the angle of arrival of the signal. If a node is positioned inside the routing pipe and closer to the destination, it assumes the forwarding duty, includes its own position in the packet, and forwards it. Otherwise, it discards the packet. Note that multiple nodes that are inside the pipe can assume the forwarding duty, resulting in VBF constructing multiple redundant paths, which makes the protocol robust against packet losses and node failures. Therefore, VBF can be regarded as a form of geographically controlled flooding.

Routing is initiated in VBF through query packets similar to the directed diffusion explained in Chapter 7. The query packets are initiated either by the sink or the source nodes themselves. The *sink-initiated queries* can be location dependent such that a group of nodes are targeted in a specific area. In this case, the same principles of VBF are used to forward these query packets to the target area. For location-independent queries, the query packets are flooded through the network. For *source-initiated queries*, the location of the sink needs to be determined. Hence, when a source has data to send, it first floods a DATA_READY packet with its location as the center of the coordinate system. This packet helps other nodes determine their positions according to the source-based coordinate system. Once the sink receives this packet, the position of the source is transformed into the sink-based coordinate system and the sink sends a location-dependent query packet to the source.

In VBF, each node in the routing pipe can forward a packet, which increases energy consumption. To minimize this effect, a localized and distributed self-adaptation algorithm is used to enhance the performance of VBF. The self-adaptation algorithm allows each node to calculate a *desirableness factor* according to its location as shown in Figure 16.18. The desirableness factor, α, of a node A is defined as

$$\alpha = \frac{p}{W} + \frac{R - d \times \cos \theta}{R} \tag{16.19}$$

where p is the projection of A to the routing vector, d is the distance between node A and the forwarder node F, and θ is the angle between the forwarder source vector and the forwarder node A vector as shown in Figure 16.18. R is the transmission range and W is the radius of the routing pipe. The lower the value of α, the more desirable it is for the node to become the forwarder. Accordingly, when a node

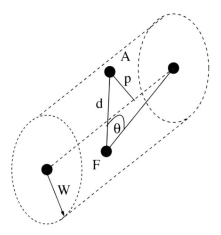

Figure 16.18   Sample topology for the calculation of desirableness factor for VBF.

inside the routing pipe receives a packet it waits for a time $T_{adaptation}$ to forward the packet. If a node receives multiple forwarded packets during this time, it recalculates the desirableness factor and discards the packet if this value is above a threshold value. The waiting time $T_{adaptation}$ is given as follows:

$$T_{adaptation} = \sqrt{\alpha} T_{delay} + \frac{R - d}{v_0} \qquad (16.20)$$

where T_{delay} is the predefined maximum delay and v_0 is the propagation speed. Accordingly, a node with a smaller desirableness factor waits less and the number of nodes that forward the packet is decreased while still providing efficient forwarding.

Routing Algorithms for Delay-Insensitive and Delay-Sensitive Applications

In this section, we explain two routing algorithms developed for delay-insensitive and delay-sensitive applications [27]. These algorithms rely on geographical routing techniques in the 3-D underwater topology to meet the requirements of delay-insensitive and delay-sensitive underwater applications. Both algorithms are characterized by efficient optimization algorithms running on various sensor nodes through a distributed routing solution.

In particular, the proposed routing solutions allow two apparently conflicting objectives to be achieved, i.e., increasing the efficiency of the channel by transmitting a *train* of short packets *back to back*, and limiting the PER by keeping the transmitted packets short. The packet-train concept allows each node to *jointly* select its best next hop, the transmitted power, and the FEC rate for each packet, with the objective of minimizing the energy consumption, taking into account the condition of the underwater channel and the application requirements. As a result, instead of sending headers for each packet, the header information is inserted at the beginning of the packet tree to minimize the overhead. We explain these two algorithms in detail next.

Delay-Insensitive Applications
The first algorithm deals with delay-insensitive applications, and tries to exploit links that guarantee a low PER, to maximize the probability that a packet is correctly decoded at the receiver, and thus minimize the number of required packet retransmissions. The objective of the algorithm is to efficiently exploit the channel and to minimize the energy consumption. The proposed algorithm relies on the

packet-train transmission scheme. In a distributed fashion, it allows each node to *jointly* select its best next hop, the transmitted power, and the FEC code rate for each packet, with the objective of minimizing the energy consumption. Furthermore, it tries to exploit those links that guarantee a low PER, in order to maximize the probability that the packet is correctly decoded at the receiver. For these reasons, the energy efficiency of the link is weighted with the number of retransmissions required to achieve link reliability, with the objective of saving energy.

Accordingly, each node i selects as the next hop that node j^* among its neighbors by solving an optimization problem that satisfies the following requirements: (1) it is closer to the surface station than i; and (2) it minimizes the link metric $E_i^{(j)*}$, which is explained next. The optimization problem can be cast as follows:

$\mathrm{P}_{\mathrm{insen}}^{\mathrm{dist}}$: **Delay-insensitive Distributed Routing Problem**

$$\text{Given:} \quad i, \ S_i, \ \mathcal{P}_i^N, \ L_P^*, \ L_P^H, \ E_{elec}^b, \ r, \ \hat{N}_{0j}, \ P_{i,\max}^{TX}$$

$$\text{Find:} \quad j^* \in S_i \cap \mathcal{P}_i^N, \ P_{ij^*}^{TX*} \leq P_{i,\max}^{TX}$$

$$\text{Minimize:} \quad E_i^{(j)} = E_{ij}^b \cdot \frac{L_P^*}{L_P^* - L_P^H - L_{P\,ij}^F} \cdot \hat{N}_{ij}^{TX} \cdot \hat{N}_{ij}^{Hop} \tag{16.21}$$

$$\text{Subject to:}$$

$$E_{ij}^b = 2 \cdot E_{elec}^b + \frac{P_{ij}^{TX}}{r}; \tag{16.22}$$

$$L_{P\,ij}^F = \Psi^{\mathcal{F}-1}\left(L_P^*, PER_{ij}, \Phi^{\mathcal{M}}\left(\frac{P_{ij}^{TX}}{\hat{N}_{0j} \cdot r \cdot TL_{ij}}\right)\right); \tag{16.23}$$

$$\hat{N}_{ij}^{TX} = \frac{1}{1 - PER_{ij}}; \quad \hat{N}_{ij}^{Hop} = \max\left(\frac{d_{iN}}{\langle d_{ij}\rangle_{iN}}, 1\right); \tag{16.24}$$

where:

- $l_P^* = L_P^H + L_{P\,ij}^F + L_{P\,ij}^N$ [bit] is the *fixed* optimal packet size, where L_P^H is the *fixed* header size of a packet, while $L_{P\,ij}^F$ is the *variable* FEC redundancy that is included in each packet transmitted from node i to j. Thus, $L_{P\,ij}^N = L_P^* - L_P^H - L_{P\,ij}^F$ is the *variable* payload size of each packet transmitted in a train on link (i, j).
- $E_{elec}^b = E_{elec}^{trans} = E_{elec}^{rec}$ [J/bit] is the *distance-independent* energy to transit 1 bit, where E_{elec}^{trans} is the energy per bit needed by transmitter electronics (PLLs, VCOs, bias currents, etc.) and digital processing, and E_{elec}^{rec} represents the energy per bit utilized by receiver electronics.
- $E_{ij}^b = 2 \cdot E_{elec}^b + P_{ij}^{TX}/r$ [J/bit] accounts for the energy to transmit 1 bit from node i to node j, when the transmitted power and the bit rate are P_{ij}^{TX} [W] and r [bps], respectively. The second term represents the *distance-dependent* portion of the energy necessary to transmit a bit.
- TL_{ij} [dB] is the transmission loss from i to j.
- $\hat{N}_{ij}^{TX}$ is the average number of transmissions of a packet sent by node i such that the packet is correctly decoded at receiver j.
- $\hat{N}_{ij}^{Hop} = \max(d_{iN}/\langle d_{ij}\rangle_{iN}, 1)$ is the estimated number of hops from node i to the surface station (sink) N when j is selected as next hop, where d_{ij} is the distance between i and j, and $\langle d_{ij}\rangle_{iN}$ (which we refer to as *advance*) is the projection of d_{ij} onto the line connecting node i with the sink.

- $BER_{ij} = \phi^{\mathcal{M}}(E_{rec}^b/\hat{N}_{0j})$ represents the bit error rate on link (i, j); it is a function of the ratio between the energy of the received bit, $E_{rec}^b = P_{ij}^{TX}/(r \cdot TL_{ij})$, and the expected noise at node j, $\hat{N}_{0j}$, and it depends on the adopted modulation scheme $\mathcal{M}$.

- $L_{P_{ij}}^F = \psi^{\mathcal{F}-1}(L_P^*, PER_{ij}, BER_{ij})$ returns the needed FEC redundancy, given the optimal packet size L_P^*, the PER and bit error rate on link (i, j), and it depends on the adopted FEC technique $\mathcal{F}$.

- S_i is the *neighbor set* of node i, while $\mathcal{P}_i^N$ is the *positive advance set*, composed of nodes closer to sink N than node i, i.e., $j \in \mathcal{P}_i^N$ iff $d_{jN} < d_{iN}$.

In summary, node i will select node j^* as its best next hop iff

$$j^* = \arg \min_{j \in S_i \cap \mathcal{P}_i^N} E_i^{(j)*} \qquad (16.25)$$

where $E_i^{(j)*}$ represents the minimum energy required to successfully transmit a payload bit from node i to the sink, taking the condition of the underwater channel into account, when i selects j as the next hop. This link metric, objective function (16.21) in $\mathbf{P_{insen}^{dist}}$, takes into account the number of packet transmissions $(\hat{N}_{ij}^{TX})$ associated with link (i, j), given the optimal packet size (L_P^*) and the optimal combination of FEC $(L_{P_{ij}}^{F*})$ and transmitted power (P_{ij}^{TX*}). Moreover, it accounts for the average hop path length $(\hat{N}_{ij}^{Hop})$ from node i to the sink when j is selected as the next hop, by assuming that the following hops will guarantee the same advance toward the surface station (sink).

The link metric $E_i^{(j)*}$ in (16.25) stands for the optimal energy per payload bit when i transmits a packet train to j using the optimal combination of power P_{ij}^{TX*} and FEC redundancy $L_{P_{ij}}^{F*}$ to achieve link reliability, jointly found by solving problem $\mathbf{P^{dist}}$. This interpretation allows node i to optimally decouple $\mathbf{P_{insen}^{dist}}$ into two *subproblems*: firstly, minimize the link metric $E_i^{(j)}$ for each of its feasible next-hop neighbors; secondly, pick as the best next hop that node j^* associated with the minimal link metric. This means that the generic node i does not have to solve a complicated optimization problem to find its best route toward a sink. Moreover, this operation does not need to be performed each time a sensor has to route a packet, but only when the channel conditions have consistently changed.

Delay-Sensitive Applications

The second algorithm is designed for delay-sensitive applications. The objective is to minimize the energy consumption while statistically limiting the end-to-end packet delay and PER by estimating at each hop the time to reach the sink and by leveraging the statistical properties of underwater links. Differently from the previous delay-insensitive routing solution, next hops are selected by also considering maximum per-packet allowed delay, while unacknowledged packets are not retransmitted to limit the delay.

Contrary to the previous algorithm, this algorithm includes two new constraints to statistically meet the delay-sensitive application requirements:

1. The end-to-end PER should be lower than an application-dependent threshold PER_{max}^{e2e}.

2. The probability that the end-to-end packet delay is over a delay bound B_{max} should be lower than an application-dependent parameter γ.

Accordingly, the optimization problem that each node will solve periodically can be cast as follows:

P_{sen}^{dist}: Delay-sensitive Distributed Routing Problem

$$\text{Given:} \quad i, \; \mathcal{S}_i, \; \mathcal{P}_i^N, \; E_{elec}^b, \; r, \; \hat{N}_{0j}, \; P_{i,max}^{TX}, \; \Delta B_i^{(m)}, \; \hat{Q}_{ij}$$

$$\text{Find:} \quad j^* \in \mathcal{S}_i \cap \mathcal{P}_i^N, \; P_{ij^*}^{TX*} \leq P_{i,max}^{TX}$$

$$\text{Minimize:} \quad E_i^{(j)} = E_{ij}^b \cdot \frac{L_P^*}{L_P^* - L_P^H - L_{P_{ij}}^F} \cdot \hat{N}_{ij}^{Hop} \tag{16.26}$$

Subject to:

$$E_{ij}^b = 2 \cdot E_{elec}^b + \frac{P_{ij}^{TX}}{r}; \tag{16.27}$$

$$L_{P_{ij}}^F = \Psi^{\mathcal{F}-1}\left(L_P^*, PER_{ij}, \Phi^{\mathcal{M}}\left(\frac{P_{ij}^{TX}}{\hat{N}_{0j} \cdot r \cdot TL_{ij}}\right)\right); \tag{16.28}$$

$$\hat{N}_{ij}^{Hop} = \max\left(\frac{d_{iN}}{\langle d_{ij}\rangle_{iN}}, 1\right); \tag{16.29}$$

$$1 - (1 - PER_{ij})^{\lceil \hat{N}_{ij}^{Hop}\rceil} \leq PER_{max}^{e2e}; \tag{16.30}$$

$$\frac{\bar{d}_{ij}}{q_{ij}} + \delta \cdot \sigma_{ij}^q \leq \min_{m=1,\dots,M}\left(\frac{\Delta B_i^{(m)}}{\hat{N}_{ij}^{Hop}}\right) - \hat{Q}_{ij} - \frac{L_P^*}{r}; \tag{16.31}$$

where:

- $M = \lfloor (L_T^* - L_T^H)/L_P^* \rfloor$ is the *fixed* number of packets transmitted in a train on each link, where L_T^* and L_P^* are the train length and packet size, respectively.
- PER_{max}^{e2e} and B_{max} [s] are the application-dependent end-to-end PER threshold and delay bound, respectively.
- $\Delta B_i^{(m)} = B_{max} - |t_{i,now}^{(m)} - t_0^{(m)}|$ [s] is the time to live of packet m arriving at node i, where $t_{i,now}^{(m)}$ is the arrival time of m at i, and $t_0^{(m)}$ is the time m was generated, which is timestamped in the packet header by its source.
- $T_{ij} = L_P^*/r + T_{ij}^q$ [s] accounts for the packet transmission delay and the propagation delay associated with link (i, j), where $T_{ij} \sim \mathcal{N}(L_P^*/r + \overline{T_{ij}^q}, \sigma_{ij}^{q2})$.
- $\overline{Q}_i$ [s] and $\overline{Q}_j$ [s] are the average queuing delays of node i (at the time the node computes its train next hop) and node j.
- $\hat{Q}_{ij}$ [s] is the network queuing delay estimated by node i when j is selected as the next hop.

The formulation of P_{sen}^{dist} is similar to P_{insen}^{dist}, except for two important differences:

1. The objective function (16.26) does not include $\hat{N}_{ij}^{TX}$ as in (16.21), since no selective packet retransmission is performed.

2. Two new constraints are included, (16.30) and (16.31), which address the two considered delay-sensitive application requirements, i.e., the end-to-end PER should be lower than an application-dependent threshold PER_{max}^{e2e}, and the probability that the end-to-end packet delay is over a delay bound B_{max} should be lower than an application-dependent parameter γ, respectively.

Note that (16.30) adjusts the PER PER_{ij} that will be experienced by packet m on link (i, j) to respect the application end-to-end PER requirement (PER_{max}^{e2e}), given the estimated number of hops to reach the sink if j is selected as the next hop ($\hat{N}_{ij}^{Hop}$). Interestingly, since the packet is assumed to be correctly forwarded up to node i, there is no need to consider the hop count number in (16.30), i.e., the number of hops of packet m from the source to the current node i. In fact, since node i is assumed to receive the packet, the conditional probability of its being correct is 1.

16.7.3   Hybrid Solutions

Resilient Routing Algorithm

In general, most routing protocols for terrestrial ad hoc networks are based on *packet switching* such that the routing function is performed separately for each single packet and paths are dynamically established. The resilient routing algorithm, on the other hand, employs *virtual circuit* routing techniques [26]. In these techniques, paths are established a priori between each source and sink and each packet from a source follows the same path. This requires centralized coordination and implies a less flexible architecture, but allows the exploitation of powerful optimization tools on a centralized manager.

The resilient routing algorithm is focused on the problem of data gathering for 3-D UWSNs by considering the interactions between the routing functions and the characteristics of the underwater acoustic channel. The algorithm consist of two phases with the objective of guaranteeing survivability of the network to node and link failures. In the first phase, energy-efficient node-disjoint primary and backup paths are optimally configured, by relying on topology information gathered by a surface station, while in the second phase paths are locally repaired in case of node failures.

In the *first phase* of the resilient routing algorithm, the network manager determines optimal *node-disjoint primary* and *backup* multi-hop data paths such that the energy consumption of the nodes is minimized. This is performed through powerful optimization tools at the centralized manager. This is needed because, unlike in terrestrial sensor networks where sensors can be redundantly deployed, the underwater environment requires minimization of the number of sensors. Hence, protection is necessary to avoid network connectivity being disrupted by node or link failures.

Once the routes are computed and communicated to the network, in the *second phase* an online distributed solution guarantees survivability of the network, by locally repairing paths in case of disconnections or failures. Moreover, in case of severe failures, the data traffic is switched to the backup paths. The emphasis on survivability is motivated by the fact that underwater long-term monitoring missions can be extremely expensive. Hence, it is crucial that the deployed network be highly reliable, so as to avoid failure of missions due to failure of single or multiple devices. The resilient routing algorithm can be classified as a hybrid solution since it consists of both centralized and distributed mechanisms to perform routing in UWSNs.

16.8   Transport Layer

The transport layer of UWSNs is a rather unexplored area. In this section, we discuss the fundamental challenges for the development of an efficient *reliable transport layer* protocol, which addresses the requirements of UWSNs.

Noticeably, in sensor networks, reliable event detection at the sink should be based on collective information provided by source nodes and not on any individual report from each single source [6]. Hence, conventional end-to-end reliability definitions and solutions can be inapplicable in the underwater sensor field, and could lead to wasting scarce sensor resources. On the other hand, the absence of a reliable transport mechanism altogether can seriously impair event detection due to the underwater challenges. Thus, the UWSN paradigm requires a new *event transport reliability* notion rather than the traditional end-to-end approaches.

A transport layer protocol is needed in UWSNs not only to achieve *reliable collective transport* of event features, but also to perform *flow control* and *congestion control*. The primary objective is to save scarce sensor resources and increase network efficiency. A reliable transport protocol should guarantee that the applications are able to correctly identify event features estimated by the sensor network. Congestion control is needed to prevent the network from being congested by excessive data with respect to the network capacity, while flow control is needed to avoid network devices with limited memory being overwhelmed by data transmissions.

Most existing TCP implementations are unsuited for the underwater environment since the flow control functionality is based on a window-based mechanism that relies on an accurate estimate of the RTT, which is twice the end-to-end delay from source to destination. The underwater RTT can be modeled as a stochastic variable with a high mean value, which reflects the sum of the high delays of the links composing the end-to-end path, and a high delay variance, which reflects the sum of the high delay variances of the composing link. This high mean, high variance RTT would affect the throughput of most TCP implementations. Furthermore, the high variability of the RTT would make it hard to effectively set the timeout of the window-based mechanism which most current TCP implementations adopt.

Rate-based transport protocols also seem unsuited for this challenging environment. In fact, although they do not adopt a window-based mechanism, they still rely on feedback control messages sent back by the destination to dynamically adapt the transmission rate, i.e., to decrease the transmission rate when packet loss is experienced or to increase it otherwise. The high delay and delay variance can thus cause instability in the feedback control.

Furthermore, due to the unreliability of the acoustic channel, it is necessary to distinguish between packet losses due to the high bit error rate of the acoustic channel and those caused by packets being dropped from the queues of sensor nodes due to network congestion. Most TCP implementations that are designed for wired networks assume that congestion is the only cause of packet loss. Due to this assumption, when a packet loss occurs, these protocols reduce the transmission rate to avoid injecting more packets into the network. Conversely, in underwater, as in terrestrial, wireless networks, it is important to discriminate between losses due to impairments of the channel and those caused by congestion. When congestion is the cause of the packet loss, the transmission rate should be decreased to avoid overwhelming the network, while in case of losses due to bad channel quality, the transmission rate should not be decreased to preserve throughput efficiency.

For these reasons, it may be necessary to devise completely new strategies to achieve underwater flow control and reliability. The transport layer in UWSNs is still a relatively unexplored field with several research challenges.

16.8.1   Open Research Issues

In order to develop a new, efficient, cross-layer reliable protocol specifically tailored to underwater acoustic sensor networks, the following issues must be studied:

- New flow control strategies need to be devised in order to tackle the high delay and delay variance of the control messages sent back by the receivers.
- New effective mechanisms tailored to the underwater acoustic channel need to be developed in order to efficiently infer the cause of packet losses.
- New event transport reliability metrics need to be devised, based on the event model and on the underwater acoustic channel model.
- Optimal update policies for the sensor reporting rate are needed, to prevent congestion and maximize the network throughput efficiency as well as the transport reliability in bandwidth-limited underwater networks.
- The effects of multiple event occurrences on the reliability and network performance require-ments must be studied, as well as efficient mechanisms to deal with it.
- It is necessary to statistically model the loss of connectivity events in order to devise mechanisms enabling delay-tolerant applications tailored to the specific underwater requirements.

- Different functionalities at the data link and transport layers such as channel access, reliability, and flow control should be jointly designed and studied. A cross-layer approach is highly recommended to optimize these mechanisms accordingly and make them adaptable to the variability of the characteristics of the underwater channel.

16.9   Application Layer

Although many application areas for UWSNs can be outlined, the definition of an application layer protocol for UWSNs remains largely unexplored.

The purpose of an application layer is multifold: (1) to provide a network management protocol that makes the hardware and software details of the lower layers transparent to management applications; (2) to provide a language for querying the sensor network as a whole; (3) to assign tasks and to advertise events and data.

No efforts in these areas have been made to date that address the specific needs of the underwater acoustic environment. A deeper understanding of the application areas and the communication problems in UWSNs is crucial for outlining some design principles on how to extend or reshape existing application layer protocols [8] for terrestrial sensor networks.

Some of the latest developments in middleware may be studied and adapted to realize a versatile application layer for UWSNs. For example, the San Diego Supercomputing Center's Storage Resource Broker (SRB) [4, 9] is client–server middleware that provides a uniform interface for connecting to heterogeneous data resources over a network and accessing replicated data sets. SRB provides a way to access data sets and resources based on their attributes and/or logical names rather than their names or physical locations.

16.10   Cross-layer Design

While the research on underwater networking so far has followed the traditional layered approach for network design, it is an increasingly accepted opinion in the wireless networking community that improved network efficiency, especially in critical environments, can be obtained with a cross-layer design approach. These techniques will entail a joint design of different network functionalities, from modem design to MAC and routing, from channel coding and modulation to source compression and the transport layer, with the objective of overcoming the shortcomings of a layered approach that lacks information sharing across protocol layers, forcing the network to operate in a suboptimal mode. Hence, the underwater environment particularly requires cross-layer design solutions that allow more efficient use of the scarce available resources. However, it is important to consider the ease of design by following a *modular design approach*. This also allows the improvement and upgrading of particular functionalities without the need to redesign the entire communication system. Next, we describe the cross-layer communication solution to support multimedia applications in UWSNs.

The cross-layer communication solution [25] is built on the distributed underwater routing protocols [27], described in Section 16.7.2, and the UW-MAC protocol [28] described in Section 16.6.2. More specifically, it relies on a distributed optimization problem to jointly control the *routing, MAC,* and *physical* functionalities. To this end, a 3-D geographical routing algorithm, a distributed CDMA/ALOHA-based scheme, and an optimized solution for the *joint* selection of modulation, FEC, and transmit power are combined in a cross-layer fashion.

The cross-layer design is aimed at maximizing the *link net rate* by jointly deciding three communication parameters. Firstly, the modulation scheme and its constellation, which affect the link raw rate, are selected. Secondly, the transmit power, which affects the bit error rate (BER), is selected. Thirdly, the FEC type and strength, which affect the PER given the BER, are determined. In addition to the physical layer properties, the MAC functionality of the cross-layer design combines Direct Sequence Code Division Multiple Access (DS-CDMA) with a simple yet effective ALOHA access for control traffic. For multi-user operation, the CDMA scheme uses locally generated *chaotic codes* to spread transmitted signals on the optimal band, which provides good auto- and cross-correlation properties. The MAC functionality also incorporates a closed-loop distributed algorithm that interacts with the physical functionality, to set the *optimal transmit power* and *code length*. The determination of the next hop is also related to the interference levels at the neighboring nodes. This captures the interactions between the routing and physical layer functionalities by trying to compose, in a distributed manner, paths using short links to exploit their higher bandwidth.

Accordingly, a node i will select j^* as its best next hop iff

$$j^* = \begin{cases} \arg\min_{j \in S_i \cap P_i^N} E_i^{(j)*} & \text{(Objective 1)} \\ \text{OR} \\ \arg\max_{j \in S_i \cap P_i^N} R_i^{(j)*} & \text{(Objective 2)} \end{cases} \qquad (16.32)$$

where $E_i^{(j)*}$ [J/bit] is the minimum energy required to successfully transmit a bit from node i to the sink, $R_i^{(j)*}$ [bps] is the maximum bit rate that can be achieved from node i, S_i is the set of neighbors of node i, and P_i^N is the *positive advance set*, which is composed of nodes closer to sink N than node i, i.e., $j \in P_i^N$ iff $d_{jN} < d_{iN}$.

The optimization problem considers different multimedia traffic classes that optimize the transmission from node i by choosing the optimal power spectral density (p.s.d.) of the transmitted signal as well as band (K^*, f_0^*, B^*), modulation (M^*), FEC (F^*, L_P^{F*}), and code length (c^*). The objective is set depending on the high-level application requirements. Accordingly, the cross-layer link optimization problem can be cast as follows:

$P_{layer}^{cross}(i, j)$: **Cross-layer Link Optimization Problem**

$$\text{Find:} \quad K_{ij}^*, \; f_{0ij}^*, \; B_{ij}^*, \; M_{ij}^*, \; F_{ij}^*, \; L_{P_{ij}}^{F*}, \; c_{ij}^*$$

$$\text{Objective 1:} \quad \text{Minimize } \mathbf{E}_i^{(j)} = \mathbf{E}_{ij}^b \cdot \Pi_{ij}^{e2e} \qquad (16.33)$$

$$\text{OR Objective 2:} \quad \text{Maximize } \mathbf{R}_i^{(j)} = \mathbf{R}_{ij}^b \cdot \Pi_{ij}^{e2e-1} \qquad (16.34)$$

Subject to:

Class-independent Constraints/Relationships

$$R_{ij}^b = \frac{\eta(M_{ij}) \cdot B_{ij}}{c_{ij}}, \quad E_{ij}^b = 2E_{elec}^b + \frac{P_{ij}}{R_{ij}^b} \qquad (16.35)$$

$$\Pi_{ij}^{e2e} = \frac{L_P^*}{L_P^* - L_P^H - L_{P_{ij}}^F} \cdot \hat{N}_{ij}^T \cdot \hat{N}_{ij}^{Hop} \qquad (16.36)$$

$$SINR_{ij} = K_{ij} \frac{\int_{f_{0ij}, B_{ij}} T L_{ij}(f)^{-1} \, df}{\int_{f_{0ij}, B_{ij}} NI_j(f) \, df} - 1 \qquad (16.37)$$

$$BER_{ij} = \Phi^{M_{ij}}(SINR_{ij}) \qquad (16.38)$$

$$PER_{ij} = \Psi^{F_{ij}}(L_P^*, L_{P_{ij}}^F, BER_{ij}) \qquad (16.39)$$

$$\hat{N}_{ij}^{Hop} = \max\left(\frac{d_{iN}}{\langle d_{ij} \rangle_{iN}}, 1\right) \qquad (16.40)$$

$$P_{ij}^{min}(c_{ij}, BER_{ij}) \le P_{ij} \le \min[P_{ij}^{max}, P_i^{max}] \qquad (16.41)$$

where

$$P_{ij} = K_{ij} \cdot B_{ij} - \int_{\langle f_{0ij}, B_{ij}\rangle} NI_j(f) \cdot TL_{ij}(f)\, df \qquad (16.42)$$

$$P_{ij}^{\min}(c_{ij}, BER_{ij}) = \frac{\int_{\langle f_{0ij}, B_{ij}\rangle} NI_j(f) \cdot TL_{ij}(f)\, df}{\alpha \cdot c_{ij} \cdot \Omega(BER_{ij})} \qquad (16.43)$$

$$P_{ij}^{\max} = \min_{k \in \mathcal{K}_i} [(\hat{R}_k - NI_k) \cdot TL_{ik}]. \qquad (16.44)$$

Class-dependent Constraints/Relationships

$$\text{Class I} = \begin{cases} \hat{N}_{ij}^T = 1 \\ 1 - (1 - PER_{ij})^{\lceil \hat{N}_{ij}^{Hop} + N_{HC}^{(m)}\rceil} \leq PER_{\max}^{e2e,(m)} \end{cases}$$

$$\text{Class II} = \left\{ \hat{N}_{ij}^T = (1 - PER_{ij})^{-1} \right.$$

$$\text{Class III} = \begin{cases} \hat{N}_{ij}^T = 1 \\ 1 - (1 - PER_{ij})^{\lceil \hat{N}_{ij}^{Hop} + N_{HC}^{(m)}\rceil} \leq PER_{\max}^{e2e,(m)} \\ \dfrac{\tilde{d}_{ij}}{q_{ij}} + \delta(\gamma) \cdot \sigma_{ij}^q \leq \left(\dfrac{\Delta D_i^{(m)}}{\hat{N}_{ij}^{Hop}}\right) - \hat{Q}_{ij} - \dfrac{L_P^*}{R_{ij}^b} \end{cases}$$

$$\text{Class IV} = \begin{cases} \hat{N}_{ij}^T = (1 - PER_{ij})^{-1} \\ \dfrac{\tilde{d}_{ij}}{q_{ij}} + \delta(\gamma) \cdot \sigma_{ij}^q \leq \left(\dfrac{\Delta D_i^{(m)}}{\hat{N}_{ij}^{Hop}}\right) - \hat{Q}_{ij} - \dfrac{L_P^*}{R_{ij}^b}. \end{cases}$$

The optimization problem includes both class-dependent and class-independent constraints. To support the requirements of various applications, four different traffic classes are defined and the optimization problem is solved for each traffic class according to the corresponding constraints. The traffic classes are: (I) delay-tolerant, loss-tolerant, (II) delay-tolerant, loss-sensitive, (III) delay-sensitive, loss-tolerant, and (IV) delay-sensitive, loss-sensitive.

The parameters used in the optimization problem are as follows:

- $L_P^* = L_P^H + L_{P\,ij}^F + L_{P\,ij}^N$ [bit] is the *fixed* packet size, where L_P^H is the header size of a packet, while $L_{P\,ij}^F$ is the *variable* FEC redundancy of each packet from i to j.

- $E_{elec}^b = E_{elec}^{trans} = E_{elec}^{rec}$ [J/bit] is the *distance-independent* energy to transit 1 bit, where E_{elec}^{trans} is the energy per bit needed by transmitter electronics and digital processing, and E_{elec}^{rec} represents the energy per bit utilized by receiver electronics.

- E_{ij}^b [J/bit] is the energy to transmit 1 bit from i to j, when the transmitted power and the bit rate are P_{ij} [W] and R_{ij}^b [bps], respectively. The second term P_{ij}/R_{ij}^b is the *distance-dependent* portion of the energy to transmit a bit.

- $P_i^{\max}$ [W] is the maximum transmitting power for node i.

- $BER = \Phi^M(SINR)$ is the BER, given the SINR and the modulation scheme $M \in \mathcal{M}$, while $\eta(M)$ is the modulation spectrum efficiency.

- $PER = \psi^F(L_P, L_P^F, BER)$ represents the link PER, given the packet size L_P, the FEC redundancy L_P^F, and the BER (BER), and it depends on the adopted FEC technique $F \in \mathcal{F}$.

- $\hat{N}_{ij}^T$ is the number of transmissions of the packet.

- $\hat{N}_{ij}^{Hop} = \max(d_{iN}/\langle d_{ij}\rangle_{iN}, 1)$ is the estimated number of hops from node i to the surface station (sink) N when j is selected as next hop and $\langle d_{ij}\rangle_{iN}$ is the advancement toward the sink.

The notations used in the class-dependent constraints/relationships are as follows:

- $PER_{max}^{e2e,(m)}$ is the maximum allowed end-to-end PER to packet m, while N_{max}^{Hop} is the maximum expected number of hops.

- $N_{HC}^{(m)}$ is the hop count, which reports the number of hops of packet m from the source to the current node.

- $\Delta D_i^{(m)} = D_{max} - (t_{i,now}^{(m)} - t_0^{(m)})$ [s] is the time to live of packet m arriving at node i, where $t_{i,now}^{(m)}$ is the arrival time of m at i, and $t_0^{(m)}$ is the time m was generated, which is timestamped in the packet header by its source, and D_{max} [s] is the maximum end-to-end delay.

- $T_{ij} = L_P^*/R_{ij}^b + T_{ij}^q$ [s] accounts for the packet transmission delay and the propagation delay associated with link (i, j); we consider a Gaussian distribution for T_{ij}, i.e., $T_{ij} \backsim \mathcal{N}(L_P^*/R_{ij}^b + \overline{T_{ij}^q}, \sigma_{ij}^{q2})$.

- $\hat{Q}_{ij}$ [s] is the network queuing delay estimated by node i when j is selected as the next hop, computed according to the information carried by incoming packets and broadcast by neighboring nodes.

The solution of the optimization problem can be achieved by decoupling the routing decision, so that, firstly, the link metric, $E_i^{(j)}$, is minimized and, secondly, the best next hop node j^* associated with the best link metric is selected. This decreases the complexity of the optimization solution. Once the solution is found, node i accesses the channel by transmitting an *extended header* (EH), which contains information about the final destination, the chosen next hop, i.e., j^*, and the parameters that node i will use to generate the *chaotic spreading code*. Immediately after transmission of the EH, the data packet is transmitted.

The extra header EH serves as a control packet for channel access using a ALOHA-like MAC protocol and the data packet is sent using CDMA techniques. This approach aims to mitigate the effects of the huge propagation delay affecting the underwater environment. With successful decoding of the EH, the receiver node adjusts its communication parameters using the locally generated chaotic code. The receiver node also sends an ACK message upon correctly receiving the packet from node i. In the case of retransmissions, node i increases the code strength to improve the probability of communication success. The cross-layer solution for UWSNs provides differentiated access to the underwater channel according to the requirements of the applications and is the first cross-layer communication protocol for the underwater medium.

References

[1] AUV Laboratory at MIT Sea Grant. Available at http://auvlab.mit.edu/.

[2] The Delay Tolerant Networking Research Group. Available at http://www.dtnrg.org.

[3] Ocean engineering at Florida Atlantic University. Available at http://www.ome.fau.edu/lab_ams.htm.

[4] SDSC storage resource broker. Available at http://www.npaci.edu/DICE/SRB/.

[5] Second field test for the AOSN program, Monterey Bay. Available at http://www.mbari.org/aosn/MontereyBay2003/MontereyBay2003Default.htm, August 2003.

[6] Ö. B. Akan and I. F. Akyildiz. Event-to-sink reliable transport in wireless sensor networks. *IEEE/ACM Transactions on Networking*, 13(5):1003–1016, October 2005.

[7] I. F. Akyildiz, D. Pompili, and T. Melodia. Underwater acoustic sensor networks: research challenges. *Ad Hoc Networks*, 3(3):257–279, March 2005.

[8] I. F. Akyildiz, W. Su, Y. Sankarasubramaniam, and E. Cayirci. Wireless sensor networks: a survey. *Computer Networks*, 38(4):393–422, March 2002.

[9] C. Baru, R. Moore, A. Rajasekar, and M. Wan. The SDSC storage resource broker. In *Proceedings of the 1998 Conference of the Centre for Advanced Studies on Collaborative research*, p. 5, Toronto, Ontario, Canada, November 1998.

[10] J. Catipovic. Performance limitations in underwater acoustic telemetry. *IEEE Journal of Oceanic Engineering*, 15:205–216, July 1990.

[11] E. Cayirci, H. Tezcan, Y. Dogan, and V. Coskun. Wireless sensor networks for underwater surveillance systems. *Ad Hoc Networks*, 4(4):431–446, April 2006.

[12] D. L. Codiga, J. A. Rice, and P. A. Baxley. Networked acoustic modems for real-time data delivery from distributed subsurface instruments in the coastal ocean: initial system development and performance. *Journal of Atmospheric and Oceanic Technology*, 21(2):331–346, February 2004.

[13] R. E. Davis, C. E. Eriksen, and C. P. Jones. Autonomous buoyancy-driven underwater gliders. In G. Griffiths, editor, *The Technology and Applications of Autonomous Underwater Vehicles*. Taylor & Francis, 2002.

[14] K. Fall. A delay-tolerant network architecture for challenged internets. In *Proceedings of ACM SIGCOMM'03*, pp. 27–34, Karlsruhe, Germany, August 2003.

[15] E. Fiorelli, N. E. Leonard, P. Bhatta, D. Paley, R. Bachmayer, and D. M. Fratantoni. Multi-AUV control and adaptive sampling in Monterey Bay. In *Proceedings of IEEE/OES Autonomous Underwater Vehicles*, pp. 134–147, Sebasco Estates, ME, USA, June 2004.

[16] L. Freitag and M. Stojanovic. Acoustic communications for regional undersea observatories. In *Proceedings of Oceanology International*, London, UK, March 2002.

[17] S. A. L. Glegg, R. Pirie, and A. LaVigne. A study of ambient noise in shallow water. In *Florida Atlantic University Technical Report*, 2000.

[18] M. Hinchey. Development of a small autonomous underwater drifter. In *Proceedings of IEEE NECEC'04*, St. John's, NF, Canada, October 2004.

[19] B. M. Howe, T. McGinnis, and H. Kirkham. Sensor networks for cabled ocean observatories. In *The 3rd International Workshop on Scientific Use of Submarine Cables and Related Technologies*, pp. 216–221, Tokyo, Japan, June 2003.

[20] J. C. Jalbert, D. R. Blidberg, and M. Ageev. Some design considerations for a solar powered AUV: energy management and its impact on operational characteristics. *Unmanned Systems*, 15(4):26–31, 1997.

[21] D. N. Kalofonos, M. Stojanovic, and J. G. Proakis. Performance of adaptive MC-CDMA detectors in rapidly fading Rayleigh channels. *IEEE Transactions on Wireless Communications*, 2(2):229–239, March 2003.

[22] D. B. Kilfoyle and A. B. Baggeroer. The state of the art in underwater acoustic telemetry. *IEEE Journal of Oceanic Engineering*, 25:4–27, January 2000.

[23] M. Molins and M. Stojanovic. Slotted FAMA: a MAC protocol for underwater acoustic networks. In *Proceedings of the MTS/IEEE Conference and Exhibition for Ocean Engineering, Science and Technology (OCEANS)*, Boston, MA, USA, September 2006.

[24] P. Ogren, E. Fiorelli, and N. E. Leonard. Cooperative control of mobile sensor networks: adaptive gradient climbing in a distributed environment. *IEEE Transactions on Automatic Control*, 49:1292–1302, August 2004.

[25] D. Pompili and I. F. Akyildiz. A cross-layer communication solution for multimedia applications in underwater acoustic sensor networks. In *Proceedings of IEEE MASS'08*, Atlanta, GA, USA, September 2008.

[26] D. Pompili, T. Melodia, and I. F. Akyildiz. A resilient routing algorithm for long-term applications in underwater sensor networks. In *Proceedings of MedHocNet'06*, Lipari, Italy, June 2006.

[27] D. Pompili, T. Melodia, and I. F. Akyildiz. Routing algorithms for delay-insensitive and delay-sensitive applications in underwater sensor networks. In *Proceedings of ACM MobiCom'06*, Los Angeles, USA, September 2006.

[28] D. Pompili, T. Melodia, and I. F. Akyildiz. A CDMA-based medium access control for underwater acoustic sensor networks. *IEEE Transactions on Wireless Communications*, 8(4):1899–1909, April 2009.

[29] J. G. Proakis, J. Rice, E. Sozer, and M. Stojanovic. Shallow water acoustic networks. In J. G. Proakis, editor, *Encyclopedia of Telecommunications*. John Wiley & Sons, Inc., 2003.

[30] A. Quazi and W. Konrad. Underwater acoustic communication. *IEEE Communications Magazine*, 20(2):24–29, March 1982.

[31] V. Ravelomanana. Extremal properties of three-dimensional sensor networks with applications. *IEEE Transactions on Mobile Computing*, 3(3):246–257, July/September 2004.

[32] V. Rodoplu and M. K. Park. An energy-efficient MAC protocol for underwater wireless acoustic networks. In *Proceedings of the MTS/IEEE Conference and Exhibition for Ocean Engineering, Science and Technology (OCEANS)*, Washington, DC, USA, September 2005.

[33] F. Salva-Garau and M. Stojanovic. Multi-cluster protocol for ad hoc mobile underwater acoustic networks. In *Proceedings of IEEE OCEANS'03*, San Francisco, USA, September 2003.

[34] N. N. Soreide, C. E. Woody, and S. M. Holt. Overview of ocean based buoys and drifters: present applications and future needs. In *Proceedings of the MTS/IEEE Conference and Exhibition OCEANS*, volume 4, pp. 2470–2472, Honolulu, HI, USA, November 2001.

[35] E. M. Sozer, J. G. Proakis, M. Stojanovic, J. A. Rice, A. Benson, and M. Hatch. Direct sequence spread spectrum based modem for underwater acoustic communication and channel measurements. In *Proceedings of OCEANS'99*, volume 1, pp. 228–233, Seattle, WA, USA, November 1999.

[36] E. M. Sozer, M. Stojanovic, and J. G. Proakis. Underwater acoustic networks. *IEEE Journal of Oceanic Engineering*, 25(1):72–83, January 2000.

[37] M. Stojanovic. Acoustic (underwater) communications. In John G. Proakis, editor, *Encyclopedia of Telecommunications*. John Wiley & Sons, Inc., 2003.

[38] M. Stojanovic, J. Catipovic, and J. Proakis. Phase coherent digital communications for underwater acoustic channels. *IEEE Journal of Oceanic Engineering*, 19(1):100–111, January 1994.

[39] M. Stojanovic, J. Proakis, and J. Catipovic. Analysis of the impact of channel estimation errors on the performance of a decision feedback equalizer in multipath fading channels. *IEEE Transactions on Communications*, 43(2/3/4):877–886, February/March/April 1995.

[40] I. Vasilescu, K. Kotay, D. Rus, M. Dunbabin, and P. Corke. Data collection, storage, and retrieval with an underwater sensor network. In *Proceedings of ACM SenSys*, San Diego, CA, USA, November 2005.

[41] G. Xie and J. H. Gibson. A network layer protocol for UANs to address propagation delay induced performance limitations. In *Proceedings of IEEE OCEANS'01*, volume 4, pp. 2087–2094, Honolulu, HI, USA, November 2001.

[42] P. Xie, J. Cui, and L. Lao. VBF: vector-based forwarding protocol for underwater sensor networks. In *Proceedings of IFIP Networking'06*, pp. 1216–1221, Coimbra, Portugal, 2006.

[43] X. Yang, K. G. Ong, W. R. Dreschel, K. Zeng, C. S. Mungle, and C. A. Grimes. Design of a wireless sensor network for long-term, in-situ monitoring of an aqueous environment. *Sensors*, 2:455–472, 2002.

[44] B. Zhang, G. S. Sukhatme, and A. A. Requicha. Adaptive sampling for marine microorganism monitoring. In *IEEE/RSJ International Conference on Intelligent Robots and Systems*, Sendai, Japan, 2004.

17

Wireless Underground Sensor Networks

Wireless underground sensor networks (WUSNs) consist of wireless devices that operate below the ground surface. These devices are either buried completely under dense soil, or placed within a bounded open underground space, such as mines and road/subway tunnels. WUSNs promise to enable a wide variety of novel applications that were not possible with current underground monitoring techniques. Compared to the current underground sensor networks, which use wired communication methods for network deployment, WUSNs have several remarkable merits, such as concealment, ease of deployment, timeliness of data, reliability, and coverage density [2].

Using sensors for underground purposes has attracted significant attention in a broad range of fields ranging from commercial agriculture and geology to security and navigation. In particular, in agricultural applications, underground sensors are used to monitor soil conditions such as water and mineral content [1]. Sensors are also successfully used to monitor the integrity of below-ground infrastructures such as plumbing [39]. Furthermore, landslide and earthquake incidents are monitored using buried seismometers [18]. The current technology for underground sensing, however, consists of deploying a buried sensor that is wired to a data-logger on the surface and writing the collected data to the data-logger. Data-loggers may be equipped with a device for wired or wireless communication with a centralized sink, but often data are manually retrieved by physically visiting the data-logger [6]. All of these existing solutions require sensor devices to be deployed at the surface and wired to a buried sensor [26]. While the usefulness of these applications of sensor network technology is clear, there remain shortcomings that can impede new and more varied uses.

Compared to the current underground sensor networks, which use wired communication methods for network deployment, WUSNs have merits such as concealment, ease of deployment, timeliness of data, reliability, and coverage density [2]. More specifically, the advantages of WUSNs are as follows:

- **Concealment:** Current underground sensing systems require data-loggers for data collection. Similarly, wireless sensor nodes can be used at the surface and wired to underground sensors [7, 26] to avoid the challenge of wireless communication underground. Therefore, above-ground devices are required at each location where an underground sensor is required. These above-ground components, however, are vulnerable to agricultural and landscaping equipment such as lawnmowers and tractors, which can damage the devices. Visible devices may also be unacceptable for performance or aesthetic reasons when monitoring sports fields or gardens. WUSNs, on the other hand, place all equipment required for sensing and transmitting underground, where it is out of sight, protected from damage by surface equipment and secure from theft or vandalism.

- **Ease of deployment:** WUSNs provide advantages also in the post-deployment phase. For traditional underground monitoring systems, the expansion of the coverage area requires deployment of additional data-loggers and underground wiring. Even if terrestrial WSN technology is used for underground monitoring [26], underground wiring must still be deployed to connect an underground sensor to a surface transceiver device. As a result, both the cost and the effort of maintaining such systems are significantly high. WUSNs provide flexibility in re-deployment since additional sensors can be deployed simply by placing them in the desired location and ensuring that they are within communication range of another device.

- **Real-time data delivery:** The existing underground sensing systems rely on data-loggers, which lack the support for real-time monitoring. Readings collected by the sensors are uploaded manually to a data-logger and then processed. Thus, real-time processing of the information is not possible. On the other hand, WUSNs provide wireless delivery of sensor readings to a central sink in real time.

- **Robustness:** Data-loggers introduce *single points of failure* for monitoring applications since all the information from tens of sensors is collected at the data-logger. Consequently, a possible failure of a data-logger could be catastrophic to a sensing application. WUSNs, however, provide a distributed service and the ability to independently forward readings. Additionally, WUSNs are self-healing. As a result, device failures can be automatically routed around, and the network operator can be alerted to device failure in real time. Consequently, the reliability of underground monitoring can be significantly improved with WUSNs compared to existing techniques.

- **Coverage density:** The coverage area and density of a traditional underground monitoring system depend on the locations of the data loggers. Since wired connections are generally required between sensors and the data logger, the sensors are usually deployed such that the distances between the data logger and the nodes are minimized. This can result in an uneven deployment, where the density is high in the vicinity of the data-logger, but low elsewhere in the environment. Furthermore, since each sensor has to be visited for data collection, the density of the network is usually limited. WUSNs allow sensors to be deployed independently of the location of a data-logger. Furthermore, since autonomous data collection is possible, the coverage area and the number of sensors in a network can be easily increased without any additional cost in data retrieval.

WUSNs provide extensive capabilities for underground applications compared to the existing methods. The realization of these networks, however, poses important and unique challenges in terms of communication channel characterization, communication protocols, as well as environment-aware networking principles. The main challenge in the realization of WUSNs is the realization of efficient and reliable underground links. The main difference between the well-established techniques in terrestrial WSNs is the communication medium. For WUSNs that are deployed in soil, the propagation characteristics of electromagnetic (EM) or magnetic waves in soil are significantly different compared to those in air. Hence, a straightforward characterization of the underground wireless channel is not possible. Similarly, for WUSNs that are deployed in underground mines and tunnels, although the EM waves propagate through air, the channel characteristics are drastically influenced by the properties of the underground space. To this end, advanced models and techniques are necessary to completely characterize the underground wireless channel and lay out the foundations for efficient communication in this environment. In addition, the previous expertise on WSNs should be leveraged to develop new cross-layer communication protocols that will allow for an efficient utilization of the underground channel.

The remainder of the chapter is organized as follows. In Section 17.1, an overview of potential applications for WUSNs is provided. Section 17.2 describes several factors that are important in the design of WUSNs and proposes possible network topologies. In Section 17.4 and Section 17.5, the underground channel model and the associated challenges are described for WUSNs in soil with EM and magnetic induction (MI) techniques, respectively. The channel model for WUSNs deployed in mines and tunnels is presented in Section 17.6. Section 17.7 examines the communication architecture of WUSNs and explains the challenges existing at each layer of the protocol stack.

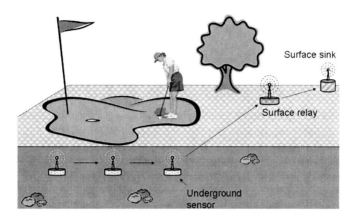

Figure 17.1   A WUSN deployed for monitoring a golf course.

17.1   Applications

WUSNs enable a vast number of applications that are in current use and have potential in the future. Generally, these applications can be categorized into four areas: *environmental monitoring*, *infrastructure monitoring*, *location determination*, and *border patrol and security monitoring*. We describe the possible applications in each area as follows:

17.1.1   Environmental Monitoring

As described above, underground sensors are being used in agriculture to monitor underground soil conditions, such as water and mineral content, and to provide data for appropriate irrigation and fertilization. A wireless underground system, however, can provide a significant refinement to the current approach for more targeted and efficient soil care. For example, since installation of WUSNs is easier than existing wired solutions, sensors can be more densely deployed to provide local detailed data. Rather than irrigating an entire field in response to broad sensor data, individual sprinklers could be activated based on local sensors. In a greenhouse setting, sensors could even be deployed within the pot of each individual plant. The concealment offered by a WUSN also makes it a more attractive and broadly viable solution than the current terrestrial agricultural WSNs. Visible and physically prominent equipment such as surface WSN devices or data-loggers would most likely be unacceptable for applications such as lawn and garden monitoring. An additional practical feature of underground sensors is that they are protected from equipment such as tractors and lawnmowers.

The concealment provided by the WUSNs makes them particularly applicable to sports field monitoring, where they can be used to monitor soil conditions at golf courses (see Figure 17.1), soccer fields, baseball fields, and grass tennis courts. For all of these sports, poor turf conditions generally create an unfavorable playing experience. Hence, soil maintenance is especially important to ensure healthy grass. On the other hand, the nature of these sports results in particular parts of the field being more damaged than others. Consequently, localized management of these sports fields, which can be provided by WUSNs, is a more efficient solution.

Monitoring the presence and concentration of various toxic substances is another important application. This is especially important for soil near rivers and aquifers, where chemical runoff could contaminate drinking water supplies. In these cases, it may be desirable to utilize a hybrid network of underground and underwater sensors.

In addition to monitoring soil *properties*, WUSNs can be used for landslide prediction by monitoring soil *movement* [33]. Current methods of predicting landslides are costly and time consuming to deploy, preventing their use in poorer regions that stand to benefit the most from such technology. Similar to terrestrial WSN devices, WUSN devices should be inexpensive, and deployment is as simple as burying each device. WUSN technology will allow for a much denser deployment of sensors so that landslides can be better predicted and residents of affected areas can be warned sufficiently early to evacuate.

Another possible application is monitoring air quality in underground coal mines. Buildup of methane and carbon monoxide is a dangerous problem that can lead to explosions or signify a fire in the mine, and the presence of these gases must be continually monitored [11]. This application would necessitate a hybrid architecture of underground open-air sensors and underground embedded sensors deployed between the surface of the ground and the roof of the mine tunnel. This would allow data from sensors in the mine to be quickly routed to surface stations vertically, rather than through the long distances of the mine tunnels.

Another mining application would include audio sensors (i.e., powerful, high-sensitivity, and low-power microphones suitable for underground environments) attached to the distributed underground sensor nodes to assist in the location and rescue of trapped miners. WUSN devices with microphones would also be useful for other applications, such as studying the noises of underground animals in their natural habitats.

Earthquake monitoring and prediction can also be facilitated by WUSN technology. Unlike landslide prediction, where soil movement near the surface is of interest, useful data for earthquakes come from multiple depths below the surface. The multi-hop nature of WUSNs will allow data to be routed back to an above-ground sink through a multi-depth topology.

17.1.2   Infrastructure Monitoring

A large amount of underground infrastructure exists, such as pipes, electrical wiring, and liquid storage tanks, where WUSNs can be used for monitoring applications. For example, fuel stations store fuel in underground tanks, which must be carefully monitored to ensure that no leaks are present and to continually determine the amount of fuel in the tank.

Homes in locations without a sewer usually have an underground septic tank, which must be monitored to prevent overflow. This is a typical application scenario for WUSNs, where monitoring can be done in real time without human interaction. WUSNs will also be useful in monitoring underground plumbing, where sensors can be deployed along the path of pipes so that leaks can be quickly localized and repaired.

Sensors may also be useful in monitoring the structural health of any underground components of a building, bridge, or dam [29]. Wireless devices could be embedded within key structural components to monitor stress, strain, and other parameters [8]. This could enable timely monitoring of wear and tear in a building and prevent catastrophic events.

17.1.3   Location Determination of Objects

Stationary underground sensor devices that are aware of their location can be used as a beacon for location-based services. One can imagine devices deployed beneath the surface of a road that communicate with a car as it drives over. A possible service would be to alert the driver to an upcoming stop sign or traffic signal. The car would receive the information about the upcoming signal and relay this to the driver.

Location information could also serve as a navigational aid for autonomous systems, e.g., an autonomous fertilizer unit, which navigates around the area to be fertilized based on underground location beacons and soil condition data from underground sensors.

WUSN technology can also be used to locate people in the event of collapse of a building. Devices could be deployed throughout a building and programmed with their physical location. The building's occupants would then carry a device on their person. In the event of a collapse, the occupant's device could be localized to a specific section of the building by communicating with the stationary devices. This could provide rescuers with a general area to search for survivors. Although this application is not strictly underground, the dense nature of rubble from a collapsed building poses challenges to wireless communication similar to soil.

17.1.4 Border Patrol and Security Monitoring

WUSNs can be used to monitor the above-ground presence and movement of people or objects. Similar to location determination, the deployed devices must be stationary and aware of their location. Unlike location determination, however, where objects announce their presence via direct communication with the embedded device, monitoring presence requires the use of sensors, such as pressure, acoustic, or magnetic sensors, to determine the presence of a person or object. This application is useful for home and commercial security, where sensors could be deployed underground around the perimeter of a building in order to detect intruders. Since the sensors are hidden, intruders would be less likely to know about them and thus take action to disable the security system.

On a larger scale, WUSNs can be very useful for border patrol. Wireless pressure sensors deployed at a shallow depth along the length of a border could be used to alert authorities to illegal crossings. Each sensor would be programmed with location information as it is deployed, allowing the exact location of an illegal crossing to be easily determined and giving a general area in which to deploy the authorities for a search. Rural areas are those needing the most security, and WUSN technology would allow a monitoring system to be easily deployed in these areas without any necessary infrastructure since they are self-powered.

17.2 Design Challenges

The realization of WUSNs requires several challenges unique to this domain to be addressed. Especially, the close interaction between the environment (soil properties, temperature, weather, location) and communication parameters results in several challenges including impaired underground channel, antenna design, and the effects of soil characteristics on communication. In this section, we describe four main topics: *energy efficiency*, *topology design*, *antenna design*, and *environmental extremes*, in detail.

17.2.1 Energy Efficiency

Depending on the intended application, WUSN devices should have a lifetime of at least several years in order to make their deployment cost efficient. This challenge is complicated by the lossy underground channel, which requires that WUSN devices have radios with greater transmission power than terrestrial WSN devices. Longer network lifetime is also crucial considering that underground sensors cannot easily be recharged or replaced. As a result, energy efficiency is a primary concern in the design of WUSNs.

As explained in the previous chapters, WSNs necessitate energy-efficient solutions to enhance the lifetime of the network that is constrained by the limited self-contained power source of each device. Like terrestrial WSNs, the lifetime of WUSNs is also limited by the energy constraints of each device. Furthermore, access to WUSN devices will be much more difficult than access to terrestrial WSN devices in most deployments, making retrieval of a device to recharge or replace its power supply less feasible, if not impossible. Deployment of new devices to replace failed ones is similarly difficult. Additionally, terrestrial WSN devices can be equipped with a solar cell [17, 42] to supplement or even replace a traditional power source, which is obviously not an option for WUSN devices. Scavenging opportunities for WUSN devices, such as converting seismic vibrations or thermal gradients to energy [28, 32, 35], do exist, but it remains to be explored whether these methods can provide sufficient energy to operate a device in the absence of a traditional power supply.

The additional peculiarities of WUSNs, compared to WSNs, make energy efficiency more pronounced in the design of these networks. While it is possible to increase the lifetime of a device by providing it with a larger stored power source, this is not necessarily desirable since it will increase the cost and size of sensor devices. Hence, energy efficiency should be achieved by utilizing power-efficient hardware and communication protocols, which is the main goal in WUSNs.

17.2.2 Topology Design

The design of an appropriate topology for WUSNs is of critical importance to network reliability and energy efficiency. WUSN topologies will likely be significantly different from their terrestrial counterparts. For example, the location of a WUSN device will usually be carefully planned, given the effort involved in the excavation necessary for deployment. Also, 3-D topologies will be common in WUSNs, with devices deployed at varying depths dictated by the sensing application.

The application of WUSNs will play an important role in dictating their topology; however, power usage minimization and deployment cost should also be considered in the design. A careful balance must be reached among these considerations to produce an optimal topology. Consequently, the design of a WUSN topology faces the following considerations:

- **Intended application:** Sensor devices must be located close to the phenomenon they are deployed to sense, which dictates the depth at which they are deployed. Some applications may require very dense deployments of sensors over a small physical area, while others may be interested in sensing phenomena over a larger physical area but with less density. Security applications, for example, will require a dense deployment of underground pressure sensors, while soil monitoring applications may need fewer devices since differences in soil properties over very small distances may not be of interest. Depending on the application type, the locations required by the application may not constitute the whole topology. Since connectivity is a major concern in network deployment, applications that require high-depth sensor deployment would require intermediate sensors at relatively lower depths to provide communication paths between the sensed phenomenon and the ground surface.

- **Power usage minimization:** Intelligent topology design can help to conserve power in WUSNs. Since attenuation is proportional to the distance between a transmitter and receiver, power usage can be minimized by designing a topology with a large number of short-distance hops rather than a smaller number of long-distance hops. Furthermore, the underground wireless channels need to be carefully studied since they exhibit significantly different characteristics than their counterparts in air. A network topology that considers these characteristics would lead to an energy-efficient network operation.

- **Cost:** Unlike terrestrial sensor devices, where deployment simply requires physically distributing devices, significant labor, and thus cost, is involved in the excavation necessary to deploy WUSNs. The deeper a sensor device is, the more the excavation required to deploy it, and the greater the cost of deploying that device. Additional costs will be incurred when the power supply of each device has been exhausted and the device must be unearthed to replace or recharge it. Thus, when cost is a factor, deeper deployment of devices should be avoided if possible, and the number of devices should be minimized. Minimizing the deployment conflicts with the dense deployment strategy suggested by power considerations, and an appropriate tradeoff must be established.

- **Soil type and condition:** An important characteristic of underground communication is its close interaction between the soil type and the condition in which the network is deployed. Depending on the soil type, the attenuation characteristics may drastically change, which significantly affects topology design. Furthermore, even for the same type of soil, the changing conditions due to humidity and weather introduce channel variability depending on the season. Therefore, WUSN topology design needs to consider the close effects of the soil that the network is deployed in for efficient and reliable communication.

17.2.3   Antenna Design

Wireless underground communication constitutes one of the major challenges in the design of WUSNs. The significant differences between WUSNs and WSNs may lead to completely different hardware for underground communication. In this respect, the design of an underground transceiver and, hence, the selection of a suitable antenna for WUSN devices are further challenging problems. The challenges facing the design are as follows:

- **Variable requirements:** Different devices may serve different communication purposes, and therefore may require antennas with differing characteristics. For example, devices deployed within several centimeters of the surface may need special consideration due to the reflection of EM radiation that will be experienced at the soil–air interface. Additionally, near-surface devices will likely act as relays between deeper devices and surface devices. Deeper devices acting as vertical relays to route data toward the surface may require antennas focused in both the horizontal and vertical directions.

- **Size:** To achieve practical transmission ranges of several meters, it is necessary for the underground transceivers to work at frequencies in the megahertz or lower ranges. However, decreasing the operating frequency necessitates larger antennas [24]. As an example, at a frequency of 100 MHz, a quarter-wavelength antenna would measure 0.75 m. While providing an acceptable transmission range, such an antenna would contradict with the goal to keep underground sensor devices small.

- **Directionality:** As explained above, WUSNs can be deployed at multiple depths to provide both coverage and connectivity in the network. This necessitates antennas that are capable of communicating in three dimensions. However, using a single omnidirectional antenna may not be feasible since the radiation patterns of these antennas contain nulls for vertical communication [24]. Therefore, underground sensors may need to be equipped with either two antennas that are oriented for both horizontal and vertical communication or a smart antenna that is capable of dynamically changing the radiation pattern according to the communication needs.

The design considerations regarding antenna design have so far focused on EM wave communication. However, as will be discussed in Section 17.4, it remains to be determined whether other technologies are better suited to this environment. Consequently, newer antenna designs may be required for underground communication.

17.2.4   Environmental Extremes

The underground environment is far from an ideal location for electronic devices. Water, temperature extremes, animals, insects, and excavation equipment all represent threats to a WUSN device, and it must be provided with adequate protection. Processors, radios, power supplies, and other components must be resilient to these factors. Additionally, the physical size of the WUSN device should be kept small, as the expense and time required for excavation increase for larger devices. Battery technology must be chosen carefully to be appropriate for the temperatures of the deployment environment while balancing environmental considerations with physical size and capacity concerns. Devices will also be subjected to pressure from people or objects moving overhead or, for deeply deployed devices, the inherent pressure of the soil above.

In addition to the physical factors that affect the lifetime of an underground sensor, the same environmental factors also affect the communication performance of WUSNs. Especially, the soil

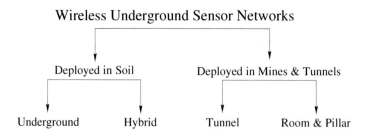

Figure 17.2   Overview of WUSN architectures.

composition and the water content of the soil affect the underground wireless channel conditions. These environmental effects on communication will be discussed in detail in Section 17.4.

17.3   Network Architecture

The peculiarities of underground communication and the close interaction between the soil content and the communication performance increase the importance of network deployment and architecture for efficient operation. According to the applications, WUSNs can be deployed in soil or in underground cavities such as mines and road or subway tunnels. Accordingly, different types of topologies for WUSNs can be deployed to address these two major types of networks. An overview of the WUSN architectures is shown in Figure 17.2. We describe next the network architecture types used for WUSNs in soil and WUSNs in mines and tunnels. The corresponding architectures also directly define the communication characteristics of each network.

17.3.1   WUSNs in Soil

Based on the network requirements, WUSNs in soil are generally deployed in two types of topologies: *underground topology* and *hybrid topology*.

Underground Topology

In many applications where concealment is important, all the sensors in the network need to be deployed underground. The only exception in this topology is the sink(s), which may be deployed underground or above ground as shown in Figure 17.3. The necessary information about the network is received by the sink through the network of underground sensors. In this particular class, the topology can be either *single depth*, i.e., all sensor devices are at the same depth, or *multi-depth*, i.e., the sensor devices are at varying depths. Next, we describe each case in detail.

Single Depth Topology: As described above, in a single depth topology sensors are deployed in the soil at almost the same depth. This deployment strategy is usually used in low-depth scenarios, where the communication between each sensor in the network and an above-ground sink is possible. Consequently, even mobile sinks can be exploited, where information can be collected from any underground sensor by a mobile above-ground entity. In this topology, deployment can be done in two ways. The nodes can be placed according to a deployment plan, where the location of each node is pre-calculated. This method can be used to monitor a particular area in the sensor field. Moreover, since deployment is carefully engineered, an optimal number of nodes can be used to *decrease* redundancy in the network. On the other hand, such a strategy would increase the deployment cost because of the additional effort

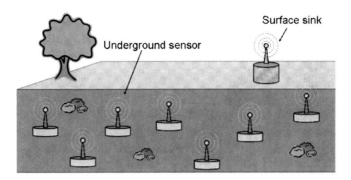

Figure 17.3　Underground topology.

in following the deployment plan for the sensor field. The other single depth deployment policy is to randomly place nodes under the surface of the soil. In this case, deployment can be relatively easier since sensors can be placed randomly at a specific depth and then covered with soil.

Data collection in a single depth topology can be performed through either fixed or mobile above-ground sinks. Consequently, underground sensors also serve as aggregators to collect information generated by the network and relay it to the sink. In this sense, network operation resembles traditional WSNs. Additionally, mobile agents (MAs) can also be used to collect and aggregate data from each node in the network. MAs can move in the field and receive data directly from sensors. When the MA moves to the communication range of a sensor, it will transfer its data to the MA.

Multi-depth Topology: In underground applications where information from a greater depth is required, multi-depth topology can be deployed as shown in Figure 17.3. Moreover, in cases where 3-D coverage of a volume of soil is necessary, multi-depth topology will be needed. Because of the limitations in transmission range for underground communication, multi-depth topology is usually necessary for high-depth applications to provide connectivity in the network. Since the sensors are located at different depths, the distances between each sensor and the ground is different. Consequently, in a multi-depth topology, the sensors at high depths may not directly communicate with a sink above the ground. Thus, intermediate nodes need to be deployed at lower depths to provide multi-hop paths between nodes located deep underground and the above-ground sinks. Although these intermediate nodes may not collect data, they serve as relays in the network. As a result, nodes at the lower depths transfer data to upper layer sensors first, and then the data are transferred in a multi-hop manner.

Similar to the single depth topology, the deployment of a multi-depth topology can also be done in two ways: through preplanning or randomly. However, considering the 3-D volume that should be covered by the network and the connectivity that should be provided, preplanning may be necessary for most of the multi-depth topology applications to decrease deployment cost.

Both communication protocols and sensor device hardware for multi-depth networks require special consideration to ensure that data can be efficiently routed to a surface sink. The depth at which devices are deployed will depend upon the application of the network, e.g., pressure sensors must be placed close to the surface, while soil-water sensors should be located deeper near the roots of plants. This topology minimizes (or eliminates, in the case of an underground sink) the above-ground equipment, providing maximum concealment of the network. Devices deployed at a shallow depth may be able to make use of a ground–air–ground path for the channel, which should produce lower path losses than a ground–ground channel.

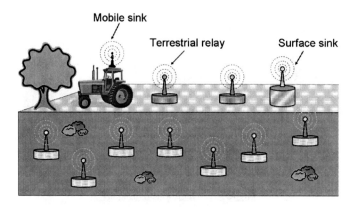

Figure 17.4   Hybrid topology.

Hybrid Topology

In addition to deploying the whole network underground, above-ground sensors can also be used for WUSNs in a hybrid topology. This is composed of a mixture of underground and above-ground sensor devices as shown in Figure 17.4. An important difference between the underground topology is that, in the former case, only sinks or MAs are above ground. Consequently, data "surface" only at the last hop. With a hybrid topology, intermediate above-ground sensors can be used to route information through the network. Since wireless signals are able to propagate through the air with lower loss than through soil, such a deployment can improve the efficiency and lifetime of the network. The above-ground sensor devices require a lower transmit power to communicate over a given distance than the underground sensor devices. A hybrid topology allows data to be routed from underground in fewer hops, thus trading power-intensive underground hops for less expensive hops in a terrestrial network. Additionally, terrestrial devices are more accessible in the event that their power supply requires replacement or recharging. Thus, given a choice, power expenditures should be made by above-ground devices rather than underground devices. The disadvantage of a hybrid topology is that the network is not fully concealed as with a strictly underground topology.

A hybrid topology could also consist of underground and above-ground sensors as well as mobile terrestrial sinks which move around the surface of the underground network deployment area and collect data from the underground sensors or terrestrial relays. Deeper devices can route their data to the nearest shallow device (which is able to communicate with both underground and above-ground devices) or a terrestrial relay, which will store the data until a mobile sink is within range. This topology should promote energy savings in the network by reducing the number of hops to reach a sink, since effectively every shallow device or terrestrial relay can act as a sink. The drawback of this topology is introduced by storing data until a mobile collector is within range. Moreover, the limited storage capabilities of sensor devices may limit the capacity of the network.

17.3.2   WUSNs in Mines and Tunnels

In addition to deployments in soil, WUSNs can also be deployed underground in mines and road/subway tunnels. Although EM waves propagate through the air in these cases, the propagation characteristics of EM waves are significantly different from those of the terrestrial wireless channels because of the restrictions caused by the structures of the mines and road/subway tunnels. In mines, multiple passageways are developed to connect the above-ground entrance and different mining areas. The structure of the mining

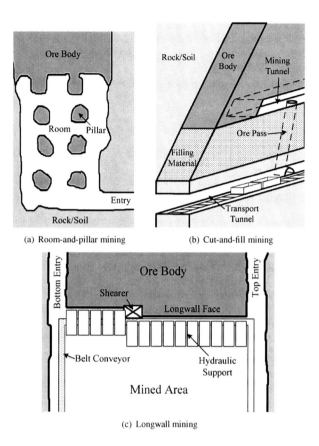

(a) Room-and-pillar mining   (b) Cut-and-fill mining

(c) Longwall mining

Figure 17.5   Mine structure of different mining methods.

area and hence the network topology is determined by mining methods, which are influenced by the shape and position of the ore body [14]:

- If the ore body is flat, room-and-pillar mining can be implemented as shown in Figure 17.5(a). The mining area can be viewed as a big room with randomly shaped pillars in it.
- If the ore body has a steep dip, cut-and-fill mining, sublevel stoping, or shrinkage stoping can be employed. Mines using these techniques have similar structures: the mining area consists of several types of tunnels, e.g. mining tunnel and transport tunnel. The sectional plan of cut-and-fill mining is shown in Figure 17.5(b).
- If the ore body has a large, thin, seam-type shape, longwall mining is preferred as shown in Figure 17.5(c). Besides the entry tunnels, the mining area near the longwall face can also be modeled as a tunnel since it is encircled by the hydraulic support and the longwall face.

17.4   Underground Wireless Channel for EM Waves

The realization of WUSNs is mainly dictated by the effects of the underground wireless channel. The characterization of the underground communication is the major challenge. Being a still developing area, the communication technology that will best suit the requirements of underground communication remains to be investigated. However, EM wave propagation through soil and rock has been studied extensively with ground-penetrating radar in the past [9, 23, 43, 44]. Moreover, existing sensor motes utilize EM waves almost exclusively. Hence, the characteristics of EM waves in soil are of particular importance for the realization of WUSNs. In this section, we describe the *properties of the underground EM channel*, the *effect of various soil properties on this channel*, and the *underground wireless channel model* in detail.

17.4.1   Underground Channel Properties

Underground communication is significantly different and challenging when compared to its terrestrial counterpart. This is mainly because of the close interaction of the soil with communication performance. In the following, five main factors that impact communication with EM waves underground are identified. These factors are *extreme path loss, reflection/refraction, multi-path fading, reduced propagation velocity*, and *noise*:

- **Extreme path loss:** Path loss due to material absorption is a major concern when using EM waves for underground communication. Path loss for underground communication is determined by both the frequency of the wave and the properties of the soil or rock through which it propagates. An important distinction to communication through air is that underground path losses are highly dependent on the soil type and water content. Soils are generally classified according to the size of their particles. In declining order of size they are *sand, silt, and clay*, or a mixture thereof [9]. Sandy soils are generally more favorable to EM wave propagation than clay soils. In addition to the soil composition, the soil-water content also significantly affects the attenuation. More specifically, any increase in soil-water content will produce a significant increase in attenuation.

  In addition to soil composition and condition, operating frequency is another factor that affects path loss. In general, over a given distance and soil condition, lower frequencies propagate underground with less attenuation than higher frequencies. Even frequencies in the megahertz range may experience attenuation on the order of over 100 dB/m depending on the soil conditions.

- **Reflection/refraction:** The soil–air interface introduces reflection and refraction because of the different attenuation characteristics of soil and air. WUSN devices deployed near the surface are able to communicate with both underground and surface devices, e.g., a surface sink, using a single radio. This implies that a communication link partially underground and partially in the air is necessary. When the propagating EM wave reaches the ground–air interface, it will be partially reflected back into the ground and partially transmitted into the air. The reverse is true for transmissions from surface devices to underground devices. Furthermore, devices that are deployed close to the surface encounter changes in the received power even if they are both deployed underground. This is because of the reflected signal from the surface. Consequently, two main paths are formed for communication, referred to as the *two-path channel* model. The reflected signal may improve or impair the communication performance depending on the distance between two entities and the deployment depth.

- **Multi-path fading:** The same mechanism described previously, whereby waves at medium transitions are partially transmitted and partially reflected, will also cause multi-path fading. This effect will especially be pronounced for sensors deployed near the surface, where the wave is close to the ground–air interface. Scattered rocks and plant roots underground, as well as varying soil properties, will act as scatterers and also produce fading.

- **Reduced propagation velocity:** EM waves propagating through a dielectric material such as soil and rock will experience a reduced propagation velocity compared to that of air. Since most soils have dielectric constants in the range of 1–80, a minimum propagation velocity of about 10% of the speed of light is implied.

- **Noise:** Even the underground channel is not immune to noise. On the contrary, underground noise levels are close to the noise levels encountered in the air [3]. Sources of underground noise include power lines, lightning, and electric motors [41]. Additionally, atmospheric noise is present underground [27, 41]. However, underground noise is generally limited to relatively low frequencies (below 1 kHz).

The above properties of the underground channel are also highly dependent on the soil properties between the transmitter and receiver. Therefore, a thorough understanding of how various soil parameters affect the channel is necessary.

17.4.2 Effect of Soil Properties on the Underground Channel

The underground wireless channel is characterized by the direct influence of soil properties on underground communication. This influence can be directly seen through the changes in attenuation of an EM signal passing through the soil, which dictates the path loss between two communicating entities. The composition of a soil, and its *volumetric water content, particle sizes, density*, and *temperature*, combine to form its complex dielectric constant ϵ. This parameter directly affects the attenuation of any EM wave passing through the soil, and it is thus useful to be able to predict its value. We now discuss in detail the effects of these and other parameters on signal attenuation:

- **Volumetric water content:** The moisture in the soil is by far the most significant factor that affects signal loss through soil. The moisture in the soil is usually referred to as the *volumetric water content* of the soil. Any increase in the volumetric water content will significantly increase the attenuation. As an example, when the volumetric water content rises from dry to 13% for a particular type of soil, the path loss increases by about 137 dB/m [44]. The effect of a rise in volumetric water content is also dependent on the type of soil. More specifically, sandy soils show less attenuation as water content increases than do clay soils [23]. As an example, at 900 MHz sandy and clay soils result in a propagation loss of 20 and 50 dB/m, respectively, with 40% volumetric water content [23].

- **Particle size:** In addition to the volumetric water content, soil composition is another factor that affects the communication performance. Soils are classified by the diameter of their particles, and are generally described as some variation of sand, silt, or clay. Sandy soils produce the least amount of loss and clay soils the most [23]. In addition, different soil particle types respond differently to changes in water content. Furthermore, increasing soil density increases path loss: the denser the soil, the greater the signal attenuation.

- **Temperature:** Increasing the temperature of soil changes its dielectric properties and will increase signal attenuation [9]. Additionally, changes in temperature will affect the dielectric properties of any water present in the soil.

In addition to the properties of soil, attenuation through soil is also dependent on the frequency being used. Lower frequencies experience less attenuation at a given volumetric water content than higher frequencies. Consequently, depending on the soil type and the volumetric water content, different operating frequencies can be used in WUSNs. These properties are summarized in Table 17.1.

17.4.3 Soil Dielectric Constant

The soil dielectric constant, ϵ, is the key factor that binds the effects of soil characteristics with the attenuation of a wave. A major challenge for predicting attenuation in an underground link is to compute the dielectric constant of the soil in which the WUSN devices are deployed. Several models are available for accurately predicting the complex dielectric constant, ϵ, for a homogeneous soil sample. However,

Table 17.1 Soil properties and their effect on signal attenuation.

Parameter	Change	Effect on signal attenuation
Water content	↑	↑
Temperature	↑	↑
Soil bulk density	↑	↑
% Sand	↑	↓
% Clay	↑	↑

predicting path losses for an underground channel is a challenge due to the inhomogeneous nature of ground. Soil makeup, density, and water content can all vary drastically over short distances [40].

In this section, we describe the Peplinski model [30], which associates key soil properties such as volumetric water content, density, and particle size with the rate of signal attenuation by the soil. The Peplinski model was constructed by taking measurements over a range of frequencies and for a variety of different soils and soil-water contents. Slightly different versions of the model are used for 0.3–1.3 GHz and 1.4–18 GHz. Both take as parameters the frequency, volumetric water content, bulk density, specific density of solid soil particles, mass fractions of sand and clay, and temperature.

Using the Peplinski model [30], the dielectric properties of soil in the 0.3–1.3 GHz band can be calculated as follows:

$$\epsilon = \epsilon' - j\epsilon'', \tag{17.1}$$

$$\epsilon' = 1.15 \left[1 + \frac{\rho_b}{\rho_s} (\epsilon_s^{\alpha'}) + m_v^{\beta'} \epsilon_{fw}'^{\alpha'} - m_v \right]^{1/\alpha'} - 0.68, \tag{17.2}$$

$$\epsilon'' = [m_v^{\beta''} \epsilon_{fw}''^{\alpha'}]^{1/\alpha'}, \tag{17.3}$$

where ϵ_m is the relative complex dielectric constant of the soil–water mixture, m_v is the water volume fraction (or volumetric water content) of the mixture, ρ_b is the bulk density in grams per cubic centimeter, $\rho_s = 2.66 \, \text{g/cm}^3$ is the specific density of the solid soil particles, $\alpha' = 0.65$ is an empirically determined constant, and β' and β'' are empirically determined constants dependent on soil type and given by

$$\beta' = 1.2748 - 0.519S - 0.152C, \quad \beta'' = 1.33797 - 0.603S - 0.166C, \tag{17.4}$$

where S and C represent the mass fractions of sand and clay, respectively. The quantities ϵ_{fw}' and ϵ_{fw}'' are the real and imaginary parts of the relative dielectric constant of free water. The Peplinski model [30] governs the value of the complex propagation constant of the EM wave in soil, which is given as $\gamma = \alpha + j\beta$, where

$$\alpha = \omega \sqrt{\frac{\mu \epsilon'}{2} \left[\sqrt{1 + \left(\frac{\epsilon''}{\epsilon'} \right)^2} - 1 \right]}, \quad \beta = \omega \sqrt{\frac{\mu \epsilon'}{2} \left[\sqrt{1 + \left(\frac{\epsilon''}{\epsilon'} \right)^2} + 1 \right]}; \tag{17.5}$$

here $\omega = 2\pi f$ is the angular frequency, μ is the magnetic permeability, and ϵ' and ϵ'' are the real and imaginary parts of the dielectric constant as given in (17.2) and (17.3), respectively.

From the above equations, it can be seen that the complex propagation constant, γ, of the EM wave in soil is dependent on the *operating frequency*, *the composition of soil in terms of sand, silt, and clay fractions*, *the bulk density*, and *the volumetric water content*. These factors also influence underground signal propagation as we will describe next.

17.4.4   Underground Signal Propagation

The unique characteristics of signal propagation in soil necessitate the derivation of path loss considering the properties of soil. The Friis equation [36] characterizes the received signal strength in free space at a distance r from the transmitter as

$$P_r = P_t + G_r + G_t - L_0 \qquad (17.6)$$

where P_t is the transmit power, G_r and G_t are the gains of the receiver and transmitter antennae, and L_0 is the path loss in free space in dB. The path loss in free space, L_0, is given by

$$L_0 = 32.4 + 20 \log(d) + 20 \log(f) \qquad (17.7)$$

where d is the distance between the transmitter and the receiver i km, and f is the operating frequency in MHz.

For the propagation in soil, a correction factor should be included in the Friis equation (17.6) to account for the effect of the soil medium. As a result, the received signal is

$$P_r = P_t + G_r + G_t - L_p \qquad (17.8)$$

where $L_p = L_0 + L_s$, and L_s stands for the additional path loss caused by the propagation in soil. L_s is calculated by considering the following differences of EM wave propagation in soil compared to that in air. The additional path loss, L_s, in soil is composed of two components:

$$L_s \text{(dB)} = L_\beta + L_\alpha \qquad (17.9)$$

where L_β is the attenuation loss due to the difference between the wavelength of the signal in soil, λ, and the wavelength in free space, λ_0, and L_α is the transmission loss caused by attenuation. Consequently,

$$L_\beta \text{(dB)} = 20 \log\left(\frac{\lambda_0}{\lambda}\right). \qquad (17.10)$$

Considering that, in soil, the wavelength is $\lambda = 2\pi/\beta$ and in free space $\lambda_0 = c/f$, where β is the phase shifting constant, $c = 3 \times 10^8$ m/s, and f is the operating frequency, Then L_β becomes

$$L_\beta = 154 - 20 \log(f) + 20 \log(\beta). \qquad (17.11)$$

The second component in (17.9), L_α, can be represented by considering the transmission loss caused by the attenuation constant, α, as $L_\alpha = e^{2\alpha d}$. In dB, L_α is given by

$$L_\alpha = 8.69\alpha d. \qquad (17.12)$$

Given that the path loss in free space is $L_0 = 20 \log(4\pi d/\lambda_0)$, the path loss of an EM wave in soil is found by combining L_0, (17.11), and (17.12) as follows:

$$L_p = 6.4 + 20 \log(d) + 20 \log(\beta) + 8.69\alpha d \qquad (17.13)$$

where distance, d, is given in meters, the attenuation constant, α, is in $1/m$, and the phase shifting constant, β, is in rad/m. The path loss, L_p in (17.13), depends on the attenuation constant, α, and the phase shifting constant, β. The values of these parameters depend on the dielectric properties of soil, which are derived in Section 17.4.3. The path loss, L_p, in soil can be found by using (17.1)–(17.5) in (17.13).

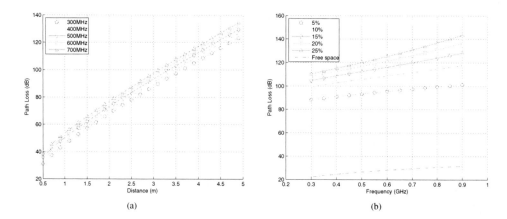

(a)   (b)

Figure 17.6   Path loss vs. (a) internode distance and operating frequency, and (b) operating frequency and soil moisture.

■ **EXAMPLE 17.1**

Consider a site where the volumetric water content is 5%, the sand particle percentage is 50%, the clay percentage is 15%, the bulk density is $1.5\,g/cm^3$, and the solid soil particle density is $2.66\,g/cm^3$. Accordingly, we evaluate the path loss in (17.13) to observe the relationship between path loss and various parameters.

In Figure 17.6(a), path loss, L_p, in (17.13) is shown in dB versus distance, d, for different values of operating frequency, f, varying from 300 to 900 MHz. It can be seen that the distance has an important effect on path loss and the path loss increases with increasing distance d as expected. Moreover, increasing operating frequency f also increases path loss, which motivates the need for lower frequencies for underground communication.

Next, the influence of soil moisture, i.e., volumetric water content, is shown in Figure 17.6(b), for values of 5%–25%. The difference between propagation in soil and that in free space is also evident in Figure 17.6(b). Since the attenuation significantly increases with higher water content, an increase of 30 dB is possible with a 20% increase in soil.

The significant effect of soil moisture on attenuation is particularly important since water content not only depends on the location of the network, but also varies during different seasons. Hence, network deployment, operation, and protocol design of WUSNs should consider this dynamic nature of the underground channel. The characteristics of path loss in soil constitute an important part of communication in soil. However, besides the attenuation in soil, various channel effects influence the performance of wireless communication. *Multi-path spreading* and *fading* are among the effects that should be considered. Moreover, the depth of the sensor nodes significantly affects the channel characteristics.

Next, the features of the underground channel are described in two main topics. Firstly, the effect of reflection from the ground surface on path loss is analyzed in Section 17.4.5. Secondly, the multi-path effects are described using a Rayleigh channel model and the bit error rate is derived for the underground wireless channel in Section 17.4.6.

17.4.5   Reflection from Ground Surface

The soil–air interface results in the waves being reflected from the ground surface. Hence, in general, underground communication results in two main paths for signal propagation as shown in Figure 17.7.

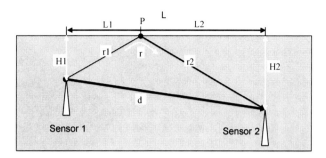

Figure 17.7 Illustration of the two-path channel model.

The first path is the direct path between two sensors and the second path is the reflection path due to the ground surface. While the direct path constitutes the main component of the received signal, the reflected path also affects communication, especially when the sensors are buried close to the surface. As a result, the signal propagation in the soil is also affected by these reflected rays in addition to the attenuation. Although this effect is mainly dependent on the depth of the sensors in soil, it should be considered in low depths as we explain next.

If the sensors are buried near the surface of the ground, i.e., *low depth*, the influence of the wave reflected by the ground surface should be considered. For such a reflection, the total path loss of the two-path channel is given by

$$L_f(\text{dB}) = L_p(\text{dB}) - V_{dB}, \tag{17.14}$$

where L_p is the path loss due to the direct path as given in (17.13) and V_{dB} is the attenuation factor due to the second path in dB, i.e., $V_{dB} = 10 \log V$.

Consider the case where two sensors are buried at a depth of H_1 and H_2, respectively, at a horizontal distance of L, and an end-to-end distance of d as illustrated in Figure 17.7. Then, the attenuation factor, V, can be found as follows:

$$V^2 = 1 + |\Gamma \cdot \exp(-\alpha \Delta(r))|^2$$

$$- 2\Gamma \exp(-\alpha \Delta(r)) \cos\left(\pi - \left(\phi - \frac{2\pi}{\lambda}\Delta(r)\right)\right) \tag{17.15}$$

where Γ and ϕ are the amplitude and phase angle of the reflection coefficient at the reflection point P, $\Delta(r) = r - d$ is the difference of the two paths, and α is the attenuation constant mentioned before.

■ **EXAMPLE 17.2**

Consider a deployment where two sensors are buried at the same depth, i.e., $H_1 = H_2 = H$, and hence $d = L$. Using (17.15) in (17.14), the path loss can be found as a function of end-to-end distance, d, operating frequency, f, buried depth, H, and volumetric water content.

In Figure 17.8(a), the path loss is shown as a function burial depth, H, for various values of operating frequency, f. It can be observed that, for the two-path model, the effect of operating frequency, f, is significant and mainly depends on the buried depth, H. For a particular operating frequency, an optimum buried depth exists such that the path loss is minimized. The effect of frequency and the volumetric water content is also shown in Figure 17.8(b), where the path loss fluctuates according to both parameters. This fluctuation is due to constructive or destructive interference from the second path based on the operating frequency and the volumetric water content.

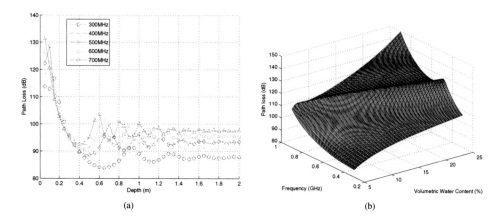

Figure 17.8 Two-path channel model: path loss vs. (a) depth for different operating frequencies, (b) operating frequency, f, and volumetric water content.

The underground channel model can be classified into two according to the buried depth of the sensors. If this depth is larger than a certain value, i.e., *high depth*, the effect of reflection is negligible and the channel is modeled through the single path model. In this case, the path loss is given in (17.13) as described in Section 17.4.4. More specifically, the underground channel exhibits a single path characteristic when the buried depth is higher than a threshold value. The results shown for Example 17.2 in Figure 17.8 reveal that if the buried depth is more than 2 m, the influence of reflection is negligible and single path model should be used. On the other hand, for low-depth deployments ($H \leq 2$ m), communication is modeled through the two-path channel model. In this case, the path loss is given in (17.14).

17.4.6 Multi-path Fading and Bit Error Rate

The two-path channel model captures the main propagation characteristics of EM waves underground. However, in fact, the underground channel exhibits additional complications. Firstly, the surface of the ground is not ideally smooth and, hence, causes not only reflection, but also refraction. Secondly, usually there are rocks or plant roots in soil and the clay of soil is generally not homogeneous. As a result of these impurities in the soil, multiple copies of the same signal can be reflected from these obstacles and received at the receiver. Hence, multi-path fading should also be considered in addition to the basic two-path channel model.

Multi-path fading has been explained in Chapter 4 for communication through air, where the random refraction due to air, the movement of objects, as well as other random effects result in the fluctuation and refraction of EM waves. Therefore, the amplitude and phase of the received signal exhibit random behavior with time. Generally, this multi-path channel character obeys the Rayleigh or log-normal probability distribution [45].

In underground communication, on the other hand, there is no random air refraction with time. This is because the channel between two transceivers is relatively stable when the composition of soil is considered. Hence, the channel is almost stable in each path with respect to time. On the other hand, considering a fixed internode distance, the received signal levels vary at different locations because the signal travels through different multiple paths. As a result, randomness is due to the locations of the nodes rather than time, which still obeys the Rayleigh probability distribution. Accordingly, the envelope of the signal from each path is modeled as an independent Rayleigh distributed random variable, χ_i, $i \in \{1, 2\}$.

Consequently, for the one-path model, the received energy per bit per noise power spectral density is given by $r = \chi^2 E_b/N_0$, which has a distribution as follows:

$$f(r) = \frac{1}{r_0} \exp\left(\frac{r}{r_0}\right) \tag{17.16}$$

where $r_0 = E[\chi^2]E_b/N_0$ and E_b/N_0 are directly related to the signal-to-noise ratio (SNR) of the channel.

Similarly, for the two-path model, a two-path Rayleigh channel is used, where χ in this channel is

$$\chi^2 = \chi_1^2 + [\chi_2 \cdot \Gamma \cdot \exp(-\alpha\Delta(r))]^2 - 2\chi_1\chi_2\Gamma \exp(-\alpha\Delta(r)) \cos\left(\pi - \left(\phi - \frac{2\pi}{\lambda}\Delta(r)\right)\right) \tag{17.17}$$

where χ_1 and χ_2 are two independent Rayleigh distributed random variables of the two paths, respectively. Γ and ϕ are the amplitude and phase angle of the reflection coefficient at the reflection point P, $\Delta(r) = r - d$ is the difference of the two paths, and α is the attenuation constant.

Based on the above model, the bit error rate (BER) characteristics of the underground channel are described next. Assuming 2PSK as the modulation method, the BER can be shown to be

$$BER = \frac{1}{2}erfc(\sqrt{SNR}) \tag{17.18}$$

where $erfc(\cdot)$ is the error function and SNR is given by

$$SNR = P_t - L_f - P_n \tag{17.19}$$

where P_t is the transmit power, L_f is the total path loss given in (17.14), and P_n is the energy of the noise. The transmit power, P_t, is usually between 10 and 30 dBm and the noise level, P_n, is found to be -103 dBm on average [3]. Although the noise, P_n, may change depending on the properties of the soil, this value is a representative value that can be used to show the properties of the BER underground.

■ EXAMPLE 17.3

Using (17.14) and (17.19) in (17.18), the BER is shown in Figure 17.9 for the single path model and in Figure 17.10 for the two-path model. The effect of transmit power, P_t, on the BER is shown in Figure 17.9(a). As the transmit power increases, the BER decreases. Moreover, as shown in Figure 17.9(b), an increase from 5% to 10% in volumetric water content results in almost an order of magnitude increase in the BER, which significantly affects the effective communication range of a sensor.

As explained earlier, for buried depths of less than 2 m, the two-path Rayleigh fading model is suitable for WUSNs. In Figure 17.10, the effect on the BER of the reflected path from the ground surface can be clearly seen. As shown in Figure 17.10(a), the communication distance can be extended to 4.5–5 m with increased transmit power of 30 dBm and low operating frequency at 400 MHz.

As discussed in Section 17.4.5 and shown in Figure 17.8(a), the path loss is significantly affected by the buried depth of sensors. This fluctuation also results in varying BER values as shown in Figure 17.10(b). As the buried depth increases, the fluctuations decrease and the BER becomes more stable. These results reveal that for a particular operating frequency, there is an optimal depth for communication where the BER is a minimum. Finally, the effect of volumetric water content in the two-path model is shown in Figure 17.10(c). Compared to the single path model results shown in Figure 17.9(b), higher volumetric water content is acceptable when the operating frequency is low. Another observation is that for a particular volumetric water content value, the BER fluctuates with changing frequency, contrary to the single path case. This is basically because the reflection from the ground surface depends greatly on the frequency of the signal.

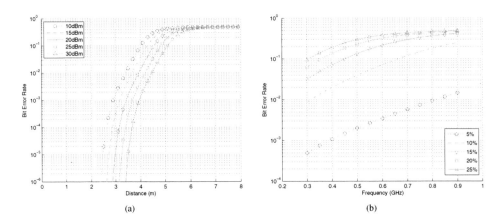

(a)

(b)

Figure 17.9 One-path channel model: BER vs. (a) internode distance with different transmit power, (b) operating frequency, f, and volumetric water content.

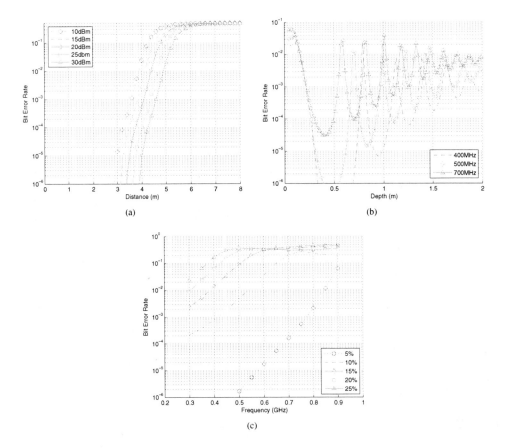

(a)

(b)

(c)

Figure 17.10 Two-path channel model: BER vs. (a) internode distance with different transmit power, (b) depth, h, and different operating frequencies, and (c) operating frequency and volumetric water content.

17.5   Underground Wireless Channel for Magnetic Induction

Traditional signal propagation techniques using EM waves encounter three major problems in the soil medium: high path loss, dynamic channel condition, and large antenna size [3]. Firstly, EM waves experience high levels of attenuation due to absorption by soil, rock, and water. Secondly, the path loss is highly dependent on numerous soil properties such as water content, soil makeup (sand, silt, or clay), and density. In addition, the path loss can change dramatically with time and space. As an example, the soil-water content increases after rainfall and the soil properties may significantly change over short distances. Therefore, the BER of the communication system also varies as a function of time and location. Thirdly, operating frequencies in the megahertz or lower ranges are necessary to achieve practical transmission range [2]. To efficiently transmit and receive signals at that frequency, the antenna size would be too large to be deployed in the soil.

An alternative signal propagation technique for underground wireless communication is *magnetic induction* (MI). The MI techniques effectively address the dynamic channel condition and large antenna size challenges of the EM wave techniques. In particular, a dense medium such as soil and water causes small variations in the attenuation rate of magnetic fields compared to that of air, since the magnetic permeabilities of the soil and air are similar [2, 16, 34]. Therefore the MI channel conditions remain constant in the soil medium. Moreover, in the MI communication, transmission and reception are accomplished with the use of a small coil of wire. Therefore, the antenna size is not a major limitation for MI techniques. However, the main challenge of MI-based communication is the attenuation: magnetic field strength falls off much faster than the EM waves [4, 5].

In this section, we first describe the path-loss expressions for the underground MI communication channel. Multiple factors are considered in the analysis, including the soil properties, coil size, the number of turns in the coil loop, coil resistance, and the operating frequency. In addition, we show how the high path loss can be reduced to extend the transmission range through MI waveguide techniques [37]. The MI waveguide has three advantages in underground wireless communication. Firstly, by carefully designing the waveguide parameters, the path loss can be greatly reduced. Secondly, the relay coils constituting the MI waveguide do not consume any energy and the cost is very small. Thirdly, the MI waveguide is not a continuous structure like a real waveguide, hence it is relatively flexible and easy to deploy and maintain.

17.5.1   MI Channel Model

Transmission and reception in MI communication are accomplished with the use of a coil of wire, as shown in Figure 17.11(a), where a_t and a_r are the radii of the transmission coil and receiving coil, respectively, r is the distance between the transmitter and the receiver, and $(90° - \alpha)$ is the angle between the axes of two coupled coils. Accordingly, the path loss for MI communication is [37]

$$\frac{P_r}{P_t} \simeq \frac{\omega^2 \mu^2 N_t N_r a_t^3 a_r^3 \sin^2 \alpha}{8r^6} \cdot \frac{1}{4R_0(2R_0 + \frac{1}{2}j\omega\mu N_t)} \tag{17.20}$$

where N_t and N_r are the number of turns of the transmitter coil and receiving coil, respectively, R_0 is the resistance of a unit length of the loop, μ is the permeability of the medium (i.e., soil), and ω is the angular frequency of the transmitted signal. If the low-resistance loop, the high signal frequency, and the large number of turns are employed ($\omega\mu N_t \gg R_0$), then the ratio can be further simplified:

$$\frac{P_r}{P_t} \simeq \frac{\omega\mu N_r a_t^3 a_r^3 \sin^2 \alpha}{16R_0 r^6}. \tag{17.21}$$

According to (17.21), the received power loss is a sixth-order function of the transmission range r. A higher signal frequency ω, larger number of turns N, lower loop resistance R_0, and larger coil size a

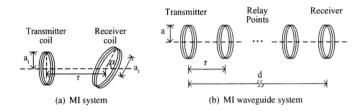

Figure 17.11   The structure of the MI transceiver and the MI waveguide.

can increase the received power. The angle between the axes of two coupled coils also affects the received power. Specifically, the smaller the angle, the higher the power received. An important characteristic of MI communication is that the received power is not affected by the environmental conditions. The only parameter related to the environment for the received power is the permeability, μ, in (17.21). Moreover, the permeability, μ, of soil and water is similar to that of air.

The expression of the path loss in (17.21) can be further compared to the Friis transmission equation for the EM wave communication, where

$$\frac{P_r}{P_t} \simeq G_t G_r \left(\frac{\lambda}{4\pi r}\right)^2 = G_t G_r \frac{\pi}{4\mu\varepsilon\omega^2 r^2}. \tag{17.22}$$

For EM waves, a higher operating frequency results in a greater path loss. On the other hand, for MI, an increased frequency leads to a lower attenuation rate. However, the received power of MI communication attenuates much faster than in the EM wave case ($1/r^6$ vs. $1/r^2$). Moreover, the permittivity ε in (17.22) is much larger in soil than in air. Furthermore, ε varies greatly at different times and locations. Hence, the path loss of EM waves is dramatically influenced by these environmental conditions. Accordingly, the MI technique has constant channel condition while the EM wave technique results in lower attenuation.

17.5.2   MI Waveguide

The MI techniques provide advantages over the EM wave techniques in terms of less dependency on the environment and antenna size. However, the received power loss is much higher than in the EM wave case. For practical applications, this can be addressed by employing relay points between the transmitter and receiver.

Contrary to the relay points used for EM waves, the MI relay point is a simple coil without any energy source or processing device. This results in a *passive* operation for the relay points, where the communication is performed without any external power source. The sinusoidal current in the transmitter coil induces a sinusoidal current in the first relay point. This sinusoidal current in the relay coil then induces another sinusoidal current in the second relay point. This continues until the destination is reached. The consecutive relay coils form an *MI waveguide* in underground environments, which acts as a waveguide to guide the so-called *MI waves*.

A typical MI waveguide structure is shown in Figure 17.11(b), where n relay coils are equally spaced along one axis between the transmitter and receiver, r is the distance between the neighbor coils, d is the distance between the transmitter and receiver, and $d = (n + 1)r$ where a is the radius of the coils. In fact, there exists mutual induction between any pair of the coils. The value of the mutual induction depends on how close the coils are to each other. For the MI waveguide model developed in this section, the distance between two relay coils is considered as 1 m and the coil radius is no more than 0.1 m. Therefore, the coils are sufficiently far from each other and only interact with their nearest neighbors. Hence, only the mutual induction between the adjacent coils needs to be taken into account.

The path loss of the MI waveguide is given as [37]

$$\frac{P_r}{P_t} \simeq \frac{\omega^2 \mu^2 N^2 a^6}{8r^6} \cdot \frac{1}{4R_0(2R_0 + \frac{1}{2}j\omega\mu N)} \cdot \left[\frac{j}{\frac{4R_0}{\omega\mu N}(\frac{r}{a})^3 + j(\frac{r}{a})^3 + \frac{\omega\mu N}{4R_0 + j\omega\mu N}(\frac{a}{r})^3} \right]^{2n}. \quad (17.23)$$

Considering the high signal frequency and large number of turns employed ($\omega\mu N \gg R_0$), (17.23) can be further simplified to

$$\frac{P_r}{P_t} \simeq \frac{\omega\mu N}{16R_0}\left(\frac{a}{r}\right)^{6n} = \frac{\omega\mu N}{16R_0}\left[\frac{a}{d}(n+1)\right]^{6n}. \quad (17.24)$$

It is shown in (17.24) that the transmission range d is divided into $n + 1$ intervals with length r. However, the path loss becomes a $6n$th-order function of the relay interval r. Hence, to reduce the path loss of the MI waveguide, the relay interval r needs to be on a par with the coil size to make the term a/r approximately 1. This means that if the coils with a radius of 0.1 m are utilized, these coils should be deployed every 0.1 m, which is infeasible in underground communication considering the deployment challenges. Consequently, using simple relay coils does not effectively reduce the path loss.

From (17.23), if the last term with exponent $2n$ is made to converge to a value around 1, then the MI waveguide path loss can be considerably reduced. This can be accomplished by adding a capacitor to each coil and determining the capacitor value, the operating frequency, and the number of turns in the coil to decrease the path loss. By assigning the capacitor C an appropriate value, the self-induction term can be neutralized. Then, the term with exponent $2n$ can be significantly diminished. More specifically, if the value of the capacitor C is selected as

$$C = \frac{2}{\omega^2 N^2 \mu\pi a} \quad (17.25)$$

then the MI waveguide path loss becomes

$$\frac{P_r}{P_t} = \frac{\omega^2 \mu^2 N^2 a^6/4r^6}{2R_0[4R_0 + (\omega^2\mu^2 N^2 a^6)/(2R_0 \cdot 4r^6)]} \cdot \left[\frac{j}{\frac{4R_0}{\omega\mu N}(\frac{r}{a})^3 + \frac{\omega\mu N}{4R_0}(\frac{a}{r})^3} \right]^{2n}. \quad (17.26)$$

Moreover, the operating frequency and the number of turns can be designed to further reduce the path loss such that

$$\frac{\omega\mu N}{4R_0}\left(\frac{a}{r}\right)^3 = 1, \quad (17.27)$$

and finally the path loss can be represented as

$$\frac{P_r}{P_t} = \frac{1}{3}\left(\frac{1}{2}\right)^{2n}. \quad (17.28)$$

As shown in (17.28), the path loss for the MI waveguide path loss is considerably smaller than that for the traditional MI and EM wave techniques. The path loss is a function of the number of relay points, n. Larger n results in higher path loss and is usually determined from the transmission distance d and the relay interval r. The value of r should be chosen as large as possible considering the constraints in (17.25) and (17.27). More specifically, in (17.27), the relay interval, r, and the coil size, a, both determine the operating frequency, ω, and the number of turns, N. In (17.25), the capacitor value C is determined by a, N, and ω. Hence, when designing the relay interval, it should be ensured that the operating frequency, the number of turns, and the capacitor value can be assigned feasible and appropriate values. If the operating frequency is considered to be several hundred megahertz and the coil radius is 0.1 m, the relay interval of 1 m can be chosen to satisfy these requirements.

17.5.3   Characteristics of MI Waves and MI Waveguide in Soil

The comparison of the path loss for the traditional EM wave technique, the current MI technique, and the improved MI waveguide technique for wireless underground communication is important in evaluating the characteristics of these techniques. In this section, we compare the techniques based on the following conditions: the volumetric water content (VWC) is 5% and the operating frequency is 300 MHz. The transmitter, receiver, and relay coil all have the same radius of 0.1 m. The coil is made of copper wire with a 0.5 mm diameter, hence the resistance of unit length R_0 is 0.216 Ω/m. The permeability of the soil medium is the same as that in air, which is $4\pi \times 10^{-7}$ H/m. The relay interval r of the MI waveguide is 1 m. The number of relay coils n is determined by the transmission distance d, where $n = \lceil d/r \rceil$. The coil capacity is calculated by (17.25), which is around 20 pF.

Low Operating Frequency and Low VWC

In Figure 17.12(a), the path loss for the three techniques is shown in dB versus the transmission distance d using a 300 MHz signal in soil with 5% VWC. It can be seen that in the very near region ($d < 1$ m), the path loss of the MI technique is smaller than that of the EM wave technique. However, for longer distances, the MI signal attenuates much faster than the EM wave signal. The difference is as high as 20 dB compared to the EM wave signal. On the other hand, the MI waveguide technique reduces the signal path loss compared to the other two techniques. Accordingly, the path loss of the MI waveguide system is less than 50% of the MI and EM wave cases at a certain range.

High Operating Frequency and Low VWC

In Figure 17.12(b), the VWC of the soil remains the same but the operating frequency is increased to 900 MHz. The path loss of the EM wave system increases slightly. Since material absorption is the major part of EM wave path loss in soil, the attenuation caused by the higher operating frequency is not significant. On the other hand, the path loss of the MI system decreases as the operating frequency increases, which can be explained through (17.21), where the operating frequency ω is in the numerator. As a result, at high operating frequency, the path loss of the EM wave system becomes higher than that of the MI system. The path loss of the MI waveguide system remains the lowest when a high operating frequency is used.

Low Operating Frequency and High VWC

In Figure 17.12(c), the influence of the VWC on the three propagation techniques is analyzed. The performance of the MI and MI waveguide systems is not affected by the environment since the permeability μ remains the same between air, water, or the soil media. According to the channel model described in Section 17.4, the path loss of the three techniques in soil with a higher water content (25% VWC) is as shown in Figure 17.12(c). As expected, the path loss of the MI and MI waveguide systems remains the same. However, the path loss of the EM wave system increases dramatically (up to 40 dB) in soil with a higher water content.

17.6   Wireless Communication in Mines and Road/Subway Tunnels

In addition to communication through soil with EM and MI techniques, WUSNs can also be deployed in underground cavities such as mines or tunnels. As explained in Section 17.3.2, different forms of underground structures exist. Accordingly, the wireless communication channels can be classified into two. The *tunnel channel model* is used to describe signal propagation in passageways and mining area tunnels. The *room-and-pillar channel model* characterizes the wireless channel of room-and-pillar

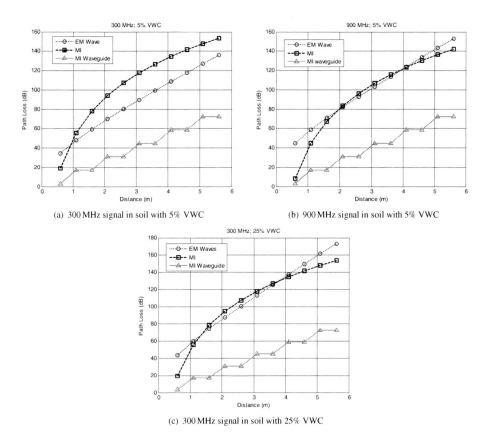

(a) 300 MHz signal in soil with 5% VWC

(b) 900 MHz signal in soil with 5% VWC

(c) 300 MHz signal in soil with 25% VWC

Figure 17.12   Path loss of the three techniques using different operating frequency in soil with different VWC.

mining area [38]. The structure of the road/subway tunnels is similar to that of underground mine tunnels and can be captured through the tunnel channel model.

In this section, we describe channel models for the tunnel and room-and-pillar environments through the *multimode model*, which lays out the foundations for reliable and efficient communication networks in mines and road tunnels [38]. For the tunnel environment, the multimode model can completely characterize the natural wave propagation in both near and far regions of the transmitter. For the room-and-pillar environment, the multimode model is combined with the shadow fading model to characterize the random effects of the pillars in underground mines. The channel models for these two environments are explained in the following.

17.6.1   Tunnel Environment

The multimode model is a multimode operating waveguide model which exploits the GO channel model [22] to analyze the EM field distribution in a tunnel cross-section. This field distribution can be considered as the weighted sum of the fields of all modes. Hence, the intensity of each mode is estimated by a mode-matching technique. Once the mode intensity is determined in the excitation plane, the propagation of each mode is mostly governed by the tunnel itself. Hence the EM field of the rest of the tunnel can be accurately predicted by summing the EM field of each mode.

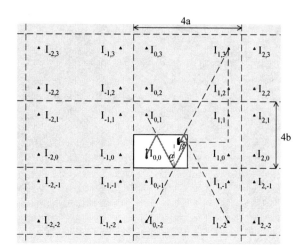

Figure 17.13   The set of images in the excitation plane in a rectangular tunnel.

Multimode Model

The excitation plane of the EM field of each mode depends on the tunnel cross-section, which is generally in the shape of a rectangle or a circle. Since the EM field distribution and attenuation of the modes in the rectangular waveguide are almost the same as the circular waveguide [21], the tunnel cross-section can be treated as an equivalent rectangle. A Cartesian coordinate system is set with its origin located at the center of the rectangular tunnel. Denoting k_v, k_h, and k_a as the complex electrical parameters of the tunnel's vertical/horizontal walls and the air in the tunnel, respectively, they are defined as

$$k_v = \varepsilon_0\varepsilon_v + \frac{\sigma_v}{j\omega}, \quad k_h = \varepsilon_0\varepsilon_h + \frac{\sigma_h}{j\omega}, \quad k_a = \varepsilon_0\varepsilon_a + \frac{\sigma_a}{j\omega} \tag{17.29}$$

where ε_v, ε_h, and ε_a are the relative permittivities for the vertical/horizontal walls and the air in the tunnel, ε_0 is the permittivity in a vacuum, σ_v, σ_h, and σ_a are their conductivities, and ω is the angular frequency of the signal. The three areas can be considered to have the same permeability μ_0. The wave number in the tunnel space is given by $k = \omega\sqrt{\mu_0\varepsilon_0\varepsilon_a}$. Accordingly, the relative electrical parameters can be defined as $\overline{k_v} = k_v/k_a$ and $\overline{k_h} = k_h/k_a$ for concise expression.

For the field analysis of the excitation plane by the GO model, the transmitter antenna is assumed to be an X-polarized electric dipole. Moreover, the total field of a point in the excitation plane is equal to the sum of ray contributions from all reflection images added to that of the source. The reflection images are located as shown in Fig 17.13 and have the following properties:

- The ray coming from image $I_{p,q}$ experiences $|p|$ times reflection from the vertical wall and $|q|$ times reflection from the horizontal wall.

- Suppose that α is the incident angle on the horizontal wall, and β is the incident angle on the vertical wall. Then α and β are complementary. These angles remain the same for all reflections of a certain ray.

Considering that the transmitter antenna is located at (x_0, y_0), the major X-polarized field at the receiver is then given by

$$E_x^{Rx} = E_x^{Tx} \cdot \sum_{p,q}\left[\frac{\exp(-jkr_{p,q})}{r_{p,q}}\right] \cdot S(\overline{k_v})^{|p|} \cdot R(\overline{k_h})^{|q|} \tag{17.30}$$

where E_x^{Tx} and E_x^{Rx} are the electric field at the transmitter and receiver, respectively, $r_{p,q}$ is the distance between image $I_{p,q}$ and the receiver, and $R(\overline{k_h})$ and $S(\overline{k_v})$ are the reflection coefficients on the horizontal and vertical walls.

The modes in a rectangular tunnel are approximately orthogonal [31, 25], i.e.,

$$\int_{-a}^{a} \int_{-b}^{b} E_{m,n}^x \cdot E_{j,k}^{x*} \, dx \, dy \simeq \begin{cases} \xi^2, & \text{if } m = j \text{ and } n = k \\ 0, & \text{otherwise} \end{cases} \tag{17.31}$$

where ξ is the norm of the modes. The field of the modes can be viewed as a basis that spans the total field. Therefore, the mode intensity C_{mn} can be calculated by projecting the field of excitation plane obtained by the GO model (E_x^{Rx} in (17.30)) on the basis function $E_{m,n}^x$:

$$C_{mn} = \int_{-a}^{a} \int_{-b}^{b} E_x^{Rx} \cdot E_{m,n}^x \, dx \, dy \tag{17.32}$$

which can be viewed as the sum of all the contributions of images: $C_{mn} = \sum_{p,q \in Z} C_{mn}^{(p,q)}$, where

$$C_{mn}^{(p,q)} = \int_{-a}^{a} \int_{-b}^{b} \left\{ E_x^{Tx} \cdot \frac{\exp(-jkr_{p,q})}{r_{p,q}} \cdot S(\overline{k_v})^{|p|} \cdot R(\overline{k_h})^{|q|} \cdot E_{m,n}^x \right\} dx \, dy. \tag{17.33}$$

By dividing the absolute mode intensity by the norm of the basis ξ with the field of reference position $E_x(r_0)$ (r_0 apart from the antenna), then the normalized mode intensity is found as

$$\overline{C_{mn}^{(p,q)}} = \frac{C_{mn}^{(p,q)}}{\xi \cdot E_x(r_0)} = \frac{C_{mn}^{(p,q)}}{\xi \cdot E_x^{Tx} \cdot \exp(-jkr_0)/r_0} \tag{17.34}$$

$$\simeq \frac{1}{\xi} \int_{-a}^{a} \int_{-b}^{b} \frac{r_0 \exp(-jkr_{p,q})}{r_{p,q} + r_0} S(\overline{k_v})^{|p|} R(\overline{k_h})^{|q|} \cdot E_{m,n}^x \, dx \, dy.$$

The closed-form solution of (17.34) is derived by composite numerical integration, which is given by

$$\overline{C_{mn}^{(p,q)}} \simeq \frac{4}{3} \frac{\sqrt{ab}}{mn} \sum_{u=0}^{m-1} \sum_{v=0}^{n-1} \left\{ \frac{r_0 \exp(-jkr_{p,q})}{r_{p,q} + r_0} S(\overline{k_v})^{|p|} R(\overline{k_h})^{|q|} \cdot (-1)^{\lfloor \frac{m+n}{2} \rfloor + 1 + u + v} \right\}. \tag{17.35}$$

The mode intensity is the summation of all contributions of the images; however, only the low-order images have significant effect. Specifically, for the X-polarized field, only images $I_{p,q}$ with subscript $p = 0, \pm 1$ and $q = 0, \pm 1, \pm 2$ are considered. In addition, to reduce the computational cost, the reflection coefficients R and S are simplified to their approximate expressions by nonlinear regression. Therefore, the normalized intensity for the mnth-order mode is

$$\overline{C_{mn}} \simeq \frac{4}{3} \frac{\sqrt{ab}}{mn} \sum_{\substack{p=0,\pm 1 \\ q=0,\pm 1,\pm 2}} \left\{ \sum_{u=0}^{m-1} \sum_{v=0}^{n-1} \left[\frac{r_0 \exp(-jkr_{p,q})}{r_{p,q} + r_0} S^{|p|} R^{|q|} \cdot (-1)^{\lfloor \frac{m+n}{2} \rfloor + 1 + u + v} \right] \right\} \tag{17.36}$$

where

$$R^{|q|} = (-1)^{|q|} \exp\left(-2|q| \cdot \frac{y_q}{r_{p,q}\sqrt{k_h}}\right)$$

$$S^{|p|} = \begin{cases} 1, & \text{if } p = 0 \\ 1 - 2 \cdot \dfrac{1}{1 + (x_p/r_{p,q})\sqrt{k_v}}, & \text{if } p = \pm 1 \end{cases}$$

$$x_p = \begin{cases} \left| 2pa - x_0 + a - \dfrac{2a}{m}\left(u + \dfrac{1}{2}\right) \right|, & \text{if } p \text{ is odd} \\ \left| 2pa + x_0 + a - \dfrac{2a}{m}\left(u + \dfrac{1}{2}\right) \right|, & \text{if } p \text{ is even} \end{cases}$$

$$y_q = \begin{cases} \left| 2qb - y_0 + b - \dfrac{2b}{n}\left(v + \dfrac{1}{2}\right) \right|, & \text{if } q \text{ is odd} \\ \left| 2qb + y_0 + b - \dfrac{2b}{n}\left(v + \dfrac{1}{2}\right) \right|, & \text{if } q \text{ is even} \end{cases}$$

$$r_{p,q} = \sqrt{x_p^2 + y_q^2}.$$

Then, the predicted field at any position (x, y, z) inside the tunnel can be obtained by summing the fields of all significant modes at that position, which is given by

$$E_x^{Rx}(x, y, z) = E_x(r_0) \sum_{m,n} \overline{C_{mn}} \cdot E_{m,n}^x(x, y) \cdot e^{-(\alpha_{mn} + j\beta_{mn}) \cdot z} \tag{17.37}$$

where α_{mn} is the attenuation coefficient and β_{mn} is the phase-shift coefficient [10, 13, 20]:

$$\alpha_{mn} = \frac{1}{a}\left(\frac{m\pi}{2ak}\right)^2 Re\frac{k_v}{\sqrt{k_v - 1}} + \frac{1}{b}\left(\frac{n\pi}{2bk}\right)^2 Re\frac{1}{\sqrt{k_h - 1}}$$

$$\beta_{mn} = \sqrt{k^2 - \left(\frac{m\pi}{2a}\right)^2 - \left(\frac{n\pi}{2b}\right)^2}. \tag{17.38}$$

Similarly, the predicted received signal power at the coordinate (x, y, z) is given by

$$P_r(x, y, z) = P_t G_t G_r \left(\frac{1}{2kr_0} \sum_{m,n} \overline{C_{mn}} \cdot E_{m,n}^x(x, y) \cdot \exp|-(\alpha_{mn} + j\beta_{mn}) \cdot z| \right)^2 \tag{17.39}$$

where P_t is the transmit power, and G_t and G_r are the antenna gains of the transmitter and receiver, respectively.

Channel Characteristics

According to the multimode model, the channel characteristics under various tunnel conditions are summarized next based on the effects of operating frequency, tunnel size and antenna polarization, antenna position, and electrical parameters.

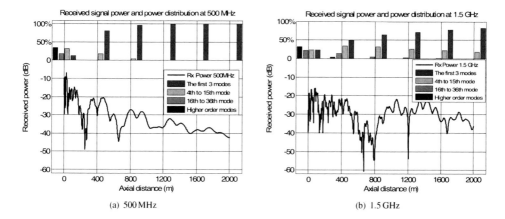

Figure 17.14 The received signal power and the power distribution among modes in tunnels at different operating frequencies.

Operating Frequency

The effects of operating frequency on the channel characteristics are illustrated in Figure 17.14. In particular, in Figure 17.14(a) and Figure 17.14(b) the signal power and the corresponding power distribution among significant modes are shown as a function of axial distance at a frequency of 500 MHz and 1.5 GHz, respectively. In the near region, the received power attenuates quickly and fluctuates very rapidly. This is attributed to the combined effect of multiple modes. On the other hand, in the far region, the decrease in the received power is gradual. This is due to the fact that the higher order modes attenuate rapidly as the distance increases. Hence, the field in the far region is governed by the few remaining low-order modes. Although the operating frequency does not significantly affect the power distribution of modes, it has an obvious influence on the propagation constants. Signals with higher frequency attenuate slower. Hence, as frequency increases, the signal attenuation decreases and the length of the rapidly fluctuating region is increased.

Tunnel Size and Antenna Polarization

The tunnel size has similar effects on the channel characteristics as the operating frequency. In tunnels with larger dimensions, the attenuation constant, α_{mn}, is smaller and more modes remain significant in the far region. Therefore, the rapidly fluctuating region is prolonged in larger tunnels, and vice versa. For horizontal polarized antennas, the tunnel width plays a more important role because the reflection coefficients on the horizontal walls are larger than those on the vertical walls. Hence, the signal attenuates slower and fluctuates longer in larger and wider tunnels for horizontally polarized antennas. Meanwhile, in larger and higher tunnels, the power of all the modes decreases slower for vertically polarized antennas.

Antenna Position

In Figure 17.15, the received power and the power distribution among modes are shown at different antenna positions. When the transmitter antenna is placed near the center of the tunnel cross-section, as shown in Figure 17.15(a), the lowest modes are effectively excited (over 50% of the total power). If the receiver is also at the center, both the signal attenuation and the fluctuation are small. If the receiver is placed near the tunnel walls, the attenuation and fluctuation are much more significant as shown in Figure 17.15(b). Near the excitation plane, higher order modes dominate (over 80% of the total power). In this case, the position of the receiver antenna does not affect the received signal.

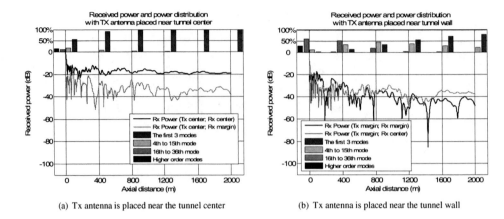

(a) Tx antenna is placed near the tunnel center   (b) Tx antenna is placed near the tunnel wall

Figure 17.15   Channel characteristics with different antenna positions and polarization.

Electrical Parameters

The electrical parameters consist of permittivity, ε, and conductivity, σ. The temperature, humidity, and pressure in a tunnel have little influence on the permittivity of the air but may affect the conductivity. However, the effect of changes on the conductivity of tunnel air may be neglected, because it is very small compared to the permittivity. Therefore, the electrical parameters of tunnel air can be considered the same as those of normal air. For the electrical parameters of the tunnel walls [10], the permittivities of tunnel materials are in the range of $5\varepsilon_0$ to $10\varepsilon_0$ (ε_0 is the permittivity in the vacuum space) and the conductivity is on the order of 10^{-2} S/m at the UHF band. In this value range, the electrical parameters of either the tunnel walls or tunnel air do not considerably influence the signal propagation inside the tunnel.

17.6.2   Room-and-Pillar Environment

The room-and-pillar environment can be regarded as a planar air waveguide superimposed with randomly distributed and random shaped pillars. A simplified multimode model is able to describe the EM wave propagation in the planar air waveguide. The randomly distributed and random shaped pillars form an environment very similar to a terrestrial metropolitan area with many buildings. Hence, the shadow fading model can be used to describe the slow fading of the signal.

Multimode Model

The GO model [22] can be utilized to analyze the excitation area. Since the planar air waveguide depends on only one coordinate, the excitation plane degenerates to a line that is perpendicular to the ceiling and floor plane and contains the point of the transmission antenna. The geometry of the cross-section is just the same as that of tunnels. The properties of the images and the reflection rays in the tunnel case are still valid.

The main differences between the multimode model for the room-and-pillar environment and that for the tunnel environment are: (1) only the y-coordinate is considered for the room-and-pillar environment; and (2) the incident angle on the ceiling and floor is a constant ($0°$). Hence, the reflection coefficient is $(1 - \sqrt{k_h})/(1 + \sqrt{k_h})$ for the X-polarized field and $(\sqrt{k_h} - 1)/(\sqrt{k_h} + 1)$ for the Y-polarized field.

The following derivation is given for a transmission antenna that is X-polarized. The result for a Y-polarized antenna can be derived in a similar way. Firstly, assume that the transmitter is located at a height y_0, and a reference point is set at the height y. The major field at the reference point is given by

$$E_x^{Rx} = E_x^{Tx} \cdot \sum_q \left[\frac{\exp(-jky_q(y))}{y_q(y)} \right] \cdot \left(\frac{1 - \sqrt{k_h}}{1 + \sqrt{k_h}} \right)^{|q|} \tag{17.40}$$

where $y_q(y)$ is the distance between image I_q and the receiver, which is given by

$$y_q(y) = \begin{cases} |2qb - y_0 - y|, & \text{if } q \text{ is odd} \\ |2qb + y_0 - y|, & \text{if } q \text{ is even.} \end{cases} \tag{17.41}$$

Secondly, the field of excitation lines obtained above is projected onto the orthogonal eigenfunctions of the planar air waveguide modes. The eigenfunctions of X-polarized modes in a planar air waveguide are given by [31]

$$E_n^x(y) = E_0^x \cdot \cos\left[\left(\frac{n\pi}{2b} - j \cdot \frac{n\pi}{2b^2 k} \frac{k_h}{\sqrt{k_h - 1}} \right) y + \varphi_y \right] \tag{17.42}$$

where $\varphi_y = \pi/2$ if n is even and $\varphi_y = 0$ if n is odd.

The mode intensity C_n is derived by projecting the field of excitation area (E_x^{Rx} in (17.40)) onto the basis function E_n^x:

$$C_n = \int_{-b}^b E_x^{Rx} \cdot E_n^x(y)\, dy$$

$$= \sum_q \left[\int_{-b}^b E_x^{Tx} \cdot \left[\frac{\exp(-jky_q(y))}{y_q(y)} \right] \cdot \left(\frac{1 - \sqrt{k_h}}{1 + \sqrt{k_h}} \right)^{|q|} \cdot E_n^x(y)\, dy \right]. \tag{17.43}$$

Using the same numerical integration technique as in the tunnel case, the normalized mode intensity $\overline{C_n}$ is

$$\overline{C_n} \simeq \frac{1}{n} \sqrt{\frac{4a}{3}} \sum_q \left\{ \sum_{v=0}^{n-1} \left[\frac{r_0 \exp(-jky_q(y))}{y_q(y) + r_0} \left(\frac{1 - \sqrt{k_h}}{1 + \sqrt{k_h}} \right)^{|q|} \cdot (-1)^{\lfloor n/2 \rfloor + 1 + v} \right] \right\}. \tag{17.44}$$

From the intensity and eigenfunction of each mode, the field at any position can be predicted for the case without pillars.

Shadow Fading Model

The shadow fading model is used to describe the slow fading caused by reflection and diffraction on the pillars. The amplitude changes caused by shadow fading are often modeled using a log-normal distribution [36]. Since one mode can be viewed as a cluster of rays with the same grasping angle, we assume that each mode experiences identically distributed and independent shadow fading when it goes through the pillars. Therefore, the predicted field at any position ($b + y$ [m] above the floor, z [m] away from the transmitter) can be obtained by summing the field of all modes, which is given by

$$E_x^{Rx}(y, z) = E_0^x \sum_n \overline{C_n} \cdot E_n^x(y) \cdot \frac{1}{2\pi z} \exp(-(\alpha_n + j\beta_n) \cdot z) \cdot \chi_n \tag{17.45}$$

where $\{\chi_n\}$ are identically distributed and independent log-normal random variables. The field is divided by $2\pi z$ because the plane wave in the room-and-pillar environment spreads in all horizontal directions; α_n is the attenuation coefficient and β_n is the phase-shift coefficient, which are given by [10, 31]

$$\alpha_n = \frac{1}{b} \left(\frac{n\pi}{2bk} \right)^2 Re \frac{1}{\sqrt{k_h - 1}}; \quad \beta_n = \sqrt{k^2 - \left(\frac{n\pi}{2b} \right)^2}. \tag{17.46}$$

The resulting received signal power is given by

$$P_r(y, z) = P_t G_t G_r \left(\frac{1}{2kr_0} \sum_n \overline{C_n} \cdot E_n^x(y) \cdot \frac{1}{2\pi z} \exp\left(-(\alpha_n + j\beta_n) \cdot z\right) \cdot \chi_n \right)^2. \qquad (17.47)$$

Channel Characteristics

The operating frequency, room height, antenna position/polarization, and electrical parameters in the room-and-pillar environment affect signal propagation in a similar way as in the tunnel case. However, their influence is much smaller. Compared to the tunnel case, signals in the room-and-pillar mining area experience extra multi-path fading caused by the pillars. Consequently, a higher path loss is experienced by the wave spreading in the room.

17.6.3   Comparison with Experimental Measurements

In this section, we discuss the evaluation of the multimode model by comparing the theoretical results to experimental measurements in both the tunnel and room-and-pillar environments [12, 21, 38].

The experiments in [12] were conducted in a straight road tunnel. The tunnel is 3.5 km long and has an equivalent rectangle (7.8 m wide and 5.3 m high, i.e. $a = 3.9$, $b = 2.65$) cross-section shape. In Figure 17.16(a), the calculated result at frequencies of 450 MHz and 900 MHz is compared to the measurements shown in [12]. The theoretical curves are intentionally displaced vertically by 75 dB and 40 dB, respectively, from the experimental curves for better comparison. The curves of the theoretical and experimental results are close to each other. The multimode model accurately predicts the attenuation velocity, fast fading in the near region, flat fading in the far region, and the effects of different operating frequency on the tunnel environment.

In [21], experiments were conducted in a room-and-pillar mining area with an average height of 6 m. In Figure 17.16(b), the calculated result at a frequency of 900 MHz is compared to the measurements shown in [21]. The comparison indicates that the theoretical result also is in good agreement with the experimental measurements in the room-and-pillar environment.

17.7   Communication Architecture

This section addresses the protocol stack of WUSNs. Figure 17.17 illustrates the classical layered protocol stack and its five layers, as well as the cross-layered power management and task management planes. The unique challenges of the underground environment cannot, however, be addressed by terrestrial WSN protocols. Therefore, it is necessary to reexamine and modify each of the layers to assure that WUSNs operate as efficiently and reliably as possible. In addition, there are many opportunities in this environment to enhance the efficiency of the protocol stack through cross-layered design. Although we promote a cross-layered design approach for WUSNs, it is nonetheless important to first understand the challenges at each layer of the traditional protocol stack. In this section, we examine each layer of the protocol stack and outline the research challenges that must be addressed to make WUSNs feasible. We then discuss cross-layered design opportunities for WUSNs.

17.7.1   Physical Layer

As explained throughout the chapter, WUSNs are characterized by the unique challenges in the physical layer because of the effects of soil on communication. EM waves propagating through soil and rock experience extreme losses. Furthermore, these losses change because of the dynamic nature of the underground environment – loss rates are highly dependent on numerous soil properties, especially

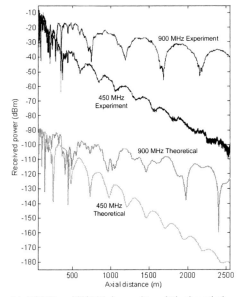

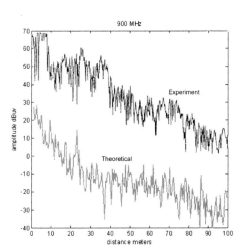

(a) 450 MHz and 900 MHz in a road tunnel (the theoretical result is shifted by 75 dB)

(b) 900 MHz in a room-and-pillar mining area (the theoretical result is shifted by 40 dB)

Figure 17.16   Experimental and theoretical received power.

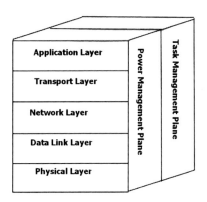

Figure 17.17   WUSN protocol stack.

volumetric water content [23], which may vary over time. Wet soils cause extreme attenuation of EM waves, even to the point of making communication impossible at any distance [44]. Losses produced by an increase in soil-water content, after rainfall for example, can last for a significant amount of time. As a result, communication techniques that account for these peculiarities of underground communication are required.

An integral part of the physical layer design for underground communication is designing an efficient antenna, given the challenging nature of EM wave propagation through soil. Embedding an antenna in a conductive medium such as soil can significantly affect its radiation and reception characteristics [19].

As discussed previously, when EM waves propagate through soil and rock, their lower frequencies experience less attenuation than higher frequencies [43], so communication at practical distances of several meters will likely only be feasible when using these lower frequencies. However, traditional EM antennas at low frequencies are too large for a WUSN device.

In addition to the antenna design, the selection of a suitable modulation scheme is another challenge considering the power constraints and the need to use low frequencies. Earlier works that have addressed underground wireless communication have focused solely on analog communication. Recently, the feasibility of QPSK, QAM-16, and QAM-32 modulation schemes has been demonstrated [41]. With a 4 kHz carrier and 10 W transmit power, a data rate of 2 kbps is possible. Apart from this work, however, modulation schemes for underground communication still remain an unexplored area.

The use of a lower carrier frequency means that less bandwidth is available for data transmission, so WUSNs will be constrained to a lower data rate than terrestrial WSNs. Extreme channel losses will also affect the data rate in WUSNs. Consequently, channel-aware error control techniques that are adaptive to intermediate communication in underground environments are necessary.

Open research issues at the physical layer are as follows:

- Although the bulk of the research in underground communication is focused on EM waves, additional analysis of EM, MI, and seismic communication underground needs to be carried out to identify the most appropriate physical layer technology. In addition, the communication characteristics for shallow nodes need to be investigated since they include both underground and surface devices. Consequently, the technologies that are suitable for each communication scenario need to be determined.

- A power-efficient modulation scheme suitable for the dynamic high-loss underground channel must be chosen. Research into varying the modulation scheme depending on underground channel conditions is needed. After rainfall, when the channel is severely impaired, for example, it may be better to trade higher data rates for a simpler modulation scheme. Modulation schemes requiring channel estimation by means of probe packets should be avoided due to the energy overhead involved in probe transmission.

- The tradeoff between reliability and capacity must be examined. Lower frequencies propagate underground with lower loss over a given distance, but also have less available bandwidth for data transmission, reducing the channel capacity. Furthermore, an informative theortical study of the capacity of underground wireless communication channels is needed.

17.7.2   Data Link Layer

The existing MAC protocols intended for terrestrial WSNs as explained in Chapter 5 will likely perform poorly in WUSNs. These protocols for terrestrial WSNs are typically either contention based or TDMA based, and focus on minimizing energy consumption by addressing four primary areas: *idle listening*, *collisions*, *control packet overhead*, and *overhearing* [15]. WUSNs require special consideration for MAC protocol development due to the characteristics of the underground wireless channel as explained in detail in Section 17.4.

Although energy conservation is the main focus of existing MAC protocols for terrestrial WSNs, energy savings are captured by reducing idle listening time [15]. While this problem still exists in WUSNs, a major problem is that higher transmit powers may be required for communication to overcome the path loss incurred through soil. Consequently, underground MAC protocols need to be designed to minimize the transmission attempts of a node. An important difference is that the traditional asymmetry between transmit and receive powers that exists in ad hoc networks may again be pronounced for WUSNs.

The type of the MAC protocol to be used is also an open research issue for WUSNs. Contention-based protocols that prevent collisions through RTS/CTS-type mechanisms may introduce additional energy consumption and overhead. On the other hand, reservation-based protocols require tight synchronization between nodes, which is much more challenging in WUSNs compared to terrestrial WSNs.

Since WUSN devices will likely report sensor data infrequently, they can operate with a low duty cycle to save power. Unfortunately, a device's clock may drift by a large amount during these periods of sleep and the network may lose its synchronization.

The significantly higher path losses also introduce errors into any communication in WUSNs. Since transmission is a major concern, ARQ schemes that try to correct errors through retransmissions may not be suitable. Instead FEC schemes or hybrid ARQ schemes may need to be developed that can adapt to the dynamic channel characteristics of underground communication.

Open research issues at the data link layer are as follows:

- Tradeoffs between the additional overhead of a TDMA-based protocol and the energy savings realized through collision elimination need to be explored to definitively determine whether a TDMA-based or contention-based MAC is most appropriate for WUSNs. A possible solution may be a hybrid MAC scheme, where, depending on network and underground wireless channel conditions, either a contention-based or a TDMA-based MAC may be optimal at a given time.

- Synchronization for very low duty cycles (on the order of minutes) needs to be explored.

- Adaptive FEC schemes need to be explored as a possible solution for the unique nature of the underground channel. When the channel is impaired by wet soil, for example, more powerful FEC schemes are necessary to overcome losses in the channel.

- Also, the optimal packet size for WUSNs needs to be determined, in particular by considering the underground channel effects in the calculations. Packet size will play an important role in both power conservation and QoS. To maximize the power efficiency, it is desirable to minimize the amount of overhead transmitted in the form of packet headers. This would suggest a larger packet size, but then larger packet sizes encounter higher error rates, which significantly affect the energy efficiency. An optimal tradeoff among these factors must be determined for WUSNs.

17.7.3   Network Layer

According to the classification in Chapter 7, three main routing categories exist: *proactive*, *reactive*, and *geographical*. Among these solutions, proactive and reactive solutions may not be applicable to WUSNs due to the excessive overhead incurred. Furthermore, these solutions usually use the same routes for long periods of time to decrease the overhead in the network. However, the topology of the WUSN can change drastically because of the very low duty cycles that can be used and the disconnections due to the soil conditions. Therefore, a route that has been created may not be reused since the nodes that were part of the route may be in sleep state or unreachable. Consequently, adaptive and low-power routing protocols are required.

Geographical routing protocols exploit the location information of each node. Therefore, knowledge of the location of sensor nodes is a major requirement. In WUSNs, geographical routing protocols can be exploited because of the careful planning required to deploy these networks for some applications. As a result, each node can be programmed with its location information at the time of the deployment.

Alternatively, a WUSN may be deployed by randomly scattering sensor devices and then covering them with soil. This could be the case when constructing a road or laying the foundation of a building since the site has already been excavated. In this scenario, location information for each device will not be known, and geographical protocols will be less useful.

In most routing protocols developed so far for terrestrial WSNs, routes are constructed by treating each node equally. An alternative to this approach is to construct energy-efficient routes by respecting the residual energy of each node. In WUSNs, however, the location of each node can affect its communication performance significantly. In large-area WUSNs, some parts of the deployment field may be affected by rain and exhibit higher attenuation. Then, the routing protocol needs to construct routes that will traverse the affected areas. Furthermore, in a hybrid topology, it would be much more energy efficient to route the data through above-ground relays. Consequently, network layer protocols must be aware of the unique challenges under ground in order to maximize the power efficiency and, thus, the network lifetime.

Open research issues at the network layer which must be addressed include the following:

- The effect of the low duty cycle of WUSNs on routing protocols must be examined. The network topology can change drastically between sensing intervals and the network layer must handle this efficiently.

- Research into routing protocols suitable for time-sensitive WUSN applications, such as presence monitoring for security by underground pressure sensors, is necessary. Protocols for these applications must be able to establish a route between an event and a sink within a short interval following the event, while still coping with the challenges of the underground channel and remaining power efficient.

- The applicability of multi-path routing algorithms on WUSNs should also be examined. These algorithms can avoid the need for complete path switching in the event of a link failure, and research is needed to make them as energy efficient as possible.

17.7.4   Transport Layer

Reliable transport, *flow control*, and *congestion control* are three main functionalities of the transport layer protocols. Some of the techniques developed so far for terrestrial WSNs as explained in Chapter 8 can be exploited in WUSNs. However, the high and variable loss rates of the underground channel require this layer to be reexamined.

The high error rates possible through underground communication make the provision of reliability a major goal for transport protocols. Especially, consideration of the several sources of error in transport protocols is crucial to correctly respond to these different cases. In addition to congestion and wireless channel errors, underground communication is also affected by the soil conditions. In the event of rainfall, the communication link between any two nodes may be lost for long periods. As a result, transport protocols need to correctly differentiate these different sources of errors and respond. As an example, treating packet drops due to disconnections as wireless channel errors may lead to crucial network resources being wasted.

Due to the low data rates of WUSNs, congestion becomes an important problem, particularly near the sink. One method of avoiding this is to route data to terrestrial relays which are capable of a higher data rate. This could be accomplished at the network layer, and points to an interesting cross-layered solution to the congestion problem. On the other hand, using above-ground relays in a hybrid underground topology may lead to new points of congestion around these relay nodes. As a result, intermediate congestion control may be required instead of end-to-end techniques.

Most existing TCP implementations are unsuited for WUSNs since the flow control functionality is based on a window-based mechanism and retransmissions. As stated before, we try to avoid/reduce the number of retransmissions in WUSNs in order to save energy. Rate-based transport protocols also seem unsuited for this challenging environment. In fact, although they do not adopt a window-based mechanism, they still rely on feedback control messages sent back by the destination to dynamically adapt the transmission rate, i.e., to decrease the transmission rate when packet loss is experienced or to increase it otherwise. The high delay and delay variance can therefore cause instability in the feedback control.

For these reasons, it may be necessary to devise completely new strategies to achieve underground flow control and reliability.

Open research issues at the transport layer which must be addressed include the following:

- New effective mechanisms tailored to the underground channel need to be developed in order to efficiently infer the cause of packet losses.
- Definitions of a new event transport reliability metric need to be proposed, based on the event model and on the underground channel model.
- Optimal update policies for the sensor reporting rate are needed, to prevent congestion and maximize the network throughput efficiency as well as the transport reliability in bandwidth-limited underground networks.
- An acceptable loss rate in WUSNs needs to be determined. This can translate directly to power savings for these severely power-constrained devices by reducing retransmissions. The acceptable loss rate is dependent on the application and the network topology, as well as on underground channel conditions.
- How best to handle the variable reporting periods of a WUSN needs to be determined. A WUSN will perform several tasks simultaneously, some of which may be more time sensitive than others. Soil-water content measurements may only be reported every hour, but the presence of any toxic substance in the soil should be reported immediately. Thus, research is needed on providing differentiated levels of service at the transport layer for different types of sensor data.

17.7.5   Cross-layer Design

Inherent in all the discussions about communication layers is that WUSNs require close coupling between the environment and protocol operation. Each individual problem discussed so far in this section highlights the importance of cross-layer interactions in communication through soil. The major reason behind this is the close interaction between communication performance and soil properties. Based on these observations, there are many challenges involving cross-layered protocol design in WUSNs. Here, we summarize some of them as follows:

- **Utilizing sensor data for channel prediction:** As described earlier, underground communication is very dependent on the soil VWC. Moreover, monitoring the VWC will be a common use of WUSNs and a large percentage of sensor devices will be equipped with moisture sensors. This argues for a cross-layer approach between the application layer, where water content readings are taken, and the lower layers, which could utilize this information for adjusting radio output power, controlling congestion, appropriately choosing routes, and selecting an appropriate adaptive FEC scheme.
- **Utilizing channel data for soil property prediction:** The opposite of the above, whereby channel properties are predicted by sensor readings, can also be accomplished. Gradually increasing losses in the channel between two devices while other devices remain reachable may be interpreted as increasing water content. This could be used to sense soil conditions in the areas between devices where no sensors are deployed, and points to an interesting interaction between the application and network layers.
- **Physical layer-based routing:** Power savings can be achieved with the use of a cross-layer MAC and routing solution. Since soil conditions can vary widely over short distances, different power levels will be necessary to communicate with a given device's neighbors. In the interest of prolonging network lifetime, routes should generally try to utilize links where lower transmit powers are necessary. The information is gathered at the physical layer, but needs to be passed on to the network layer. Additionally, soil-water content readings from surrounding devices can be processed to form a map of water content over the network's deployment terrain, allowing packets to be routed through dry areas where the soil produces less attenuation.
- **Opportunistic MAC scheduling:** Opportunistic scheduling at the MAC layer can be accomplished with the help of application layer sensor data. For example, if a device detects continually increasing soil-water content, it may try for a period to send packets at a higher power level to

overcome the additional losses incurred, followed by a period of silence where it caches outbound packets, waiting for a decrease in soil-water content in order to conserve power. Waiting for water content to decrease means that a device will need fewer retransmissions and a lower transmit power.

- **Cross layer between link and transport layers:** Transport layer functionalities can be tightly integrated with data link layer functionalities in a cross-layer integrated module. The purpose of such an integrated module is to make information about the condition of the variable underground channel available at the transport layer also. In fact, usually the state of the channel is known only at the physical and channel access sublayers, while the design principle of layer separation makes this information transparent to the higher layers. This integration allows the efficiency of the transport functionalities to be maximized, and the behavior of data link and transport layer protocols can be dynamically adapted to the variability of the underground environment.

In this chapter, the concept of WUSNs was introduced, where sensor devices are deployed completely below ground. There are existing applications of underground sensing, such as soil monitoring for agriculture. However, WUSNs provide benefits over current sensing solutions including: complete network concealment, ease of deployment, and increased timeliness of data. These benefits enable a new and wider range of underground sensing applications, from sports field and garden monitoring, where surface sensors could impede sports activity or are unsightly, to military applications such as border monitoring, where sensors should be hidden to avoid detection and deactivation. The underground environment is a particularly difficult one for wireless communication and poses several research challenges for WUSNs. An important characteristic is that the underground channel depends on the properties of the soil or rock in which devices are deployed, particularly the volumetric water content. Additionally, low frequencies are able to propagate with lower losses under ground and frequencies used by traditional terrestrial WSNs are infeasible for this environment. The use of low frequencies, however, severely restricts the bandwidth available for data transmission in WUSNs. The close interactions between communication performance and environmental effects necessitate cross-layer solutions for WUSNs, which is still an evolving field.

References

[1] Advanced Aeration Systems, Inc. Rz-aer tech sheet.

[2] I. F. Akyildiz and E. P. Stuntebeck. Wireless underground sensor networks: Research challenges. *Ad Hoc Networks*, 4:669–686, July 2006.

[3] I. F. Akyildiz, M. C. Vuran, and Z. Sun. Signal propagation techniques for wireless underground communication networks. *Physical Communication Journal*, 2(3):167–183, September 2009.

[4] R. Bansal. Near-field magnetic communication. *IEEE Antennas and Propagation Magazine*, 46(2):114–115, April 2004.

[5] C. Bunszel. Magnetic induction: a low-power wireless alternative. *RF Design*, 24(11):78–80, November 2001.

[6] Campbell Scientific, Inc. Soil science brochure.

[7] R. Cardell-Oliver, K. Smettem, M. Kranz, and K. Mayer. A reactive soil moisture sensor network: design and field evaluation. *International Journal of Distributed Sensor Networks*, 1(2):149–162, April–June 2005.

[8] S. Cheekiralla. Development of a wireless sensor unit for tunnel monitoring. Master's thesis, Massachusetts Institute of Technology, 2004.

[9] D. Daniels. *Surface-Penetrating Radar*. IEE, 1996.

[10] P. Delogne. *Leaky Feeders and Subsurface Radio Communications*. Peter Peregrinus, 1982.

[11] T. H. Dubaniewicz, J. E. Chilton, and H. Dobroski. Fiber optics for atmospheric mine monitoring. In *Proceedings of the IEEE Industry Applications Society Annual Meeting*, volume 2, pp. 1243–1249, Dearborn, MI, USA, September 1991.

[12] D. G. Dudley, M. Lienard, S. F. Mahmoud, and P. Degauque. Wireless propagation in tunnels. *IEEE Antenna and Propagation Magazine*, 49(2):11–26, April 2007.

[13] A. G. Emslie, R. L. Lagace, and P. F. Strong. Theory of the propagation of UHF radio waves in coal mine tunnels. *IEEE Transactions on Antennas and Propagation*, 23(2):192–205, March 1975.

[14] R. E. Gertsch and R. L. Bullock. *Techniques in Underground Mining: Selections from Underground Mining Methods Handbook*. Society for Mining, Metallurgy, and Exploration, 1998.

[15] K. Kredo II and P. Mohapatra. Medium access control in wireless sensor networks. *Computer Networks*, 51(4):961–994, 2007.

[16] N. Jack and K. Shenai. Magnetic induction IC for wireless communication in RF-impenetrable media. In *Proceedings of the IEEE Workshop on Microelectronics and Electron Devices (WMED'07)*, pp. 47–48, Boise, ID, USA, April 2007.

[17] X. Jiang, J. Polastre, and D. Culler. Perpetual environmentally powered sensor networks. In *Proceedings of the IEEE Workshop on Sensor Platform, Tools and Design Methods for Networked Embedded Systems (SPOTS'05)*, Los Angeles, USA, April 2005.

[18] S. G. Prejean K. Imanishi, and W. L. Ellsworth. Earthquake source parameters determined by the SAFOD pilot hole seismic array. *Geophysical Research Letters*, 31(12):L12S09.1–L12S09.5, May 2004.

[19] R. W. P. King and G. S. Smith. *Antennas in Matter: Fundamentals, Theory, and Applications*. MIT Press, 1981.

[20] K. D. Laakmann and W. H. Steier. Waveguides: characteristic modes of hollow rectangular dielectric waveguides. *Applied Optics*, 15(5):1334–1340, May 1976.

[21] M. Lienard and P. Degauque. Natural wave propagation in mine environments. *IEEE Transactions on Antennas and Propagation*, 48(9):1326–1339, September 2000.

[22] S. F. Mahmoud and J. R. Wait. Geometrical optical approach for electromagnetic wave propagation in rectangular mine tunnels. *Radio Science*, 9(12):1147–1158, December 1974.

[23] T. W. Miller, B. Borchers, J. M. H. Hendrickx, S. Hong, L. W. Dekker, and C. J. Ritsema. Effects of soil physical properties on GPR for landmine detection. In *Proceedings of the 5th International Symposium on Technology and the Mine Problem*, Monterey, CA, USA, 2002.

[24] T. A. Milligan. *Modern Antenna Design*, 2nd edition. IEEE Press, 2005.

[25] J. M. Molina-Garcia-Pardo, M. Lienard, P. Degauque, D. G. Dudley, and L. Juan-Llacer. Interpretation of MIMO channel characteristics in rectangular tunnels from modal theory. *IEEE Transactions on Vehicular Technology*, 57(3):1974–1979, May 2008.

[26] R. Musaloiu, A. Terzis, K. Szlavecz, A. Szalay, J. Cogan, and J. Gray. Life under your feet: a wireless soil ecology sensor network. In *Proceedings of the 3rd IEEE Conference on Embedded Networked Sensors*, Cambridge, MA, USA, 2006.

[27] E. F. Neuenschwander and D. F. Metcalf. A study of electrical earth noise. *Geophysics*, 7(1):69–77, January 1942.

[28] J. Pan, B. Xue, and Y. Inoue. A self-powered sensor module using vibration-based energy generation for ubiquitous systems. In *6th International Conference On ASIC*, volume 1, pp. 443–446, Briançon, France, 2005.

[29] C. Park, Q. Xie, P. H. Chou, and M. Shinozuka. Duranode: wireless networked sensor for structural health monitoring. In *Proceedings of IEEE Sensors'05*, pp. 277–280, Irvine, CA, USA, 2005.

[30] N. R. Peplinski, F. T. Ulaby, and M. C. Dobson. Dielectric properties of soils in the 0.3–1.3-GHz range. *IEEE Transactions on Geoscience and Remote Sensing*, 33(3):803–807, 1995.

[31] D. Porrat. Radio propagation in hallways and streets for UHF communications. PhD thesis, Stanford University, 2002.

[32] S. Roundy, P. K. Wright, and J. Rabaey. A study of low level vibrations as a power source for wireless sensor nodes. *Computer Communications*, 26(11):1131–1144, July 2003.

[33] A. Sheth, K. Tejaswi, P. Mehta, C. Parekh, R. Bansal, S. Merchant, T. Singh, U. B. Desai, C. A. Thekkath, and K. Toyama. Senslide: a sensor network based landslide prediction system. In *SenSys'05: Proceedings of the 3rd International Conference on Embedded Networked Sensor Systems*, pp. 280–281, San Diego, CA, USA, 2005.

[34] J. J. Sojdehei, P. N. Wrathall, and D. F. Dinn. Magneto-inductive (MI) communications. In *Proceedings of the MTS/IEEE Conference and Exhibition (OCEANS 2001)*, volume 1, pp. 513–519, Honolulu, HI, USA, November 2001.

[35] M. Stordeur and I. Stark. Low power thermoelectric generator-self-sufficient energy supply for micro systems. In *International Conference on Thermoelectrics*, pp. 575–577, Melbourne, Australia, 1997.

[36] G. L. Stuber. *Principles of Mobile Communication*, 2nd edition. Kluwer Academic, 2001.

[37] Z. Sun and I. F. Akyildiz. Underground wireless communication using magnetic induction. In *Proceedings of IEEE ICC 2009*, Dresden, Germany, June 2009.

[38] Z. Sun and I. F. Akyildiz. Channel modeling of wireless networks in tunnels. In *Proceedings of IEEE GLOBECOM'08*, New Orleans, USA, November 2008.

[39] *US Water News*. Street spies detect water leakage, December 1998.

[40] R. L. Van Dam, B. Borchers, and J. M. H. Hendrickx. Methods for prediction of soil dielectric properties: a review. In *Detection and Remediation Technologies for Mines and Minelike Targets X*, volume 5794, pp. 188–197, 2005.

[41] J. Vasquez, V. Rodriguez, and D. Reagor. Underground wireless communications using high-temperature superconducting receivers. *IEEE Transactions on Applied Superconductivity*, 14(1):46–53, 2004.

[42] T. Voigt, H. Ritter, and J. Schiller. Utilizing solar power in wireless sensor networks. In *IEEE Local Computer Networks, 2003*, pp. 416–422, 2003.

[43] J. Wait and J. Fuller. On radio propagation through earth. *IEEE Transactions on Antennas and Propagation*, 19(6):796–798, 1971.

[44] T. P. Weldon and A. Y. Rathore. Wave propagation model and simulations for landmine detection. Technical report, University of North Carolina, Charlotte, NC, 1999.

[45] M. Zuniga and B. Krishnamachari. Analyzing the transitional region in low power wireless links. In *Proceedings of IEEE SECON'04*, pp. 517–526, Santa Clara, CA, USA, October 2004.

18

Grand Challenges

The vast number of solutions that have driven the research community over the years have made the WSN phenomenon a reality. The communication solutions and existing deployments explained throughout this book provide an extensive knowledge base of WSNs. However, their proliferation has so far been limited to the research community with just a limited number of commercial applications. To effectively transform these lessons into practical solutions, several *grand challenges* still exist.

As discussed extensively throughout the chapters of this book, the major challenge for the proliferation of WSNs is

ENERGY
ENERGY
ENERGY

Extremely energy-efficient solutions are required for each aspect of WSN design to deliver the potential advantages of the WSN phenomenon. Therefore, in both existing and future solutions for WSNs, energy efficiency is *the* grand challenge.

In addition to energy efficiency, there still exist several other grand challenges. In this chapter, we discuss these challenges and highlight open research issues for addressing them.

18.1   Integration of Sensor Networks and the Internet

The evolution of wireless technology has enabled the realization of various network architectures for different applications such as cognitive radio networks [11], mesh networks [9], and WSNs [6]. In order to extend the applicability of these architectures and provide useful information anytime and anywhere, their integration with the Internet is very important. So far, research has progressed in each of these areas separately, but realization of these networks will require tight integration and interoperability. In this respect, it is crucial to develop location- and spectrum-aware cross-layer communication protocols as well as heterogeneous network management tools for the integration of WSNs, cognitive radio networks, mesh networks, and the Internet.

As we explained in Chapter 1, the 6LoWPAN [1] standard has been developed to integrate the IPv6 standard with low-power sensor nodes. Accordingly, the IPv6 packet header is compressed to sizes that are suitable for sensor motes. This provides efficient integration for communication between an IPv6-based device and a sensor mote. However, significant challenges in seamless integration between WSNs and the Internet still exist at the higher layers of the protocol stack. The coexistence of WLANs and WSNs is a major challenge at the MAC layer since they both operate in the same spectrum range. End-to-end routing between a sensor node and an Internet device is not feasible using existing solutions. Similarly, existing transport layer solutions for WSNs are not compatible with the TCP and UDP protocols, which are extensively used in the Internet. In most sensor deployment scenarios, the sink is usually assumed to reside within or very near to the sensor field, which makes it part of the multi-hop communication in receiving the sensor readings. However, it would be desirable to be able to reach the sensor network from a distant monitoring or management node residing in the wireless Internet. Therefore, new adaptive transport protocols must be developed to provide the seamless reliable transport of event features throughout the WSN and next-generation wireless Internet. Moreover, Internet protocols are generally prone to energy and memory inefficiency since these performance metrics are not of interest. Instead, WSN protocols are tailored to provide high energy and memory efficiency. The fundamental differences between the design principles for each domain may necessitate novel solutions that require significant modifications in each network to provide seamless operation.

18.2   Real-Time and Multimedia Communication

The majority of the developed solutions in WSNs focus on energy efficiency. These provide efficient monitoring applications where strict delay guarantees are not employed. However, recent developments in robotics and the need to perform specialized tasks such as acting on environments based on sensed events have led to the development of wireless sensor and actor networks (WSANs) [10] and, eventually, cyber-physical systems (CPS) [17], as explained in Chapter 14. In these networks, real-time operation guarantees need to be provided through the networking protocols. This requires integration of the existing solutions from the real-time systems community [23] with the WSN phenomenon. Considering the non-deterministic nature of communication due to wireless channel errors and traffic characteristics, probabilistic analysis of network performance is crucial to provide quality of service (QoS) guarantees.

In WSANs, for sensor–actor coordination, algorithms that can provide ordering, synchronization, and elimination of the redundancy of actions need to developed. For actor–actor coordination, there is a need to provide a unified framework that can be exploited by different applications to always select the best networking paradigm available, according to the events sensed and to the operation to be performed, so as to provide efficient actor–actor communication. In addition to specific coordination issues, there is a need for an analytic framework to characterize the three planes, i.e., the management, coordination, and communication planes.

In WSANs, the application, transport, routing, MAC, and physical layers have common requirements and are highly dependent on each other. Hence, leveraging a cross-layer approach can provide much more effective sensing, data transmission, and acting in WSANs. Several cross-layer integration issues among the communication layers should be investigated to improve the overall efficiency of WSANs.

In addition to energy efficiency, WSNs have so far been designed to carry scalar data, e.g., temperature, humidity, acceleration, etc. Recent progress in CMOS technology has, however, enabled the development of single chip camera modules that could easily be embedded into inexpensive transceivers. Moreover, microphones have for long been used as an integral part of wireless sensor nodes. Consequently, wireless multimedia sensor networks (WMSNs) became the focus of research in a wide variety of areas including digital signal processing, communication, networking, control, and statistics in recent years [3], as explained in Chapter 15. Aggregation and fusion of both inter-media and intra-media are necessary considering the fundamental differences between WMSNs and WSNs. Especially, the computational burden exposed through complex multimedia operations prevents well-established distributed processing schemes that have been developed for WSNs being used in this context. Further, the tradeoffs between compression at end-nodes and communicating raw data have yet to be clearly analyzed. This leads to fundamental architecture decisions such as whether this processing can be done on sensor nodes (i.e., a flat architecture of multifunctional sensors that can perform any task), or if the need for specialized devices, e.g. *computation hubs*, arises.

At the physical layer, UWB communication for point-to-point links has been successfully accomplished. However, multi-hop communication through the UWB technology is still an open research issue. Although there exist recent efforts in this direction [14, 20], multi-user channel access and multi-hop communication techniques are yet to be defined for UWB networks. Moreover, comprehensive analytical models are needed to quantitatively compare different variants of UWB and determine the tradeoffs in their applicability to high-data-rate and low-power-consumption devices such as multimedia sensors. A promising research direction may also be to integrate UWB with advanced cognitive radio [7] techniques to increase utilization of the spectrum. For example, UWB pulses could be adaptively shaped to occupy portions of the spectrum that are subject to lower interference.

Although dedicated communication slots are possible in ZigBee, the low data rate limits its applicability for multimedia applications. The standard describes a self-organizing network but heterogeneous nodes necessitate some form of topology control in order to derive optimum ratios of FFD and RFD devices. Such a ratio will depend on the region being monitored and the desired coverage accuracy, among others.

18.3   Protocol Stack

The MAC solutions explained in Chapter 5 are tailored to the unique challenges of the WSN paradigm. Although these solutions address many of the challenges in WSNs, there still exist many open research issues for MAC protocols in WSNs.

Most MAC protocols are tailored to topologies where the nodes are static. However, developments in MEMS and robotics technology have enabled the production of mobile sensor nodes at low cost. Hence, mobility support at the MAC layer is also required for WSN applications. Neither the contention-based nor the reservation-based MAC protocols try to provide low-delay medium access. Moreover, the access latency is usually traded off against energy conservation. However, in order for WSNs to provide real-time support for delay-crucial applications, low-latency MAC protocols are required.

At the transport layer, some sensor applications and multimedia information such as target images, acoustic signals, and even video of a moving target need to be transported as we discussed in Chapter 15. However, the multimedia traffic has significantly different characteristics and, hence, different reliability and congestion notions are required to address these challenges. Therefore, new transport layer solutions, which address the requirements of multimedia delivery over WSNs, must be developed. Due to the severe processing, memory, and energy limitations of sensor nodes, it is imperative that communication is achieved with maximum efficiency. In this respect, cross-layer optimization of transport, routing, link, and physical layers must be investigated and the theoretical bounds should be identified to develop new cross-layer communication protocols for reliable transport in WSNs.

18.4   Synchronization and Localization

The synchronization protocols developed for WSNs provide a common reference frame for various applications while addressing the unique challenges of the WSN paradigm as discussed in Chapter 11. While several aspects of synchronization have been addressed by these protocols, there still exist many open research issues. As an example, the existing synchronization protocols result in an accuracy that is acceptable for most WSN applications with scalar data delivery. However, this accuracy is not acceptable for the evolving WMSN paradigm [3]. Real-time communication constraints and the strict timing requirements of multimedia necessitate precise synchronization methods with a low overhead. As an example, in wireless video sensor networks, low-cost cameras distributed throughout the network need to be synchronized to perform distributed image processing algorithms and communicate with each other. This requires multi-hop techniques where multiple distant nodes are synchronized.

The synchronization protocols developed for WSNs assume a mostly static topology, where message exchanges can be used for time offset calculations. While this assumption is mostly true, recent developments in embedded system design have enabled the realization of mobile WSNs. Consequently, the dynamic topology changes and the effect of mobility on timing measurements need to be addressed to develop synchronization protocols for these networks. Furthermore, most synchronization protocols rely on simulations and experiments to determine the performance of multi-hop synchronization. In addition, novel analytical tools are necessary to model the factors that affect synchronization in high-density, multi-hop WSNs.

Due to the multi-hop nature of WSNs, the synchronization protocols are closely coupled with routing. The path that a packet traverses also affects its timing information. Especially, the synchronization protocols that rely on time translation rather than network-wide synchronization require the routes to include translation nodes. Furthermore, MAC protocols significantly affect the non-deterministic delay between a pair of nodes. Based on these cross-layer interactions between several layers of the protocol stack, cross-layer synchronization protocols are necessary.

As explained in Chapter 12, the developed localization mechanisms provide several capabilities for location estimation through both range-based and range-free techniques. There is, however, still many issues to be addressed for efficient localization protocols. Generally, localization protocols assume a static topology for WSNs. By exploiting localized beacon nodes, several techniques have been developed to estimate the locations of the remaining nodes. In mobile WSNs, however, most of the assumptions made for the existing protocols do not hold true. Since the locations of the nodes change frequently in mobile WSNs, velocity estimation algorithms should be integrated to provide dynamic localization. Furthermore, the effects of mobility on ranging techniques should be investigated. While the theory for static localization through robust quadrilaterals is well established, necessary and sufficient conditions for localization in mobile WSNs require further investigation.

In addition to mobility support, novel ranging techniques are also required. Existing ranging techniques such as received signal strength and time-of-arrival techniques suffer severely from non-line-of-sight factors. Moreover, the accuracy of these techniques is limited. While angle-of-arrival measurements provide higher accuracy, higher cost and more severe vulnerability from non-line-of-sight conditions limit the applicability of these solutions. Hence, low-cost and robust ranging techniques are necessary to improve the accuracy of localization algorithms without high computation demands.

The effects of non-line-of-sight operation make some of the ranging measurements erroneous for localization. Hence, efficient methods should be developed such that these situations are determined and the information from unreliable nodes is disregarded. Moreover, similar to the synchronization protocols, the cross-layer interactions between the various layers of the communication stack and the localization algorithms need to be addressed for more efficient solutions. Since localization accuracy also depends on communication success, MAC and routing solutions should be designed considering the requirements of the localization protocols. This leads to more deterministic ranging measurements and efficient distributed localization algorithms.

18.5   WSNs in Challenging Environments

The promising advantages of the WSN phenomenon have opened up new application areas in challenging environments such as under water and under ground, thus requiring communication protocols that are able to cope with the characteristics and impairments of the propagation medium and the environmental characteristics of such environments. Underwater wireless sensor networks (UWSNs), explained in Chapter 16, are envisioned to enable applications for a wide variety of purposes such as oceanographic data collection, pollution monitoring, offshore exploration, disaster prevention, assisted navigation, and tactical surveillance [4]. Similarly, the realization of wireless underground sensor networks (WUSNs), explained in Chapter 17, will lead to potential applications in the fields of intelligent irrigation, border patrol, assisted navigation, sports field maintenance, intruder detection, and infrastructure monitoring [5]. This is possible by exploiting real-time information of soil condition from a network of underground sensors and enabling localized interaction with the soil. In both these fields, however, RF signals have limited applicability. As explained in Chapter 16 and Chapter 17, acoustic communication techniques have been utilized for UWSNs [4, 15, 18] and magnetic induction techniques can be used in conjunction with RF radios for WUSNs [5, 8]. Therefore, novel communication protocols are required that are designed considering the particular challenges in these environments.

For UWSNs, it is necessary to develop inexpensive transmitter/receiver modems for underwater communications. Accordingly, the design of low-complexity suboptimal filters characterized by rapid convergence is necessary to enable real-time underwater communication with decreased energy expenditure. In this respect, there is a need to overcome the stability problems in the coupling between the phase-locked loop (PLL) and the decision feedback equalizer (DCE). To enable data link layer solutions specifically tailored to underwater acoustic sensor networks, extensive research is necessary. For CDMA-based solutions, access codes with high auto-correlation and low cross-correlation properties are necessary to achieve minimum interference among users. This needs to be achieved even when the transmitting and receiving nodes are not synchronized.

UWSNs require efficient routing solutions to address the unique challenges of the underwater environment. For delay-tolerant applications, there is a need to develop mechanisms to handle loss of connectivity without provoking immediate retransmissions. Strict integration with transport and data link layer mechanisms may be advantageous to this end. Moreover, routing algorithms are required to be robust with respect to the intermittent connectivity of acoustic channels. Algorithms and protocols need to be developed that detect and deal with disconnections due to failures, unforeseen mobility of nodes, or battery depletion. These solutions should be local so as to avoid communication with the surface station and global reconfiguration of the network, and should minimize the signaling overhead. Furthermore, local route optimization algorithms are needed to react to consistent variations in the metrics describing the energy efficiency of the underwater channel. These variations can be caused by increased bit error rates due to acoustic noise and relative displacement of communicating nodes due to variable currents. In addition to sensor nodes, mechanisms are needed to integrate AUVs in underwater networks and to enable communication between sensors and AUVs. In particular, all the information available to sophisticated AUV devices (trajectory, localization) could be exploited to minimize the signaling needed for reconfiguration.

Energy efficiency is one of the most important factors in the design of WSNs [6]. This is even more pronounced in WUSNs because of the environmental effects. Firstly, the higher attenuation in underground communication requires higher transmit powers to establish reasonable communication ranges. Secondly, changing batteries is even harder, if not impossible, in WUSNs compared to their terrestrial counterparts. Consequently, an energy-efficient design is crucial for the design of WUSNs. There are many possible methods to save energy in WUSNs. Although the underground channel exhibits multi-path effects with higher attenuation, it is stable for a particular pair of nodes in the short term. As a result, simple modulation schemes and error control techniques may be sufficient for efficient underground communication. This in turn decreases the overhead incurred by these techniques and provides a potential way of saving energy.

The results for maximum attainable communication range illustrate that the underground environment is much more limited compared to the terrestrial one for WSNs. In particular, at the operating frequency of 300–400 MHz, the communication range can be extended up to 5 m. This suggests that multi-hop communication is essential in WUSNs. Consequently, in the design of WUSN topology, multi-hop communication should be emphasized.

Another important factor is the direct influence of soil properties on the communication performance. It is clear that any increase in water content significantly hampers the communication quality. The network topology should be designed to be robust to such changes in channel conditions. Furthermore, soil composition at a particular location should be carefully investigated to tailor the topology design according to specific characteristics of the underground channel at that location. Underground communication is also affected by changes according to depth. As a result, different ranges of communication distance can be attained at different depths. This requires a topology structure that is adaptive to the 3-D effects of the channel. Optimum strategies for providing connectivity and coverage should be developed considering these peculiarities.

In WUSNs, communication quality is directly related to the environmental conditions. Besides the effect of soil type, seasonal changes result in the variation of volumetric water content, which significantly affects the communication performance. Therefore, in the protocol design for WUSNs, the dynamics of the environment need to be considered. This implies an *environment-aware protocol* design paradigm that can adjust the operating parameters according to the surroundings. Furthermore, the dynamic nature of the physical layer and its direct influence on communication quality call for novel cross-layer design techniques that are adaptive to environmental changes for WUSNs.

18.6   Practical Considerations

WSNs are characterized by the *cross-layer* interactions at each level of the system. As an example, experimental evaluations reveal that the MAC layer collisions and wireless channel impurities have significant impacts on a simple flooding protocol, which is not evident through the *unit disc graph model* used in several solutions [16]. Similarly, it has been found that the state-of-the-art protocols can only reach half of the theoretical throughput in both single channel and multi-channel communication and, more importantly, the reason was found to be the packet copying latency between the microcontroller and the transceiver [21], which is almost never considered in the development of a communication protocol. Moreover, it has been found that intrinsic limitations of existing motes significantly affect the application limits, e.g., the sampling rate is bounded by the memory access latency of MicaZ [22]. These empirical findings show that the state-of-the-art analytical models as well as simulation tools cannot capture these cross-layer interactions. Moreover, performing real-life experiments at large scales are costly in terms of time and budget and may not be feasible at each step of the development cycle. This calls for a comprehensive set of tools in terms of both analytical models and simulation environments that can effectively capture these intrinsic relations so that application- and platform-specific solutions can be developed efficiently.

18.7   Wireless Nano-sensor Networks

The recent improvements in nanotechnology have enabled the realization of various components at the nano-scale. Nanotechnology mainly consists of the processing, separation, consolidation, and deformation of materials by one atom or by one molecule [25]. With the help of this technology, *nano-machines* can be developed. Nano-machines are tiny components consisting of an arranged set of molecules which are able to perform very simple sensing, actuation, computation, storage, and communication tasks [2]. The small sizes reduce the power requirements of the nano-machines significantly, which could be operated by small batteries or can be solar powered [19]. The nano-machines are envisioned to constitute the building blocks for wireless nano-sensor networks.

Nanotechnology can be instrumental in the development and implementation of several components of a sensor node at the nano-scale. Nanocrystalline materials and nanotubes have been successfully used to develop *nanobatteries*, which have significantly larger energy density compared to state-of-the-art batteries [24]. Accordingly, energy levels that are required for the operation of nano-machines can be easily provided for long periods of time. In addition, carbon nanotube (CNT) technology can be utilized to develop *supercapacitors* that can store as much as eight times more energy compared to state-of-the-art capacitors with a small form factor.

Nanotechnology is also used to develop nano-sensors that can convey biological, chemical, or physical information at the nano-scale. Bio-sensors can be used to detect DNA and other biomaterials to an accuracy that was not possible before. Similarly, chemical substances can be detected to enable the sensors to *smell*. CNTs are also used to develop processors that are 500 times smaller than their micro-counterparts and that can provide significantly higher processing speeds at a fraction of the power with negligible heat dissipation compared to state-of-the-art processors [24]. Nanotechnology memory (NRAM) units will also be available that can store a trillion bits per square inch for the development of a stand-alone microprocessor unit (MCU).

These advances enable completely nano-scale sensor nodes, which can autonomously sense, process, and store data, to be developed. Communication between these autonomous entities to relay this information and create networks similar to traditional WSNs is a grand challenge in this area. Nano-scale RF antennas are possible [13, 12]. However, the characteristics of these antennas are significantly different to their micro-counterparts due to their smaller size and high inductance. Moreover, nano-scale operations allow molecular communication that opens up a wide array of possibilities for communication in addition to RF waves [2].

Molecular communication is realized through nano-scale transmitters and receivers that are capable of encoding/decoding information on various molecules. Accordingly, the transmitter encodes the message onto molecules and inserts it into the medium. The medium can be wet or dry depending on the application, e.g., body implants or environmental monitoring. The message can also be attached to molecular carriers, which are able to transport chemical signals or molecular structures containing the information. The message then propagates from the transmitter to the receiver. This propagation can take times that are orders of magnitude larger than those encountered in RF communication and is highly dependent on the medium. Once the molecules reach the receiver, they are detected and decoded.

While the communication structure is analogous to that in WSNs, the particular characteristics of nano-communication are drastically different. Depending on the application scenario, various encoding/decoding techniques, propagation medium, and carriers can be used. Hence, efficient communication models are required for each particular situation to better understand the features of nano-communication. This requires nano-scale measurement components, e.g., smaller counterparts of signal generators, oscilloscopes, etc. Moreover, development of communication components, testbeds, and simulators is a major challenge for accurate evaluation. Furthermore, the theory of nano-communication is still in its infancy and requires substantial amounts of research.

Once the basic nano-network components are built, the transmission is controlled, and the propagation is understood, advanced networking knowledge can be applied to design and realize more complex nano-networks [2]. The design, development, and implementation of networking protocols require unique approaches that consider the peculiarities of nano-machines as well as nano-communication. For nano-scale RF communication, interference characteristics should be carefully studied for efficient MAC protocols. On the other hand, molecular communication provides additional tools that do not exist in traditional wireless networking. Since molecules are used for communication between parties, interference issues may not be severe and the delay in communication dominates medium access challenges. Besides point-to-point communication challenges, creating multi-hop routes between nano-machines is a major undertaking. This requires addressing schemes at the nano-scale using the molecular properties of transmitters and receivers. Assuming multi-hop paths can be created in this domain, end-to-end reliability requirements of certain applications should be guaranteed. The significant per-hop delay in molecular communication poses an important challenge for the development of end-to-end solutions. Finally, cross-layer solutions are required that exploit the advantages of nano-machines and nano-communication for extremely energy-efficient wireless nano-sensor networks.

References

[1] IPv6 over low power WPAN working group. http://tools.ietf.org/wg/6lowpan/.

[2] I. F. Akyildiz, F. Brunetti, and C. Blazquez. NanoNetworking: A new communication paradigm. *Computer Networks*, 52:2260–2279, June 2008.

[3] I. F. Akyildiz, T. Melodia, and K. Chowdury. A survey on wireless multimedia sensor networks. *Computer Networks*, 51(4):921–960, March 2007.

[4] I. F. Akyildiz, D. Pompili, and T. Melodia. Underwater acoustic sensor networks: research challenges. *Ad Hoc Networks*, 3(3):257–279, March 2005.

[5] I. F. Akyildiz and E. P. Stuntebeck. Wireless underground sensor networks: research challenges. *Ad Hoc Networks*, 4:669–686, July 2006.

[6] I. F. Akyildiz, W. Su, Y. Sankarasubramaniam, and E. Cayirci. Wireless sensor networks: a survey. *Computer Networks*, 38(4):393–422, March 2002.

[7] I. F. Akyildiz, M. C. Vuran, and Ö. B. Akan. A cross layer protocol for wireless sensor networks. In *Proceedings of the Conference on Information Sciences and Systems (CISS'06)*, pp. 1102–1107, Princeton, NJ, USA, March 2006.

[8] I. F. Akyildiz, M. C. Vuran, and Z. Sun. Signal propagation techniques for wireless underground communication networks. *Physical Communication Journal*, 2(3):167–183, September 2009.

[9] I. F. Akyildiz and X. Wang. A survey on wireless mesh networks. *IEEE Communications Magazine*, 43(9):S23–S30, September 2005.

[10] I. F. Akyildiz and I. H. Kasimoglu. Wireless sensor and actor networks: Research challenges. *Ad Hoc Networks*, 2(4):351–367, October 2004.

[11] I. F. Akyildiz, W. -Y. Lee, M. C. Vuran, and S. Mohanty. NeXt generation/dynamic spectrum access/cognitive radio wireless networks: a survey. *Computer Networks*, 50(13):2127–2159, September 2006.

[12] P. J. Burke, S. Li, and Z. Yu. Quantitative theory of nanowire and nanotube antenna performance. *IEEE Transactions on Nanotechnology*, 5(4):314–334, 2006.

[13] P. J. Burke. An RF circuit model for carbon nanotubes. *IEEE Transactions on Nanotechnology*, 2(1):55–58, March 2003.

[14] F. Cuomo, C. Martello, A. Baiocchi, and F. Capriotti. Radio resource sharing for ad-hoc networking with UWB. *IEEE Journal on Selected Areas in Communications*, 20(9):1722–1732, December 2002.

[15] M. C. Domingo. Overview of channel models for underwater wireless communication networks. *Physical Communication*, 1(3):163–182, March 2008.

[16] D. Ganesan, B. Krishnamachari, A. Woo, D. Culler, D. Estrin, and S. Wicker. An empirical study of epidemic algorithms in large scale multihop wireless networks. Technical report IRB-TR-02-003, Intel Research, March 2002.

[17] E. A. Lee. Cyber physical systems: design challenges. Technical report UCB/EECS-2008-8, EECS Department, University of California, Berkeley, January 2008.

[18] D. Lucani, M. Medard, and M. Stojanovic. Underwater acoustic networks: channel models and network coding based lower bound to transmission power for multicast. *IEEE Journal on Selected Areas in Communications*, 26(9):1708–1719, December 2008.

[19] M. U. Mahfuz and K. M. Ahmed. A review of micro-nano-scale wireless sensor networks for environmental protection: prospects and challenges. *Science and Technology of Advanced Materials*, 6(3–4):302–306, April–May 2005.

[20] R. Merz, J. Widmer, J.-Y. Le Boudec, and B. Radunovic. A joint PHY/MAC architecture for low-radiated power TH-UWB wireless ad-hoc networks. *Wireless Communications and Mobile Computing*, 5(5):567–580, July 2005.

[21] F. Österlind and A. Dunkels. Approaching the maximum 802.15.4 multi-hop throughput. In *Proceedings of the ACM Workshop on Embedded Networked Sensors (HotEmNets'08)*, Charlottesville, VA, USA, June 2008.

[22] J. Paek, K. Chintalapudi, J. Cafferey, R. Govindan, and S. Masri. A wireless sensor network for structural health monitoring: performance and experience. In *Proceedings of the 2nd IEEE Workshop on Embedded Networked Sensors (EmNetS-II)*, Sydney, Australia, May 2005.

[23] L. Sha, T. Abdelzaher, K.-E. Ârzén, A. Cervin, T. Baker, A. Burns, G. Buttazzo, M. Caccamo, J. Lehoczky, and A. K. Mok. Real time scheduling theory: a historical perspective. *Real-Time Systems*, 28(2–3):101–155, 2004.

[24] J. P. M. She and J. T. W. Yeow. Nanotechnology-enabled wireless sensor networks: from a device perspective. *IEEE Sensors Journal*, 6(5):1331–1339, October 2006.

[25] N. Taniguchi. On the basic concept of nano-technology. In *Proceedings of the International Conference on Production Engineering*, Tokyo, Japan, 1974.

Index

CPSIA information can be obtained at www.ICGtesting.com
Printed in the USA
LVOW05*1205090714

393483LV00011B/98/P